Advances in Computational Management Science

Volume 12

Series editors

Hans M. Amman, Eindhoven, The Netherlands
Berc Rustem, London, UK

More information about this series at http://www.springer.com/series/5594

Dmitrii Lozovanu · Stefan Pickl

Optimization of Stochastic Discrete Systems and Control on Complex Networks

Computational Networks

Springer

Dmitrii Lozovanu
Institute of Mathematics and Informatics
Academy of Sciences of Moldova
Chisinau
Moldova

Stefan Pickl
Institut für Theoretische Informatik
Mathematik und Operations Research
Universität der Bundeswehr München
Munchen
Germany

ISSN 1388-4301
ISBN 978-3-319-11832-1 ISBN 978-3-319-11833-8 (eBook)
DOI 10.1007/978-3-319-11833-8

Library of Congress Control Number: 2014951665

Springer Cham Heidelberg New York Dordrecht London

Printed on acid-free paper

Springer is part of Springer Science+Business Media (www.springer.com)

Preface

Complex Networks are the heart of Operations Research and Operations Research is in the core of Complex Networks.

Panos Pardalos

This book comprises the recent results on studying stochastic dynamic programming models and solving optimal control problems on networks. The main attention is addressed to stochastic versions of classical discrete control problems, Markov decision processes, and a game-theoretical approach to these stochastic dynamic models. The authors included in the book their own results concerning the determination of the optimal solution of discrete optimal control problems on networks and solving the game variants of Markov decision problems in the sense of computational networks.

Investigations on elaborations of efficient methods and algorithms for modern stochastic dynamic programming problems represent a relevant topic in Operations Research. With respect to computational networks, stochastic dynamic optimization models are widely used for studying and solving many practical decision problems from diverse areas such as ecology, economics, engineering, communications systems, etc. Therefore, in the book an attempt is made to systematize the most important existing methods for the mentioned class of problems as well as to describe new algorithms for solving different classes of the stochastic dynamic programming problems.

The book consists of four chapters.

The first chapter studies the finite state space of Markov processes and gives a review of existing methods and algorithms for determining the main characteristics in Markov chains. New approaches based on dynamic programming and combinatorial methods for determining the state-time probabilities, the matrix of limiting probabilities, and differential matrices in a Markov chain are proposed and formulated. Based on a classical numerical methods in this chapter new polynomial-time algorithms for determining the limiting probabilities and differential matrices are described.

Along with well-known numerical methods for determining the basic characteristics in finite Markov processes, new calculation procedures and algorithms for their finding in the case of stationary and nonstationary discrete processes are presented.

Asymptotic behavior of the average and expected total rewards (costs) in Markov processes with rewards (costs) are analyzed and the corresponding asymptotic formulae are derived. Computational complexity aspects of the problems of determining the average and the discounted expected total rewards (costs) are analyzed and the corresponding procedures for calculating these characteristics are described.

Chapter 2 is dedicated to infinite horizon stochastic discrete optimal control models and Markov decision problems with average and expected total discounted optimization criteria. Necessary and sufficient conditions for determining the optimal stationary strategies in such problems are formulated and algorithms for determining their solutions are developed. The stochastic control problems are formulated on networks and the corresponding algorithms for determining their optimal solutions using a linear programming approach are proposed. Furthermore, a relationship between the class of stationary Markov decision problems and the class of stochastic control problems on networks is analyzed and procedures how to reduce one class of problems to another are described. The most important results of Chap. 2 are related to a linear programming approach for unichain and multichain Markov decision problems with average and expected total discounted costs optimization criteria. Based on such an approach new algorithms for solving problems in general form are derived. Afterwards, efficient iterative procedures for determining the optimal solutions of Markov decision problems and stochastic control problems on networks are described. Additionally, a new class of Markov decision problems and stochastic control problems with stopping states are introduced and the corresponding mathematical tool for solving this class of problems is developed. Some extensions and generalizations of the stochastic decision models are considered and their possible applications for solving classical optimization problems are suggested.

In Chap. 3 a special game-theoretical approach to Markov decision processes and stochastic discrete optimal control problems is developed. An essentially new class of stochastic positional games is formulated and studied applying the game-theoretical concept to Markov decision problems with average and expected total discounted costs optimization criteria. To formulate this class of games we assume that Markov decision processes may be controlled by several actors (players).

The set of states of the system in such processes is divided into several disjoint subsets which represent the corresponding position sets of the players. Each player has to determine which action should be taken in each state of his position set in order to minimize (or maximize) his own average cost per transition or the expected total discounted cost. For the stochastic discrete optimal control problems with infinite time horizon this approach is developed in a similar way and a new class of stochastic positional games on networks is obtained.

The main results we describe in this chapter are related to the existence of Nash equilibria for a considered class of stochastic positional games and an elaboration of the algorithms for determining the optimal stationary strategies of the players. We formulate and prove Nash equilibria conditions for the stochastic positional games with average and discounted payoff functions of the players and develop algorithms for determining the optimal strategies for different classes of games. These results are specified for antagonistic stochastic positional games and algorithms for determining the optimal strategies of the players are gained. In the following, we show that the obtained results generalize the well-known results for deterministic positional games and new conditions for determining the solutions of the problems can be derived. Moreover, we show that the considered class of stochastic positional games can be used for studying cyclic games and Shapley stochastic games. New polynomial-time algorithms for deterministic antagonistic positional games are described. The algorithms for determining the optimal strategies of the players in deterministic cases are developed for a more general class of positional games on networks. Additionally, the multi-criteria decision problems with Pareto and Stackelberg optimization principles are formulated and some approaches for determining the solutions of these problems are suggested.

Chapter 4 is devoted to finite horizon stochastic control problems and Markov decision processes. In this chapter dynamic programming algorithms for determining the optimal solutions of the problems using the network representation of the finite discrete processes are developed. We show that the solutions for the considered stochastic horizon decision problems can be efficiently found using a backward dynamic induction principle. The algorithms are described in general form for stationary and nonstationary cases of the problems. Moreover, the algorithms are developed for the case with varying time of states' transitions of the dynamical systems. These algorithms are a contribution to the important field of computational network theory.

January 2015

Dmitrii Lozovanu
Stefan Pickl

Contents

Notations

A	Set of actions in a Markov decision process
a	An action in a Markov decision process
$A(x)$	Set of actions in the state x
$\mathbb{C}$	Complex space
$C = (c_{x,y})$	Matrix of transition costs in a finite state space Markov process
$c_{x,y}$	Cost of the system's transition from the state x to the state y
$D_x(t)$	Variance of total cost for the time-discrete system during t transitions with a starting state x
E	Set of edges of the graph of states' transitions of the system for control problems on a network
$e = (x, y)$	A directed edge of the graph of the states' transitions in control problems on a network
$E(x)$	Set of directed edges originated in vertex x
$F(z)$	z-Transform for a nondecreasing time-discrete function $f(t)$
$G = (X, E)$	Graph of the states' transitions of the system for control problems on a network
$G_p = (X, E_p)$	Graph induced by the transition probability function in a Markov chain or in a Markov decision process
I	Identity matrix
$K(z)$	Characteristic polynomial of the transition probability matrix in a Markov chain
$\mathbb{L}$	Time-discrete system
L	Length of input data of a decision problem
$P = (p_{x,y})$	Matrix of transition probabilities for a finite state space Markov process
P^t	t-th Power of matrix P
$P^s = (p^s_{x,y})$	Matrix of transition probability induced by a stationary strategy s in a Markov decision process

$P(t)$	Dynamic stochastic matrix
$\overline{P}(t)$	P^t
$P_x(y,t)$	Probability that the system will reach the state y using t transitions if it starts transitions in the state x
$Q=(q_{x,y})$	Matrix of limiting probabilities for a finite state space Markov chain
$\mathbb{R}^m$	m-Dimensional real space
$\mathbb{S}$	Set of stationary strategies for a finite state space control problem and Markov decision process
S	Set of strategies for a continuous control model and Markov decision problem
s	A strategy from $\mathbb{S}$ or from S
$T(t)$	$P^t - Q$
t	Time-moment for the state in a discrete dynamical system
$U_t(x(t))$	Set of feasible controls for a time-discrete system in the state $x(t)$ at the time moment t
$u(t)$	Control vector from $U_t(x(t))$
X	Finite set of states for a time-discrete system and a finite state space Markov process
X_C	Set of controllable states for a control problem on a network
X_N	Set of uncontrollable states for a control problem on a network
$X(x)$	Set of neighbor vertices for the vertex x in the graph of states' transitions $G=(X,E)$
$x(t)$	State of the dynamical system at the moment of time t
Z	Set of dynamical states for the discrete control problem, i.e., $Z=\{x(t)=(x,t)\,\vert\, x\in X,\ t=0,\ 1,2,\ldots\}$
Z^C	Set of controllable dynamical states for the control problem
Z^N	Set of dynamical states for the discrete control problem
γ	Discount rate in discrete Markov decision models
ϵ	A small quantity
μ_x	Immediate cost in the state x for a Markov process with transition costs; it is expressed as $\sum_{y\in X} p_{x,y}c_{x,y}$
π	Vector of the limiting state probabilities in an ergodic Markov chain
$\pi(\tau)$	Vector of the state probabilities of the system in a Markov chain at the moment of time τ
$\sigma_x(t)$	Expected total cost for the time-discrete system during t transitions if it starts transitions in the state x

σ_x^γ	Expected total discounted cost with discount factor γ in a Markov chain for the system if it starts transitions in the state x
$\sigma_x^\gamma(t)$	Expected total discounted cost with discount factor γ in a Markov chain for the system during t transitions if it starts transitions in the state x
τ	Time counter in discrete systems
τ_e	Transition time through a directed edge $e = (x, y)$ in the graph of states' transitions for the control problem on a network
$\tau_{x(t_j)}\big(u(t_j)\big)$	Transition time from the state $x(t_j)$ to the state $x(t_{j+1})$ induced by the control vector $u(t) \in U_t(x(t_j))$
ω_x	Expected cost per transition in the Markov chain for the system if it starts transitions in the state x
$\omega_x(t)$	Expected cost per transition in the Markov chain for the system during t transitions if it starts transitions in the state x
$*$	Superscript denoting an optimal strategy, value or decision rule in the problems
argmax	An element or a subset of elements at which the maximum of a function is obtained
argmin	An element or a subset of elements at which the minimum of a function is obtained
0	Scalar or vector zero
$\times$	Cartesian product
$\square$	Completion of the proof
$\Longrightarrow$	Implication token
(G, X_C, X_N, c, p)	Decision network for an infinite horizon stochastic control problem with a finite set of states $X = X_C \cup X_N$
$(G, Z^C, Z^N, c, p, \bar{t}, z_0, z_f)$	Time-expanded network for a finite horizon stochastic control problem
(X, A, c, p)	Markov process with the set of states X, the set of actions A, the transition cost function $c : X \times X \times A \rightarrow \mathbb{R}$ and the transition probability function $p : X \times X \times A \rightarrow [0, 1]$
$\Gamma = (\{\mathbb{S}i\}_{i=\overline{1,m}}, \{F_i\}_{i=\overline{1,m}})$	Noncooperative game of m players with the sets of strategies $\mathbb{S}_i$ and the payoff functions F_i, $i = 1, 2, \ldots, m$
$(X, A, \{X_i\}_{i=\overline{1,m}}, \{c^i\}_{i=\overline{1,m}}, p, x_0)$	Stochastic positional game of m players with average payoff functions
$(X, A, \{X_i\}_{i=\overline{1,m}}, \{c^i\}_{i=\overline{1,m}}, p, \gamma, x_0)$	Stochastic positional game of m players with discounted payoff functions

Chapter 1
Discrete Stochastic Processes, Numerical Methods for Markov Chains and Polynomial Time Algorithms

In this chapter we consider stochastic discrete systems with a finite set of states. We study stationary and non-stationary discrete Markov processes. The main attention we address to the problems of determining the state-time probabilities, calculation of the limiting and the differential probability matrices, determining the expected total cost and calculation of the average and the expected total discounted costs in the finite-state space of Markov processes. These problems arise in many practical decision problems from different areas and need efficient numerical methods and algorithms for their solving. Our aim is to study the efficiency of the existing algorithms for solving these problems, to analyze the algorithms from a computational complexity point of view and to describe some new approaches for solving the considered problems. Especially, new algorithms and calculation procedures are proposed for determining the state-time probabilities and the limiting matrix in Markov chains. Mainly, we describe and formulate two classes of algorithms.

The first class is based on the properties of generating functions and classical numerical methods; the second one mainly uses the combinatorial properties of the graphs of transition probabilities in Markov chains and a dynamic programming approach for such processes. In a detailed form we analyze the dynamic programming procedures for determining the state-time probabilities in finite Markov processes. We study the asymptotic behavior of these procedures and show how to use them for calculating the matrix of limiting state probabilities in Markov chains. Beside that we describe new algorithms for determining the differential matrices for the generating functions based on classical numerical methods. We analyze especially asymptotic behavior of the basic characteristics in Markov chains using the z-transformation and illustrate the calculation procedures for determining these characteristics. In addition, computational complexity aspects of the considered problems and of the proposed algorithms are analyzed and discussed.

D. Lozovanu and S. Pickl, *Optimization of Stochastic Discrete Systems and Control on Complex Networks*, Advances in Computational Management Science 12,
DOI 10.1007/978-3-319-11833-8_1

1.1 The Basic Definitions and Some Preliminary Results

The basic definitions related to finite-state space of Markov processes and the main results concerning the calculation of the state-time probabilities of the dynamical system in such processes can be found in [26, 39, 48–50, 52, 55, 56, 114, 140]. In this section we introduce some necessary notions and describe some preliminary results related to Markov processes that we use in this book.

1.1.1 Discrete Markov Processes and Determining the State-Time Probabilities of the System

Consider a *discrete Markov process* that models the evolution of the stochastic dynamical system $\mathbb{L}$ with the finite set of states $X = \{x_1, x_2, \ldots, x_n\}$. Assume that at the moment of time $t = 0$ the system is in the state $x_{i_0} \in X$ and for an arbitrary state $x \in X$ the probabilities $p_{x,y}$ of the systems' transitions from x to another state $y \in X$ are given, i.e.,

$$\sum_{y \in X} p_{x,y} = 1, \quad \forall x \in X; \quad p_{x,y} \geq 0, \quad \forall x, y \in X.$$

Here, the probabilities $p_{x,y}$ do not depend on time, i.e. we have a *stationary Markov process* determined by the *stochastic matrix* $P = (p_{x,y})$ and a given starting state x_{i_0}. So, we consider a stationary finite-state space *Markov chain* where the transition time from one state to another is constant and it is equal to 1 [56, 114, 115, 140].

For the dynamical system $\mathbb{L}$ we denote by $P_{x_{i_0}}(x, t)$ the probability to reach the state x at the moment of time t if it starts transitions at the moment of time $t = 0$ in the state x_{i_0}. We consider the probability $P_{x_{i_0}}(x, t)$ at the discrete moment of time $t = 0, 1, 2, \ldots$. Following [47] we define and calculate $P_{x_{i_0}}(x, t)$ using a recursive formula

$$P_{x_{i_0}}(x, \tau + 1) = \sum_{y \in X} P_{x_{i_0}}(y, \tau) p_{y,x}, \quad \tau = 0, 1, 2, \ldots, t - 1, \tag{1.1}$$

where $P_{x_{i_0}}(x_{i_0}, 0) = 1$ and $P_{x_{i_0}}(x, 0) = 0$ for $x \in X \backslash \{x_{i_0}\}$. We call these probabilities *state-time probabilities of the system* $\mathbb{L}$. Formula (1.1) can be represented in the matrix form

$$\pi(\tau + 1) = \pi(\tau) P, \quad \tau = 0, 1, 2, \ldots, t - 1. \tag{1.2}$$

Here $\pi(\tau) = (\pi_1(\tau), \pi_2(\tau), \ldots, \pi_n(\tau))$ is a vector, where an arbitrary component i expresses the probability of the system $\mathbb{L}$ to reach from the state x_{i_0} to the state x_i using τ transitions, i.e., $\pi_i(\tau) = P_{x_{i_0}}(x_i, \tau)$.

At the starting moment of time $\tau = 0$ the vector $\pi(\tau)$ is given and its components are defined as follows: $\pi_{i_0}(0) = 1$ and $\pi_i(0) = 0$ for arbitrary $i \neq i_0$. It is easy to observe that if for a given starting vector $\pi(0)$ we apply formula (1.2) for $\tau = 0, 1, 2, \dots, t-1$ then we obtain

$$\pi(t) = \pi(0)P^t, \tag{1.3}$$

where $P^t = P \times P \times \cdots \times P$. So, an arbitrary element $p^{(t)}_{x_i,x_j}$ of this matrix represents the probability of the system $\mathbb{L}$ to reach the state x_j from x_i by using t transitions.

It is easy to see that if the starting state of the system $\pi(0)$ is fixed then

$$\sum_{i=1}^{n} \pi_i(\tau) = 1, \quad \tau = 0, 1, 2, \dots. \tag{1.4}$$

The correctness of this relation can be proved easily by using the induction principle with respect to τ. Indeed, for $\tau = 0$ the equality (1.4) holds according to the definition. If we assume that (1.4) takes place for every $\tau \leq t$ then for $\tau = t + 1$ we obtain

$$\sum_{i=1}^{n} \pi_i(t+1) = \sum_{i=1}^{n} \sum_{x_j \in X} p_{x_j,x_i} \pi_j(t) = \sum_{x_j \in X} \pi_j(t) \sum_{i=1}^{n} p_{x_j,x_i} = \sum_{x_j \in X} \pi_j(t) = 1.$$

So, the relation (1.4) holds.

Let us assume that for the matrix P there exists the limit

$$\lim_{t\to\infty} P^t = Q.$$

Then for an arbitrary starting state in the case of a large number of transitions the vector of state-time probability $\pi(t)$ can be approximated with the corresponding row vector of the matrix Q. Indeed, if in (1.3) we take the limit then we have

$$\pi = \lim_{t\to\infty} \pi(t) = \pi(0) \lim_{t\to\infty} P^t = \pi(0)Q.$$

An arbitrary component π_j of $\pi = (\pi_1, \pi_2, \dots, \pi_n)$ can be treated as the probability that the system $\mathbb{L}$ will occupy the state x_j after a large number of transitions if it starts transitions in x_{i_0}. The vector π is called the *vector of limiting state probabilities* for the dynamical system $\mathbb{L}$ with a given starting state x_{i_0}.

Based on the property mentioned above we may conclude that

$$\sum_{j=1}^{n} \pi_j = 1$$

for an arbitrary $\pi(0)$.

This means that the elements $q_{x,y}$ of the matrix Q satisfy the condition

$$\sum_{y \in X} q_{x,y} = 1, \quad \forall x \in X$$

where $q_{x,y} \geq 0$, $\forall x, y \in X$, i.e., $Q = (q_{x,y})$ is a stochastic matrix. An arbitrary element $q_{x,y}$ of this matrix expresses the limiting probability of the system $\mathbb{L}$ to occupy the state $y \in X$ if the system starts transitions in x. The matrix Q is called the *matrix of limiting state probabilities* [47]; shortly we call this matrix the *limit matrix*. The Markov chain for which there exists $\lim_{t \to \infty} P^t = Q$ is called *aperiodic Markov chain* [114, 140].

If the limit matrix Q possesses the property that all its rows are the same then the corresponding Markov process is called *Markov unichain* [1, 47]. In [1, 47, 114] such processes are called *ergodic Markov processes*. Some authors use the notion of *ergodic Markov process* in the sense of an aperiodic Markov chain where all limiting probabilities are different from zero [37, 56]. Here we shall use the terminology from [114, 140]. The Markov process for which the limit matrix contains at least two different rows is called *Markov multichain* [56, 114, 140]. A more detailed classification of a Markov chain with respect to values of limiting probabilities is given in the next section.

So, for a Markov unichain we have $q_{x_i,x_j} = q_{x_k,x_j} = \pi_j, \quad \forall x_i, x_j, x_k \in X$, i.e., the limiting state probabilities π_j, $j = 1, 2, \ldots, n_j$ do not depend on the state in which the system starts transitions. For such processes the vector π of limiting state probabilities can be found by solving the system of linear equations

$$\pi(I - P) = 0; \quad \sum_{j=1}^{n} \pi_j = 1, \tag{1.5}$$

where I is the identity matrix. The first equation in (1.5) corresponds to the condition $\pi = \pi P$ that can be obtained from (1.2) if $\tau \to \infty$. The second equation in (1.5) reflects the property that after a large number of transitions the dynamical system will occupy one of the states $x_j \in X$. It is well known that the rank of the matrix $(I - P)$ for a Markov unichain is equal to $n - 1$ and the system (1.5) has a unique solution [98, 140]. In the case of a Markov multichain the rank of the matrix $(I - P)$ is less than $n - 1$.

It is easy to check that the Markov chain with the following probability transition matrix

$$P = \begin{pmatrix} \frac{1}{2} & \frac{1}{2} \\ \frac{2}{5} & \frac{3}{5} \end{pmatrix}$$

is ergodic and there exists $\lim_{t \to \infty} P^t = Q$.

The vector $\pi = (\pi_1, \pi_2)$ can be found by solving the system of linear equations

$$\begin{cases} \frac{1}{2}\pi_1 - \frac{2}{5}\pi_2 = 0, \\ \frac{1}{2}\pi_1 - \frac{2}{5}\pi_2 = 0, \\ \pi_1 + \pi_2 = 1. \end{cases}$$

If we solve this system then we obtain $\pi_1 = 4/9$, $\pi_2 = 5/9$, i.e., $\pi = (4/9, 5/9)$ and

$$Q = \begin{pmatrix} \frac{4}{9} & \frac{5}{9} \\ \frac{4}{9} & \frac{5}{9} \end{pmatrix}.$$

In general, the system (1.5) may also have a unique solution if the limit $\lim_{t\to\infty} P^t$ does not exist. This case corresponds to a *periodic Markov chain* [114]. In such a process a component π_j of the vector π that satisfies (1.5) can be treated as the probability of the system $\mathbb{L}$ to occupy the state x_j at random moments of times during a large number of transitions. An example for which $\lim_{t\to\infty} P^t$ does not exist and the system (1.5) has a unique solution is represented below by the following probability transition matrix

$$P = \begin{pmatrix} 0 & 1 \\ 1 & 0 \end{pmatrix}.$$

It is easy to check that

$$P^{2t} = \begin{pmatrix} 1 & 0 \\ 0 & 1 \end{pmatrix}, \quad P^{2t+1} = \begin{pmatrix} 0 & 1 \\ 1 & 0 \end{pmatrix}, \quad \forall t \geq 0,$$

i.e., we have 2−periodic Markov chains. In general, for a *k−periodic Markov chain* $(k > 1)$ for the matrix P there exists t_0 such that for $t > t_0$ it holds $P^{k+t} = P^t$. The system of linear equations (1.5) for the given example of 2−periodic Markov chains is the following

$$\begin{cases} \pi_1 - \pi_2 = 0, \\ \pi_2 - \pi_1 = 0, \\ \pi_1 + \pi_2 = 1. \end{cases}$$

If we solve this system then we obtain $\pi_1 = 1/2$, $\pi_2 = 1/2$. In such a way we obtain the matrix

$$Q = \begin{pmatrix} \frac{1}{2} & \frac{1}{2} \\ \frac{1}{2} & \frac{1}{2} \end{pmatrix}.$$

A more detailed characterization of the periodic and aperiodic Markov chains can be found in [33, 56, 114, 127]. In the following we show how the notion of the limit matrix Q can be extended for an arbitrary Markov chain, i.e., for the case that $\lim_{t\to\infty} P^t$ does not exist. We can see that in this case the elements of the matrix Q have a different interpretation.

1.1.2 Definition of the Limit Matrix and the Classification of the States in Markov Chains

The limit matrix Q for an arbitrary transition probability matrix P can be defined by using the following limit

$$\lim_{t\to\infty} \frac{1}{t} \sum_{\tau=0}^{t-1} P^{\tau}.$$

This limit is called *Cesaro limit* [33, 114, 127, 140]. In [20] it has been proven that this limit exists for an arbitrary stochastic matrix P, i.e.,

$$\lim_{t\to\infty} \frac{1}{t} \sum_{\tau=0}^{t-1} P^{\tau} = Q$$

where $P^0 = I$ is the identity matrix. Moreover, in [20] it is shown that if $\lim_{t\to\infty} P^t$ exists then

$$\lim_{t\to\infty} P^t = \lim_{t\to\infty} \frac{1}{t} \sum_{\tau=0}^{t-1} P^{\tau} = Q.$$

In addition, in [20] it is shown that for an arbitrary stochastic matrix P the corresponding limit matrix Q satisfies the condition

$$PQ = QP = Q. \tag{1.6}$$

Based on this property we can give the interpretation of the solution of the Eq. (1.5) for the periodic Markov process. In the general case an element $q_{x,y}$ of the limit matrix Q in the Markov chain can be interpreted as the long-run time fraction that the system occupies the state y if it starts transitions in x. For an aperiodic Markov chain an arbitrary element $q_{x,y}$ can be treated as the steady-state probability that the system will be in the state y after a large number of transitions if it starts transitions in x.

We give a classification of the states in a Markov chain using the structure properties of the matrices P and Q. We say that the state y is *accessible* from the state x if there exists $t > 0$ such that $P_x(y, t) > 0$, i.e., y is accessible from x if y can be reached from x with a positive probability $P_x(y, t)$. A subset Y of X is called *closed set* if no state outside Y is accessible from any state in Y.

If a closed set X does not contain a proper closed subset then it is called *irreducible set*. An irreducible set in X induces a distinct Markov chain which we call *irreducible Markov chain*. The irreducible set consisting of a single state is said to be an *absorbing state*.

Let $Q = (q_{x,y})$ be the limit matrix of the Markov process with a given transition probability matrix $P = (p_{x,y})$. Two states $x, y \in X$ are called *equivalent* if in the matrix Q the rows that correspond to x and y are the same, i.e., the states $x, y \in X$ are equivalent if $q_{x,z} = q_{y,z}$, $\forall z \in X$. Thus, the set of states X can be divided into several subsets $X_1, X_2, \ldots, X_k$ $\big(X = \bigcup_{i=1}^{k} X_i, X_i \cap X_j = \emptyset,\ i \neq j\big)$ such that each subset contains only equivalent states and there are no equivalent states that belong to different subsets. A state $x \in X$ is called *positive recurrent state* if $q_{x,x} > 0$; a state $x \in X$ for which $q_{x,x} = 0$ is called *transient state*.

Let $X^+ = \{x \in X \mid q_{x,x} > 0\}$, $X^0 = \{x \in X \mid q_{x,x} = 0\}$. Then the set X can be represented as follows

$$X = X_1^+ \cup X_2^+ \cup \cdots \cup X_{k'}^+ \cup X^0,$$

where each X_i^+ consists of positive recurrent states, i.e., $X_i^+ = \{x \in X_i \mid q_{x,x} > 0\}$, $i \in \{1, 2, \ldots, k'\}$, and X^0 is the set of transient states. Taking into account that two arbitrary states $x, y \in X_i^+$ are equivalent we have $q_{x,y} > 0$ and $q_{y,x} > 0$. It is easy to observe that X_i^+ in X represents the maximal set with mutual accessible states and that no state outside of X_i is accessible from X_i. Thus, each $X_i^+ \subseteq X$ represents an irreducible set in X.

So, for arbitrary two states $x, y \in X_i$, $i = 1, 2, \ldots, k'$, it holds $q_{x,y} > 0$ and for arbitrary two states $x, y \in X$ that belong to different irreducible sets X_i^+ and X_j^+, it holds $q_{x,y} = 0$ and $p_{x,y} = 0$. Based on these properties we may conclude that the states $x \in X$ can be relabeled in such a way that the matrices P and Q can be expressed as follows

$$P = \begin{pmatrix} P_1 & 0 & \ldots & 0 & 0 \\ 0 & P_2 & \ldots & 0 & 0 \\ \cdot & \cdot & \ldots & \cdot & \cdot \\ \cdot & \cdot & \ldots & \cdot & \cdot \\ \cdot & \cdot & \ldots & \cdot & \cdot \\ 0 & 0 & \ldots & P_{k'} & 0 \\ W_1 & W_2 & \ldots & W_{k'} & W_{k'+1} \end{pmatrix}, \qquad Q = \begin{pmatrix} Q_1 & 0 & \ldots & 0 & 0 \\ 0 & Q_2 & \ldots & 0 & 0 \\ \cdot & \cdot & \ldots & \cdot & \cdot \\ \cdot & \cdot & \ldots & \cdot & \cdot \\ \cdot & \cdot & \ldots & \cdot & \cdot \\ 0 & 0 & \ldots & Q_{k'} & 0 \\ R_1 & R_2 & \ldots & R_{k'} & 0 \end{pmatrix}.$$

In these matrices P_i and Q_i correspond to transitions between states in X_i^+; the matrices W_i and R_i correspond to transitions from states in X^0 to X_i and $W_{k'+1}$ is the matrix that corresponds to transitions between states in X^0. The matrix P represented in such a form is called the *transition probability matrix in canonical form*.

1.1.3 A Decomposition Algorithm for Determining the Limit Matrix

If the matrix P of transition probabilities is given in canonical form then for determining the limit matrix Q a decomposition algorithm which is described in [37, 115] can be applied. Each component P_i corresponds to a *irreducible Markov chain* and, therefore, all rows of the matrix Q_i are identical. Each matrix Q_i is determined according to the formula

$$Q_i = \overline{e}_i \pi^i,$$

where $\overline{e}_i$ is the column vector, where each component of which is equal to one and π^i is a row vector with the component that represents the solution of the system

$$\pi^i = \pi^i P_i, \quad \sum_j \pi^i_j = 1.$$

The matrices R_i can be calculated by using the formula

$$R_i = (I - W_{k'+1})^{-1} W_i Q_i, \quad i = 1, 2, \ldots, k'. \tag{1.7}$$

The correctness of this formula can be proved if we represent, respectively, the transition matrix P as

$$P = \begin{pmatrix} \overline{P} & 0 \\ \overline{W} & W_{k'+1} \end{pmatrix},$$

the limit matrix Q as

$$Q = \begin{pmatrix} \overline{Q} & 0 \\ \overline{R} & 0 \end{pmatrix},$$

and then apply the property (1.6). Indeed, since $PQ = Q$, we obtain

$$\overline{W}\,\overline{Q} + W_{k'+1}\overline{R} = \overline{R}.$$

Thus,

$$W_i Q_i + W_{k'+1} R_i = R_i, \quad i = 1, 2, \ldots, k'.$$

Taking into account that in the case of a Markov multichain the inverse matrix $(I - W_{k'+1})^{-1}$ for $(I - W_{k'+1})$ exists (see [7, 21]) we obtain from this formula (1.7).

So, to determine the limit matrix Q it is sufficient to represent the matrix P in the canonical form and then to determine $Q_1, Q_2, \ldots, Q_{k'}$ and $R_1, R_2, \ldots, R_{k'}$.

An algorithm for relabeling the states in Markov chains which transform the matrix P of probability transitions into its canonical form is described in [37, 62, 115]. This algorithm uses $O(n^2)$ elementary operations. In general, to determine the irreducible sets X_i we can use the *graph of probability transitions* $G_p = (X, E_p)$ for Markov chains.

The *adjacency matrix* of this graph is obtained from the matrix P by changing the nonzero probabilities with units. Then the strongly connected components which do not contain directed edges to other strongly connected components correspond to the irreducible sets X_i. Therefore, if we apply an algorithm for finding the strongly connected components of G_p we obtain an algorithm for finding the irreducible sets X_i.

Example Consider the problem of determining the limit matrix Q for a Markov chain with the probability matrix

$$P = \begin{pmatrix} 1 & 0 & 0 & 0 \\ 0 & \frac{1}{2} & \frac{1}{2} & 0 \\ 0 & \frac{2}{5} & \frac{3}{5} & 0 \\ \frac{1}{4} & \frac{1}{4} & \frac{1}{4} & \frac{1}{4} \end{pmatrix}.$$

In Fig. 1.1 the graph of probability transitions $G_p = (X, E_p)$ of a Markov chain is given. The strongly connected components of this graph are determined by the subsets

$$X_1 = \{1\}, \quad X_2 = \{2, 3\}, \quad X_3 = \{4\}.$$

The irreducible sets correspond to strongly connected components that do not contain outgoing directed edges to other strongly connected components. Thus, we have two irreducible sets $X_1 = \{1\}$ and $X_2 = \{2, 3\}$. Therefore, the *transition probability matrix* mentioned above has a canonical form, where

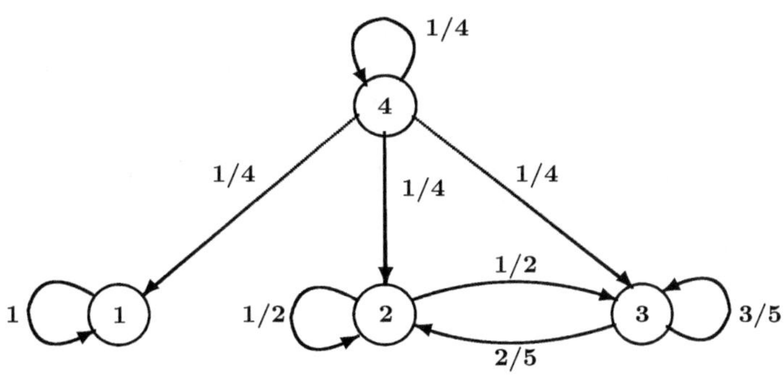

Fig. 1.1 The graph of probability transitions

$$P_1 = (1), \; P_2 = \begin{pmatrix} \frac{1}{2} & \frac{1}{2} \\ \frac{2}{5} & \frac{3}{5} \end{pmatrix}, \; W_1 = \begin{pmatrix} \frac{1}{4} \end{pmatrix}, \; W_2 = \begin{pmatrix} \frac{1}{4} & \frac{1}{4} \end{pmatrix}, \; W_3 = \begin{pmatrix} \frac{1}{4} \end{pmatrix}.$$

Therefore, we solve the system of linear equations $\pi^i = \pi^i P_i$, $\sum_j \pi^i_j = 1$ for $i = 1, 2$ and determine $\pi^1 = (1)$, $\pi^2 = (4/9 \; 5/9)$. This means that

$$Q_1 = (1), \quad Q_2 = \begin{pmatrix} \frac{4}{9} & \frac{5}{9} \\ \frac{4}{9} & \frac{5}{9} \end{pmatrix}.$$

Now we calculate $R_i = (I - W_3)^{-1} W_i Q_i$, $i = 1, 2$ and obtain

$$R_1 = \begin{pmatrix} \frac{3}{4} \end{pmatrix}^{-1} (1) \begin{pmatrix} \frac{1}{4} \end{pmatrix} = \begin{pmatrix} \frac{1}{3} \end{pmatrix};$$

$$R_2 = \begin{pmatrix} \frac{3}{4} \end{pmatrix}^{-1} \begin{pmatrix} \frac{1}{4} & \frac{1}{4} \end{pmatrix} \begin{pmatrix} \frac{4}{9} & \frac{5}{9} \\ \frac{4}{9} & \frac{5}{9} \end{pmatrix} = \begin{pmatrix} \frac{8}{27} & \frac{10}{27} \end{pmatrix}.$$

In such a way we obtain the limit matrix

$$Q = \begin{pmatrix} 1 & 0 & 0 & 0 \\ 0 & \frac{4}{9} & \frac{5}{9} & 0 \\ 0 & \frac{4}{9} & \frac{5}{9} & 0 \\ \frac{1}{3} & \frac{8}{27} & \frac{10}{27} & 0 \end{pmatrix}.$$

It is easy to check that in the worst case this algorithm uses $O(|X|^3)$ elementary operations. We can see that along to this algorithm other variations for determining the limit matrix in a Markov chain can be used. In the following we analyze an approach for determining the limit matrix that allows us to study the asymptotic behavior of the state-time probability in Markov chains. This approach is based on the properties of the generating functions [33, 47]. In [47] the generating functions are called z-transforms.

1.1.4 The z-Transform and the Asymptotic Behavior of State-Time Probabilities

For the analysis of the *asymptotic behavior of state-time probabilities in Markov processes* we shall use the apparatus of the generating functions [33, 47, 56, 127, 140]. In [47] the generating function is called z-transform. In this section we describe shortly the results from [47] concerning the application of the z-transform to an asymptotic estimation of the state-time probability in a Markov chain. Then in the next section we will develop numerical algorithms for the calculation of the limit matrix and the state-time probabilities using the z-transform.

We consider the z-transform for a time-discrete function $f(t)$, $t = 0, 1, 2, \ldots$ with the values $f(0), f(1), (2), \ldots$ that do not increase in magnitude faster than a geometric sequence. In this case the z-transform

$$F(z) = \sum_{t=0}^{\infty} f(t)z^t$$

can be defined uniquely.

Based on the definition of the z-transform we can easily obtain the following elementary properties:

(1) $F(z) = F_1(z) + F_2(z)$ if $f(t) = f_1(t) + f_2(t)$,
where $F_1(z) = \sum_{t=0}^{\infty} f_1(t)z^t$ and $F_2(z) = \sum_{t=0}^{\infty} f_2(t)z^t$;

(2) $F(z) = kF'(z)$ if $f(t) = kf'(t)$,
where $F'(z) = \sum_{t=0}^{\infty} f'(t)z^t$;

(3) $F(z) = z^{-1}[F'(z) - f'(0)]$ if $f(t) = f'(t+1)$,
where $F'(z) = \sum_{t=0}^{\infty} f'(t)z^t$;

(4) $F(z) = zF'(z)$ if $f(t) = f'(t-1)$,
where $F'(z) = \sum_{t=0}^{\infty} f'(t)z^t$;

(5) $F(z) = \dfrac{1}{1-z}$ if $f(t) = \begin{cases} 1, & t = 0, 1, 2, \ldots, \\ 0, & t < 0; \end{cases}$

(6) $F(z) = \dfrac{1}{1-\alpha z}$ if $f(t) = \alpha^t$;

(7) $F(z) = \dfrac{\alpha z}{(1-\alpha z)^2}$ if $f(t) = t\alpha^t$;

(8) $F(z) = \dfrac{z}{(1-z)^2}$ if $f(t) = t$;

(9) $F(z) = F'(\alpha z)$ if $f(t) = \alpha^t f'(t)$,
where $F'(z) = \sum_{t=0}^{\infty} f'(t)z^t$.

In a more detailed form the properties of the z-transform for different classes of time-discrete functions are described in [33, 47, 56, 127]. Note that the z-transform can be applied to the vectors and matrices by applying the z-transform to each component of the array. If we denote the z-transform of the vector $\pi(t)$ by $F(z)$ and if we apply the z-transform to the relationship

$$\pi(t+1) = \pi(t)P$$

then we obtain

$$F(z+1) = F(z)P.$$

Based on property (3) this formula can be written as follows

$$z^{-1}(F(z) - F(0)) = F(z)P,$$

i.e.,

$$F(z)(I - zP) = F(0)$$

were I is the identity matrix. Taking into account that $F(0) = \pi(0)$ we finally obtain

$$F(z) = \pi(0)(I - zP)^{-1}.$$

So, the z-transform of the vector of state-time probabilities is equal to the initial vector of probabilities $\pi(0)$ multiplied by the matrix $(I - zP)^{-1}$. This mean that the solution of the transient problem is contained in the matrix $(I - zP)^{-1}$. We can obtain the probabilities of our transient problem in an analytical form if we weight the rows of the matrix $(I - zP)^{-1}$ by the initial state probabilities, sum them, and then take the inverse transformation. In such a way we can obtain the matrix $\overline{P}(t)$ which represents the inverse transformation of the matrix $(I - zP)^{-1}$. An example which illustrates how to obtain the matrix $\overline{P}(t)$ for the Markov process with the following stochastic matrix of probability transitions is given below

$$P = \begin{pmatrix} \frac{1}{2} & \frac{1}{2} \\ \frac{2}{5} & \frac{3}{5} \end{pmatrix}.$$

To determine the matrix $\overline{P}(t)$ we form the matrix

$$(I - zP) = \begin{pmatrix} 1 - \frac{1}{2}z & -\frac{1}{2}z \\ -\frac{2}{5}z & 1 - \frac{3}{5}z \end{pmatrix}$$

and find the inverse matrix

$$(I - zP)^{-1} = \frac{1}{\Delta(z)} \begin{pmatrix} 1 - \frac{3}{5}z & \frac{1}{2}z \\ \frac{2}{5}z & 1 - \frac{1}{2}z \end{pmatrix}$$

where $\Delta(z) = (1 - z)(1 - z/10)$.

Finally, the inverse matrix can be represented as follows

$$(I - zP)^{-1} = \frac{1}{1 - z} \begin{pmatrix} \frac{4}{9} & \frac{5}{9} \\ \frac{4}{9} & \frac{5}{9} \end{pmatrix} + \frac{1}{1 - \frac{1}{10}z} \begin{pmatrix} \frac{5}{9} & -\frac{5}{9} \\ -\frac{4}{9} & \frac{4}{9} \end{pmatrix}.$$

If we take the inverse transform of $(I - zP)^{-1}$, then we obtain

$$\overline{P}(t) = \begin{pmatrix} \frac{4}{9} & \frac{5}{9} \\ \frac{4}{9} & \frac{5}{9} \end{pmatrix} + \frac{1}{10^t} \begin{pmatrix} \frac{5}{9} & -\frac{5}{9} \\ -\frac{4}{9} & \frac{4}{9} \end{pmatrix}.$$

From this formula in the case $t \to \infty$ we obtain

$$\overline{P}(t) \to Q = \begin{pmatrix} \frac{4}{9} & \frac{5}{9} \\ \frac{4}{9} & \frac{5}{9} \end{pmatrix}.$$

Note that the example given above corresponds to an aperiodic Markov chain and for the matrix P there exists the limit $\lim_{t \to \infty} P^t$. An example for which this limit does not exist is given by the following matrix

$$P = \begin{pmatrix} 0 & 1 \\ 1 & 0 \end{pmatrix}.$$

In a similar way we can find the matrix $(I - zP)^{-1}$ of the z-transform

$$(I - zP)^{-1} = \frac{1}{1-z}\begin{pmatrix} \frac{1}{2} & \frac{1}{2} \\ \frac{1}{2} & \frac{1}{2} \end{pmatrix} + \frac{1}{1+z}\begin{pmatrix} \frac{1}{2} & -\frac{1}{2} \\ -\frac{1}{2} & \frac{1}{2} \end{pmatrix}.$$

Based on property (6) of the z-transform we obtain the matrix

$$\overline{P}(t) = \begin{pmatrix} \frac{1}{2} & \frac{1}{2} \\ \frac{1}{2} & \frac{1}{2} \end{pmatrix} + (-1)^t \begin{pmatrix} \frac{1}{2} & -\frac{1}{2} \\ -\frac{1}{2} & \frac{1}{2} \end{pmatrix}.$$

We can see that in this case for the matrix P the limit $\lim_{t\to\infty} \overline{P}(t)$ does not exist. However, the matrix Q can be found by using the z-transform.

Thus, the vector $\pi(t)$ can be calculated by using the following formula

$$\pi(t) = \pi(0)\overline{P}(t).$$

By comparison of this formula with formula $\pi(t) = \pi(0)P^t$ we conclude that $\overline{P}(t) = P^t$. This formula can be used for the asymptotic analysis of the transient problem. An important property of the matrix $\overline{P}(t)$ which can be obtained on the bases of the z transform is the following: Among the component matrices in the representation of $\overline{P}(t)$ there exists at least one matrix that is stochastic and it corresponds in the term $(I - zP)^{-1}$ with the coefficient $1/(1 - z)$. This follows from the fact that the determinant of $(I - zP)$ is equal to zero for $z = 1$, i.e., the stochastic matrix P always has at least one characteristic value equal to 1. In the ergodic case the Markov process $\overline{P}(t)$ will contain exactly one such stochastic matrix. Moreover, in the ergodic case the rows of the stochastic components are identical and correspond to the vector of limiting-state probabilities of the process. This component does not depend on time and we denote it by Q. The remaining terms of $\overline{P}(t)$ express the transient behavior of the process and represent matrices multiplied by coefficients of the form $\lambda_k^t, t\lambda_k^t, t^2\lambda_k^t, \ldots$, where $|\lambda_k| \leq 1$; for the ergodic processes $|\lambda_k| < 1$. In the following we can see that λ_k represents the proper values of the characteristic polynomial of the matrix P. The transient component of $\overline{P}(t)$ that depends on t is denoted by $T(t)$; this component vanishes as t becomes very large. The matrices of the component $T(t)$ possess the property that the sum of the elements in each row is equal to zero. This property for transient components holds because it expresses the perturbations from the limiting state probabilities. The matrices with such properties are called *differential matrices*. So, the matrix $\overline{P}(t)$ for the Markov process can be represented as follows

$$\overline{P}(t) = Q + T(t),$$

where Q is a stochastic matrix where the rows represent the vector of limiting state probabilities and $T(t)$ is the sum of the differential matrices with the geometric coefficients which for the ergodic case tends to zero if $t \to \infty$.

The z-transform shows that the matrix Q exists for an arbitrary transition probability matrix P and $T(t)$ can be expressed as the sum of differential matrices multiplied with the coefficients $t^i \lambda_k^t$, where λ_k are the roots of the characteristic polynomial of the matrix P. In the next two sections we show how to determine Q and the corresponding components (differential matrices) of the matrix $T(t)$.

1.2 An Algorithm for Determining the Limit Matrix Based on the z-Transform and Classical Numerical Methods

In this section we develop a numerical method for determining the limit matrix in discrete Markov processes based on the z-transform and classical numerical methods [64, 72]. We propose and formulate a polynomial time algorithm for determining the limit matrix Q.

The running time of the algorithm is $O(n^4)$ where n is the number of the states of the dynamical system. In the following we show that the proposed method can be extended for calculating the differential matrices.

1.2.1 The Main Approach and the General Scheme of the Algorithm

Let $\mathbb{C}$ be a complex space and denote by $\mathbb{M}(\mathbb{C})$ the set of complex matrices with n rows and n columns. We consider the function $A : \mathbb{C} \to \mathbb{M}(\mathbb{C})$, where

$$A(z) = I - zP, \quad z \in \mathbb{C}.$$

We denote the elements of the matrix $A(z)$ by $a_{i,j}(z)$, $i, j = 1, 2, \ldots, n$, i.e.,

$$a_{i,j}(z) = \delta_{i,j} - zp_{i,j} \in \mathbb{C}[z],$$

where

$$\delta_{i,j} = \begin{cases} 1 & \text{if} \quad i = j, \\ 0 & \text{if} \quad i \neq j, \end{cases} \qquad i, j = 1, 2, \ldots, n.$$

It is evident that the determinant $\Delta(z)$ of the matrix $A(z)$ is a polynomial of degree less or equal to n, $(\deg(\Delta(z)) \leq n, \Delta(z) \in \mathbb{C}[z])$. Therefore, if we denote $\mathcal{D} = \{z \in \mathbb{C} \mid \Delta(z) \neq 0\}$ then we obtain that $|\mathbb{C} \backslash \mathcal{D}| \leq \deg(\Delta(z)) \leq n$ and for an arbitrary $z \in \mathcal{D}$

there exists the inverse matrix of $A(z)$. So, we can define the function $F:\mathcal{D} \to \mathbb{M}(\mathbb{C})$ where

$$F(z) = (A(z))^{-1}.$$

Then the elements $F_{i,j}(z)$, $i, j = 1, 2, \ldots, n$ of $F(z)$ can be found as follows

$$F_{i,j}(z) = \frac{M_{j,i}(z)}{\Delta(z)}, \quad i, j = 1, 2, \ldots, n,$$

where

$$M_{i,j}(z) = (-1)^{i+j} A_{i,j}(z)$$

and $A_{i,j}(z)$ is a determinant of the matrix obtained from $A(z)$ by deleting the row i and the column j, $i, j = 1, 2, \ldots, n$. Therefore,

$$M_{j,i}(z) \in \mathbb{C}[z], \ \deg(M_{j,i}(z)) \leq n - 1, \quad i, j = 1, 2, \ldots, n.$$

Note that $\Delta(1) = 0$ because for the matrix $A(1)$ the following property holds

$$\sum_{j=1}^{n}(\delta_{i,j} - p_{i,j}) = \sum_{j=1}^{n} \delta_{i,j} - \sum_{j=1}^{n} p_{i,j} = \delta_{i,i} - 1 = 0, \quad i = 1, 2, \ldots, n.$$

This means that $1 \in \mathbb{C}\backslash\mathcal{D}$ and $\Delta(z)$ can be factorized by $z - 1$.

Taking into account that $F_{i,j}(z)$ is a rational fraction with the denominator $\Delta(z)$ we can represent $F_{i,j}(z)$ uniquely in the following form

$$F_{i,j}(z) = B_{i,j}(z) + \sum_{y \in \mathbb{C}\backslash\mathcal{D}} \sum_{k=1}^{m(y)} \frac{\alpha_{i,j,k}(y)}{(z - y)^k}, \quad i, j = 1, 2, \ldots, n, \tag{1.8}$$

where $m(z)$ is the order of the root of the polynomial $\Delta(z)$, $z \in \mathbb{C}\backslash\mathcal{D}$, $\alpha_{i,j,k}(y) \in \mathbb{C}$, $\forall y \in \mathbb{C}\backslash\mathcal{D}$, $k = 1, 2, \ldots, m(y)$, $i, j = 1, 2, \ldots, n$. In the representation of $F_{i,j}(z)$ given above the degree of the polynomial $B_{i,j}(z) \in \mathbb{C}[z]$ satisfies the condition

$$\deg(B_{i,j}(z)) = \deg(M_{j,i}(z)) - \deg(\Delta(z)),$$

where $\deg(M_{j,i}(z)) \geq \deg(\Delta(z))$, otherwise $B_{i,j}(z) = 0$.

To represent (1.8) in a more convenient form we shall use the expressions of series expansion for the functions $\nu_k(z) = 1/(1-z)^k$, $k = 1, 2, \ldots$. First of all we observe that for these functions there exist a series expansion. In particular for $k = 1$ the function $\nu_1(z)$ can be represented by $\nu_1(z) = \sum_{t=0}^{\infty} z^t$. In the general case (for an arbitrary $k > 1$) the following recursive relation holds $\nu_{k+1}(z) = \mathrm{d}\nu_k(z)/(k\mathrm{d}z)$, $k = 1, 2, \ldots$. Using these properties and the induction principle we obtain the series expansion of the function $\nu_k(z)$, $\forall k \geq 1$: $\nu_k(z) = \sum_{t=0}^{\infty} H_{k-1}(t)z^t$, where $H_{k-1}(t)$ is a polynomial of degree less or equal to $(k-1)$.

Based on the properties mentioned above we can make the following transformation in (1.8):

$$\begin{aligned}
F_{i,j}(z) &= B_{i,j}(z) + \sum_{y \in \mathbb{C}\setminus\mathcal{D}} \sum_{k=1}^{m(y)} \frac{\left(-\frac{1}{y}\right)^k \alpha_{i,j,k}(y)}{\left(1 - \frac{1}{y}z\right)^k} \\
&= B_{i,j}(z) + \sum_{y \in \mathbb{C}\setminus\mathcal{D}} \sum_{k=1}^{m(y)} \left(-\frac{1}{y}\right)^k \alpha_{i,j,k}(y)\nu_k\left(\frac{z}{y}\right) \\
&= B_{i,j}(z) + \sum_{y \in \mathbb{C}\setminus\mathcal{D}} \sum_{k=1}^{m(y)} \left(-\frac{1}{y}\right)^k \alpha_{i,j,k}(y) \sum_{t=0}^{\infty} H_{k-1}(t)\left(\frac{z}{y}\right)^t \\
&= B_{i,j}(z) + \sum_{t=0}^{\infty} \left(\frac{z}{y}\right)^t \sum_{y \in \mathbb{C}\setminus\mathcal{D}} \sum_{k=0}^{m(y)-1} \left(-\frac{1}{y}\right)^{k+1} \alpha_{i,j,k+1}(y)H_k(t).
\end{aligned}$$

We can observe that in the relation above the expression

$$\sum_{k=0}^{m(y)-1} \left(-\frac{1}{y}\right)^{k+1} \alpha_{i,j,k+1}(y)H_k(t)$$

represents a polynomial of degree less or equal to $m(y)-1$ and we can write it in the form $\sum_{k=0}^{m(y)-1} \beta_{i,j,k}(y)t^k$, where $\beta_{i,j,k}(y)$ represent the corresponding coefficients of this polynomial. Therefore, if in this expression we substitute $\sum_{k=0}^{m(y)-1}(-1/y)^{k+1}$ $\alpha_{i,j,k+1}(y)H_k(t)$ by $\sum_{k=0}^{m(y)-1} \beta_{i,j,k}(y)t^k$ then we obtain

$$\begin{aligned}
F_{i,j}(z) &= B_{i,j}(z) + \sum_{t=0}^{\infty} z^t \sum_{y \in \mathbb{C}\setminus\mathcal{D}} \sum_{k=0}^{m(y)-1} \frac{t^k}{y^t}\beta_{i,j,k}(y) \\
&= W_{i,j}(z) + \sum_{t=1+\deg(B_{i,j}(z))}^{\infty} z^t \sum_{y \in \mathbb{C}\setminus\mathcal{D}} \sum_{k=0}^{m(y)-1} \frac{t^k}{y^t}\beta_{i,j,k}(y), \\
&\qquad i, j = 1, 2, \ldots, n,
\end{aligned} \tag{1.9}$$

where $\beta_{i,j,k}(y) \in \mathbb{C}$, $\forall y \in \mathbb{C}\setminus\mathcal{D}$, $k = 0, 1, 2, \ldots, m(y)-1$, $i, j = 1, 2, \ldots, n$, and $W_{i,j}(z) \in \mathbb{C}[z]$ is a polynomial, where the degree satisfies the condition $\deg(W_{i,j}(z)) = \deg(B_{i,j}(z))$, $i, j = 1, 2, \ldots, n$.

In addition we observe that the norm of the matrix P satisfies the condition $\|P\| = \max_{i=1,2,\ldots,n} \sum_{j=1}^{n} p_{i,j} = 1$, and, therefore, $\|zP\| = |z|\|P\| = |z|$.

Let $|z| < 1$. Thus, for $F(z)$ we have

$$F(z) = (I - zP)^{-1} = \sum_{t=0}^{\infty} P^t z^t.$$

This means that

$$F_{i,j}(z) = \sum_{t=0}^{\infty} \overline{p}_{i,j}(t) z^t, \quad i, j = 1, 2, \ldots, n. \tag{1.10}$$

From the definition of the z-transform and from (1.9), (1.10) we obtain

$$\overline{p}_{i,j}(t) = \sum_{y \in \mathbb{C} \setminus \mathcal{D}} \sum_{k=0}^{m(y)-1} \frac{t^k}{y^t} \beta_{i,j,k}(y), \quad \forall t > \deg(B_{i,j}(z)), \quad i, j = 1, 2, \ldots, n.$$

Since $0 \le \overline{p}_{i,j}(t) \le 1, i, j = 1, 2, \ldots, n, \ \forall t \ge 0$, we have

$$|y| \ge 1, \ \forall y \in \mathbb{C} \setminus \mathcal{D}, \quad \beta_{i,j,k}(1) = 0, \ \forall k \ge 1.$$

This implies $\alpha_{i,j,k}(1) = 0, \ \forall k \ge 2$.

Now let us assume that $\Delta(z) = (z-1)^{m(1)} H(z)$, where $H(1) \ne 0$. Then the relation (1.8) is represented as follows:

$$\begin{aligned} F_{i,j}(z) &= \frac{\alpha_{i,j,1}(1)}{z-1} + B_{i,j}(z) + \sum_{y \in (\mathbb{C} \setminus \mathcal{D}) \setminus \{1\}} \sum_{k=1}^{m(y)} \frac{\alpha_{i,j,k}(y)}{(z-y)^k} \\ &= \frac{\alpha_{i,j,1}(1)}{z-1} + \frac{Y_{i,j}(z)}{H(z)}, \quad i, j = 1, 2, \ldots, n, \end{aligned}$$

where $Y_{i,j}(z) \in \mathbb{C}[z]$ and

$$\begin{aligned} \deg(Y_{i,j}(z)) &= \deg(B_{i,j}(z)) + \deg(T(z)) = \deg(B_{i,j}(z)) + \deg(\Delta(z)) - m(1) \\ &= \deg(M_{j,i}(z)) - m(1) \le n - 1 - m(1) \le n - 2, \quad i, j = 1, 2, \ldots, n. \end{aligned}$$

If we denote

$$Y(z) = (Y_{i,j}(z))_{i,j=1,2,\ldots,n}, \quad \alpha_1(1) = (\alpha_{i,j,1}(1))_{i,j=1,2,\ldots,n}$$

then the matrix $F(z)$ can be represented as follows:

$$F(z) = \frac{1}{z-1} \alpha_1(1) + \frac{1}{H(z)} Y(z). \tag{1.11}$$

From this formula and from the definition of the limiting state matrix Q we have

$$Q = -\alpha_1(1), \tag{1.12}$$

i.e., in the representation of the inverse matrix of $(I - zP)$ the limit matrix Q will correspond to the term with the coefficient $1/(1-z)$.

From (1.11) and (1.12) we obtain the formula

$$Q = \lim_{z \to 1} (1-z)(I - zP)^{-1}.$$

In the following we show how to determine the polynomial $\Delta(z)$ and the function $F(z)$ in the matrix form.

1.2.2 The Calculation of the Coefficients of the Characteristic Polynomial

Let us consider the characteristic polynomial

$$K(\lambda) = |P - \lambda I| = \sum_{k=0}^{n} \nu_k \lambda^k.$$

It is easy to observe that $\nu_n = |-I| = (-1)^n \neq 0$. This means that $\deg(K(\lambda)) = n$ and the characteristic polynomial may be written in the following form

$$K(\lambda) = (-1)^n(\lambda^n - \alpha_1\lambda^{n-1} - \alpha_2\lambda^{n-2} - \cdots - \alpha_n).$$

If we assume that $\alpha_0 = -1$ then the coefficients ν_k can be represented as follows

$$\nu_k = (-1)^{n+1}\alpha_{n-k}, \quad k = 0, 1, 2, \ldots, n.$$

In [44] it is shown that the coefficients α_k can be calculated by using $O(n^3)$ elementary operations based on Leverrier's method. This method can be applied for determining the coefficients α_k by the following way:

(1) Determine the matrices

$$P^k = (p_{i,j}^{(k)}), \quad i, j = 1, 2, \ldots, n, \quad k = 1, 2, \ldots, n$$

where $P^k = P \times P \times \cdots \times P$;

(2) Find the traces of the matrices P^k:

$$\overline{s}_k = \operatorname{tr}(P^k) = \sum_{j=1}^{n} p_{j,j}^{(k)}, \quad k = 1, 2, \ldots, n;$$

(3) Calculate the coefficients

$$\alpha_k = \frac{1}{k}\left(\overline{s}_k - \sum_{j=1}^{k-1} \alpha_j \overline{s}_{k-j}\right), \quad k = 1, 2, \ldots, n.$$

Thus, if the coefficients α_k are known then we can determine the coefficients of the polynomial $\Delta(z) = \sum_{k=0}^{n} \beta_k z^k$. Indeed, if $z \in \mathbb{C}\setminus\{0\}$ then

$$\begin{aligned}
\Delta(z) = |I - zP| &= (-z)^n \left|P - \frac{1}{z} I\right| = (-1)^n z^n K\left(\frac{1}{z}\right) \\
&= (-1)^n z^n \sum_{k=0}^{n} \nu_k \frac{1}{z^k} = (-1)^n \sum_{k=0}^{n} \nu_k z^{n-k} = \sum_{k=0}^{n} (-1)^n \nu_{n-k} z^k \\
&= \sum_{k=0}^{n} (-1)^n (-1)^{n+1} \alpha_k z^k = \sum_{k=0}^{n} (-\alpha_k) z^k.
\end{aligned}$$

For $z = 0$ we have

$$\Delta(0) = |I| = 1 = -\alpha_0.$$

Therefore, finally we obtain

$$\Delta(z) = \sum_{k=0}^{n} (-\alpha_k) z^k, \quad \forall z \in \mathbb{C}.$$

This means that

$$\beta_k = -\alpha_k, \quad k = 0, 1, 2, \ldots, n.$$

So, the coefficients $\beta_k, k = 0, 1, 2, \ldots, n$ can be calculated using a similar recursive formula

$$\beta_k = -\alpha_k = -\frac{1}{k}\left(\overline{s}_k - \sum_{j=1}^{k-1} \alpha_j s_{k-j}\right) = -\frac{1}{k}\left(\overline{s}_k + \sum_{j=1}^{k-1} \beta_j \overline{s}_{k-j}\right), \quad k = 1, 2, \ldots, n,$$

$$\beta_0 = -\alpha_0 = 1.$$

Based on the result described above we can propose the following algorithm for determining the coefficients β_k:

Algorithm 1.1 Determining the Coefficients of the Characteristic Polynomial

(1) Calculate the matrices $P^k = \left(p_{i,j}^{(k)}\right), \quad i, j = 1, 2, \ldots, n, \quad k = 1, 2, \ldots, n;$

(2) Determine the traces of the matrices P^k :

$$\overline{s}_k = \operatorname{tr}(P^k) = \sum_{j=1}^{n} p_{j,j}^{(k)}, \quad k = 1, 2, \ldots, n;$$

(3) Find the coefficients

$$\beta_0 = 1, \quad \beta_k = -\frac{1}{k}\left(\overline{s}_k + \sum_{j=1}^{k-1} \beta_j \overline{s}_{k-j}\right), \quad k = 1, 2, \ldots, n.$$

1.2.3 Determining the z-Transform Function

Now let us show how to determine the function $F(z)$. Consider

$$H'(z) = (z-1)H(z)$$

and denote $N = \deg(H'(z)) = n - (m(1) - 1)$. We have already shown that the function $F(z)$ can be represented in the following matrix form:

$$F(z) = \frac{1}{\delta(z)} \sum_{k=0}^{N-1} z^k R^{(k)},$$

where

$$(z-1)^{m(1)-1} \sum_{k=0}^{N-1} z^k R_{i,j}^{(k)} = M_{j,i}, \quad i, j = 1, 2, \ldots, n.$$

We shall use the identity relation $I = (I - zP)(I - zP)^{-1}$ and will make in it some elementary transformations:

$$\begin{aligned}\delta(z)I &= (I - zP)\sum_{k=0}^{N-1} z^k R^{(k)} = \sum_{k=0}^{N-1} z^k R^{(k)} - \sum_{k=0}^{N-1} z^{k+1}(PR^{(k)}) \\ &= R^{(0)} + \sum_{k=1}^{N-1} z^k (R^{(k)} - PR^{(k-1)}) - z^N (PR^{(N-1)}).\end{aligned}$$

Let $H'(z) = \sum_{k=0}^{N} \beta_k^* z^k$ and let us substitute this expression in the formula above. Then we obtain the following formula for determining the matrices $R^{(k)}$, $k = 0, 1, 2, \ldots, N-1$:

$$R^{(0)} = \beta_0^* I; \quad R^{(k)} = \beta_k^* I + PR^{(k-1)}, \quad k = 1, 2, \ldots, N-1. \tag{1.13}$$

So, we have

$$F(z) = \left(\frac{V_{i,j}(z)}{\delta(z)}\right)_{i,j=1,2,\ldots,n},$$

where

$$V_{i,j}(z) = \sum_{k=0}^{N-1} R_{i,j}^{(k)} z^k, \quad i, j = 1, 2, \ldots, n.$$

Based on these formulae we can propose the following algorithm for determining the matrix Q.

1.2.4 An Algorithm for Calculating the Limit Matrix

Consider

$$H(z) = \sum_{k=0}^{N-1} \gamma_k z^k; \quad Y(z) = \sum_{k=0}^{N-2} y^{(k)} z^k; \quad y^* = \alpha_1(1).$$

Then according to relation (1.11) we obtain

$$\frac{V_{i,j}(z)}{\delta(z)} = F_{i,j}(z) = \frac{y_{i,j}^*}{z-1} + \frac{\sum_{k=0}^{N-2} y_{i,j}^{(k)} z^k}{H(z)}, \quad i, j = 1, 2, \ldots, n.$$

This implies

$$\begin{aligned} V_{i,j}(z) &= \sum_{k=0}^{N-1} R_{i,j}^{(k)} z^k = y_{i,j}^* H(z) + (z-1) \sum_{k=0}^{N-2} y_{i,j}^{(k)} z^k \\ &= y_{i,j}^* \sum_{k=0}^{N-1} \gamma_k z^k + \sum_{k=0}^{N-2} y_{i,j}^{(k)} z^{k+1} - \sum_{k=0}^{N-2} y_{i,j}^{(k)} z^k \\ &= \sum_{k=0}^{N-1} \gamma_k y_{i,j}^* z^k + \sum_{k=1}^{N-1} y_{i,j}^{(k-1)} z^k - \sum_{k=0}^{N-2} y_{i,j}^{(k)} z^k \\ &= \left(\gamma_0 y_{i,j}^* - y_{i,j}^{(0)}\right) + \sum_{k=1}^{N-2} \left(\gamma_k y_{i,j}^* + y_{i,j}^{(k-1)} - y_{i,j}^{(k)}\right) z^k \\ &\quad + \left(\gamma_{N-1} y_{i,j}^* + y_{i,j}^{(N-2)}\right) z^{N-1}, \qquad i, j = 1, 2, \ldots, n. \end{aligned}$$

If we equate the corresponding coefficients of the variable z with the same exponents then we obtain the following system of linear equations:

$$\begin{cases} R_{i,j}^{(0)} = \gamma_0 y_{i,j}^* - y_{i,j}^{(0)}, \\ R_{i,j}^{(k)} = \gamma_k y_{i,j}^* + y_{i,j}^{(k-1)} - y_{i,j}^{(k)}, \quad k = 1, 2, \ldots, N-2, \quad i, j = 1, 2, \ldots, n, \\ R_{i,j}^{(N-1)} = \gamma_{N-1} y_{i,j}^* + y_{i,j}^{(N-2)}. \end{cases}$$

This system is equivalent to the following system:

$$\begin{cases} y_{i,j}^{(0)} = \gamma_0 y_{i,j}^* - R_{i,j}^{(0)}, \\ y_{i,j}^{(k)} = \gamma_k y_{i,j}^* + y_{i,j}^{(k-1)} - R_{i,j}^{(k)}, \quad k = 1, 2, \ldots, N-2, \quad i, j = 1, 2, \ldots, n, \\ y_{i,j}^{(N-2)} = -\gamma_{N-1} y_{i,j}^* + R_{i,j}^{(N-1)}. \end{cases}$$

Here we can observe that there exist the coefficients $u_{i,j}^{(k)}, v_{i,j}^{(k)} \in \mathbb{C}, k = 0, 1, 2, \ldots, N-2, i, j = 1, 2, \ldots, n$, such that

$$y_{i,j}^{(k)} = u_{i,j}^{(k)} y_{i,j}^* + v_{i,j}^{(k)}, \quad k = 0, 1, 2, \ldots, N-2, \quad i, j = 1, 2, \ldots, n.$$

From the first equation we obtain

$$u_{i,j}^{(0)} = \gamma_0, \quad v_{i,j}^{(0)} = -R_{i,j}^{(0)}, \quad i, j = 1, 2, \ldots, n.$$

From the next $N-2$ equations we obtain

$$\begin{aligned} y_{i,j}^{(k)} &= \gamma_k y_{i,j}^* + y_{i,j}^{(k-1)} - R_{i,j}^{(k)} \\ &= \gamma_k y_{i,j}^* + u_{i,j}^{(k-1)} y_{i,j}^* + v_{i,j}^{(k-1)} - R_{i,j}^{(k)} \\ &= \left(\gamma_k + u_{i,j}^{(k-1)}\right) y_{i,j}^* + \left(v_{i,j}^{(k-1)} - R_{i,j}^{(k)}\right), \\ &\quad k = 1, 2, \ldots, N-2, \quad i, j = 1, 2, \ldots, n, \end{aligned}$$

which involve the recursive equations

$$u_{i,j}^{(k)} = u_{i,j}^{(k-1)} + \gamma_k, \quad v_{i,j}^{(k)} = v_{i,j}^{(k-1)} - R_{i,j}^{(k)}, \quad k = 1, 2, \ldots, N-2, \quad i, j = 1, 2, \ldots, n.$$

We obtain the direct formula for the calculation of the coefficients:

$$u_{i,j}^{(k)} = \sum_{r=0}^{k} \gamma_r, \quad v_{i,j}^{(k)} = -\sum_{r=0}^{k} R_{i,j}^{(r)}, \quad k = 0, 1, 2, \ldots, N-2, \quad i, j = 1, 2, \ldots, n.$$

If we introduce these coefficients in the last equation of the system then we obtain

$$\begin{aligned} & u_{i,j}^{(N-2)} y_{i,j}^* + v_{i,j}^{(N-2)} = -\gamma_{N-1} y_{i,j}^* + R_{i,j}^{(N-1)}, \quad i, j = 1, 2, \ldots, n \\ \Leftrightarrow \; & y_{i,j}^* \sum_{r=0}^{N-1} \gamma_r = \sum_{r=0}^{N-1} R_{i,j}^{(r)}, \quad i, j = 1, 2, \ldots, n \\ \Leftrightarrow \; & y_{i,j}^* = \frac{\sum\limits_{r=0}^{N-1} R_{i,j}^{(r)}}{\sum\limits_{r=0}^{N-1} \gamma_r} = \frac{R_{i,j}}{H(1)}, \quad i, j = 1, 2, \ldots, n, \end{aligned}$$

where $R_{i,j} = \sum_{r=0}^{N-1} R_{i,j}^{(r)}$, $i, j = 1, 2, \ldots, n$. Finally, if we denote $R = (R_{i,j})_{n\times n}$ then we obtain:

$$Q = -\frac{1}{H(1)}R. \tag{1.14}$$

Based on the result described above we can describe the algorithm for determining the matrix Q.

Algorithm 1.2 The Determination of the Limit Matrix Q

(1) Find the coefficients of the characteristic polynomial $\Delta(z) = \sum_{k=0}^{n} \beta_k z^k$ using Algorithm 1.1;
(2) Divide $m(1)$ times the polynomial $\Delta(z)$ by $z - 1$, using Horner's scheme and find the polynomial $H(z)$ that satisfies the condition $H(1) \neq 0$. At the same time we preserve the coefficients β_k^*, $k = 0, 1, 2, \ldots, N$ of the polynomial $(z-1)H(z)$ obtained at the previous step of Horner's scheme;
(3) Determine $H(1)$ according to the rule described above;
(4) Find the matrices $R^{(k)}$, $k = 0, 1, 2, \ldots, N-1$, according to (1.13);
(5) Determine the matrix $R = \sum_{k=0}^{N-1} R^{(k)}$;
(6) Calculate the matrix Q according to formula (1.14).

It is easy to check that the running time of Algorithm 1.2 is $O(|X|^4)$. Indeed, step (1) and step (4) of the algorithm use $O(|X|^4)$ elementary operations and each of the remaining steps uses in the worst case $O(|X|^3)$ elementary operations.

Below we give some numerical examples which illustrate the main details of the algorithm described above.

Example 1 Consider the discrete Markov process with the stochastic matrix of probability transitions

$$P = \begin{pmatrix} 0 & 1 \\ 1 & 0 \end{pmatrix}.$$

We have already noted that

$$P^{2t} = \begin{pmatrix} 1 & 0 \\ 0 & 1 \end{pmatrix}, \quad P^{2t+1} = \begin{pmatrix} 0 & 1 \\ 1 & 0 \end{pmatrix}, \quad \forall t \geq 0,$$

i.e., this Markov chain is 2−periodic. So, for the considered example $\lim_{t\to\infty} P^t$ does not exist.

Nevertheless we can see that the matrix Q exists and it can be determined by using the algorithm described above. If we apply this algorithm then we obtain:

(1) Find

$$P = \begin{pmatrix} 0 & 1 \\ 1 & 0 \end{pmatrix}, \quad P^2 = \begin{pmatrix} 1 & 0 \\ 0 & 1 \end{pmatrix};$$

$$\overline{s}_1 = \mathrm{tr}(P) = 0, \quad \overline{s}_2 = \mathrm{tr}\big(P^2\big) = 2;$$

$$\beta_0 = 1, \ \beta_1 = -\overline{s}_1 = 0, \quad \beta_2 = -\frac{1}{2}(\overline{s}_2 + \beta_1\overline{s}_1) = -1.$$

(2) Divide the polynomial $\beta_2 z^2 + \beta_1 z + \beta_0$ by $z - 1$ using Horner's scheme

	−1	0	1
1	−1	−1	0
1	−1	−2	

and obtain $m(1) = 1$, $N = 2$; $\beta_0^* = 1$, $\beta_1^* = 0$, $\beta_2^* = -1$; $\gamma_0 = -1$, $\gamma_1 = -1$.

(3) Determine

$$H(1) = \gamma_0 + \gamma_1 = -2.$$

(4) Calculate

$$R^{(0)} = \beta_0^* I = \begin{pmatrix} 1 & 0 \\ 0 & 1 \end{pmatrix}, \quad R^{(1)} = \beta_1^* I + PR^{(0)} = \begin{pmatrix} 0 & 1 \\ 1 & 0 \end{pmatrix}.$$

(5) Find

$$R = R^{(0)} + R^{(1)} = \begin{pmatrix} 1 & 0 \\ 0 & 1 \end{pmatrix} + \begin{pmatrix} 0 & 1 \\ 1 & 0 \end{pmatrix} = \begin{pmatrix} 1 & 1 \\ 1 & 1 \end{pmatrix}.$$

(6) Determine

$$Q = -\frac{1}{H(1)} R = \frac{1}{2}\begin{pmatrix} 1 & 1 \\ 1 & 1 \end{pmatrix} = \begin{pmatrix} 0.5 & 0.5 \\ 0.5 & 0.5 \end{pmatrix}.$$

In such a way we obtain the limit matrix

$$Q = \begin{pmatrix} 0.5 & 0.5 \\ 0.5 & 0.5 \end{pmatrix}.$$

Note that the considered process is not ergodic in the sense of the definition from [114, 140] because the matrix P^t contains zero elements $\forall t \geq 0$.

However, the limit matrix exists and the rows of this matrix are identical. As we have already shown the vector of the limiting probabilities $\pi^* = (0.5,\ 0.5)$ can also be found by solving the system of linear equations (1.5).

Example 2 Consider a Markov process with the stochastic matrix

$$P = \begin{pmatrix} 0.5 & 0.5 \\ 0.4 & 0.6 \end{pmatrix}.$$

It easy to observe that here we have an ergodic Markov chain. We can determine the matrix Q by using our algorithm.

In the same way we determine

$$P = \begin{pmatrix} 0.5 & 0.5 \\ 0.4 & 0.6 \end{pmatrix}, \quad P^2 = \begin{pmatrix} 0.45 & 0.55 \\ 0.44 & 0.56 \end{pmatrix};$$

$$\overline{s}_1 = \text{tr}(P) = 0.5 + 0.6 = 1.1, \quad \overline{s}_2 = \text{tr}(P^2) = 0.45 + 0.56 = 1.01;$$

$$\beta_0 = 1,\ \beta_1 = -\overline{s}_1 = -1.1, \quad \beta_2 = -\frac{1}{2}(\overline{s}_2 + \beta_1 \overline{s}_1) = 0.1.$$

Using Horner's scheme

	0.1	−1.1	1
1	0.1	−1	0
1	0.1	−0.9	

we obtain

$$\beta_0^* = 1,\ \beta_1^* = -1.1,\ \beta_2^* = 0.1; \ \ \gamma_0 = -1,\ \gamma_1 = 0.1; \ \ H(1) = \gamma_0 + \gamma_1 = -0.9;$$

$$R^{(0)} = \beta_0^* I = \begin{pmatrix} 1 & 0 \\ 0 & 1 \end{pmatrix}, \quad R^{(1)} = \beta_1^* I + P R^{(0)} = \begin{pmatrix} -0.6 & 0.5 \\ 0.4 & -0.5 \end{pmatrix};$$

$$R = R^{(0)} + R^{(1)} = \begin{pmatrix} 0.4 & 0.5 \\ 0.4 & 0.5 \end{pmatrix}; \quad Q = -\frac{1}{H(1)} R = \frac{1}{9} \begin{pmatrix} 4 & 5 \\ 4 & 5 \end{pmatrix}.$$

Finally we obtain

$$Q = \begin{pmatrix} \frac{4}{9} & \frac{5}{9} \\ \frac{4}{9} & \frac{5}{9} \end{pmatrix}.$$

The rows of this matrix are the same and all elements of the matrix P^t are non zero if $t \to \infty$. So, this is an ergodic Markov chain with the vector of limiting probabilities $\pi^* = (4/9,\ 5/9)$. As we have shown this vector can be found by solving system (1.5).

Example 3 We consider a Markov multichain with the stochastic matrix of probability transitions

$$P = \begin{pmatrix} 1 & 0 & 0 \\ 0 & 1 & 0 \\ \frac{1}{3} & \frac{1}{3} & \frac{1}{3} \end{pmatrix}.$$

In this case the solution of the system of linear equations (1.5) is not unique. If we apply the proposed algorithm we can determine the matrix Q. According to this algorithm we obtain:

$$P = \begin{pmatrix} 1 & 0 & 0 \\ 0 & 1 & 0 \\ \frac{1}{3} & \frac{1}{3} & \frac{1}{3} \end{pmatrix}, \quad P^2 = \begin{pmatrix} 1 & 0 & 0 \\ 0 & 1 & 0 \\ \frac{4}{9} & \frac{4}{9} & \frac{1}{9} \end{pmatrix}, \quad P^3 = \begin{pmatrix} 1 & 0 & 0 \\ 0 & 1 & 0 \\ \frac{13}{27} & \frac{13}{27} & \frac{1}{27} \end{pmatrix};$$

$$\overline{s}_1 = \mathrm{tr}(P) = \frac{7}{3}, \quad \overline{s}_2 = \mathrm{tr}(P^2) = \frac{19}{9}, \quad \overline{s}_3 = \mathrm{tr}(P^3) = \frac{55}{27}; \quad \beta_0 = 1,$$

$$\beta_1 = -\overline{s}_1 = -\frac{7}{3}, \quad \beta_2 = -\frac{\overline{s}_2 + \beta_1 \overline{s}_1}{2} = \frac{5}{3}, \quad \beta_3 = -\frac{\overline{s}_3 + \beta_1 \overline{s}_2 + \beta_2 \overline{s}_1}{3} = -\frac{1}{3}.$$

If we apply Horner's scheme

	$-\frac{1}{3}$	$\frac{5}{3}$	$-\frac{7}{3}$	1
1	$-\frac{1}{3}$	$\frac{4}{3}$	-1	0
1	$-\frac{1}{3}$	1	0	
1	$-\frac{1}{3}$	$\frac{2}{3}$		

then we obtain

$$\beta_0^* = -1, \ \beta_1^* = \frac{4}{3}, \ \beta_2^* = -\frac{1}{3}; \quad \gamma_0 = 1, \ \gamma_1 = -1/3; \quad H(1) = \gamma_0 + \gamma_1 = \frac{2}{3};$$

$$R^{(0)} = \beta_0^* I = \begin{pmatrix} -1 & 0 & 0 \\ 0 & -1 & 0 \\ 0 & 0 & -1 \end{pmatrix}, \quad R^{(1)} = \beta_1^* I + P R^{(0)} = \begin{pmatrix} 1/3 & 0 & 0 \\ 0 & 1/3 & 0 \\ -1/3 & -1/3 & 1 \end{pmatrix};$$

$$R = R^{(0)} + R^{(1)} = \begin{pmatrix} -2/3 & 0 & 0 \\ 0 & -2/3 & 0 \\ -1/3 & -1/3 & 0 \end{pmatrix}; \quad Q = -\frac{1}{H(1)} R = \begin{pmatrix} 1 & 0 & 0 \\ 0 & 1 & 0 \\ 1/2 & 1/2 & 0 \end{pmatrix}.$$

So, finally we have

$$Q = \begin{pmatrix} 1 & 0 & 0 \\ 0 & 1 & 0 \\ 1/2 & 1/2 & 0 \end{pmatrix}.$$

In this case all rows of the matrix Q are different. It is easy to observe that for the considered example $\lim_{t\to\infty} P^t = Q$ exists.

Finally, we can remark that the proposed approach also allows to determine the differential components of the matrix $T(t)$.

1.3 An Algorithm for Determining the Differential Matrices in Markov Chains

Now we show how to use the method from the previous section for determining the differential components of the matrix $T(t)$. We propose a simple modification of this method that allows us to calculate all components of the transient matrix in the case if the roots of the characteristic polynomial are known. So, if there exist efficient algorithms for determining the roots of the characteristic polynomial then the differential components of the matrix $T(t)$ can be determined efficiently by using algorithms similar to the algorithms from the previous section. We show that if the roots of the characteristic polynomial are known then the differential matrices of the matrix $T(t)$ can be calculated by using $O(|X|^4)$ elementary operations.

1.3.1 Determining the Differential Matrices Based on the z-Transform

To formulate the algorithm for determining the differential matrices we shall use the relationship between the coefficients of the matrix $\overline{P}(t) = P^t$ and the corresponding coefficients of $F(z)$ in the formulae (1.9), (1.10). An arbitrary element $p_{i,j}(t)$ of the matrix $\overline{P}(t) = P^t$ represents the probability of the system to reach the state x_j from x_i using t transitions. The corresponding coefficients in the formulae (1.9), (1.10) have the same sense and, therefore, we obtain that an arbitrary element $\overline{p}_{i,j}(t)$ of the matrix $\overline{P}(t)$ can be expressed by the formula

$$\overline{p}_{i,j}(t) = \sum_{y\in\mathbb{C}\setminus\mathcal{D}} \sum_{k=0}^{m(y)-1} \frac{t^k}{y^t}\beta_{i,j,k}(y), \quad \forall t > \deg(B_{i,j}(z)),\ i, j = 1, 2, \ldots, n,$$

where $\mathcal{D} = \{z \in \mathbb{C} \mid |I - zP| \neq 0\}$, $\beta_{i,j,k}(y) \in \mathbb{C}$, $\forall y \in \mathbb{C}\setminus\mathcal{D}$, $k = 0, 1, 2, \ldots, m(y) - 1$, $m(y)$—is the order of the root y of the polynomial $\Delta(z) = |I - zP|$ and $B_{i,j}(z)$ is a polynomial of degree less or equal to $n-1$, $i, j = 1, 2, \ldots, n$.

If we denote $\beta_k(y) = (\beta_{i,j,k}(y))_{i,j=\overline{1,n}}$, $\forall y \in \mathbb{C}\backslash\mathcal{D}$, $k = 0, 1, 2, \dots, m(y) - 1$, then

$$\overline{P}(t) = \sum_{y \in \mathbb{C}\backslash\mathcal{D}} \sum_{k=0}^{m(y)-1} \frac{t^k}{y^t} \beta_k(y), \quad \forall t \geq n. \tag{1.15}$$

As we have shown $\mathbb{C}\backslash\mathcal{D}$ consists of the set of inverses of the nonzero proper values of the matrix P, where the order of each element is the same as the order of the corresponding proper value. This means that (1.15) gives the components of the matrix $\overline{P}(t)$ with respect to proper values of the matrix P. Thus, if we determine these components then we find the limit and differential matrices in a Markov chain.

We have already described the algorithm for determining the stationary components of the matrix $\overline{P}(t)$ in the previous section. In the following we show how to determine all matrices $\beta_k(y)$ from (1.15). To develop such an algorithm we shall use some properties of the linear recurrent equations.

1.3.2 Linear Recurrent Equations and its Main Properties

Consider an arbitrary set K on which the operations of summation and multiplication are defined. On this set we consider the following relation

$$a_n = \sum_{k=0}^{m-1} q_k a_{n-1-k}, \quad \forall n \geq m, \tag{1.16}$$

where q_k are given elements from K. A sequence $a = \{a_n\}_{n=0}^{\infty}$ is called the *linear* m*-recurrence on* K if there exists a vector $q = (q_k)_{k=0}^{m-1} \in K^m$ such that (1.16) holds. Here we call q the *generating vector* and we call $I_m^{[a]} = (a_n)_{n=0}^{m-1}$ the *initial value of the sequence* a. The sequence a is called the *linear recurrence on* K if there exists $m \in \mathbb{N}^*$ ($\mathbb{N}^* = \{1, 2, \dots\}$) such that the sequence a is a linear m-recurrence on K. If $q_{m-1} \neq 0$ then the sequence a is called non-degenerated; otherwise it is called degenerated.

Denote:

$Rol[K][m]$—the set of non-degenerated linear m-recurrences on K;

$Rol[K]$—the set of non-degenerated recurrences on K;

$G[K][m](a)$—the set of generating vectors of length m of the sequence $a \in Rol[K][m]$;

$G[K](a)$—the set of generating vectors of the sequence $a \in Rol[K]$.

In the following we will consider K as a subfield of the field of complex numbers $\mathbb{C}$ and $a = \{a_n\}_{n=0}^{\infty} \subseteq \mathbb{C}$.

We call the function $G^{[a]} : \mathbb{C} \to \mathbb{C}$, $G^{[a]}(z) = \sum_{n=0}^{\infty} a_n z^n$, the *generating function of the sequence* $a = (a_n)_{n=0}^{\infty} \subseteq \mathbb{C}$ and we call the function $G_t^{[a]} : \mathbb{C} \to \mathbb{C}$,

$G_t^{[a]}(z) = \sum_{n=0}^{t-1} a_n z^n$ the *partial generating function of order t of the sequence* $a = (a_n)_{n=0}^{\infty} \subseteq \mathbb{C}$.

Let $a \in Rol[K][m]$, $q \in G[K][m](a)$. For this sequence we will consider the *characteristic polynomial* $\Delta_m^{[q]}(z) = 1 - zG_m^{[q]}(z)$ and the characteristic equation $\Delta_m^{[q]}(z) = 0$. For an arbitrary $\alpha \in K^*$ we also call the polynomial $\Delta_{m,\alpha}^{[q]}(z) = \alpha\Delta_m^{[q]}(z)$ characteristic polynomial of the sequence a. We introduce the following notations:

$\Delta[K][m](a)$—the set of characteristic polynomials of degree m of the sequence $a \in Rol[K]$;
$\Delta[K](a)$—the set of characteristic polynomials of the sequence $a \in Rol[K]$.

If we operate with arbitrary recurrence (not obligatory non-degenerated) then for the corresponding set we will use similar notations and will specify this with the mark "*", i.e., we will denote respectively: $Rol^*[K][m]$, $Rol^*[K]$, $G^*[K][m](a)$, $G^*[K](a)$, $\Delta^*[K][m](a)$, $\Delta^*[K](a)$. We shall use the following well known properties:

(1) Let: $a \in Rol[K][m]$, $q \in G[K][m](a)$, $\Delta_{m,\alpha}^{[q]}(z) = \prod_{k=0}^{p-1}(z - z_k)^{s_k}$, $z_i \neq z_j \ \forall i \neq j$.
Then $a_n = I_m^{[a]} \cdot ((B^{[a]})^T)^{-1} \cdot (\beta_n^{[a]})^T$, $\forall n \in \mathbb{N}$ $(\mathbb{N} = \{0, 1, 2, \dots\})$,
where $\beta_i^{[a]} = \left(\frac{\tau_{i,j}}{z_k^i}\right)_{k=\overline{0,p-1},\, j=\overline{0,s_k-1}}$,
$$\tau_{i,j} = \begin{cases} i^j & \text{if } i^2 + j^2 \neq 0, \\ 1 & \text{if } i = j = 0, \end{cases} \quad i \in \mathbb{N}, \quad B^{[a]} = (\beta_i^{[a]})_{i=0}^{m-1}.$$

(2) If a is a matrix sequence, $a \in Rol[M_n(K)][m]$ and $q \in G[M_n(K)][m](a)$, then $a \in Rol^*[K][mn]$ and $|I - zG_m^{[q]}(z)| \in \Delta^*[K][mn](a)$.

1.3.3 The Main Results and an Algorithm

Consider a matrix sequence $a = (\overline{P}(t))_{t=0}^{\infty}$. Then it is easy to observe that the following recurrent relation $a_t = Pa_{t-1}$, $\forall t \geq 1$ holds. So, a belongs to $Rol[M_n(\mathbb{R})][1]$ with the generating vector $q = (P) \in G[M_n(\mathbb{R})][1](a)$. Therefore, according to property (2) from Sect. 1.3.2 we have $a \in Rol^*[\mathbb{R}][n]$ and $\Delta(z) \in \Delta^*[\mathbb{R}][n](a)$.

Let $r = \deg \Delta(z)$ and consider the subsequence $\overline{a} = (\overline{P}(t))_{t=n-r}^{\infty}$ of the sequence a. We have $\overline{a} \in Rol[\mathbb{R}][r]$ and $\Delta(z) \in \Delta[\mathbb{R}][r](\overline{a})$. For the corresponding elements this relation can be expressed as follows: $\overline{a}_{i,j} \in Rol[\mathbb{R}][r]$, $\Delta(z) \in \Delta[\mathbb{R}][r](\overline{a}_{i,j})$, $i, j = 1, 2, \dots, n$.

According to property (1) from Sect. 1.3.2 we obtain

$$\begin{aligned} \overline{p}_{i,j}(t) = a_{i,j}(t) &= \overline{a}_{i,j}(t - n + r) \\ &= I_r^{[\overline{a}_{i,j}]}(B^T)^{-1}(\beta_{t-n+r})^T, \quad i, j = 1, 2, \dots, n, \ \forall t \geq n - r, \end{aligned} \tag{1.17}$$

where

$$\beta_t = \left(\frac{t^k}{y^t}\right)_{k=\overline{0,m(y)-1};\, y\in\mathbb{C}\backslash\mathcal{D}}, \quad B = (\beta_j)_{j=\overline{0,r-1}}, \quad 0^0 \equiv 1 \tag{1.18}$$

for $t \geq 0$. Therefore, we can determine the initial values of the subsequences $\overline{a}_{i,j}$, $i, j = 1, 2, \ldots, n$:

$$I_r^{[\overline{a}_{i,j}]} = (\overline{a}_{i,j}(t))_{t=0}^{r-1} = (a_{i,j}(t))_{t=n-r}^{n-1} = (\overline{p}_{i,j}(t))_{t=n-r}^{n-1}, \quad i, j = 1, 2, \ldots, n. \tag{1.19}$$

If for $y \in \mathbb{C}\backslash\mathcal{D}$ we denote

$$I_r^{[\overline{a}_{i,j}]}(B^T)^{-1} = (\gamma_{i,j,l}(y))_{l=\overline{0,m(y)-1}}, \quad i, j = 1, 2, \ldots, n, \tag{1.20}$$

then formula (1.17) can be represented in the following form:

$$\begin{aligned}
\overline{p}_{i,j}(t) &= \sum_{y\in\mathbb{C}\backslash\mathcal{D}} \sum_{s=0}^{m(y)-1} \frac{(t-n+r)^l}{y^{t-n+r}} \gamma_{i,j,s}(y) \\
&= \sum_{y\in\mathbb{C}\backslash\mathcal{D}} \sum_{s=0}^{m(y)-1} \sum_{k=0}^{l} C_l^k (r-n)^{l-k} y^{n-r} \gamma_{i,j,l}(y) \frac{t^k}{y^t} \\
&= \sum_{y\in\mathbb{C}\backslash\mathcal{D}} \sum_{k=0}^{m(y)-1} \frac{t^k}{y^t} \sum_{l=k}^{m(y)-1} C_l^k (r-n)^{l-k} y^{n-r} \gamma_{i,j,l}(y) \\
&= \sum_{y\in\mathbb{C}\backslash\mathcal{D}} \sum_{k=0}^{m(y)-1} \frac{t^k}{y^t} \beta_{i,j,k}(y), \quad i, j = 1, 2, \ldots, n, \quad \forall t \geq n-r,
\end{aligned} \tag{1.21}$$

where

$$\begin{aligned}
\beta_{i,j,k}(y) &= y^{n-r} \sum_{l=k}^{m(y)-1} C_l^k (r-n)^{l-k} \gamma_{i,j,l}(y), \\
&\forall y \in \mathbb{C}\backslash\mathcal{D}, \quad k = 0, 1, 2, \ldots, m(y)-1, \quad i, j = 1, 2, \ldots, n.
\end{aligned} \tag{1.22}$$

If we rewrite the relations (1.21) in the matrix form then we obtain the representation (1.15) of the matrices $\beta_k(y)$ ($y \in \mathbb{C}\backslash\mathcal{D}$, $k = 0, 1, 2, \ldots, m(y) - 1$), i.e., these matrices can be determined according to formula (1.22).

Based on the results above we can describe the following algorithm for the decomposition of the transient matrix:

Algorithm 1.3 The Decomposition of the Transient Matrix

Input Data: The matrix of the transition probability P.

Output Data: The matrices $\beta_k(y)$ ($y \in \mathbb{C}\backslash\mathcal{D}$, $k = 0, 1, 2, m(y) - 1$).

1. Calculate the coefficients of the polynomial $\Delta(z)$ for the matrix P using the algorithm from Sect. 1.2 (the algorithm based on Leverrier's method [44]);
2. Solve the equation $\Delta(z) = 0$ and find all roots of this equations in $\mathbb{C}$; then determine $\mathbb{C}\backslash\mathcal{D}$;
3. Determine the order of each root $m(y)$ of the polynomial $\Delta(z)$ (the order of each root can be found using Horner's scheme; $m(y)$ is equal to the number of successive factorizations of the polynomial $\Delta(z)$ by $(z - y)$, $\forall y \in \mathbb{C}\backslash\mathcal{D}$);
4. Calculate the matrix B using formula (1.18);
5. Determine the matrix $(B^T)^{-1}$;
6. Calculate the values C_l^k, $l = 0, 1, 2, \ldots, \max_{y\in\mathbb{C}\backslash\mathcal{D}} m(y) - 1$, $k = 0, 1, 2, \ldots, l$, according to Pascal's triangle rule: $C_l^0 = C_l^l = 1$, $C_l^k = C_{l-1}^{k-1} + C_{l-1}^k$ $(k = 1, 2, \ldots, l - 1)$;
7. Find recursively $(r - n)^l$, $l = 0, 1, 2, \ldots, \max_{y\in\mathbb{C}\backslash\mathcal{D}} m(y) - 1$;
8. For every $i, j = 1, 2, \ldots, n$ do the following steps:

   a. Find the initial value $I_r^{[\overline{a}_{i,j}]}$ according to formula (1.19);
   b. Calculate the values $\gamma_{i,j,l}(y)$, $y \in \mathbb{C}\backslash\mathcal{D}$, $l = 0, 1, 2, \ldots, m(y)-1$, according to (1.20);
   c. For arbitrary $y \in \mathbb{C}\backslash\mathcal{D}$, $k = 0, 1, 2, \ldots, m(y) - 1$, determine the coefficients $\beta_{i,j,k}(y)$ of the matrix $\beta_k(y)$ using formula (1.22) and the parameters calculated in the steps 6, 7.

1.3.4 Comments on the Computational Complexity of the Algorithm

The proposed algorithm can be used for determining the differential matrices in the case if the characteristic values of the matrix P are known. Therefore, the computational complexity of the algorithm depends on the computational complexity of the algorithm for determining the characteristic values of the matrix P.

If the set of characteristic values of the matrix P is known then it is easy to observe that the running time of the algorithm for determining the differential matrices is $O(n^4)$. We obtain this estimation of the algorithm if we estimate the number of elementary operations at the steps 3–8 in the worst case.

Note that the matrix $\beta_0(1)$ corresponds to the limit probability matrix Q of the Markov process and, therefore, this matrix can be calculated using $O(n^4)$ elementary operations.

So, based on the results described above we may conclude that the matrix $\overline{P}(t)$ can be represented as follows

$$\overline{P}(t) = \sum_{y\in\mathbb{C}\backslash\mathcal{D}} \sum_{k=0}^{m(y)-1} \beta_k(y)\frac{t^k}{y^t}, \quad \forall t \geq n - r.$$

For $t = 0, 1, 2, \ldots, n - r - 1$ this formula can be expressed in the form

$$\overline{P}(t) = L(t) + \sum_{y \in \mathbb{C} \backslash \mathcal{D}} \sum_{k=0}^{m(y)-1} \beta_k(y) \frac{t^k}{y^t}, \tag{1.23}$$

where $L(t)$ is a matrix that depends only on t. If the matrices $\beta_k(y)$, $\forall y \in \mathbb{C}\backslash\mathcal{D}$, $k = 0, 1, 2, m(y) - 1$, are known then we can determine the matrices $L(t)$ from (1.23) taking into account that $\overline{P}(t) = P^t$, $\forall t \geq 0$.

In [47] it is noted that the matrices $L(t)$, $t = 0, 1, 2, \ldots, n-r-1$, and $\beta_k(y)$, for each $y \in (\mathbb{C}\backslash\mathcal{D})\backslash\{1\}$, $k = 0, 1, 2, \ldots, m(y) - 1$, are differential matrices, i.e., the sum of elements of each row is equal to zero. The unique non-differential component in the representation (1.23) is the matrix $\beta_0(1)$; the remaining matrices $\beta_k(1)$, $k = 1, 2, \ldots, m(1) - 1$, are null (see [72]).

1.4 An Algorithm for Determining the Limit and the Differential Matrices

The results from the previous sections can be used for a simultaneous calculation of the limit and the differential matrices in a Markov process. We propose a modification of the algorithms from the Sects. 1.2 and 1.3.3 that allows us to determine the limit and the differential matrices in the case if the proper values of the characteristic polynomial are known.

1.4.1 Some Auxiliary Results Concerning the Representation of the z-Transform

Let us consider the method for determining the matrix $F(z) = (I - zP)^{-1}$ from the previous section with a simple modification: In the calculation procedure we will not divide $F(z)$ by $(z-1)^{m(1)-1}$.

Then it easy to observe that

$$F(z) = \frac{1}{\Delta(z)} \sum_{k=0}^{n-1} R^{(k)} z^k, \tag{1.24}$$

where the matrix-coefficients $R^{(k)}$, $k = 0, 1, 2, \ldots, n-1$, are determined recursively according to the formula

$$R^{(0)} = \beta_0 I; \quad R^{(k)} = \beta_k I + P R^{(k-1)}, \quad k = 1, 2, \ldots, n-1; \tag{1.25}$$

and the values β_k, $k = 0, 1, 2, \ldots, n$, represent the coefficients of the polynomial $\Delta(z)$ calculated according to the algorithm described above.

As we have shown in Sect. 1.2 the elements of the matrix $F(z)$ can be represented by the following formula

$$F_{i,j}(z) = B_{i,j}(z) + \sum_{y \in \mathbb{C} \setminus \mathcal{D}} \sum_{k=1}^{m(y)} \frac{\alpha_{i,j,k}(y)}{(z-y)^k}, \quad i,j = 1,2,\ldots,n. \tag{1.26}$$

In the general form the relation (1.26) can be written as follows

$$F_{i,j}(z) = B_{i,j}(z) + \frac{Q_{i,j}(z)}{\Delta(z)}, \quad i,j = 1,2,\ldots,n, \tag{1.27}$$

where $Q_{i,j}(z) \in \mathbb{C}[z]$ and $\deg(Q_{i,j}(z)) < \deg(\Delta(z)) = r$, $i,j = 1,2,\ldots,n$.

If we write the equality (1.24) for each element and after that we make the corresponding substitutions in (1.27) then we obtain the formula

$$\sum_{k=0}^{n-1} R_{i,j}^{(k)} z^k = B_{i,j}(z)\Delta(z) + Q_{i,j}(z), \quad i,j = 1,2,\ldots,n.$$

So, $B_{i,j}(z) = \sum_{k=0}^{n-1-r} b_{i,j,k} z^k$ and $Q_{i,j}(z) = \sum_{k=0}^{r-1} q_{i,j,k} z^k$ represent the quotient and the rest, respectively, after the division of the polynomial $\sum_{k=0}^{n-1} R_{i,j}^{(k)} z^k$ by $\Delta(z)$. Therefore, the polynomials $B_{i,j}(z)$ and $Q_{i,j}(z)$ can be found by using the procedure described below.

Calculation procedure for determining
the polynomials $B_{i,j}(z)$ and $Q_{i,j}(z)$, $i,j = 1,2,\ldots,n$:

- For $i,j = 1,2,\ldots,n$ do:
  - $q_{i,j,k} = R_{i,j}^{(k)}$, $k = 0,1,2,\ldots,n-1$;
- For $k = n-1, n-2, \ldots, r$ do:
  - $b_{i,j,k-r} = \dfrac{q_{i,j,k}}{\beta_r}$;
  - $q_{i,j,k-t} = q_{i,j,k-t} - b_{i,j,k-r}\beta_{r-t}$, $t = 0,1,2,\ldots,r$.

1.4.2 *Expansion of the z-Transform with Respect to Nonzero Characteristic Values*

Let $\mu \in \mathbb{C} \setminus \mathcal{D}$, $m(\mu) = m$ (μ^{-1} be a nonzero characteristic value of the matrix P and assume that the order of this characteristic value is m). According to the formulae (1.26), (1.27), for the separated root μ we have

$$\frac{Q_{i,j}(z)}{\Delta(z)} = \sum_{k=1}^{m} \frac{\alpha_{i,j,k}(\mu)}{(z-\mu)^k} + \sum_{y \in (\mathbb{C} \setminus \mathcal{D}) \setminus \{\mu\}} \sum_{k=1}^{m(y)} \frac{\alpha_{i,j,k}(y)}{(z-y)^k}, \quad i,j = 1,2,\ldots,n. \tag{1.28}$$

Let $\Delta(z) = (z-\mu)^m D(z)$, $D(z) = \sum_{k=0}^{r-m} d_k z^k$ and denote $\deg(D(z)) = M$. The relation (1.28) can be written as follows:

$$\frac{Q_{i,j}(z)}{\Delta(z)} = \frac{G_{i,j}(z)}{(z-\mu)^m} + \frac{E_{i,j}(z)}{D(z)}, \quad i,j = 1,2,\ldots,n,$$

where $E_{i,j}(z) = \sum_{k=0}^{M-1} e_{i,j,k} z^k$, $G_{i,j}(z) = \sum_{k=0}^{m-1} g_{i,j,k} z^k \in \mathbb{C}[z]$, $i,j = 1,2,\ldots,n$.

Making an elementary transformation we obtain

$$Q_{i,j}(z) = G_{i,j}(z) D(z) + E_{i,j}(z)(z-\mu)^m, \quad i,j = 1,2,\ldots,n.$$

By the expansion $(z-\mu)^m = \sum_{k=0}^{m} C_m^k (-\mu)^{m-k} z^k$ and then introducing the notation $\xi(k) = C_m^k (-\mu)^{-k}$, $k = 0,1,2,\ldots,m$ we have

$$(z-\mu)^m = \sum_{k=0}^{m} C_m^k (-\mu)^{m-k} z^k = (\xi(m))^{-1} \sum_{k=0}^{m} \xi(k) z^k.$$

Now for our relation we make the following transformations:

$$\begin{aligned}
\sum_{t=0}^{r-1} q_{i,j,t} z^t &= \sum_{k=0}^{m-1} g_{i,j,k} z^k \sum_{l=0}^{M} d_l z^l + (\xi(m))^{-1} \sum_{l=0}^{M-1} e_{i,j,l} z^l \sum_{k=0}^{m} \xi(k) z^k \\
&= \sum_{k=0}^{m-1} \sum_{l=0}^{M} g_{i,j,k} d_l z^{k+l} + (\xi(m))^{-1} \sum_{k=0}^{m} \sum_{l=0}^{M-1} \xi(k) e_{i,j,l} z^{k+l} \\
&= \sum_{t=0}^{r-1} z^t \left[\sum_{\substack{k+l=t \\ 0\le k\le m-1 \\ 0\le l\le M}} g_{i,j,k} d_l + (\xi(m))^{-1} \sum_{\substack{k+l=t \\ 0\le k\le m \\ 0\le l\le M-1}} \xi(k) e_{i,j,l} \right].
\end{aligned}$$

Equating the corresponding coefficients in this formula we obtain

$$\begin{aligned}
q_{i,j,t} &= \sum_{\substack{0\le k\le m-1 \\ 0\le t-k\le M}} d_{t-k} g_{i,j,k} + (\xi(m))^{-1} \sum_{\substack{0\le l\le t \\ t-m\le l\le M-1}} \xi(t-l) e_{i,j,l} \\
&= \sum_{k=0}^{m-1} d_{t-k} I_{\{0\le x\le M\}}(t-k) g_{i,j,k} \\
&\quad + (\xi(m))^{-1} \sum_{l=0}^{t} \xi(t-l) I_{\{t-m\le x\le M-1\}}(l) e_{i,j,l},
\end{aligned}$$

where $I_A(x)$ is the index of the set A: $I_A(x) = 1,\ \forall x \in A$ and $I_A(x) = 0,\ \forall x \notin A$.

Now we observe that for $t \leq M - 1$ the formula above can be written in the following form:

$$\begin{aligned} q_{i,j,t} &= \sum_{k=0}^{m-1} d_{t-k} I_{\{x \leq t\}}(k) g_{i,j,k} + (\xi(m))^{-1} \sum_{l=0}^{t} \xi(t-l) I_{\{x \geq t-m\}}(l) e_{i,j,l} \\ &= \sum_{k=0}^{m-1} d_{t-k} I_{\{x \leq t\}}(k) g_{i,j,k} \\ &\quad + (\xi(m))^{-1} \left(e_{i,j,t} + \sum_{l=0}^{t-1} \xi(t-l) I_{\{x \geq t-m\}}(l) e_{i,j,l} \right). \end{aligned}$$

This involves

$$\begin{aligned} e_{i,j,t} = \xi(m) \Bigg[q_{i,j,t} &- \sum_{k=0}^{m-1} d_{t-k} I_{\{x \leq t\}}(k) g_{i,j,k} \\ &- (\xi(m))^{-1} \sum_{l=0}^{t-1} \xi(t-l) I_{\{x \geq t-m\}}(l) e_{i,j,l} \Bigg]. \end{aligned}$$

So, finally we obtain the following expression

$$e_{i,j,t} = w_{i,j,t} + \sum_{k=0}^{m-1} x_{t,k}\, g_{i,j,k}, \quad t = 0, 1, 2, \ldots, M-1, \quad i, j = 1, 2, \ldots, n.$$

In the following we will determine the coefficients $w_{i,j,t}$ and $x_{t,k}$ from the expression above.

Then we have

$$\begin{aligned} e_{i,j,t} = w_{i,j,t} + \sum_{k=0}^{m-1} x_{t,k}\, g_{i,j,k} &= \xi(m) q_{i,j,t} - \sum_{k=0}^{m-1} \xi(m) d_{t-k} I_{\{x \leq t\}}(k) g_{i,j,k} \\ &- \sum_{l=0}^{t-1} \xi(t-l) I_{\{x \geq t-m\}}(l) \left[w_{i,j,l} + \sum_{k=0}^{m-1} x_{l,k}\, g_{i,j,k} \right] \\ = \left[\xi(m) q_{i,j,t} - \sum_{l=0}^{t-1} \xi(t-l) I_{\{x \geq t-m\}}(l) w_{i,j,l} \right] & \\ + \sum_{k=0}^{m-1} g_{i,j,k} \left[-\xi(m) d_{t-k} I_{\{x \leq t\}}(k) \right. &\left. - \sum_{l=0}^{t-1} \xi(t-l) I_{\{x \geq t-m\}}(l) x_{s,k} \right]. \end{aligned}$$

So, we obtain

$$\begin{cases} x_{t,k} = -\xi(m) d_{t-k} I_{\{x \le t\}}(k) - \sum\limits_{l=\max\{0,\, t-m\}}^{t-1} \xi(t-l) x_{l,k}, \\ \qquad\qquad k = 0, 1, 2, \dots, m-1; \\ w_{i,j,t} = \xi(m) q_{i,j,t} - \sum\limits_{l=\max\{0,\, t-m\}}^{t-1} \xi(t-l) w_{i,j,l}, \\ \qquad\qquad t = 0, 1, 2, \dots, M-1, \quad i, j = 1, 2, \dots, n. \end{cases} \tag{1.29}$$

In the case $t \ge M$ we obtain the following transformations

$$\begin{aligned} q_{i,j,t} &= \sum_{k=0}^{m-1} d_{t-k} I_{\{0 \le x \le M\}}(t-k) g_{i,j,k} \\ &\quad + (\xi(m))^{-1} \sum_{l=0}^{M-1} \xi(t-l) I_{\{x \ge t-m\}}(l) \left[w_{i,j,l} + \sum_{k=0}^{m-1} x_{l,k}\, g_{i,j,k} \right] \\ &= (\xi(m))^{-1} \sum_{l=0}^{M-1} \xi(t-l) I_{\{x \ge t-m\}}(l) w_{i,j,l} \\ &\quad + \sum_{k=0}^{m-1} g_{i,j,k} \left[d_{t-k} I_{\{0 \le x \le M\}}(t-k) \right. \\ &\quad \left. + (\xi(m))^{-1} \sum_{l=0}^{M-1} \xi(t-l) I_{\{x \ge t-m\}}(l) x_{l,k} \right]. \end{aligned}$$

This involves

$$\sum_{k=0}^{m-1} r_{t,k}\, g_{i,j,k} = s_{i,j,t}, \quad t = M, M+1, \dots, r-1, \quad i, j = 1, 2, \dots, n, \tag{1.30}$$

where

$$\begin{cases} r_{t,k} = d_{t-k} I_{\{0 \le x \le M\}}(t-k) + (\xi(m))^{-1} \sum\limits_{l=\max\{0,\, t-m\}}^{M-1} \xi(t-l) x_{l,k}; \\ s_{i,j,t} = q_{i,j,t} - (\xi(m))^{-1} \sum\limits_{l=\max\{0,\, t-m\}}^{M-1} \xi(t-l) w_{i,j,l}, \ k = 0, 1, 2, \dots, m-1. \end{cases} \tag{1.31}$$

The results described above allow us to determine the values $\alpha_{i,j,k}(\mu)$, $k = 1, 2, \dots, m$, $i, j = 1, 2, \dots, n$. Indeed, according to formula (1.28) we have

$$\frac{G_{i,j}(z)}{(z-\mu)^m} = \sum_{k=1}^{m} \frac{\alpha_{i,j,k}(\mu)}{(z-\mu)^k} = \frac{1}{(z-\mu)^m} \sum_{k=1}^{m} \alpha_{i,j,k}(\mu)(z-\mu)^{m-k}.$$

If we express this formula in a more detailed form then we obtain

$$\begin{aligned}\sum_{l=0}^{m-1} g_{i,j,l} z^l &= \sum_{k=1}^{m} \alpha_{i,j,k}(\mu)(z-\mu)^{m-k} = \sum_{k=0}^{m-1} \alpha_{i,j,m-k}(\mu)(z-\mu)^k \\ &= \sum_{k=0}^{m-1} \alpha_{i,j,m-k}(\mu) \sum_{l=0}^{k} C_k^l (-\mu)^{k-l} z^l \\ &= \sum_{l=0}^{m-1} z^l \sum_{k=l}^{m-1} \alpha_{i,j,m-k}(\mu) C_k^l (-\mu)^{k-l}.\end{aligned}$$

So, we have

$$g_{i,j,l} = \sum_{k=s}^{m-1} C_k^l (-\mu)^{k-l} \alpha_{i,j,m-k}(\mu), \ \ l = 0, 1, 2, \ldots, m-1, \ \ i, j = 1, 2, \ldots, n.$$

If we substitute the expression $g_{i,j,l}$ in (1.30) then we obtain

$$\begin{aligned}s_{i,j,t} &= \sum_{k=0}^{m-1} r_{t,k} \sum_{l=k}^{m-1} C_l^k (-\mu)^{l-k} \alpha_{i,j,m-l}(\mu) \\ &= \sum_{l=0}^{m-1} \alpha_{i,j,m-l}(\mu) \sum_{k=0}^{l} C_l^k (-\mu)^{l-k} r_{t,k} \\ &= \sum_{l=1}^{m} \alpha_{i,j,l}(\mu) \sum_{k=0}^{m-l} C_{m-l}^k (-\mu)^{m-l-k} r_{t,k} \\ &= \sum_{l=1}^{m} r_{t,l}^* \alpha_{i,j,l}(\mu), \quad t = M, M+1, \ldots, r-1, \ \ i, j = 1, 2, \ldots, n,\end{aligned}$$

where

$$r_{t,l}^* = \sum_{k=0}^{m-l} C_{m-l}^k (-\mu)^{m-l-k} r_{t,k}, \ \ t = M, M+1, \ldots, r-1, \ \ s = 1, 2, \ldots, m. \tag{1.32}$$

The solution of the system is

$$\alpha_{i,j}(\mu) = (R^*)^{-1} S_{i,j}, \quad i, j = 1, 2, \ldots, n, \tag{1.33}$$

where

$$\alpha_{i,j}(\mu) = ((\alpha_{i,j,l}(\mu))_{l=\overline{1,m}})^T, \quad S_{i,j} = ((s_{i,j,t})_{t=\overline{M,r-1}})^T$$

and

$$R^* = (r^*_{t,l})_{t=\overline{M,r-1},\ \ l=\overline{1,m}}.$$

1.4.3 The Main Conclusion and the Description of the Algorithm

In Sect. 1.2.1 we have introduced the complex functions $\nu_k(z) = (1-z)^{-k}, \forall k \geq 1$ that satisfy the recurrent relation $\nu_{k+1}(z) = \mathrm{d}\nu_k(z)/(k\mathrm{d}z)$, $\forall k \geq 1$. In addition, we have shown that $\nu_k(z) = \sum_{t=0}^{\infty} H_{k-1}(t)z^t$, $\forall k \geq 1$, where the coefficient $H_{k-1}(t)$ is a polynomial of degree less or equal to $k-1$. Moreover, the calculation formula for the elements $\beta_{i,j,k}(y)$ and the corresponding matrices

$$W_{i,j}(y,t) = \sum_{k=0}^{m(y)-1} (-y)^{-k-1}\alpha_{i,j,k+1}(y)H_k(t), \quad \forall y \in \mathbb{C}\backslash\mathcal{D}, \ i,j = 1,2,\ldots,n$$

have been obtained.

Let $H_k(t) = \sum_{l=0}^{k} u_s^{(k)} t^l$, $\forall k \geq 0$. Then

$$\nu_{k+1}(z) = \frac{\mathrm{d}}{k\mathrm{d}z}\sum_{t=0}^{\infty} H_{k-1}(t)z^t = \frac{1}{k}\sum_{t=1}^{\infty} tH_{k-1}(t)z^{t-1} = \frac{1}{k}\sum_{t=0}^{\infty}(t+1)H_{k-1}(t+1)z^t$$

and we obtain

$$\begin{aligned} H_k(t) &= \frac{1}{k}(t+1)H_{k-1}(t+1) = \frac{1}{k}(t+1)\sum_{l=0}^{k-1} u_l^{(k-1)}(t+1)^l \\ &= \frac{1}{k}\sum_{l=0}^{k-1} u_l^{(k-1)}(t+1)^{l+1} = \frac{1}{k}\sum_{l=0}^{k-1} u_l^{(k-1)} \sum_{l=0}^{l+1} C_{l+1}^l t^l \\ &= \frac{1}{k}\sum_{l=0}^{k-1} u_l^{(k-1)}\left(1+\sum_{l=1}^{l+1} C_{l+1}^l t^l\right) \\ &= \frac{1}{k}\sum_{l=0}^{k-1} u_l^{(k-1)} + \frac{1}{k}\sum_{l=1}^{k} t^l \sum_{l=l-1}^{k-1} u_l^{(k-1)} C_{l+1}^l. \end{aligned}$$

This means that

$$\begin{cases} u_0^{(0)} = 1, \\ u_0^{(k)} = \frac{1}{k}\sum_{l=0}^{k-1} u_l^{(k-1)}, \ u_j^{(k)} = \frac{1}{k}\sum_{l=j-1}^{k-1} C_{l+1}^r u_l^{(k-1)}, \ \forall k \geq 1, \ j = 1, 2, \ldots, k. \end{cases} \tag{1.34}$$

We obtain a formula for calculating the elements of the matrices for the expressions

$$\begin{cases} \beta_{i,j,k}(y) = \sum_{l=k+1}^{m(y)} (-y)^{-l} \alpha_{i,j,l}(y) u_k^{(l-1)}, \\ y \in \mathbb{C} \setminus \mathcal{D}, \ k = 0, 1, 2, \ldots, m(y) - 1, \ i, j = 1, 2, \ldots, n. \end{cases} \tag{1.35}$$

Based on the results described above we can use the following algorithm for determining the limit and the differential matrices in a Markov chain:

Algorithm 1.4 Determining the Limit and Differential Matrices

Input Data: The matrix of the probability transitions P.

Output Data: The matrices $\beta_k(y)$ $(y \in \mathbb{C} \setminus \mathcal{D}, \ k = 0, 1, 2, \ldots, m(y) - 1)$.

1–3. Do steps 1–3 of Algorithm 1.3;

4. Calculate the matrices $R^{(k)}, k = 0, 1, 2, \ldots, n-1$, according to formula (1.25);
5. Find the values $q_{i,j,k}$, $k = 0, 1, 2, \ldots, r - 1$, $i, j = 1, 2, \ldots, n$, using the calculation procedure described in Sect. 1.4.1;
6. Calculate C_s^k, $s = 1, 2, \ldots, \max_{y \in \mathbb{C} \setminus \mathcal{D}} m(y)$, $k = 0, 1, 2, \ldots, s$, using Pascal's triangle rule;
7. Determine $u_j^{(k)}$, $k = 0, 1, 2, \ldots, \max_{y \in \mathbb{C} \setminus \mathcal{D}} m(y) - 1$, $j = 0, 1, 2, \ldots, k$, using formula (1.34);
8. For every $\mu \in \mathbb{C} \setminus \mathcal{D}$ do the items *a–g*:
   a. Determine the values $\xi(k) = C_m^k(-\mu)^{-k}, k = 0, 1, 2, \ldots, m \ \ (m = m(\mu))$;
   b. Determine the coefficients d_k, $k = 0, 1, 2, \ldots, r - m$, using Horner's scheme;
   c. Calculate the values $x_{t,k}$, $t = 0, 1, 2, \ldots, M - 1, k = 0, 1, 2, \ldots, m - 1$, according to (1.29);
   d. Calculate the values $r_{t,k}, t = M, M + 1, \ldots, r - 1, k = 0, 1, 2, \ldots, m - 1$, using formula (1.31);
   e. Determine the elements of the matrix R^* according to the relation (1.32);
   f. Determine the matrix $(R^*)^{-1}$ using known numerical algorithms;
   g. For $i, j = 1, 2, \ldots, n$ do the items g_1–g_4 :
      g_1. Calculate the values $w_{i,j,t}, t = 0, 1, 2, \ldots, M-1$, according to formula (1.29);
      g_2. Calculate the values $s_{i,j,t}$, $t = M, M + 1, \ldots, r - 1$, using formula (1.31);

g3. Determine the vector $\alpha_{i,j}(\mu)$ according to relation (1.33);

g4. Calculate the elements $\beta_{i,j,k}(\mu)$ of the matrix $\beta_k(\mu)$, $k = 0, 1, 2, , \dots, m(\mu) - 1$, according to formula (1.35).

It is easy to observe that the computational complexity of this algorithm is similar to the computational complexity of the algorithm from the previous section. If the characteristic values of the matrix P are known then the algorithm finds the limit and the differential matrices using $O(n^4)$ elementary operations. Note that this algorithm can be applied if the set (or a subset) of characteristic values of the matrix P is known. In this case we use the set (or the subset) $\mathbb{C} \setminus \mathbb{D}$ of the inverse nonzero characteristic values; the algorithm will determine the corresponding matrices which correspond to these characteristic values. The computational complexity of the algorithm in the case of unknown characteristic values depends on the complexity of the algorithm for determining the roots of the characteristic polynomial.

1.4.4 Numerical Examples

We will illustrate the details of the proposed algorithms for periodic as well as for aperiodic Markov chains.

Example 1 Let a Markov chain with the following transition probability matrix be given

$$P = \begin{pmatrix} 1 & 0 & 0 \\ 0 & 0.5 & 0.5 \\ 0.5 & 0 & 0.5 \end{pmatrix}$$

and consider the problem of determining the limit matrix and the differential components of the matrix $\overline{P}(t)$.

At first we apply Algorithm 1.3.

Step 1
Calculate the coefficients of the characteristic polynomial. Thus, we find

$$P^2 = \begin{pmatrix} 1 & 0 & 0 \\ 0.25 & 0.25 & 0.5 \\ 0.75 & 0 & 0.25 \end{pmatrix}, \quad P^3 = \begin{pmatrix} 1 & 0 & 0 \\ 0.5 & 0.125 & 0.375 \\ 0.875 & 0 & 0.125 \end{pmatrix};$$

$$\overline{s}_1 = \mathrm{tr} P = 2, \quad \overline{s}_2 = \mathrm{tr} P^2 = 1.5, \quad \overline{s}_3 = \mathrm{tr} P^3 = 1.25$$

and determine

$$\beta_0 = 1, \quad \beta_1 = -\overline{s}_1 = -2, \quad \beta_2 = -(\overline{s}_2 + \overline{\beta}_1 \overline{s}_1)/2 = 1.25,$$
$$\beta_3 = -(\overline{s}_3 + \beta_1 \overline{s}_2 + \beta_2 \overline{s}_1)/3 = -0.25.$$

Steps 2–3
Find the roots of the equation $\Delta(z) = 0$ and the set $\mathbb{C}\backslash\mathcal{D}$:

$$\Delta(z) = \sum_{k=0}^{3} \beta_k z^k = 1 - 2z + 1.25z^2 - 0.25z^3 = (1-z)(1-0.5z)^2,$$
$$\mathbb{C}\backslash\mathcal{D} = \{z \in \mathbb{C} \mid \Delta(z) = 0\} = \{1, 2\}, \quad m(1) = 1, \quad m(2) = 2, \quad r = n = 3.$$

Step 4
Find the matrix B:

$$\overline{\beta}_0 = (1, 1, 0), \quad \overline{\beta}_1 = (1, 0.5, 0.5),$$
$$\overline{\beta}_2 = (1, 0.25, 0.5) \implies B = \begin{pmatrix} 1 & 1 & 0 \\ 1 & 0.5 & 0.5 \\ 1 & 0.25 & 0.5 \end{pmatrix}.$$

Step 5
Calculate $(B^T)^{-1}$:

$$(B^T)^{-1} = \begin{pmatrix} 1 & 0 & -2 \\ -4 & 4 & 6 \\ 4 & -4 & -4 \end{pmatrix}.$$

Steps 6–7
Find the coefficients C_s^k using Pascal's triangle rule:

$$C_0^0 = C_1^0 = C_1^1 = 1, \quad (r-n)^0 = 0^0 = 1, \quad (r-n)^1 = 0^1 = 0.$$

Steps 8a–8b
Determine $I_r^{[\overline{a}_{i,j}]}$ and $\gamma_{i,j,l}(y)$:

$$\Gamma_{1,1} = I_3^{[\overline{a}_{1,1}]}(B^T)^{-1} = (1, 1, 1) \begin{pmatrix} 1 & 0 & -2 \\ -4 & 4 & 6 \\ 4 & -4 & -4 \end{pmatrix} = (1, 0, 0),$$
$$\Gamma_{1,2} = (0, 0, 0)(B^T)^{-1} = (0, 0, 0),$$
$$\Gamma_{1,3} = (0, 0, 0)(B^T)^{-1} = (0, 0, 0),$$
$$\Gamma_{2,1} = (0, 0, 0.25)(B^T)^{-1} = (1, -1, -1),$$
$$\Gamma_{2,2} = (1, 0.5, 0.25)(B^T)^{-1} = (0, 1, 0),$$
$$\Gamma_{2,3} = (0, 0.5, 0.5)(B^T)^{-1} = (0, 0, 1),$$
$$\Gamma_{3,1} = (0, 0.5, 0.75)(B^T)^{-1} = (1, -1, 0),$$
$$\Gamma_{3,2} = (0, 0, 0)(B^T)^{-1} = (0, 0, 0),$$
$$\Gamma_{3,3} = (1, 0.5, 0.25)(B^T)^{-1} = (0, 1, 0).$$

Step 8c
Find the coefficients $\beta_{i,j,k}(y)$ for the limit and the differential matrices by using the formula:

$$\beta_{i,j,k}(y) = \sum_{l=k}^{m(y)-1} 0^{l-k}\gamma_{i,j,l}(y) = \gamma_{i,j,k}(y)$$

for $y \in \mathbb{C}\backslash\mathcal{D}$, $k = 0, 1, 2, \ldots, m(y) - 1$, $i, j = 1, 2, 3$.

Based on this formula we obtain

$$\beta_0(1) = \begin{pmatrix} 1 & 0 & 0 \\ 1 & 0 & 0 \\ 1 & 0 & 0 \end{pmatrix}, \quad \beta_0(2) = \begin{pmatrix} 0 & 0 & 0 \\ -1 & 1 & 0 \\ -1 & 0 & 1 \end{pmatrix}, \quad \beta_1(2) = \begin{pmatrix} 0 & 0 & 0 \\ -1 & 0 & 1 \\ 0 & 0 & 0 \end{pmatrix}.$$

So, the matrix $\overline{P}(t)$ can be represented as follows:

$$\overline{P}(t) = \begin{pmatrix} 1 & 0 & 0 \\ 1 & 0 & 0 \\ 1 & 0 & 0 \end{pmatrix} + \begin{pmatrix} 0 & 0 & 0 \\ -1 & 1 & 0 \\ -1 & 0 & 1 \end{pmatrix}\left(\frac{1}{2}\right)^t + \begin{pmatrix} 0 & 0 & 0 \\ -1 & 0 & 1 \\ 0 & 0 & 0 \end{pmatrix} t\left(\frac{1}{2}\right)^t \quad (1.36)$$

because $\lambda = 1/2$.

If we apply Algorithm 1.4 for Example 1, then we obtain: Steps 1–3
Repeat the steps 1–3 of Algorithm 1.3 and find:

$$P^2 = \begin{pmatrix} 1 & 0 & 0 \\ 0.25 & 0.25 & 0.5 \\ 0.75 & 0 & 0.25 \end{pmatrix}, \quad P^3 = \begin{pmatrix} 1 & 0 & 0 \\ 0.5 & 0.125 & 0.375 \\ 0.875 & 0 & 0.125 \end{pmatrix};$$

$$\beta_0 = 1, \quad \beta_1 = -2, \quad \beta_2 = 1.25, \quad \beta_3 = -0.25;$$

$$\mathbb{C}\backslash\mathcal{D} = \{1, 2\}, \quad m(1) = 1, \quad m(2) = 2, \quad r = n = 3.$$

Steps 4–5
Find $R^{(k)}$ and $q_{i,j,k}$:

$$R^{(0)} = \beta_0 I = \begin{pmatrix} 1 & 0 & 0 \\ 0 & 1 & 0 \\ 0 & 0 & 1 \end{pmatrix}, \quad R^{(1)} = \beta_1 I + PR^{(0)} = \begin{pmatrix} -1 & 0 & 0 \\ 0 & -1.5 & 0.5 \\ 0.5 & 0 & -1.5 \end{pmatrix},$$

$$R^{(2)} = \beta_2 I + PR^{(1)} = \begin{pmatrix} 0.25 & 0 & 0 \\ 0.25 & 0.5 & -0.5 \\ -0.25 & 0 & 0.5 \end{pmatrix};$$

$$q_{i,j,k} = R_{i,j}^{(k)}, \quad i, j = 1, 2, 3, \quad k = 0, 1, 2.$$

Steps 6–7

Calculate C_s^k and $u_l^{(k)}$:

$$C_1^0 = C_1^1 = C_2^0 = C_2^2 = 1, \quad C_2^1 = C_1^0 + C_1^1 = 2; \quad u_0^{(0)} = u_0^{(1)} = u_1^{(1)} = 1.$$

Step 8

For $\mu = 1$ and $\mu = 2$ do the items *a–g*: If we fix $\mu = 1$ then we have

$$8') \qquad m = m(\mu) = 1, \quad M = r - m = 2, \quad xi(0) = 1, \quad xi(1) = -1.$$

Based on Horner's scheme we obtain

	−0.25	1.25	−2	1
1	−0.25	1	−1	0

$\Rightarrow d_0 = -1, \quad d_1 = 1, \quad d_2 = -0.25.$

Then we calculate

$$x_{0,0} = -\xi(1)d_0 = -1, \quad x_{1,0} = -\xi(1)d_1 - \xi(1)x_{0,0} = 0;$$
$$r_{2,0} = d_2 - \xi(1)x_{1,0} = -0.25, \quad r_{2,1}^* = r_{2,0} = -0.25;$$
$$R^* = (-0.25); \quad (R^*)^{-1} = (-4);$$
$$w_{i,j,0} = -q_{i,j,0} = -R_{i,j}^{(0)}, \; w_{i,j,1} = -q_{i,j,1} + w_{i,j,0} = -R_{i,j}^{(0)} - R_{i,j}^{(1)}, \; i, j = 1, 2, 3;$$
$$(s_{i,j,2})_{3\times 3} = (q_{i,j,2} - w_{i,j,1})_{3\times 3} = (R_{i,j}^{(0)} + R_{i,j}^{(1)} + R_{i,j}^{(2)})_{3\times 3} = \begin{pmatrix} 0.25 & 0 & 0 \\ 0.25 & 0 & 0 \\ 0.25 & 0 & 0 \end{pmatrix};$$
$$(\alpha_{i,j}(1))_{3\times 3} = -4(s_{i,j,2})_{3\times 3} = \begin{pmatrix} -1 & 0 & 0 \\ -1 & 0 & 0 \\ -1 & 0 & 0 \end{pmatrix};$$
$$(\beta_{i,j,0}(1))_{3\times 3} = (-\alpha_{i,j}(1))_{3\times 3} = \begin{pmatrix} 1 & 0 & 0 \\ 1 & 0 & 0 \\ 1 & 0 & 0 \end{pmatrix} \Rightarrow \beta_0(1) = \begin{pmatrix} 1 & 0 & 0 \\ 1 & 0 & 0 \\ 1 & 0 & 0 \end{pmatrix}.$$

If we fix $\mu = 2$ then we have

$$8'') \qquad m = m(\mu) = 2, \quad M = r - m = 1; \quad \xi(0) = 1, \quad \xi(1) = -1, \quad \xi(2) = 0.25\,.$$

Find the coefficients d_i using Horner's scheme

	−0.25	1.25	−2	1
2	−0.25	0.75	−0.5	0
2	−0.25	0.25	0	

$\Rightarrow d_0 = 0.25, \; d_1 = -0.25.$

Then we calculate

$$x_{0,0} = -\xi(2)d_0 = -0.0625,\ x_{0,1} = 0;\ r_{1,0} = 0,\ r_{1,1} = 0.25,\ r_{2,0} = -0.0625,$$
$$r_{2,1} = -0.25 \Rightarrow r^*_{1,1} = 0.25, \quad r^*_{1,2} = 0, \quad r^*_{2,1} = -0.125, \quad r^*_{2,2} = -0.0625$$
$$\Longrightarrow R^* = \begin{pmatrix} 0.25 & 0 \\ -0.125 & -0.0625 \end{pmatrix} \Longrightarrow (R^*)^{-1} = \begin{pmatrix} 4 & 0 \\ -8 & -16 \end{pmatrix};$$
$$w_{i,j,0} = 0.25 q_{i,j,0} = 0.25 R^{(0)}_{i,j}; \quad s_{i,j,1} = q_{i,j,1} + 4w_{i,j,0} = R^{(0)}_{i,j} + R^{(1)}_{i,j},$$
$$s_{i,j,2} = q_{i,j,2} - w_{i,j,0} = R^{(2)}_{i,j} - 0.25 R^{(0)}_{i,j} \Longrightarrow S_{i,j} = \begin{pmatrix} R^{(0)}_{i,j} + R^{(1)}_{i,j} \\ R^{(2)}_{i,j} - 0.25 R^{(0)}_{i,j} \end{pmatrix}$$
$$\Longrightarrow \alpha_{i,j}(2) = (R^*)^{-1} S_{i,j} = \begin{pmatrix} 4R^{(0)}_{i,j} + 4R^{(1)}_{i,j} \\ -4R^{(0)}_{i,j} - 8R^{(1)}_{i,j} - 16\, R^{(2)}_{i,j} \end{pmatrix}, \ i,\, j = 1, 2, 3;$$
$$\beta_{i,j,0}(2) = -0.5\alpha_{i,j,1}(2) + 0.25\alpha_{i,j,2}(2) = -3R^{(0)}_{i,j} - 4R^{(1)}_{i,j} - 4R^{(2)}_{i,j},$$
$$\beta_{i,j,1}(2) = 0.25\alpha_{i,j,2}(2) = -R^{(0)}_{i,j} - 2R^{(1)}_{i,j} - 4R^{(2)}_{i,j},\ i,\, j = 1, 2, 3;$$
$$\Longrightarrow \beta_0(2) = \begin{pmatrix} 0 & 0 & 0 \\ -1 & 1 & 0 \\ -1 & 0 & 1 \end{pmatrix}, \quad \beta_1(2) = \begin{pmatrix} 0 & 0 & 0 \\ -1 & 0 & 1 \\ 0 & 0 & 0 \end{pmatrix}.$$

So, we obtain formula (1.36).

Example 2 Let a 2-periodic Markov process determined by the matrix of probability transition

$$P = \begin{pmatrix} 0 & 1 \\ 1 & 0 \end{pmatrix} \quad \text{be given.}$$

Consider the problem of determining the limit and the differential components of the matrix $\overline{P}(t)$.

At first we apply Algorithm 1.3.

Steps 1–3
Find the characteristic polynomial $\Delta(z) = 0$ and the set $\mathbb{C}\backslash\mathcal{D}$ in a similar way as in Example 1:

$$P^2 = \begin{pmatrix} 1 & 0 \\ 0 & 1 \end{pmatrix}, \quad \overline{s}_1 = \mathrm{tr} P = 0, \quad \overline{s}_2 = \mathrm{tr} P^2 = 2 \Longrightarrow \beta_0 = 1, \quad \beta_1 = -\overline{s}_1 = 0,$$
$$\beta_2 = -(\overline{s}_2 + \beta_1 \overline{s}_1)/2 = -1 \Longrightarrow \Delta(z) = \sum_{k=0}^{2} \beta_k z^k = 1 - z^2 = (1-z)(1+z),$$

i.e.,

$$\mathbb{C}\backslash\mathcal{D} = \{z \in \mathbb{C} \mid \Delta(z) = 0\} = \{1, -1\}, \quad m(1) = m(-1) = 1, \; r = n = 2.$$

Steps 4–5
Find the matrices B and $(B^T)^{-1}$:

$$\overline{\beta}_0 = (1, 1), \qquad ov\beta_1 = (1, -1) \Rightarrow B = \begin{pmatrix} 1 & 1 \\ 1 & -1 \end{pmatrix},$$

$$(B^T)^{-1} = \begin{pmatrix} 0.5 & 0.5 \\ 0.5 & -0.5 \end{pmatrix}.$$

Steps 6–8
Determine C_s^k, $I_r^{[\overline{a}_{i,j}]}$ and $\gamma_{i,j,l}(y)$:

$$C_0^0 = 1, \;\; ,(r-n)^0 = 0^0 = 1; \;\; \Gamma_{1,1} = \Gamma_{2,2} = (1, 0)\begin{pmatrix} 0.5 & 0.5 \\ 0.5 & -0.5 \end{pmatrix} = (0.5,\, 0.5),$$

$$\Gamma_{1,2} = \Gamma_{2,1} = (0, 1)\begin{pmatrix} 0.5 & 0.5 \\ 0.5 & -0.5 \end{pmatrix} = (0.5,\, -0.5)$$

$$\Longrightarrow \beta_0(1) = \begin{pmatrix} 0.5 & 0.5 \\ 0.5 & 0.5 \end{pmatrix}, \quad \beta_0(-1) = \begin{pmatrix} 0.5 & -0.5 \\ -0.5 & 0.5 \end{pmatrix}.$$

So, we obtain the following formula for the matrix $P(t)$:

$$\overline{P}(t) = \begin{pmatrix} 0.5 & 0.5 \\ 0.5 & 0.5 \end{pmatrix} + \begin{pmatrix} 0.5 & -0.5 \\ -0.5 & 0.5 \end{pmatrix}(-1)^t. \tag{1.37}$$

Now we apply Algorithm 1.4 for Example 2.

Steps 1–3
Repeat the steps 1–3 of Algorithm 1.3 and find

$$P^2 = \begin{pmatrix} 1 & 0 \\ 0 & 1 \end{pmatrix}, \quad \beta_0 = 1, \quad \beta_1 = 0, \quad \beta_2 = -1$$
$$\Longrightarrow \Delta(z) = 1 - z^2 = (1-z)(1+z)$$
$$\Longrightarrow \mathbb{C}\backslash\mathcal{D} = \{1, -1\}, \quad m(1) = m(-1) = 1, \quad r = n = 2.$$

Step 4
Calculate

$$R^{(0)} = \beta_0 I = \begin{pmatrix} 1 & 0 \\ 0 & 1 \end{pmatrix}, \quad R^{(1)} = \beta_1 I + P R^{(0)} = \begin{pmatrix} 0 & 1 \\ 1 & 0 \end{pmatrix}.$$

Steps 5–7
Find

$$q_{i,j,k} = R_{i,j}^{(k)}, \quad i, j = 1, 2, \ k = 0, 1; \quad C_1^0 = C_1^1 = 1, \quad u_0^0 = 1.$$

Step 8
For $\mu = 1$ and $\mu = -1$ do the items *a–g*:
If we fix $\mu = 1$ then we have

$$8') \qquad m = m(\mu) = 1, \quad M = r - m = 1, \quad \xi(0) = 1, \quad \xi(1) = -1;$$

then determine d_0 and d_1 using Horner's scheme

	-1	0	1
1	-1	-1	0

$\Rightarrow d_0 = -1, \quad d_1 = -1.$

This means that

$$x_{0,0} = -\xi(1)d_0 = -1; \quad r_{1,0} = d_1 - \xi(1)x_{0,0} = -2 \Longrightarrow r_{1,1}^* = -2$$

$$\Longrightarrow R^* = (-2), \ (R^*)^{-1} = (-0.5); \quad w_{i,j,0} = \xi(1)q_{i,j,0} = -R_{i,j}^{(0)}, \ i, j = 1, 2;$$

$$s_{i,j,1} = q_{i,j,1} + \xi(1)w_{i,j,0} = R_{i,j}^{(0)} + R_{i,j}^{(1)}$$

$$\Longrightarrow \alpha_{i,j}(1) = (-0.5)(s_{i,j,1}) = -0.5R_{i,j}^{(0)} - 0.5R_{i,j}^{(1)}, \ i, j = 1, 2;$$

$$\beta_{i,j,0}(1) = -\alpha_{i,j,1}(1) = 0.5R_{i,j}^{(0)} + 0.5R_{i,j}^{(1)}, \ i, j = 1, 2;$$

$$\Longrightarrow \beta_0(1) = \begin{pmatrix} 0.5 & 0.5 \\ 0.5 & 0.5 \end{pmatrix}.$$

If we fix $\mu = -1$ then we have

$$8'') \qquad m = m(\mu) = 1, \quad M = r - m = 1, \quad \xi(0) = 1, \quad \xi(1) = 1;$$

then determine d_0 and d_1 using Horner's scheme

	-1	0	1
-1	-1	1	0

$\Rightarrow d_0 = 1, \quad d_1 = -1.$

This means that

$$x_{0,0} = -\xi(1)d_0 = -1; \quad r_{1,0} = d_1 + \xi(1)x_{0,0} = -2 \Longrightarrow r_{11}^* = -2$$

$$\Longrightarrow R^* = (-2), \ (R^*)^{-1} = (-0.5); \quad w_{i,j,0} = \xi(1)q_{i,j,0} = R_{i,j}^{(0)}, \ i, j = 1, 2;$$

$$s_{i,j,1} = q_{i,j,1} - \xi(1)w_{i,j,0} = R_{i,j}^{(1)} - R_{i,j}^{(0)}$$

$$\Longrightarrow \alpha_{i,j}(-1) = (-0.5)(s_{i,j,1}) = -0.5R_{i,j}^{(1)} + 0.5R_{i,j}^{(0)}, \ i, j = 1, 2;$$

$$\beta_{i,j,0}(-1) = \alpha_{i,j,0}(-1) = -0.5R_{i,j}^{(1)} + 0.5R_{i,j}^{(0)},\ i,\ j = 1, 2;$$

$$\Longrightarrow \beta_0(-1) = \begin{pmatrix} 0.5 & -0.5 \\ -0.5 & 0.5 \end{pmatrix}.$$

So, we obtain formula (1.37).

In the examples given above the roots of the characteristic polynomials are real values. Below we consider an example where the characteristic polynomial contains complex roots. For this example the calculations in the algorithms are similar as for the case with real roots, however, at the final stage of the algorithms in order to obtain the real representation of $\overline{P}(t)$, it is necessary to make some additional elementary transformations that eliminate the imaginary component of $T(t)$. We illustrate these transformations on the example below.

Example 3 Let a Markov chain with the matrix of probability transitions

$$P = \begin{pmatrix} \frac{1}{2} & 0 & \frac{1}{2} \\ \frac{1}{2} & \frac{1}{4} & \frac{1}{4} \\ \frac{1}{4} & \frac{1}{2} & \frac{1}{4} \end{pmatrix}$$

be given and consider the problem of determining the limit matrix and the differential components of the matrix $\overline{P}(t)$.

If we apply Algorithm 1.3 then we obtain:

Step 1

Calculate the coefficients of the characteristic polynomial. Thus, we find

$$P^2 = \begin{pmatrix} \frac{3}{8} & \frac{1}{4} & \frac{3}{8} \\ \frac{7}{16} & \frac{3}{16} & \frac{3}{8} \\ \frac{7}{16} & \frac{1}{4} & \frac{5}{16} \end{pmatrix}, \quad P^3 = \begin{pmatrix} \frac{13}{32} & \frac{1}{4} & \frac{11}{32} \\ \frac{13}{32} & \frac{15}{64} & \frac{23}{64} \\ \frac{27}{64} & \frac{7}{32} & \frac{23}{64} \end{pmatrix};$$

$$\overline{s}_1 = \mathrm{tr}P = 1, \quad \overline{s}_2 = \mathrm{tr}P^2 = \frac{7}{8}, \quad \overline{s}_3 = \mathrm{tr}P^3 = 1$$

and determine

$$\beta_0 = 1, \quad \beta_1 = -\overline{s}_1 == -1, \quad \beta_2 = -(\overline{s}_2 + \overline{\beta}_1\overline{s}_1)/2 = \frac{1}{16},$$

$$\beta_3 = -(\overline{s}_3 + \beta_1\overline{s}_2 + \beta_2\overline{s}_1)/3 = -\frac{1}{16}.$$

Steps 2–3

Find the roots of the equation $\Delta(z) = 0$ and the set $\mathbb{C}\backslash\mathcal{D}$:

$$\Delta(z) = \sum_{k=0}^{3} \beta_k z^k = 1 - z + \frac{1}{16}z^2 - \frac{1}{16}z^3 = -\frac{1}{16}(z-1)(z-4i)(z+4i),$$

$\mathbb{C}\backslash\mathcal{D} = \{z \in \mathbb{C} \mid \Delta(z) = 0\} = \{1, -4i, 4i\}$; $m(1) = m(-4i) = m(4i) = 1$, $r = n = 3$.

Step 4
Find the matrix B: $\overline{\beta}_0 = (1, 1, 1)$, $\overline{\beta}_1 = (1, i/4, -i/4)$, $\overline{\beta}_2 = (1, -1/16, -1/16)$, i.e.,

$$B = \begin{pmatrix} 1 & 1 & 1 \\ 1 & \frac{i}{4} & -\frac{i}{4} \\ 1 & -\frac{1}{16} & -\frac{1}{16} \end{pmatrix}.$$

Step 5
Calculate $(B^T)^{-1}$:

$$(B^T)^{-1} = \begin{pmatrix} \frac{1}{17} & \frac{8}{17} + \frac{2}{17}i & \frac{8}{17} - \frac{2}{17}i \\ 0 & -2i & 2i \\ \frac{16}{17} & -\frac{8}{17} + \frac{32}{17}i & -\frac{8}{17} - \frac{32}{17}i \end{pmatrix}.$$

Steps 6–7
Find the coefficients C_s^k using Pascal's triangle rule:

$$C_0^0 = C_1^0 = C_1^1 = 1, \quad (r-n)^0 = 0^0 = 1, \quad (r-n)^1 = 0^1 = 0.$$

Steps 8a–8b
Determine $I_r^{[\overline{a}_{i,j}]}$ and $\gamma_{i,j,l}(y)$:

$$\Gamma_{1,1} = I_3^{[\overline{a}_{1,1}]}(B^T)^{-1} = \left(1, \frac{1}{2}, \frac{3}{8}\right)(B^T)^{-1} = \left(\frac{7}{17}, \frac{5}{17} - \frac{3}{17}i, \frac{5}{17} + \frac{3}{17}i\right),$$

$$\Gamma_{1,2} = \left(0, 0, \frac{1}{4}\right)(B^T)^{-1} = \left(\frac{4}{17}, -\frac{2}{17} + \frac{8}{17}i, -\frac{2}{17} - \frac{8}{17}i\right),$$

$$\Gamma_{1,3} = \left(0, \frac{1}{2}, \frac{3}{8}\right)(B^T)^{-1} = \left(\frac{6}{17}, -\frac{3}{17} - \frac{5}{17}i, -\frac{3}{17} + \frac{5}{17}i\right),$$

$$\Gamma_{2,1} = \left(0, \frac{1}{2}, \frac{7}{16}\right)(B^T)^{-1} = \left(\frac{7}{17}, -\frac{7}{34} - \frac{3}{17}i, -\frac{7}{34} + \frac{3}{17}i\right),$$

$$\Gamma_{2,2} = \left(1, \frac{1}{4}, \frac{3}{16}\right)(B^T)^{-1} = \left(\frac{4}{17}, \frac{13}{34} - \frac{i}{34}, \frac{13}{34} + \frac{i}{34}\right),$$

$$\Gamma_{2,3} = \left(0,\ \frac{1}{4},\ \frac{3}{8}\right)(B^T)^{-1} = \left(\frac{6}{17},\ -\frac{3}{17}+\frac{7}{34}i,\ -\frac{3}{17}-\frac{7}{34}i\right),$$

$$\Gamma_{3,1} = \left(0,\ \frac{1}{4},\ \frac{7}{16}\right)(B^T)^{-1} = \left(\frac{7}{17},\ -\frac{7}{34}+\frac{11}{34}i,\ -\frac{7}{17}-\frac{11}{34}i\right),$$

$$\Gamma_{3,2} = \left(0,\ \frac{1}{2},\ \frac{1}{4}\right)(B^T)^{-1} = \left(\frac{4}{17},\ -\frac{2}{17}-\frac{9}{17}i,\ -\frac{2}{17}+\frac{9}{17}i\right),$$

$$\Gamma_{3,3} = \left(1,\ \frac{1}{4},\ \frac{5}{16}\right)(B^T)^{-1} = \left(\frac{6}{17},\ \frac{11}{34}+\frac{7}{17}i,\ \frac{11}{17}-\frac{7}{17}i\right).$$

Step 8c

Find the coefficients $\beta_{i,j,k}(y)$ for the limit and the differential matrices using the formula

$$\beta_{i,j,k}(y) = \sum_{l=k}^{m(y)-1} 0^{l-k}\gamma_{i,j,l}(y) = \gamma_{i,j,k}(y)$$

for $y \in \mathbb{C}\backslash\mathcal{D}$, $k = 0, 1, 2, \ldots, m(y)-1$, $i, j = 1, 2, 3$.

Based on this formula we obtain

$$\beta_0(1) = \begin{pmatrix} \frac{7}{17} & \frac{4}{17} & \frac{6}{17} \\ \frac{7}{17} & \frac{4}{17} & \frac{6}{17} \\ \frac{7}{17} & \frac{4}{17} & \frac{6}{17} \end{pmatrix},$$

$$\beta_0(-4i) = \begin{pmatrix} \frac{5}{17}-\frac{3}{17}i & -\frac{2}{17}+\frac{8}{17}i & -\frac{3}{17}-\frac{5}{17}i \\ -\frac{7}{34}-\frac{3}{17}i & \frac{13}{34}-\frac{1}{34}i & -\frac{3}{17}+\frac{7}{34}i \\ -\frac{7}{34}+\frac{11}{34}i & -\frac{2}{17}-\frac{9}{17}i & \frac{11}{34}+\frac{7}{34}i \end{pmatrix},$$

$$\beta_0(4i) = \begin{pmatrix} \frac{5}{17}+\frac{3}{17}i & -\frac{2}{17}-\frac{8}{17}i & -\frac{3}{17}+\frac{5}{17}i \\ -\frac{7}{34}+\frac{3}{17}i & \frac{13}{34}+\frac{1}{34}i & -\frac{3}{17}-\frac{7}{34}i \\ -\frac{7}{34}-\frac{11}{34}i & -\frac{2}{17}+\frac{9}{17}i & \frac{11}{34}-\frac{7}{34}i \end{pmatrix}.$$

So, the matrix $\overline{P}(t)$ can be represented as follows:

$$\overline{P}(t) = \beta_0(1) + \lambda^t \beta_0(-4i) + \overline{\lambda}^t \beta_0(4t), \tag{1.38}$$

where $\lambda = 1/(-4i) = i/4$ and $\overline{\lambda} = 1/(4i) = -i/4$.

If we set $i = \cos(\pi/2) + i\sin(\pi/2)$ then

$$\lambda^t = \left(\frac{i}{4}\right)^t = \left(\frac{1}{4}\right)^t \left(\cos\frac{t\pi}{2} + i\sin\frac{t\pi}{2}\right).$$

Here $\beta_0(4i)$ is the conjugate matrix of $\beta_0(-4i)$, i.e., $\beta_0(4i) = \overline{\beta}_0(-4i)$.

Therefore,

$$\lambda^t \beta_0(-4i) + \overline{\lambda}^t \beta_0(4t) = 2\Big(\mathrm{Re}(\beta_0(-4i))\mathrm{Re}(\lambda^t) - \mathrm{Im}(\beta_0(-4i))\mathrm{Im}(\lambda^t)\Big).$$

If we take these properties into account then we obtain

$$\overline{P}(t) = \begin{pmatrix} \frac{7}{17} & \frac{4}{17} & \frac{6}{17} \\ \frac{7}{17} & \frac{4}{17} & \frac{6}{17} \\ \frac{7}{17} & \frac{4}{17} & \frac{6}{17} \end{pmatrix} + 2\left(\frac{1}{4}\right)^t \cos\frac{t\pi}{2} \begin{pmatrix} \frac{5}{17} & -\frac{2}{17} & -\frac{3}{17} \\ -\frac{7}{34} & \frac{13}{34} & -\frac{3}{17} \\ -\frac{7}{34} & -\frac{2}{17} & \frac{11}{34} \end{pmatrix}$$

$$+ 2\left(\frac{1}{4}\right)^t \sin\frac{t\pi}{2} \begin{pmatrix} \frac{3}{17} & -\frac{8}{17} & \frac{5}{17} \\ \frac{3}{17} & \frac{1}{34} & -\frac{7}{34} \\ -\frac{11}{34} & \frac{9}{17} & -\frac{7}{34} \end{pmatrix}.$$

Formula (1.38) can be obtained by using Algorithm 1.4. The calculation procedure according to this algorithm is similar to the calculation procedures for the previous examples.

1.5 A Dynamic Programming Approach for Discrete Markov Processes and Combinatorial Algorithms for Determining the Limit Matrix

In this section we develop new dynamic calculation procedures for finite Markov processes. We consider the problem of determining the probability of systems' transitions from a given starting state to a given final state if the final state should be reached at the moment of time from a given interval. We can see that the asymptotic behavior analysis of the proposed calculation of the dynamic programming procedures for this problem will allow us to formulate several polynomial time algorithms for calculating the limit matrix in Markov chains [88, 110], including also the algorithm from Sect. 1.4.

1.5.1 Calculation of the State-Time Probability of the System with the Restriction on the Number of Transitions

We consider the problem of determining the probability of systems' transitions from the state x_{i_0} to the state x with the condition that the state x should be reached at the time moment $t(x)$ such that $t_1 \leq t(x) \leq t_2$ where t_1 and t_2 are given. So, we need to calculate the probability that the system $\mathbb{L}$ will reach the state x at least at one of the moments in time $t_1, t_1 + 1, \ldots, t_2$. We denote this probability by $P_{x_{i_0}}(x,\ t_1 \leq t(x) \leq t_2)$. Some reflections on this definition allow us to write the following formula

$$\begin{aligned} &P_{x_{i_0}}(x,\ t_1 \leq t(x) \leq t_2) \\ &\quad = P_{x_{i_0}}(x,\ 0 \leq t(x) \leq t_2) - P_{x_{i_0}}(x,\ 0 \leq t(x) \leq t_1 - 1). \end{aligned} \tag{1.39}$$

Further we describe some results that can be used for calculating the probability $P_x(y,\ 0 \leq t(y) \leq t)$ for $x, y \in X$ and $t = 1, 2, \ldots$. For this reason we shall give the graphical interpretation of the Markov processes using the graph of probability transitions $G_p = (X, E_p)$. In this graph each vertex $x \in X$ corresponds to a state of the dynamical system and a possible transition of the system from one state x to another state y with positive probability $p_{x,y}$ is represented by the directed edge $e = (x, y) \in E_p$ from x to y; the corresponding probabilities $p_{x,y}$ are associated to directed edges $(x, y) \in E_p$ in G_p. It is evident that in the graph G_p each vertex x contains at least one leaving edge (x, y) and $\sum_{y \in X} p_{x,y} = 1$. In general this graph can be treated as a random graph [10] with some additional properties. As example the graph of states' transitions $G_p = (X, E_p)$ for the Markov process with the stochastic matrix of probabilities

$$P = \begin{pmatrix} 0.3 & 0.3 & 0.4 & 0 \\ 0.5 & 0 & 0.5 & 0 \\ 0 & 0.4 & 0 & 0.6 \\ 0 & 0.3 & 0.5 & 0.2 \end{pmatrix}$$

is represented in Fig. 1.2.

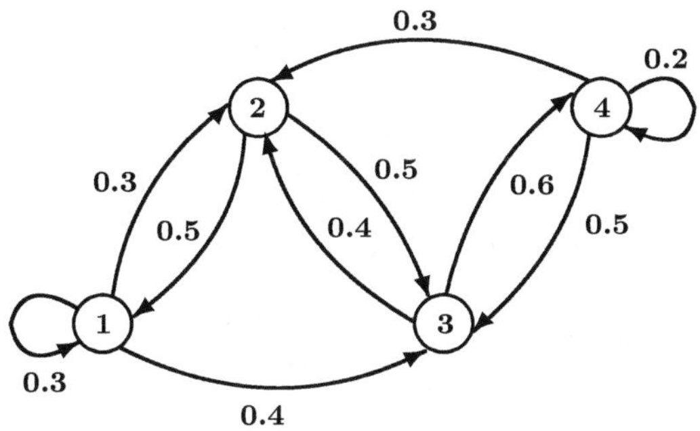

Fig. 1.2 The graph of states' transitions

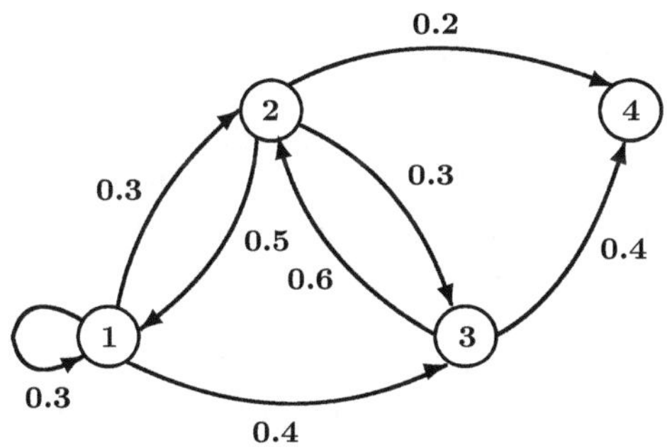

Fig. 1.3 The graph of the process with a deadlock vertex

In the following we will consider also the stochastic process which may stop if one of the states from a given subset of states of the dynamical system is reached. This means that the graph of such a process may contain the deadlock vertices.

So, we consider the stochastic process for which the graph may contain the *deadlock vertices* $y \in X$ and $\sum_{z \in X} p_{x,z} = 1$ for the vertices $x \in X$ which contain at least one leaving directed edge. As example in Fig. 1.3 a graph $G_p = (X, E_p)$ which contains deadlock vertices is represented.

This graph corresponds to the stochastic process with the following matrix of states' transitions

$$P = \begin{pmatrix} 0.3 & 0.3 & 0.4 & 0 \\ 0.5 & 0 & 0.3 & 0.2 \\ 0 & 0.6 & 0 & 0.4 \\ 0 & 0 & 0 & 0 \end{pmatrix}.$$

Such a graph does not correspond to a Markov process and the matrix of the transition probability P contains rows with zero components. Nevertheless in this case the probabilities $P_{x_0}(x, t)$ can be calculated on the basis of the recursive formula (1.2).

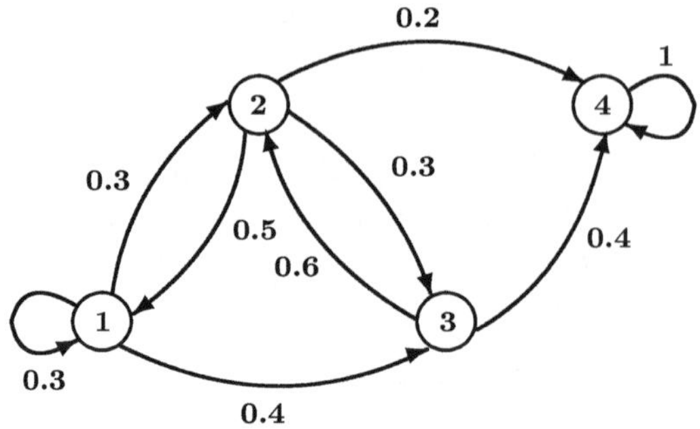

Fig. 1.4 The graph of the process with an absorbing state

Note that the matrix P can easily be transformed into a stochastic matrix changing the probabilities $p_{y,y} = 0$ for the deadlock states $y \in X$ by the probabilities $p_{y,y} = 1$. This transformation leads to a new graph which corresponds to a Markov process because the obtained graph contains a new directed edges $e = (y, y)$ with $p_e = 1$ for $y \in X$. In this graph the vertices $y \in X$ contain the loops and the corresponding states of the dynamical system in a new Markov process that represents the absorbing states. So, the stochastic process which may stop in a given set of states can be represented either by the graph with deadlock vertices or by a graph with absorbing vertices. In Fig. 1.4 the graph is represented with absorbing vertex $y = 4$ for the Markov process defined by the matrix P given below.

$$P = \begin{pmatrix} 0.3 & 0.3 & 0.4 & 0 \\ 0.5 & 0 & 0.3 & 0.2 \\ 0 & 0.6 & 0 & 0.4 \\ 0 & 0 & 0 & 1.0 \end{pmatrix}.$$

It is easy to see that the stochastic matrix P in this example is obtained from the previous one by changing $p_{4,4} = 0$ by $p_{4,4} = 1$. The corresponding graph with the absorbing vertex $y = 4$ in this case is obtained from the graph in Fig. 1.3 by adding the directed edge $e = (4, 4)$ with $p_{4,4} = 1$.

We shall use the graph with absorbing vertices for the calculation of the probabilities $P_x(y, 0 \le t(y) \le t)$.

Lemma 1.5 *Let a Markov process for which the graph $G_p = (X, E_p)$ contains an absorbing vertex $y \in X$ be given. Then for an arbitrary state $x \in X$ the following recurrence formula holds:*

$$P_x(y, 0 \le t(y) \le \tau + 1) = \sum_{z \in X} p_{x,z} P_z(y, 0 \le t(y) \le \tau),$$

$$\tau = 0, 1, 2, \ldots, \qquad (1.40)$$

where $P_x(y, 0 \le t(y) \le 0) = 0$ *if* $x \ne y$ *and* $P_y(y, 0 \le t(y) \le 0) = 1$.

Proof It is easy to observe that for $\tau = 0$ the lemma holds. Moreover, we can see that here the condition that y is an absorbing state is essential; otherwise for $x = y$ the recursive formula (1.40) fails to hold. For $t \ge 1$ the correctness of formula (1.40) follows from the definition of the probabilities $P_x(y, 0 \le t(y) \le t+1)$, $P_z(y, 0 \le t(z) \le t)$ and the induction principle on τ. □

The recursive formula from this lemma can be written in matrix form as follows

$$\pi'(\tau+1) = P\pi'(\tau), \quad \tau = 0, 1, 2, \ldots. \tag{1.41}$$

Here P is the stochastic matrix of the Markov process with the absorbing state $y \in X$ and

$$\pi'(\tau) = \begin{pmatrix} \pi'_1(\tau) \\ \pi'_2(\tau) \\ \vdots \\ \pi'_n(\tau) \end{pmatrix}, \quad \tau = 0, 1, 2, \ldots$$

are the column vectors, where an arbitrary component $\pi'_i(\tau)$ expresses the probability of the dynamical system to reach the state y from x_i by using not more than τ units of times, i.e., $\pi'_i(\tau) = P_{x_i}(y, 0 \le t(y) \le \tau)$. At the starting moment of time $\tau = 0$ the vector $\pi'(0)$ is given:

All components are equal to zero except the component corresponding to the absorbing vertex which is equal to one, i.e.,

$$\pi'_i(0) = \begin{cases} 0, & \text{if } x_i \ne y, \\ 1, & \text{if } x_i = y. \end{cases}$$

If we apply this formula for $\tau = 0, 1, 2, \ldots, t-1$ then we obtain

$$\pi'(t) = P^t\pi'(0), \quad t = 1, 2, \ldots, \tag{1.42}$$

where P^t is the t-th power of the matrix P. So, if we denote by j_y the index of the column of the matrix P^t which corresponds to the absorbing state y then an arbitrary element $p^{(t)}_{i,j_y}$ of this column expresses the probability of the system $\mathbb{L}$ to reach the state y from x_i by using not more than t units of time, i.e., $p^{(t)}_{i,j_y} = P_{x_i}(y, 0 \le t(x) \le t)$. This allows us to formulate the following lemma:

Lemma 1.6 *Let a discrete Markov process with the absorbing state* $y \in X$ *be given. Then:*

(a) $P_{x_i}(y, t) = P_{x_i}(y, 0 \le t(y) \le t), \quad for \quad x_i \in X \setminus \{y\},\ t = 1, 2, \ldots;$
(b) $P_{x_i}(y, t_1 \le t(y) \le t_2) = p^{(t_2)}_{i,j_y} - p^{(t_1-1)}_{i,j_y}$,

where $p_{i,j_y}^{(t_2)}$, $p_{i,j_y}^{(t_1-1)}$ *represent the corresponding elements of the matrices* P^{t_2} *and* P^{t_1-1}.

Proof The condition (a) in this lemma holds because

$$P_{x_i}(y,t) = p_{i,j_y}^{(t)} = P_{x_i}(y,\ 0 \le t(y) \le t).$$

The condition (b) follow from Lemma 1.5 and the following properties

$$P_{x_i}(y,\ 0 \le t(y) \le t_2) = p_{i,j_y}^{(t_2)}, \quad P_{x_i}(y,\ 0 \le t(y) \le t_1 - 1) = p_{i,j_y}^{(t_1-1)}. \qquad \square$$

So, to calculate $P_{x_i}(y,\ t_1 \le t(y) \le t_2)$ it is sufficient to find the matrices P^{t_1-1}, P^{t_2} and then to apply the formula from Lemma 1.6.

Below we give an example which illustrates the calculation procedure of the state probabilities on the bases of the recursive formula above for the stationary case of the Markov process.

Example 1 Let the Markov process with the stochastic matrix P which corresponds to the graph of transition probabilities represented in Fig. 1.3 be given. It is easy to see that the state $y = 4$ is an absorbing state. We consider the problem of finding the probabilities $P_{x_i}(y, 4)$ and $P_{x_i}(y,\ 2 \le t(x) \le 4)$, where $x_i = 2$. To determine this probability we shall use the probability matrices $P^1 = P$ and P^4:

$$P^1 = \begin{pmatrix} 0.3 & 0.3 & 0.4 & 0 \\ 0.5 & 0 & 0.3 & 0.2 \\ 0 & 0.6 & 0 & 0.4 \\ 0 & 0 & 0 & 1.0 \end{pmatrix}.$$

All probabilities $P_{x_i}(y, 4)$, $i = 1, 2, 3$ can be derived from the last column of the following matrix:

$$P^4 = \begin{pmatrix} 0.1701 & 0.1881 & 0.1542 & 0.4876 \\ 0.1455 & 0.1584 & 0.1335 & 0.5626 \\ 0.1260 & 0.0990 & 0.0954 & 0.6796 \\ 0 & 0 & 0 & 1.0 \end{pmatrix}.$$

The probability $P_{x_i}(y,\ 2 \le t(x) \le 4)$ for $x_i = 2$ we find on the bases of Lemma 1.6. According to condition (a) of this lemma we have

$$P_2(4,\ 0 \le t(4) \le 4) = P_2(4, 4) = p_{2,4}^{(4)} = 0.5626;$$
$$P_2(4,\ 0 \le t(4) \le 1) = P_2(4, 1) = p_{2,4}^{(1)} = p_{2,4} = 0.2$$

and according to the condition (b) we obtain

$$P_2(4,\ 2 \le t(4) \le 4) = p_{2,4}^{(4)} - p_{2,4}^{(1)} = 0.5626 - 0.2 = 0.3626\,.$$

The procedure of calculating the probabilities $P_x(y,\ 0 \le t(y) \le t)$ in the case of the Markov process without absorbing states can easily be reduced to the procedure of calculating the probabilities in the Markov process with the absorbing state y by using the following transformation of the stochastic matrix P. We put $p_{i_y,j} = 0$ if $j \ne i_y$ and $p_{i_y,i_y} = 1$. It is easy to see that such a transformation of the matrix P does not change the probabilities $P_x(y,\ 0 \le t(y) \le t)$. After such a transformation we obtain a new stochastic matrix for which the recursive formula from the lemma above can be applied. In general, for the Markov processes without absorbing states these probabilities can be calculated by using the algorithm which operates with the original matrix P without changing its elements. Below such an algorithm is described.

Algorithm 1.7 Determining the State-Time Probabilities of the System with a Restriction on the Number of Transitions

Preliminary step (Step 0): Put $P_x(y,\ 0 \le t(y) \le 0) = 0$ for every $x \in X \setminus \{y\}$ and $P_y(y,\ 0 \le t(x) \le 0) = 1$.
General step (Step $\tau + 1$, $\tau \ge 0$): For every $x \in X$ calculate

$$P_x(y,\ 0 \le t(x) \le \tau + 1) = \sum_{z \in X} p_{x,z} P_z(y,\ 0 \le t(y) \le \tau) \tag{1.43}$$

and then put

$$P_y(y,\ 0 \le t(y) \le \tau + 1) = 1. \tag{1.44}$$

If $\tau < t - 1$ then go to the next step, i.e., $\tau = \tau + 1$; otherwise STOP.

Theorem 1.8 *Algorithm 1.7 correctly finds the probabilities $P_x(y,\ 0 \le t(x) \le \tau)$ for $x \in X$, $\tau = 1, 2, \ldots, t$. The running time of the algorithm is $O(|X|^2 t)$.*

Proof It is easy to see that at each general step of the algorithm the probabilities $P_x(y,\ 0 \le t(x) \le \tau + 1)$ are calculated on the bases of formula (1.43) which takes condition (1.44) into account. This calculation procedure is equivalent with the calculation of the probabilities $P_x(y,\ 0 \le t(x) \le \tau + 1)$ with the condition that the state y is an absorbing state. So, the algorithm correctly finds the probabilities $P_x(y,\ 0 \le t(x) \le \tau)$ for $x \in X$, $\tau = 1, 2, \ldots, t$. To estimate the running time of the algorithm it is sufficient to estimate the number of elementary operations at the general step of the algorithm. It is easy to see that at the iteration τ the algorithm uses $O(|X|^2)$ elementary operations. So, the running time of the algorithm is $O(|X|^2 t)$. □

If we use in Algorithm 1.7 the same notation $\pi'_i(\tau) = P_{x_i}(y,\ 0 \le t(y) \le \tau)$, $\pi_{i_y}(\tau) = P_y(y,\ 0 \le t(y) \le \tau)$ as in formula (1.40) then we obtain the following simple description in matrix form:

Algorithm 1.9 Calculation of the State-Time Probabilities of the System in the Matrix Form (Stationary Case)

Preliminary step (Step 0): Fix the vector $\pi'(0) = (\pi'_1(0), \pi'_2(0), \dots, \pi'_n(0))$, where $\pi'_i(0) = 0$ for $i \neq i_y$ and $\pi'_{i_y}(0) = 1$.

General step (Step $\tau + 1$, $\tau \geq 0$): For a given τ calculate

$$\pi'(\tau + 1) = P\pi'(\tau)$$

and then put

$$\pi'_{i_y}(\tau + 1) = 1. \tag{1.45}$$

If $\tau < t - 1$ then go to the next step, i.e., $\tau = \tau + 1$; otherwise STOP.

Note that in the algorithm the condition (1.45) allows us to preserve the value $\pi'_{i_y}(t) = 1$ at every moment of time t in the calculation process. This condition reflects the property that the system remains in the state y at every time-step t if the state y is reached. We can modify this algorithm for determining the probability $P_x(y, 0 \leq t(y) \leq 0)$ in a more general case if we assume that the system will remain at every time step t in the state y with the probability $\pi'_{i_y}(t) = q(y)$, where $q(y)$ may differ from 1, i.e., $q(y) \leq 1$. In the following we can see that this modification allows us to elaborate a new polynomial-time algorithm for determining the matrix of the limit probabilities in a stationary Markov process.

If $q(y)$ is known then we can use the following algorithm for calculating the state probabilities of the system with a given restriction on the number of its transitions.

Algorithm 1.10 Calculation of the State-Time Probabilities of the System with Known Probability of its Remaining in the Final State (Stationary Case)

Preliminary step (Step 0): Fix the vector $\pi'(0) = (\pi'_1(0), \pi'_2(0), \dots, \pi'_n(0))$, where $\pi'_i(0) = 0$ for $i \neq i_y$ and $\pi'_{i_y}(0) = q(y)$.

General step (Step $\tau + 1$, $\tau \geq 0$): For a given τ calculate

$$\pi'(\tau + 1) = P\pi'(\tau)$$

and then put

$$\pi'_{i_y}(\tau + 1) = q(y). \tag{1.46}$$

If $\tau < t - 1$ then go to the next step, i.e., $\tau = \tau + 1$; otherwise STOP.

Remark 1.11 The results and algorithms described above are valid for an arbitrary stochastic process and for an arbitrary graph for which $\sum_{z \in X} p_{x_i,z} = r(x_i) \leq 1$.

An example which illustrates the calculation procedure in this algorithm is given below.

Example 2 Consider the Markov process with the stochastic matrix P which corresponds to the graph of transition probabilities $G_p = (X, E_p)$ represented in Fig. 1.4. Fix $y = 4$ and let us calculate the probabilities $\pi_{i_x} = P_x(y,\ 0 \le t(y) \le t)$ for $x \in \{1, 2, 3\}$ and $t = 3$. If we use this algorithm in the case $q(y) = 1$, i.e., we apply Algorithm 1.9, then we obtain:

Step 0

$$P_1(4,\ 0 \le t(4) \le 0) = 0; \quad P_2(4,\ 0 \le t(4) \le 0) = 0;$$
$$P_3(4,\ 0 \le t(4) \le 0) = 0; \quad P_4(4,\ 0 \le t(4) \le 0) = 1.$$

Step 1

$$P_1(4,\ 0 \le t(4) \le 1) = 0; \quad P_2(4,\ 0 \le t(4) \le 1) = 0;$$
$$P_3(4,\ 0 \le t(4) \le 1) = 0.6; \quad P_4(4,\ 0 \le t(4) \le 1) = 1.$$

Step 2

$$P_1(4,\ 0 \le t(4) \le 2) = 0.4 \cdot 0.6 = 0.24;$$
$$P_2(4,\ 0 \le t(4) \le 2) = 0.5 \cdot 0.6 = 0.30;$$
$$P_3(4,\ 0 \le t(4) \le 2) = 0.6;$$
$$P_4(4,\ 0 \le t(4) \le 2) = 1.$$

Step 3

$$P_1(4,\ 0 \le t(4) \le 3) = 0.3 \cdot 0.24 + 0.3 \cdot 0.30 + 0.4 \cdot 0.6 = 0.402;$$
$$P_2(4,\ 0 \le t(4) \le 3) = 0.5 \cdot 0.24 + 0.5 \cdot 0.6 = 0.42;$$
$$P_3(4,\ 0 \le t(4) \le 3) = 0.4 \cdot 0.30 + 0.6 = 0.72;$$
$$P_4(4,\ 0 \le t(4) \le 3) = 1.$$

If we put in this algorithm $q(y) = 0.7$ then we obtain:

Step 0

$$P_1(4,\ 0 \le t(4) \le 0) = 0; \quad P_2(4,\ 0 \le t(4) \le 0) = 0;$$
$$P_3(4,\ 0 \le t(4) \le 0) = 0; \quad P_4(4,\ 0 \le t(4) \le 0) = 0.7.$$

Step 1

$$P_1(4,\ 0 \le t(4) \le 1) = 0; \quad P_2(4,\ 0 \le t(4) \le 1) = 0;$$
$$P_3(4,\ 0 \le t(4) \le 1) = 0.42; \quad P_4(4,\ 0 \le t(4) \le 1) = 0.7.$$

Step 2

$$P_1(4,\ 0 \le t(4) \le 2) = 0.4 \cdot 0.42 = 0.168;$$
$$P_2(4,\ 0 \le t(4) \le 2) = 0.5 \cdot 0.42 = 0.21;$$
$$P_3(4,\ 0 \le t(4) \le 2) = 0.6 \cdot 0.7 = 0.42;$$
$$P_4(4,\ 0 \le t(4) \le 2) = 0.7\,.$$

Step 3

$$P_1(4,\ 0 \le t(4) \le 3) = 0.3 \cdot 0.168 + 0.3 \cdot 0.21 + 0.4 \cdot 0.42 = 0.2814;$$
$$P_2(4,\ 0 \le t(4) \le 3) = 0.5 \cdot 0.168 + 0.5 \cdot 0.42 = 0.294;$$
$$P_3(4,\ 0 \le t(4) \le 3) = 0.4 \cdot 0.21 + 0.6 \cdot 0.7 = 0.504;$$
$$P_4(4,\ 0 \le t(4) \le 3) = 0.7\,.$$

1.5.2 Determining the Limiting State Probabilities in Markov Chains Based on Dynamic Programming

Now let us show that the algorithms from Sect. 1.5.1 allow us to elaborate an algorithm for determining the matrix of the limiting probabilities for Markov chains which is similar to the algorithm from Sect. 1.4. To characterize this algorithm we will analyze the algorithms from Sect. 1.5.1 in the case of a large number of iterations, i.e., if $\tau \to \infty$. We can see that in the case $\tau \to \infty$ these give the conditions for determining the limit states probabilities in the Markov chain.

Denote by $Q = (q_{i,j})$ the limit matrix for the Markov chain induced by the stochastic matrix $P = (p_{x_i,x_j})$. We denote the column vectors of the matrix Q by

$$\overline{q}^j = \begin{pmatrix} q_{1,j} \\ q_{2,j} \\ \vdots \\ q_{n,j} \end{pmatrix}, \quad j = 0, 1, 2, \ldots, n,$$

and the row vectors of the matrix Q are denoted by $\overline{q}_i = (q_{i,1}, q_{i,2}, \ldots, q_{i,n})$, $i = 1, 2, \ldots, n$. To characterize algorithms for finding the limit matrix Q for an arbitrary Markov chain we need to analyze the structure of the corresponding graph of transition probabilities $G_p = (X, E_p)$ and to study the behavior of the algorithms from the previous section in the case $t \to \infty$. First of all we note that for the ergodic Markov chain with positive recurrent states the graph G_p is strongly connected and all row vectors $\overline{q}_i$, $i = 1, 2, \ldots, n$ are the same. In this case the limit state probabilities can be derived by solving the system of linear equations

$$\pi = \pi P, \quad \sum_{J=1}^{n} \pi_j = 1,$$

i.e., $\overline{q}_i = \pi, i = 1, 2, \ldots, n$. In general such an approach can be used for an arbitrary Markov unichain.

If we have a Markov multichain then the graph G_p consists of several strongly connected components $G^1 = (X^1, E^1)$, $G^2 = (X^2, E^2)$, ..., $G^k = (X^k, E^k)$ where $\bigcup_{i=1}^{k} X^i = X$. Additionally, among these components, there are such strongly connected components $G^{i_r} = (X^{i_r}, E^{i_r})$, $r = 1, 2, \ldots, k'$ $(k' < k)$ which do not contain a leaving directed edge $e = (x, y)$ where $x \in X^{i_r}$ and $y \in X \setminus X^{i_r}$. We call such components G^{i_r} *deadlock components* in G_p. In the following we shall use these deadlock components for the characterization of the structure of the Markov multichains.

Lemma 1.12 *If $G^{i_r} = (X^{i_r}, E^{i_r})$ is a strongly connected deadlock component in G_p then X^{i_r} is a positive recurrent class of Markov chains (irreducible chains); if $x \in X \setminus \bigcup_{r=1}^{k'} X^{i_r}$ then x is a transient state of the Markov chain.*

Lemma 1.12 reflects the well known properties of Markov chains from [47, 56, 114, 127] expressed in the terms of the graphs of probability transitions.

Now we analyze some properties that can be derived from the algorithms from the previous section in the case $t \to \infty$.

Let a Markov process with finite set of states X be given. For an arbitrary state $x_j \in X$ we denote by X_j the subset of states $x_i \in X$ for which in G_p there exists at least a directed path from x_i to x_j. In addition we denote $N = \{1, 2, \ldots, n\}$, $I(X_j) = \{i \mid x_i \in X_j\}$.

Lemma 1.13 *Let a Markov process with a finite set of states X be given and let us assume that x_j is an absorbing state. Furthermore, let $\pi'(x_j)$ be a solution of the following system of linear equations*

$$\pi'(x_j) = P\pi'(x_j); \quad \pi_{j,j} = 1; \quad \pi_{i,j} = 0 \quad for \quad i \in N \setminus I(X_j), \tag{1.47}$$

where

$$\pi'(x_j) = \begin{pmatrix} \pi_{1,j} \\ \pi_{2,j} \\ \vdots \\ \pi_{n,j} \end{pmatrix}.$$

Then $\pi'(x_j) = \overline{q}^j$, i.e., $\pi_{i,j} = q_{i,j}$, $i = 1, 2, \ldots, n$. If x_j is a unique absorbing state in the graph G_p of the Markov chain and if x_j in G_p is attainable from every $x_i \in X$ (i.e. $I(X_j) = N$) then $\pi_{i,j} = q_{i,j} = 1$, $i = 1, 2, \ldots, n$.

Proof We apply Algorithm 1.9 with respect to a given absorbing state x_j, $(y_j = x_j)$ if $t \to \infty$. Then $\pi'(t) \to \pi'(x_j)$ and, therefore, we obtain $\pi'(x_j) = P\pi'(x_j)$ where $\pi_{j,j} = 1$ and $\pi_{i,j} = 0$ for $i \in N \setminus I(X_j)$. The correctness of the second part of the lemma corresponds to the case if $I(X_j) = N$ and, therefore, we obtain that the vector π^j with the components $\pi_{i,j} = 1, i = 1, 2, \dots, n$ is the solution of the system $\pi'(x_j) = P\pi'(x_j)$, $\pi_{j,j} = 1$. So, the lemma holds. □

Remark 1.14 If x_j is not an absorbing state then Lemma 1.13 may fail to hold.

Remark 1.15 Lemma 1.13 can be extended to the case if $\sum_{y \in X} p_{x_i,y} = q(x_i) \le 1$ for some states $x_i \in X$. In this case the solution of the system (1.47) also can be treated as the limiting probabilities of the system to reach the state x_j. In such processes there exists always at least one component $\pi_{i,j}$ of the vector $\pi'(x_j)$ that is less then 1 even if $X_j = X$.

Let us show that the result formulated above allows us to find the vector of limit probabilities $\overline{q}^j$ of the matrix Q if the diagonal elements $q_{j,j}$ of Q are known. We consider the subset of states $Y^+ = \{x_j \mid q_{j,j} \ge 0\}$. It is easy to observe that $Y^+ = \bigcup_{r=1}^{k'} X^{i_r}$; we denote the corresponding set of indices of this set by $I(Y^+)$. For each $j \in I(Y^+)$ we define the set X_j in the same way as above.

Lemma 1.16 *If a non zero diagonal element $q_{j,j}$ of the limit matrix Q in the Markov multichain is known, i.e., $q_{j,j} = q(x_j)$, then the corresponding vector $\overline{q}^j$ of the matrix Q can be found by solving the following systems of linear equations:*

$$\overline{q}^j = P\overline{q}^j; \quad q_{j,j} = q(x_j); \quad q_{i,j} = 0 \quad for \quad i \in N \setminus I(X_j).$$

Proof We apply Algorithm 1.10 with respect to the fixed final state $y_j = x_j \in X$ with $q(y_j) = q_{j,j}$ if $t \to \infty$. Then for given $y_j = x$ we have $\pi(t)' \to \overline{q}^j$ and, therefore, we obtain $\overline{q}^j = P\overline{q}^j$ where $q(y_j) = q_{j,j}$ and $q_{i,j} = 0$ for $i \in N \setminus I(X_j)$. So, the lemma holds. □

Based on this lemma and Algorithm 1.10 we can prove the following result.

Theorem 1.17 *The limit state matrix Q for Markov chains can be derived by using the following algorithm:*

(1) *For each ergodic class X^{i_r} solve the system of linear equations*

$$\pi^{i_r} = \pi^{i_r} P^{(i_r)}, \quad \sum_{j \in I(X^{i_r})} \pi_j^{i_r} = 1,$$

where π^{i_r} is the row vector with components $\pi_j^{i_r}$ for $j \in I(X^{i_r})$ and $P^{(i_r)}$ is the submatrix of P induced by the class X^{i_r}. Then for every $j \in I(X^{i_r})$ put $q_{j,j} = \pi_j^{i_r}$; for each $j \in I\left(X \setminus \bigcup_{r=1}^{k'} X^{i_r}\right)$ set $q_{j,j} = 0$;

(2) *For every* $j \in I(Y^+), Y^+ = \bigcup_{r=1}^{k'} X^{i_r}$ *solve the system of linear equations*

$$\overline{q}^j = P\overline{q}^j; \quad q_{j,j} = \pi_{j,j}; \quad q_{i,j} = 0, \quad \forall i \in N \setminus I(X_j)$$

and determine the vector $\overline{q}^j$. *For every* $j \in I(X \setminus Y^+)$ *set* $\overline{q}^j = \mathbf{0}$, *where* $\mathbf{0}$ *is the column vector with zero components.*

The algorithm determines the matrix Q *using* $O(n^3)$ *elementary operations.*

Proof Let us show that the algorithm finds correctly the limit matrix Q. Item (1) of the algorithm determines the limit probabilities $q_{j,j}$. This item is based on Lemma 1.12 and on the conditions which each ergodic class X^{i_r} and each transient state $x \in X \setminus Y^+$ should satisfy. So, item (1) correctly finds the limit probabilities $q_{i,j}$ for $j \in N$. Item (2) of the algorithm is based on Lemma 1.16 and, therefore, correctly determines the vectors $\overline{q}^j$ of the matrix Q wif the diagonal elements $q_{j,j}$ are known. So, the algorithm correctly determines the limit matrix Q of the Markov multichain.

The running time of the algorithm is $O(n^3)$. We obtain this estimation if we estimate the number of elementary operations at each step of the algorithm. At step 1 of the algorithm we solve $n' \leq n$ system of equations where each system contains $|X^{i_r}|$ variables and $\sum_{r=1}^{k'} |X_{i_r}| \leq n$. Therefore, as a whole the solutions of these systems can be obtained using $O(n^3)$ elementary operations. At step 2 of the algorithm we solve n systems of linear equations for determining the vectors $\overline{q}^j$. However, these systems have the same left part and can be solved simultaneously using Gaussian elimination. Therefore, we obtain $O(n^3)$ elementary operations. □

An example that illustrates the details of the algorithm described in this theorem is given below. As we have noted, in the worst case the running time of the algorithm is $O(n^3)$, however, intuitively it is clear that the upper bound of this estimation cannot be reached. Practically, this algorithm efficiently finds the limit matrix Q. In the next section we will show how the proposed algorithm can be modified such that the solution can be found in a more suitable form.

Example Consider the problem of determining the limit matrix of probabilities Q for the Markov multichain determined by the following probability transition matrix

$$P = \begin{pmatrix} 1 & 0 & 0 & 0 \\ 0.25 & 0.25 & 0.25 & 0.25 \\ 0 & 0 & 0.5 & 0.5 \\ 0 & 0 & 0.5 & 0.5 \end{pmatrix}.$$

The corresponding graph $G_p = (X, E_p)$ of this Markov multichain is represented in Fig. 1.5. We apply the algorithm from Theorem 1.17.

(1) Find the deadlock components X^{i_r}, $r = 1, 2, \ldots, k'$ in the graph G_p and for each of them solve the system of linear equations

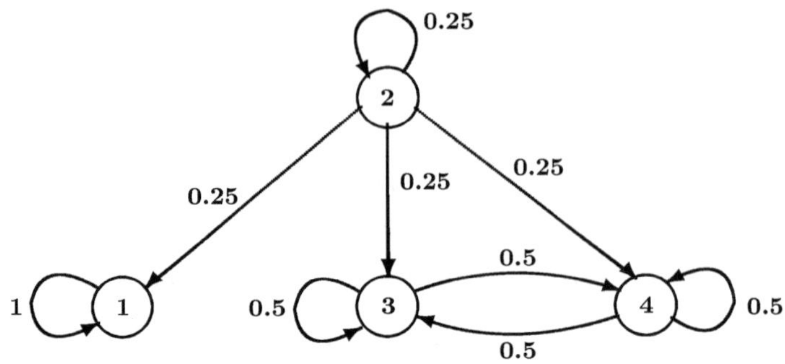

Fig. 1.5 The graph of the Markov multichain

$$\pi^{i_r} = \pi^{i_r} P^{(i_r)}, \quad \sum_{j \in I(X^{i_r})} \pi_j^{i_r} = 1.$$

In our case we have two deadlock components $X^1 = \{1\}$, $X^2 = \{3, 4\}$ and, therefore, we have to solve the following two systems of linear equations

$$\pi^1 = \pi^1 P^{(1)}, \quad \pi_1^1 = 1;$$
$$\pi^2 = \pi^2 P^{(2)}, \quad \pi_3^2 + \pi_4^2 = 1$$

where

$$\pi^1 = (\pi_1^1); \qquad P^{(1)} = (1);$$
$$\pi^2 = (\pi_3^2, \pi_4^2); \quad P^{(2)} = \begin{pmatrix} 0.5 & 0.5 \\ 0.5 & 0.5 \end{pmatrix}.$$

So, we have to solve the following systems

$$\begin{cases} \pi_1^1 = \pi_1^1, \quad \pi_1^1 = 1; \\ \pi_3^2 = 0.5\pi_3^2 + 0.5\pi_4^2, \\ \pi_4^2 = 0.5\pi_3^2 + 0.5\pi_4^2, \\ \pi_3^2 + \pi_4^2 = 1. \end{cases}$$

and we obtain $\pi_1^1 = 1$; $\pi_3^2 = 0.5$; $\pi_4^2 = 0.5$.

This means that the diagonal elements $q_{j,j}$ of the matrix Q are:

$$q_{1,1} = 1; \quad q_{3,3} = 0.5; \quad q_{4,4} = 0.5; \quad q_{2,2} = 0.$$

Here the vertex $x = 2$ corresponds to a transient state and, therefore, $q_{2,2} = 0$. At the next step of the algorithm we obtain the vectors $\overline{q}^j$, $j = 1, 2, 3, 4$.

(2) Fix $j = 1$. For this case $q_{1,1} = 1$, $X_1 = \{1,\ 2\}$, $N \setminus X_1 = \{3,\ 4\}$. Therefore, we have to solve the system of linear equations

$$\overline{q}^1 = P\overline{q}^1; \quad q_{1,1} = 1; \quad q_{1,3} = 0; \quad q_{1,4} = 0.$$

This system can be written as follows

$$\begin{cases} q_{1,1} = q_{1,1}; \\ q_{2,1} = 0.25q_{1,1} + 0.25q_{2,1} + 0.25q_{3,1} + 0.25q_{4,1}; \\ q_{3,1} = \qquad 0.5\ q_{3,1} + 0.5\ q_{4,1}; \\ q_{4,1} = \qquad 0.5\ q_{3,1} + 0.5\ q_{4,1}; \\ q_{1,1} = 1, \quad q_{3,1} = 0, \quad q_{4,1} = 0 \end{cases}$$

and we determine

$$\overline{q}^1 = \begin{pmatrix} 1 \\ 0.33(3) \\ 0 \\ 0 \end{pmatrix}.$$

For $j = 2$ we have $q_{2,2} = 0$, therefore,

$$\overline{q}^2 = \begin{pmatrix} 0 \\ 0 \\ 0 \\ 0 \end{pmatrix},$$

because the state $x = 2$ is a transient state.

For $j = 3$ we have $q_{3,3} = 0.5$, $X_3 = \{2,\ 3,\ 4\}$, $N \setminus X_3 = \{1\}$.
In this case it is necessary to solve the system of linear equations

$$\overline{q}^3 = P\overline{q}^3; \quad q_{3,3} = 0.5; \quad q_{1,3} = 0$$

Thus, if we solve the system of linear equations

$$\begin{cases} q_{1,3} = q_{1,3}; \\ q_{2,3} = 0.25q_{1,3} + 0.25q_{2,3} + 0.25q_{3,3} + 0.25q_{3,4}; \\ q_{3,3} = \qquad 0.5\ q_{3,3} + 0.5\ q_{3,4}; \\ q_{4,3} = \qquad 0.5\ q_{3,3} + 0.5\ q_{3,4}; \\ q_{3,3} = 0.5, \quad q_{1,3} = 0. \end{cases}$$

then we obtain

$$\overline{q}^3 = \begin{pmatrix} 0 \\ 0.33(3) \\ 0.5 \\ 0.5 \end{pmatrix}.$$

For $j = 4$ we have $q_{4,4} = 0.5$, $X_4 = \{2,\ 3,\ 4\}$, $N \setminus X_4 = \{1\}$. In this case it is necessary to solve the system of the linear equations

$$\overline{q}^4 = P\overline{q}^4; \quad q_{4,4} = 0.5; \quad q_{1,4} = 0,$$

If we solve the system of linear equations

$$\begin{cases} q_{1,4} = q_{1,4}; \\ q_{2,4} = 0.25q_{1,4} + 0.25q_{2,4} + 0.25q_{3,4} + 0.25q_{4,4}; \\ q_{3,4} = \qquad\qquad\qquad\qquad 0.25q_{3,4} + 0.25q_{4,4}; \\ q_{4,4} = \qquad\qquad\qquad\qquad 0.25q_{3,4} + 0.25q_{4,4}; \\ q_{4,4} = 0.5, \quad q_{1,4} = 0 \end{cases}$$

then we find

$$\overline{q}^4 = \begin{pmatrix} 0 \\ 0.33(3) \\ 0.5 \\ 0.5 \end{pmatrix}.$$

So, the limit matrix is

$$Q = \begin{pmatrix} 1 & 0 & 0 & 0 \\ 0.33(3) & 0 & 0.33(3) & 0.33(3) \\ 0 & 0 & 0.5 & 0.5 \\ 0 & 0 & 0.5 & 0.5 \end{pmatrix}.$$

1.5.3 An Algorithm for the Calculation of the Limit Matrix in Markov Chains with Running Time $O(n^3)$

In the following we propose a modification of the algorithm for calculating the limit matrix from the previous section. The theoretical estimation of the number of elementary operations of the modified algorithm is also $O(n^3)$. However, we can see that in this algorithm we can solve less then $|Y^+|$ systems of linear equations of the form

$$P\overline{q}^j = \overline{q}^j; \quad q_{j,j} = 1; \quad q_{i,j} = 0, \quad \forall i \in N \setminus I(X_j).$$

Moreover, we show that these systems can be simplified. We describe and characterize the proposed modification for calculating the limit matrix Q using the structure properties of the graph of the probability transitions $G_p = (X, E_p)$ and its strongly connected deadlock components $G^{i_r} = (X^{i_r}, E^{i_r})$, $r = 1, 2, \dots, k'$.

Algorithm 1.18 Determining the Limit State Matrix for a Markov Multichain The algorithm consists of two parts. The first part determines the limit probabilities $q_{x,y}$ for $x \in \bigcup_{r=1}^{k'} X^{i_r}$ and $y \in X$. The second procedure calculates the limit probabilities $q_{x,y}$ for $x \in X \setminus \bigcup_{r=1}^{k'} X^{i_r}$ and $y \in X$.

Procedure 1:

1. For each ergodic class X^{i_r} we solve the system of linear equations:

$$\pi^{i_r} = \pi^{i_r} P^{(i_r)}, \quad \sum_{y \in X^{i_r}} \pi^{i_r}_y = 1$$

   where $P^{(i_r)}$ is the matrix of probability transitions corresponding to the ergodic class X^{i_r}, i.e., $P^{(i_r)}$ is a submatrix of P, and π^{i_r} is a row vector with the components $\pi^{i_r}_y$ for $y \in X^{i_r}$. If $\pi^{i_r}_y$ are known then $q_{x,y}$ for $x \in X^{i_r}$ and $y \in X$ can be calculated as follows:

   Set $q_{x,y} = \pi^{i_r}_y$ if $x, y \in X^{i_r}$ and $q_{x,y} = 0$ if $x \in X^{i_r}$, $y \in X \setminus X^{i_r}$.

Procedure 2:

1. We construct an auxiliary directed graph $GA = (XA, EA)$ which is obtained from the graph $G_p = (X, E_p)$ by using the following transformations:
   We contract each set of vertices X^{i_r} into one vertex z^{i_r} where X^{i_r} is a set of vertices of a strongly connected deadlock component $G^{i_r} = (X^{i_r}, E^{i_r})$. If the obtained graph contains parallel directed edges $e^1 = (x, z)$, $e^2 = (x, z)$, $\dots$, $e^{m'} = (x, z)$ with the corresponding probabilities $p^1_{x,z}, p^2_{x,z}, \dots, p^{m'}_{x,z}$ then we change them by one directed edge $e = (x, z)$ with the probability $p_{x,z} = \sum_{i=1}^{m'} p^i_{x,z}$; after this transformation we associate to each vertex z^i_r a directed edge $e = (z^r, z^r)$ with the probability $p'_{z^r,z^r} = 1$.
2. We fix the directed graph $GA = (XA, EA)$ obtained by the construction described in step 1 where $XA = \left(X \setminus (\bigcup_{r=1}^{k'} X^{i_r})\right) \cup Z^r$, $Z^r = (z^1, z^2, \dots, z^{k'})$. In addition we fix a new probability matrix $P' = (p'_{x,y})$ which corresponds to this graph GA.

3. For each $x \in XA$ and every $z^i \in Z^r$ we find the probability $\pi'_x(z^i)$ of the system's transition from the state x to the state z^r. The probabilities $\pi'_x(z^i)$ can be found by solving the following k' systems of linear equations:

$$\begin{array}{llllll} P'\pi'(z^1) = \pi'(z^1), & \pi'_{z^1}(z^1) = 1, & \pi'_{z^2}(z^1) = 0, & \ldots, & \pi'_{z^{k'}}(z^1) = 0; \\ P'\pi'(z^2) = \pi'(z^2), & \pi'_{z^1}(z^2) = 0, & \pi'_{z^2}(z^2) = 1, & \ldots, & \pi'_{z^{k'}}(z^2) = 0; \\ \vdots & \vdots & \vdots & & \vdots \\ P'\pi'(z^{k'}) = \pi'(z^{k'}), & \pi'_{z^1}(z^{k'}) = 0, & \pi'_{z^2}(z^{k'}) = 0, & \ldots, & \pi'_{z^{k'}}(z^{k'}) = 1, \end{array}$$

   where $\pi'(z^i), i = 1, 2, \ldots, k'$ are the column vectors with components $\pi'_x(z^i)$ for $x \in XA$. So, each vector $\pi'_x(z^i)$ gives the probabilities of the systems' transitions from the states $x \in XA$ to the ergodic classes X^i.
4. We put $q_{x,y} = 0$ for every $x, y \in X \setminus \bigcup_{r=1}^{k'} X^{i_r}$ and $q_{x,y} = \pi'_x(z^r)\pi_y^{i_r}$ for every $x \in X \setminus \bigcup_{r=1}^{r} X^{i_r}$ and $y \in X^{i_r}$, $X^{i_r} \subset X$. If $x \in X^{i_r}$ and $y \in X \setminus X^{i_r}$ then fix $q_{x,y} = 0$.

In the algorithm $\pi(z^i)'$ can be treated as the vector, where each component $\pi_x(z^i)'$ expresses the probability that the system will be in the positive recurrent class X^i after a large number of states' transitions of the system if it starts transitions in the state $x \in X$. Therefore, $q_{x,y} = \pi'_x(z^i)\pi^i_y$, if $y \in X^i$.

By comparison this algorithm with the previous one we observe that here we solve the system of linear equations

$$P'\pi(z^r) = \pi'_{z^r}, \quad \pi'_{z^r} = 1, \quad \pi'_{z^i} = 0, \ i = 1, 2, \ldots k' \quad (i \neq r), \ r = 1, 2, \ldots, k'$$

instead of the system of equations

$$P\overline{q}^j = \overline{q}^j; \quad q_{j,j} = 1; \quad q_{i,j} = 0, \quad \forall i \in N \setminus I(X_j)$$

for $j = 1, 2, \ldots, k'$.

Theorem 1.19 *The algorithm correctly finds the limit state matrix Q and the running time of the algorithm is $O(|X|^3)$.*

Proof The correctness of Procedure 1 of the algorithm follows from the definition of the ergodic Markov class (positive recurrent Markov chain). So, Procedure 1 finds the probabilities $q_{x,y}$ for $x \in \bigcup_{r=1}^{k'} X^{i_r}$ and $y \in X$. Let us show that Procedure 2 finds correctly the remaining elements $q_{x,y}$ of the matrix Q. Indeed, each vertex $x \in X \setminus \bigcup_{r=1}^{r} X^{i_r}$ in GA corresponds to transient states of the Markov chain and, therefore, we have $q_{x,y} = 0$ for every $x, y \in X \setminus \bigcup_{r=1}^{k'} X^{i_r}$. If $x \in X^{i_r}$ then the system cannot reach a state $y \in X \setminus X^{i_r}$ and, therefore, for such two arbitrary states x, y we have $q_{x,y} = 0$. Finally, we show that the algorithm determines correctly the limit probability $q_{x,y}$ if $x \in X \setminus \bigcup_{r=1}^{r} X^{i_r}$ and $y \in X^{i_r}$. In this case the limit probability

$q_{x,y}$ is equal to the limit probability of the system to reach the ergodic class X^{i_r} multiplied by the limit probability of the system to remain in the state $y \in X^{i_r}$, i.e., $q_{x,y} = \pi_x(z^r)\pi_y^{i_r}$. Here $\pi_y^{i_r}$ is the probability of the system to remain in the state $y \in X^{i_r}$ and $\pi_x(z^{i_r})$ is the limit probability of the system to reach the absorbing state z_{i_r} in GA. According to the construction of the auxiliary graph GA the value $\pi_x(z^{i_r})$ coincides with the limit probability of the system to reach the ergodic class X^{i_r}. The correctness of this fact can be easily obtained from Lemma 1.12 and Theorem 1.17. According to Lemma 1.12 the probabilities $\pi_x(z^r)$ for $x \in X \setminus \bigcup_{r=1}^{k'} X^{i_r}$ can be found by solving the following system of linear equations

$$P'\pi'(z^r) = \pi'(z^r),$$
$$\pi'_{z^1}(z^r) = 0, \quad \pi'_{z^2}(z^r) = 0, \quad \dots, \quad \pi'_{z^r}(z^r) = 1, \quad \dots, \quad \pi'_{z^{k'}}(z^r) = 0,$$

which them determine correctly. So, the algorithm correctly finds the limit state matrix Q.

Now let us show that the running time of the algorithm is $O(n^3)$. We obtain this estimation from the items 3 and 4. In item 3 we solve $k' \leq n$ systems of linear equations $\pi^{i_r} = \pi^{i_p}P^{(i_r)}$, $\sum_{y \in X^{i_r}} \pi_y^{i_r} = 1$, where $\sum_{r=1}^{k'} |X_{i_r}| \leq n$. Therefore, here we determine the solution of these systems using $O(n^3)$ elementary operations. In item 4 we also solve $k' \leq n$ systems of linear equations. Each of these systems contains not more then n variables. All these systems have the same left part and, therefore, can be solved simultaneously applying the Gaussian elimination. The simultaneous solution of these r systems with the same left part by using the Gaussian elimination method uses $O(n^3)$ elementary operations. □

As we have shown in the proof of the theorem each component $\pi'_x(z^i)$ of the vector $\pi'(z^i)$ represents the probability that the system will occupy a state of the recurrent class X^i after a large number of states' transitions if the system starts transitions in the state $x \in X$, i.e., the vector $\pi'(z^i)$ gives the limiting probabilities $\pi'_x(z^i)$ for every $x \in X$ and the arbitrary recurrent class X^i. Therefore, $q_{x,y} = \pi'_x(z^i)\pi_y^i$, if $y \in X^i$.

It is easy to observe that if the subgraph $G' = \left(X \setminus \bigcup_{r=1}^{k'} X^{i_r}, E'\right)$ of GA induced by the subset of vertices $X \setminus \bigcup_{r=1}^{k'} X^{i_r}$ is an acyclic graph then

$$\pi_x(z^i)' = P_x(z^i, 0 \leq t(z^i) \leq |XA|),$$

where $P_x(z^i, 0 \leq t(z^i) \leq |XA|$ is the probability of the system to reach z^i in GA from x using $T(z)$ transitions such that $0 \leq t(z^i) \leq |XA|$. These probabilities can be calculated using the algorithms from Sect. 1.5. Thus, in the case that the subgraph G' is acyclic Procedure 2 can be modified by introducing the calculation of the vectors $\pi_x(z^i)$ on the bases of the algorithms from Sect. 1.5.

Below an example that illustrates the calculation procedure in Algorithm 1.18 is given.

Example Consider the problem of determining the limit matrix of the probabilities Q for the example from the previous section, i.e., the Markov chain is determined by the stochastic matrix of probabilities

$$P = \begin{pmatrix} 1 & 0 & 0 & 0 \\ 0.25 & 0.25 & 0.25 & 0.25 \\ 0 & 0 & 0.5 & 0.5 \\ 0 & 0 & 0.5 & 0.5 \end{pmatrix}.$$

The graph $G_p = (X, E_p)$ of the Markov process is represented in Fig. 1.5.

If we apply Procedure 1 of the algorithm, then we find the probabilities

$$\pi_1^1 = 1, \quad \pi_3^2 = 0.5, \quad \pi_4^2 = 0.5$$

which represent the solutions of the system:

$$\pi^1 = \pi^1 P^{(1)}, \quad \pi_1^1 = 1;$$
$$\pi^2 = \pi^2 P^{(2)}, \quad \pi_3^2 + \pi_4^2 = 1.$$

The first system represents the ergodicity condition for the recurrent class that corresponds to the deadlock component $X^1 = \{1\}$ and the second one represents the ergodicity condition that corresponds to the deadlock component $X^2 = \{3, 4\}$, i.e.,

$$P^{(1)} = (1); \quad P^{(2)} = \begin{pmatrix} 0.5 & 0.5 \\ 0.5 & 0.5 \end{pmatrix}.$$

In such a way we determine

$$\begin{array}{llll} q_{1,1} = 1, & q_{1,2} = 0, & q_{1,3} = 0, & q_{1,4} = 0, \\ q_{3,3} = 1, & q_{3,4} = 0, & q_{3,1} = 0, & q_{3,2} = 0, \\ q_{4,4} = 0.5, & q_{4,3} = 0.5, & q_{4,1} = 0, & q_{4,2} = 0. \end{array}$$

After that we apply Procedure 2. To apply this procedure we construct the auxiliary graph $GA = (XA, EA)$. This graph is represented in Fig. 1.6. Graph GA is obtained from G_p where the components $X^{i_1} = \{1\}$ and X^{i_2} are contracted respectively into vertices $Z^1 = 1'$ and $Z^2 = 3'$.

The matrix P' for this graph is given by

$$P' = \begin{pmatrix} 1 & 0 & 0 \\ 0.25 & 0.25 & 0.5 \\ 0 & 0 & 1 \end{pmatrix}.$$

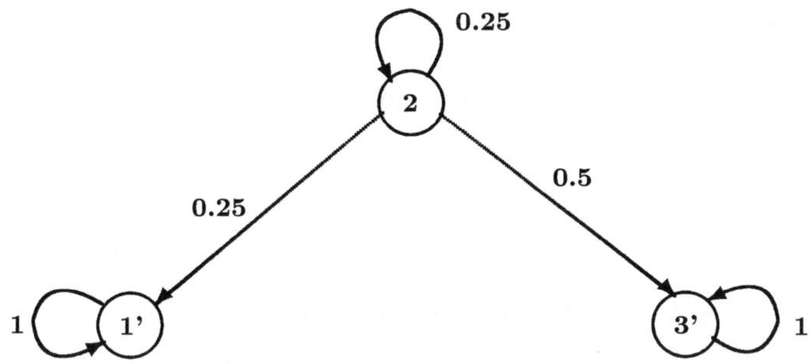

Fig. 1.6 The graph GA obtained from G_p

We solve the following two systems of linear equations:

$$P'\pi'(1') = \pi'(1'), \quad \pi'_{1'}(1') = 1, \quad \pi'_{3'}(3') = 0;$$
$$P'\pi'(3') = \pi'(3'), \quad \pi'_{3'}(1') = 0, \quad \pi'_{1'}(3') = 0,$$

where

$$\pi'(1') = \begin{pmatrix} \pi'_{1'}(1') \\ \pi'_{2'}(1') \\ \pi'_{3'}(1') \end{pmatrix}, \quad \pi'(3') = \begin{pmatrix} \pi'_{1'}(3') \\ \pi'_{2'}(3') \\ \pi'_{3'}(3') \end{pmatrix}.$$

The first system of equations can be written in the following form

$$\begin{cases} \pi'_{1'}(1') = \pi'_{1'}(1'), \\ 0.25\pi'_{1'}(1') + 0.25\pi'_{2'}(1') + 0.5\pi'_{3'}(1') = \pi'_{2'}(1'), \\ \qquad\qquad\qquad\qquad\qquad\qquad \pi'_{3'}(1') = \pi'_{3'}(1'), \\ \pi'_{1'}(1') = 1, \quad \pi'_{3'}(1') = 0 \end{cases}$$

and we obtain

$$\pi'_{1'}(1') = 1; \quad \pi'_{2'}(1') = 0.33(3); \quad \pi'_{3'}(1') = 0.$$

The second system of equations can be represented as follows

$$\begin{cases} \pi'_{1'}(3') = \pi'_{1'}(3'), \\ 0.25\pi'_{1'}(3') + 0.25\pi'_{2'}(3') + 0.5\pi'_{3'}(3') = \pi'_{2'}(3'), \\ \qquad\qquad\qquad\qquad\qquad\qquad \pi'_{3'}(3') = \pi'_{3'}(3'), \\ \pi'_{3'}(3') = 1, \quad \pi'_{1'}(3') = 0 \end{cases}$$

and we obtain

$$\pi'_{3'}(3') = 1; \quad \pi'_{2'}(3') = 0.66(6); \quad \pi'_{1'}(3') = 0.$$

After that we calculate

$$\begin{aligned} q_{2,1} &= \pi'_2(1') \cdot \pi_1^1 = 0.33(3), \qquad\qquad q_{2,2} = 0, \\ q_{2,3} &= \pi'_2(3') \cdot \pi_3^2 = 0.66(6) \cdot 0.5 = 0.33(3), \\ q_{2,4} &= \pi'_2(3') \cdot \pi_4^2 = 0.66(6) \cdot 0.5 = 0.33(3). \end{aligned}$$

Thus, we obtain the limit matrix

$$Q = \begin{pmatrix} 1 & 0 & 0 & 0 \\ 0.33(3) & 0 & 0.33(3) & 0.33(3) \\ 0 & 0 & 0.5 & 0.5 \\ 0 & 0 & 0.5 & 0.5 \end{pmatrix}.$$

1.5.4 An Algorithm for Determining the Limit Probabilities Based on the Ergodicity Condition

The main idea of the algorithm described in this section is based on the ergodicity condition for Markov chains and on the structure properties of the graph G_p of probability transitions. We consider the problem of finding the limit probabilities $q_{i_0,j}$, $j = 1, 2, \dots, n$ in the Markov chain with a fixed starting state $x(0) = x_{i_0}$ and a given stochastic matrix P. We show that this problem can be reduced to an auxiliary problem of finding the limit probabilities in a new Markov chain for which the corresponding graph of probability transitions $G^0 = (X, E^0)$ is strongly connected. Finally, for a new Markov chain we obtain the problem of determining the limiting probability

$$q^0_{i_0,j} = \pi^0_j, \quad j = 1, 2, \dots, n$$

where $\pi^0 = (\pi^0_1, \pi'_2, \dots, \pi^0_n)$ can be found by solving the system of linear equations

$$\pi^0 = \pi^0 P^0, \quad \sum_{j=1}^{n} \pi^0_j = 1. \tag{1.48}$$

Here P^0 is the stochastic matrix of the auxiliary irreducible Markov chain with the graph of probability transitions $G^0 = (X, E^0)$. Afterwards we show that each component $q_{i_0,j}$ can be obtained from the corresponding π_j^0, $j \in \{1, 2, \ldots, n\}$ using a special approximation procedure from [58]. We define the auxiliary graph $G^0 = (X, E^0)$ and the corresponding auxiliary Markov chain in the following way:

Let $G^1 = (X^1, E^1)$, $G^2 = (X^2, E^2)$, ..., $G^k = (X^k, E^k)$ be the strongly connected components of the graph $G_p = (X, E_p)$ where $\bigcup_{i=1}^k X^i = X$ and $k > 1$. Denote by $G^{i_r} = (X^{i_r}, E^{i'_n})$, $r = 1, 2, \ldots, k'$ the deadlock components of graph Gp and assume that $x_{i_0} \in X \setminus \bigcup_{r=1}^{k'} X^{i_r}$ and G_p satisfy the condition that each vertex $x \in X$ is attainable from x_{i_0}. If G_p contains vertices x_j which could not be reached from x_{i_0} then we set $q_{i_0,j} = 0$ and delete these vertices from G_p. Our problem can easily be solved also in the case that x_{i_0} belongs to a deadlock component G^{i_r}. In this case the limit probabilities also can be determined easily: We put $q_{i_0,j} = 0$ for $x_j \in X \setminus X^{i_r}$ and determine the nonzero components $q_{i_0,j}$ by solving the system of linear equations $\pi^{i_r} = \pi^{i_p} P^{(i_r)}$, $\sum_{y \in X^{i_r}} \pi_y^{i_r} = 1$ (see Procedure 1 of the algorithm from the previous section).

The strongly connected graph $G^0 = (X, E^0)$ of the auxiliary Markov unichain is obtained from the graph $G_p = (X, E_p)$ using the following construction:

- graph G^0 contains the same set of vertices X as the graph G_p;
- the set of directed edges E^0 of the graph G^0 consists of the set of edges E and the new directed edges $e = (x, x_{i_0})$ oriented from $x \in \bigcup_{r=1}^{k'} X^{i_r}$ to x_{i_0}, i.e., $E^0 = E \cup E'$ where $E' = \left\{(x, x_{i_0}) \,\middle|\, x \in \bigcup_{r=1}^{k'} X^{i_r}\right\}$;
- we define the probabilities $p_{x,y}^0$ for $(x, y) \in E^0 (x \in X_2)$ in G^0 using the following rules:

  (a) $p_{x,y}^0 = p_{x,y}$ for $(x, y) \in E^0$ if $x \in X \setminus \bigcup_{r=1}^{k'} X^{i_r}$;
  (b) $p_{x,x_{i_0}}^0 = \epsilon$ for $(x, x_{i_0}) \in E'$, $x \in \bigcup_{r=1}^{k'} X^{i_r}$, where ϵ is a small positive value in comparison with p_e for $e \in E$;
  (c) $p_{x,y}^0 = p_{x,y}(1 - \epsilon)$ for $(x, y) \in E$ if $x, y \in X \setminus \bigcup_{r=1}^{k'} X^{i_r}$.

Graph G^0 is strongly connected and, therefore, for the corresponding Markov chain with the new probability transition matrix P^0 there exists the limit matrix Q^0 whose rows are all identical. The vector $\pi^0 = (\pi_1^0, \pi_2^0, \ldots, \pi_n^0)$ with the components $\pi_j^0 = q_{i,j}^0$, $i, j = 1, 2, \ldots, n$ can be found by solving the system of linear equations (1.48). Moreover, based on the algorithm from the previous section we may conclude that for a small ϵ the solution of this system (1.48) will represent the approximation values for the components of the vector of the limiting probabilities in the initial Markov chain with given starting state i_0 and probability matrix P. The probabilities $p_{x,y}^0$ for $(x, y) \in E^0$ are defined according to the items a)–c) that allow us to preserve in the auxiliary Markov chain the small values of the limiting state probabilities π_j^0 for the states $x_j \in X \setminus \bigcup_{r=1}^{k'} X^{i_r}$. In addition, we can see that the condition c) in the auxiliary problem allows us to preserve an appropriate proportion of the limiting states probabilities in each irreducible set X^{i_r} and between different ergodic classes as the proportion of the limiting probabilities in the initial problem.

Using the results from [58] we can see that the exact values of the limiting probabilities $\pi_j = q_{i_0,j}$, $j = 1, 2, \ldots, n$ can be found from the corresponding π_j^0 using a special approximation procedure from [58] if ϵ is a suitable small value.

Indeed let us assume that the probabilities $p_{i,j}$ are given in the form of irreducible decimal fractions $p_{i,j} = a_{i,j}/b_{i,j}$, $i, j = 1, 2, \ldots, n$ where the numerators as well as the denominators are integer numbers. Then the values π_j can be found from π_j^0 using the roundoff procedure from [58] if ϵ satisfies the following condition:

$$\epsilon \leq 2^{-L},$$

where

$$L = \sum_{(x,y)\in E} \log(a_{x,y} + 1) + \sum_{(x,y)\in E} \log(b_{x,y} + 1) + 2\log(n) + 1.$$

Here L is the length of the binary-coded representation of the elements of the matrix P where each probability $p_{x,y}$ is given by the integer couple $a_{x,y}$, $b_{x,y}$. If for given ϵ the values π_j^0, $j = 1, 2, \ldots, n$ are known then each π_j^0 can be represented as a convergent continued fraction, and we may find a unique rational fraction A_j/B_j which satisfies the condition $\left|\pi_j^0 - A_j/B_j\right| \leq 2^{-2L-2}$. After that we fix $\pi_j = A_j/B_j$. In such a way we find the exact limiting probabilities π_j, $j = 1, 2, \ldots, n$. In general we can see that the probabilities π_j^0 can be expressed as functions that depend on ϵ, i.e.,

$$\pi_j^0 = \pi_j^0(\epsilon), \quad j = 1, 2, \ldots, n.$$

This means that the limiting probabilities π_j for our initial problem can be found as follows:

$$\pi_j = \lim_{\epsilon \to 0} \pi_j^0(\epsilon), \quad j = 1, 2, \ldots, n.$$

Below an example that illustrate the algorithm described above is given.

Example Again we use the data for a Markov chain from the previous example and consider the problem of determining the limiting state probabilities in the case if the starting state of the system is $x_{i_1} = 2$ and the stochastic matrix of probabilities is

$$P = \begin{pmatrix} 1 & 0 & 0 & 0 \\ 0.25 & 0.25 & 0.25 & 0.25 \\ 0 & 0 & 0.5 & 0.5 \\ 0 & 0 & 0.5 & 0.5 \end{pmatrix}.$$

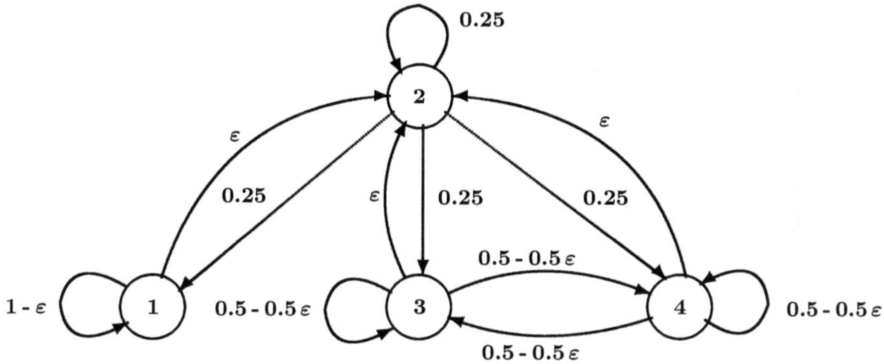

Fig. 1.7 The auxiliary graph $G^0 = (X, E^0)$

The corresponding graph $G_p = (X, E_p)$ of this Markov chain is represented in Fig. 1.5.

We apply the algorithm described above for determining the vector of limiting-state probabilities $\pi = (\pi_1, \pi_2, \pi_3)$, where $\pi_1 = q_{2,1}$, $\pi_2 = q_{2,2}$, $\pi_3 = q_{2,3}$, $\pi_4 = q_{2,4}$.

The corresponding strongly connected auxiliary graph $G^0 = (X, E^0)$ is represented in Fig. 1.7 and the corresponding matrix of probabilities P' for the auxiliary Markov chain is given below

$$P^o = \begin{pmatrix} 1-\epsilon & \epsilon & 0 & 0 \\ 0.25 & 0.25 & 0.25 & 0.25 \\ 0 & \epsilon & 0.5-0.5\epsilon & 0.5-0.5\epsilon \\ 0 & \epsilon & 0.5-0.5\epsilon & 0.5-0.5\epsilon \end{pmatrix}.$$

For the matrix P we have $p_{1,1} = 1$, $p_{1,2} = p_{1,3} = p_{1,4} = 0$; $p_{2,1} = p_{2,2} = p_{2,3} = p_{2,4} = 1/4$; $p_{3,1} = p_{3,2} = 0$, $p_{3,3} = p_{3,4} = 1/2$; $p_{4,1} = p_{4,2} = 0$, $p_{4,3} = p_{4,4} = 1/2$. Therefore, we can take $L \geq 35$. This implies that we can set $\epsilon = 0.0001$. We consider the system of linear equations

$$\pi^0 = \pi^0 P^0; \quad \sum_{j=1}^{4} \pi_j^0 = 1$$

and determine the solution of this system. So, if for $\epsilon = 0.0001$ we solve the system of linear equations

$$\begin{cases} \pi_1^0 = (1-\epsilon)\pi_1^0 + 0.25\pi_2^0; \\ \pi_2^0 = \epsilon\pi_1^0 + 0.25\pi_2^0 + \epsilon\pi_3^0 + \epsilon\pi_4^0; \\ \pi_3^0 = 0.25\pi_2^0 + (0.5-0.5\epsilon)\pi_3^0 + (0.5-0.5\epsilon)\pi_4^0; \\ \pi_4^0 = 0.25\pi_2^0 + (0.5-0.5\epsilon)\pi_3^0 + (0.5-0.5\epsilon)\pi_4^0; \\ 1 = \pi_1^0 + \pi_2^0 + \pi_3^0 + \pi_4^0 \end{cases}$$

then we obtain the solution

$$\pi_1^0 = 0.3333; \quad \pi_2^0 = 0.0001; \quad \pi_3^0 = 0.3333; \quad \pi_4^0 = 0.3333.$$

If for each π_j^0, $j = 1, 2, 3, 4$ we find the approximate rational fraction then we determine

$$\pi_1 = \frac{1}{3}; \quad \pi_2 = 0; \quad \pi_3 = \frac{1}{3}; \quad \pi_4 = \frac{1}{3}$$

that satisfies the conditions

$$\left|\pi_1^0 - \frac{1}{3}\right| \le 0.0001; \quad |\pi_2^0 - 0| \le 0.0001;$$

$$\left|\pi_3^0 - \frac{1}{3}\right| \le 0.0001; \quad \left|\pi_4^0 - \frac{1}{3}\right| \le 0.0003.$$

Therefore, finally we obtain

$$q_{2,1} = \frac{1}{3}; \quad q_{2,2} = 0; \quad q_{2,3} = \frac{1}{3}; \quad q_{2,4} = \frac{1}{3}.$$

The system above can be solved in general form with respect to ϵ and the following representation of the solution in parametrical form can be obtained:

$$\pi_1^0(\epsilon) = \frac{1}{3+4\epsilon}; \qquad \pi_2^0(\epsilon) = \frac{4\epsilon}{3+4\epsilon};$$

$$\pi_3^0(\epsilon) = \frac{2+3\epsilon}{(2+2\epsilon)(3+4\epsilon)}; \quad \pi_4^0(\epsilon) = \frac{2+\epsilon}{(2+2\epsilon)(3+4\epsilon)}.$$

If after that we find the corresponding limits when $\epsilon \to 0$ then we obtain

$$\pi_1 = \lim_{\epsilon\to 0} \pi_1^0(\epsilon) = \frac{1}{3}; \quad \pi_2 = \lim_{\epsilon\to 0} \pi_2^0(\epsilon) = 0;$$

$$\pi_3 = \lim_{\epsilon\to 0} \pi_3^0(\epsilon) = \frac{1}{3}; \quad \pi_4 = \lim_{\epsilon\to 0} \pi_4^0(\epsilon) = \frac{1}{3},$$

i.e., we obtain the limiting probabilities for our problem.

1.5.5 *Determining the First Hitting Limiting Probability of the State in a Markov Chain*

In this section we consider Markov processes in which the system may stop transitions as soon as a given state $y \in X$ is reached. We call such processes *Markov processes with stopping state*. We show that the dynamic programming algorithms from Sect. 1.5 can be used for determining the probability of first hitting the stopping state from an arbitrary starting state.

Let a Markov chain with the corresponding matrix of probability transitions P be given. Fix a state $y \in X$ and consider the problem of determining the probabilities π_i of first hitting the state y when the system starts transition in a state $x_i \in X$.

It is evident that if we have a Markov unichain and y as a absorbing state then the probability of first hitting the state y from a given state $x_i \in X$ is equal to the limiting probability $q_{x,y}$ from x to y where $q_{x,y} = 1$ for every $x \in X$. If we apply Algorithm 1.10 and if we take into account that $\tau \to \infty$ then we obtain the following system of linear equations

$$\pi' = P\pi', \quad \pi'_{i_y} = 1,$$

where π' is the column vector with components π'_{i_x}, i.e.,

$$\pi' = \begin{pmatrix} \pi'_1 \\ \pi'_2 \\ \vdots \\ \pi'_n \end{pmatrix}.$$

So, this system for Markov unichains with absorbing state always has a unique solution $p_i = 1, i = 1, 2, \ldots, n$ and $q_{x_i,y} = p_i = 1, \forall x_i \in X$.

If we have an arbitrary Markov multichain that contains an absorbing state $y \in X$ then the limiting probabilities $q_{x_i,y}$ from $x_i \in X$ to y also coincide with the limiting probabilities π_i of first hitting the state $y \in X$. However, here these probabilities may be different from 1 and some of them may be equal to zero. The zero components of the vector π can easily be determined from the graph $G_p = (X, E_p)$ of this Markov process. Indeed, let X_y be the set of vertices $x \in X$ for which in G_p there exists a directed path from x to y, we consider $I_y = \{i \in \{1, 2, \ldots, n\} \mid x_i \in X_y\}$. Then it is evident that $\pi_i = 0$ if and only if $i \in N \setminus I_y$, where $N = \{1, 2, \ldots, n\}$. Therefore, the probabilities π_i for $x_i \in X$ can be found by solving the following system of linear equations

$$\pi' = P\pi', \quad \pi'_{i_y} = 1 \quad \text{and} \quad \pi_i = 0, \quad \forall\, i \in N \setminus I_y.$$

We obtain this system of equations from Algorithm 1.10 if $\tau \to \infty$.

In the general case for an arbitrary Markov chain if y is not an absorbing state the probability of first hitting the state y from $x \in X$ can be found by the following way: We consider a new matrix of probability transitions P' which is obtained from P by changing the elements $p_{y,z}$ with the new elements $p'_{y,z}$, where $p'_{y,y} = 1$ and $p_{y,z} = 0, \forall z \in X \setminus \{y\}$. After that we obtain a new Markov chain with a new matrix of probability transitions P' where y is an absorbing state and we determine the limiting probabilities of first hitting the state y from $x \in X$ using the procedure described above. So, in order to determine the vector of first hitting probabilities we have to solve the following system of linear equations

$$\pi' = P'\pi'; \quad \pi'_{i_y} = 1, \quad \pi'_i = 0, \quad \forall i \in N \setminus I_y.$$

Thus, for an arbitrary starting state $x \in X$ and an arbitrary positive recurrent state y in a Markov unichain the first hitting probability $\pi'_{x,y}$ is equal to 1, i.e., $\pi'_{x,y} = 1$, $\forall x \in X$.

Example Consider the problem of determining the probabilities of first hitting for the Markov process with the matrix of probability transitions

$$P = \begin{pmatrix} 0.3 & 0.3 & 0.4 & 0 \\ 0.5 & 0 & 0.3 & 0.2 \\ 0 & 0.6 & 0 & 0.4 \\ 0 & 0 & 0 & 1.0 \end{pmatrix}$$

and a given stopping state $y = 3$.
To determine $\pi'_1, \pi'_2, \pi'_3, \pi'_4$ we form the matrix

$$P' = \begin{pmatrix} 0.3 & 0.3 & 0.4 & 0 \\ 0.5 & 0 & 0.3 & 0.2 \\ 0 & 0 & 1.0 & 0 \\ 0 & 0 & 0 & 1.0 \end{pmatrix}$$

and find the set $X_y = \{4\}$ for the corresponding graph $G'_p = (X, E'_p)$. This means that we have to solve the system of linear equations

$$\pi' = P'\pi', \quad \pi'_3 = 1, \quad \pi'_4 = 0.$$

The solution of this system is

$$\pi'_1 = 0.8909(09), \quad \pi'_2 = 0.745(45), \quad \pi'_3 = 1, \quad \pi'_4 = 0.$$

It is evident that the vector of the first hitting limiting probabilities π' for different fixed stopping states $y \in X$ may be different, i.e., it depends on the stopping state y. Therefore, in the following we will denote this vector for a given stopping state $y \in X$ by $\pi'(y)$, where

$$\pi'(y) = \begin{pmatrix} \pi'_{1,j_y} \\ \pi'_{2,j_y} \\ \vdots \\ \pi'_{n,j_y} \end{pmatrix}.$$

If we find the vectors of first hitting probabilities $\pi'(y)$ for every stopping state $y \in X$ then we determine the matrix $\pi' = (\pi_{i,j})'$, where an arbitrary element $\pi'_{i,j}$ represents the limiting probability of first hitting the state x_j if the system starts transitions in the state x_i.

It easy to observe that the following relationship between the elements of the limit matrix $Q = (q_{i,j})$ and the elements of the matrix of the first hitting probabilities π' in the Markov chain holds

$$q_{i,j} = \pi'_{i,j} q_{j,j}, \quad i, j = 1, 2, \ldots, n.$$

Using this property and the property of the limit matrix we obtain the system of linear equations

$$\sum_{j=1}^{n} \pi'_{i,j} q_{j,j} = 1, \quad i = 1, 2, \ldots, n.$$

In the general case the rank of this system may be less then n and, therefore, the values $q_{j,j}$ cannot be determined uniquely if the matrix π' is known. This system has a unique solution with respect to $q_{j,j}$ only in the case if in the Markov chain each positive recurrent state is an absorbing state.

1.6 Determining the State-Time Probabilities of the System for Non-stationary Markov Processes

We consider Markov processes for which the probabilities of systems' transitions from one state to another depend on time. Such processes are called non-stationary Markov processes. In this case the process is defined by a dynamic matrix $P(t) = (p_{x,y}(t))$ and a given starting state x_{i_0}, where the dynamic matrix $P(t)$ is assumed to be stochastic for every discrete moment of time $t = 1, 2, \ldots$. The state-time probabilities $P_{x_{i_0}}(x, t)$ for non-stationary Markov processes are defined and calculated in a similar way as for stationary processes by using the recursive formula

$$P_{x_{i_0}}(x, \tau+1) = \sum_{y \in X} P_{x_{i_0}}(y, \tau) p_{y,x}(\tau), \quad \tau = 0, 1, 2, \ldots, t-1$$

where $P_{x_{i_0}}(x_{i_0}, 0) = 1$ and $P_{x_{i_0}}(x_{i_0}, 0) = 0$ for $x \in X \setminus \{x_{i_0}\}$. It is evident that if $p_{x,y}(t)$ does not depend on time; then this formula becomes the formula from Sect. 1.5. The matrix form of this formula can be represented as follows

$$\pi(\tau+1) = \pi(\tau)P(\tau), \quad \tau = 0, 1, 2, \ldots, t-1$$

where $\pi(\tau) = (\pi_1(\tau), \pi_2(\tau), \ldots, \pi_n(\tau))$ is the vector with the components $\pi_i(\tau) = P_{x_{i_0}}(x_i, \tau)$. At the starting moment of time $\tau = 0$ the vector $\pi(\tau)$ is given in the same way as for stationary processes, i.e., $\pi_{i_0}(0) = 1$ and $\pi_i(0) = 0$ for arbitrary $i \neq i_0$. If we apply this formula for a given starting vector $\pi(0)$ and $\tau = 0, 1, 2, \ldots, t-1$ then we obtain

$$\pi(t) = \pi(0)P(0)P(1)P(2) \cdots P(t-1).$$

So, an arbitrary element $\overline{p}_{x_i,x_j}(t)$ of the matrix $\overline{P}(t) = P(0)P(1)P(2)\cdots P(t-1)$ expresses the probability of the system $\mathbb{L}$ to reach the state x_j from x_i by using t units of time.

Now let us show how to calculate the probability $P_{x_{i_0}}(y,\ t_1 \leq t(y) \leq t_2)$ in the case of non-stationary Markov processes. In the same way as for the stationary case we consider the non-stationary Markov process with a given absorbing state $y \in X$. So, we assume that the dynamic matrix $P(t)$ which is stochastic for every $t = 0, 1, 2, \ldots$ and $p_{y,y}(t) = 1$ for arbitrary t is given. Then the probabilities $P_x(y,\ 0 \leq t(y) \leq t)$ for $x \in X$ can be determined if we tabulate the values $P_x(y,\ t-\tau \leq t(y) \leq t)$, $\tau = 0, 1, 2, \ldots, t$ using the following recursive formula:

$$P_x(y,\ t-\tau-1 \leq t(y) \leq t) = \sum_{z \in X} p_{x,z}(t-\tau-1) P_z(y,\ t-\tau \leq t(y) \leq t)$$

where for $\tau = 0$ we fix

$$P_x(y,\ t \leq t(y) \leq t) = 0 \quad \text{if} \quad x \neq y \quad \text{and} \quad P_y(y,\ t \leq t(y) \leq t) = 1.$$

We can represent this recursive formula in the following matrix form

$$\pi''(\tau+1) = P(t-\tau-1)\pi''(\tau), \quad \tau = 0, 1, 2, \ldots, t-1.$$

At the starting moment of time $t = 0$ the vector $\pi''(0)$ is given: All components are equal to zero except the component corresponding to the absorbing state which is equal to one, i.e.,

$$\pi_i''(0) = \begin{cases} 0, & \text{if } x_i \neq y, \\ 1, & \text{if } x_i = y. \end{cases}$$

If we apply this formula to $\tau = 0, 1, 2, \ldots, t-1$ then we obtain

$$\pi''(t) = P(0)P(1)P(2)\cdots P(t-1)\pi''(0), \quad t = 1, 2, \ldots.$$

So, if we consider the matrix $\overline{P}(t) = P(0)P(1)P(2)\cdots P(t-1)$ then an arbitrary element $\overline{p}_{i,j_y}(t)$ of the column j_y in the matrix $\overline{P}(t)$ expresses the probability of the system $\mathbb{L}$ to reach the state y from x_i by using not more than t units of time, i.e., $\overline{p}_{i,j_y}(t) = P_{x_i}(y,\ 0 \le t(y) \le t)$.

Here the matrix $P(t)$ is a stochastic matrix for $t = 0, 1, 2, \ldots$ where $p_{y,y}(t) = 1$ for $t = 1, 2, \ldots$ and

$$\pi''(\tau) = \begin{pmatrix} \pi''_1(\tau) \\ \pi''_2(\tau) \\ \vdots \\ \pi''_n(\tau) \end{pmatrix}, \quad \tau = 0, 1, 2, \ldots$$

is the column vector, where an arbitrary component $\pi''_i(\tau)$ expresses the probability of the dynamical system to reach the state y from x_i by using not more than τ units of time if the system starts transitions in the state x at the moment of time $t - \tau$, i.e., $\pi''_i(\tau) = P_{x_i}(y,\ t - \tau \le t(y) \le t)$.

This means that in the case that y is an absorbing state the probability $P_x(y,\ t_1 \le t(y) \le t_2)$ can be found in the following way:

(a) Find the matrices

$$\overline{P}(t_1 - 1) = P(0)P(1)P(2)\cdots P(t_1 - 2)$$

and

$$\overline{P}(t_2) = P(0)P(1)P(2)\cdots P(t_2 - 1);$$

(b) Calculate

$$\begin{aligned} P_x(y,\ t_1 \le t(y) \le t_2) &= P_x(y,\ 0 \le t(y) \le t_2) - P_x(y,\ 0 \le t(y) \le t_1 - 1) \\ &= \overline{p}_{i_x,j_y}(t_2) - \overline{p}_{i_x,j_y}(t_1 - 1), \end{aligned}$$

where $\overline{p}_{i_x,j_y}(t_1 - 1)$ and $\overline{p}_{i_x,j_y}(t_2)$ represent the corresponding elements of the matrices $\overline{P}(t_1 - 1)$ and $\overline{P}(t_2)$.

The results described above allow us to formulate algorithms for the calculation of the probabilities $P_x(y,\ 0 \le t(y) \le t)$ for an arbitrary non-stationary Markov process. Such algorithms can be obtained if in the general steps of the Algorithms 1.9 and 1.10 we change the matrix P by the matrix $P(t - \tau - 1)$ and $\pi'(\tau)$ by $\pi''(\tau)$.

Below we describe these algorithms. They can be derived in an analogues way as the algorithms from the previous section.

Algorithm 1.20 Calculation of the State-Time Probabilities of the System in the Matrix Form (Non-stationary Case)

Preliminary step (Step 0): Fix the vector $\pi''(0) = (\pi''_1(0), \pi''_2(0), \dots, \pi''_n(0))$, where $\pi''_i(0) = 0$ for $i \neq i_y$ and $\pi''_{i_y}(0) = 1$.

General step (Step $\tau + 1, \quad \tau \geq 0$): For a given τ calculate

$$\pi''(\tau + 1) = P(t - \tau - 1)\pi''(\tau)$$

and then put

$$\pi''_{i_y}(\tau + 1) = 1.$$

If $\tau < t - 1$ then go to the next step, i.e., $\tau = \tau + 1$; otherwise STOP.

Algorithm 1.21 Calculation of the State-Time Probabilities of the System with Known Probability of its Remaining in the Final State (Non-stationary Case)

Preliminary step (Step 0): Fix the vector $\pi''(0) = (\pi''_1(0), \pi''_2(0), \dots, \pi''_n(0))$, where $\pi''_i(0) = 0$ for $i \neq i_y$ and $\pi''_{i_y}(0) = 1$.

General step (Step $\tau + 1, \quad \tau \geq 0$): For a given τ calculate

$$\pi''(\tau + 1) = P(t - \tau - 1)\pi''(\tau)$$

and then put

$$\pi''_{i_y}(\tau + 1) = q(y).$$

If $\tau < t - 1$ then go to the next step, i.e., $\tau = \tau + 1$; otherwise STOP.

Note that Algorithm 1.21 finds the probabilities $P_x(y, \ 0 \leq t(y) \leq t)$ when the value $q(y)$ is given. We treat this value as the probability of the system to remain in the state y; for the case $q(y) = 1$ this algorithm coincides with the previous one.

To calculate the probability $P_{x_{i_0}}(x, \ t_1 \leq t(x) \leq t_2)$ for $x \in X$ in the case of non-stationary Markov processes we shall use also the following auxiliary result.

Lemma 1.22 *Let a Markov process determined by the stochastic matrix of probabilities $P = (p_{x,y})$ and the starting state x_{i_0} be given. Then the following formula holds:*

$$\begin{aligned} P_{x_{i_0}}(x, \ t_1 \leq t(x) \leq t_2) = P_{x_{i_0}}(x, t_1) &+ P^{t_1}_{x_{i_0}}(x, t_1 + 1) \\ &+ P^{t_1, t_1+1}_{x_{i_0}}(x, t_1 + 2) + \dots \\ &+ P^{t_1, t_1+1, \dots, t_2-1}_{x_{i_0}}(x, t_2) \end{aligned} \tag{1.49}$$

where $P_{x_{i_0}}^{t_1,t_1+1,\ldots,t_1+i-1}(x, t_1+i)$, $i = 1, 2, \ldots, t_2 - t_1$, *is the probability of the dynamical system to reach the state* x *from* x_0 *by using* $t_1 + i$ *transitions such that it does not pass through* x *at the moments of time* $t_1, t_1+1, t_1+2, \ldots, t_1+i-1$.

Proof Taking into account that $P_{x_{i_0}}(x,\ t_1 \le t(x) \le t_1+i)$ expresses the probability of the system $\mathbb{L}$ to reach from x_0 the state x at least at one of the moments of time $t_1, t_1+1, \ldots, t_1+i$ we can use the following recursive formula

$$P_{x_{i_0}}(x,\ t_1 \le t(x) \le t_1+i) = P_{x_{i_0}}(x,\ t_1 \le t(x) \le t_1+i-1) + P_{x_{i_0}}^{t_1,t_1+1,\ldots,t_1+i-1}(x, t_1+i). \tag{1.50}$$

Applying formula (1.50) $t_2 - t_1$ times to $i = 1, 2, \ldots, t_2 - t_1$ we obtain the equality (1.49). □

Note that formulae (1.49) and (1.50) could not be used directly for the calculation of the probability $P_{x_0}(x,\ t_1 \le t(x) \le t_2)$. Nevertheless we can see that such a representation of the probability $P_{x_{i_0}}(x,\ t_1 \le t(x) \le t_2)$ in the time-expanded network method will allow us to formulate a suitable algorithm for calculating this probability and to develop new algorithms for solving the problems from Chap. 4.

Corollary 1.23 *If the state* x *of the dynamical system* $\mathbb{L}$ *in the graph of probability transitions* $G_p = (X, E_p)$ *corresponds to a deadlock vertex then*

$$P_{x_{i_0}}(x,\ t_1 \le t(x) \le t_2) = \sum_{t=t_1}^{t_2} P_{x_{i_0}}(x, t). \tag{1.51}$$

Let X_f be a subset of X and assume that at the moment of time $t = 0$ the dynamical system $\mathbb{L}$ is in the state x_0. Denote by $P_{x_{i_0}}(X_f,\ t_1 \le t(X_f) \le t_2)$ the probability that at least one of the states $x \in X_f$ will be reached at the time moment $t(x)$ such that $t_1 \le t(x) \le t_2$. Then the following corollary holds:

Corollary 1.24 *If the subset of states* $X_f \subset X$ *of the dynamical system* $\mathbb{L}$ *in the graph* $G_p = (X, E_p)$ *corresponds to the subset of deadlock vertices then for the probability* $P_{x_{i_0}}(X_f,\ t_1 \le t(X_f) \le t_2)$ *the following formula holds*

$$P_{x_{i_0}}(X_f,\ t_1 \le t(X_f) \le t_2) = \sum_{x \in X_f} \sum_{t=t_1}^{t_2} P_{x_{i_0}}(x, t). \tag{1.52}$$

It is easy to see that formula (1.52) generalizes formula (1.51).

1.7 Markov Processes with Rewards and Transition Costs

Markov processes with rewards have been introduced in [1, 3, 4, 47]. The main problems related to such processes and approaches for their solving can be found in [47, 114, 140]. In this section we consider Markov processes with rewards and analyze some basic notions that we will relate to stochastic control problems in the following. For a Markov processes with rewards there are given the probability matrix $P = (p_{x,y})$, a starting state and the matrix $C = (c_{x,y})$ where an arbitrary element $c_{x,y}$ expresses the reward if the system makes a transition from the state x to the state y. Thus, when the system makes transitions from one state to another we obtain a sequence of rewards. This sequence of rewards is a random variable with a probability distribution induced by the probability relations of the Markov process and for which the *expected total reward* during T state-time transitions of the system can be defined. In this book for stochastic control problems we will treat the rewards as the costs, i.e., we consider Markov processes with transition costs. Thus, in the following we call the *expected total reward* the *expected total cost*.

1.7.1 Definition and Calculation of the Expected Total Cost

Consider a Markov process with the transition probability matrix $P = (p_{x,y})$ and a given starting state of the dynamical system. In addition, assume that the matrix $C = (c_{x,y})$ is defined where $c_{x,y}$ expresses the cost of the system's transition from the state x to the state y. This Markov process generates a sequence of costs which is a random variable induced by transition probability distributions in the states. Therefore, the expected total cost during t transitions when the system starts transitions in a given state $x \in X$ can be defined in an analogues way as for Markov processes with rewards. We denote this expected total cost by $\sigma_x(t)$. The values $\sigma_x(t)$ for $x \in X$ are defined and calculated on the bases of the following recursive formula

$$\sigma_x(\tau + 1) = \sum_{y \in X} p_{x,y}(c_{x,y} + \sigma_y(\tau)), \quad \tau = 0, 1, 2, \dots, t - 1 \tag{1.53}$$

where $\sigma_x(0) = 0$ for every $x \in X$. Formula (1.53) can be treated in a similar way as the formula for calculating the total earning in Markov processes with rewards [47]. The expression $c_{x,y} + \sigma_y(\tau)$ means that if the system makes a transition from the state x to the state y then it spends the amount $c_{x,y}$ plus the amount it expects to spend during the next $\tau - 1$ transitions if the system starts transitions in the state y at the moment of time $\tau = 0$. Taking into account that in the state x the system makes transitions in a random way (with the probability distribution $p_{x,y}$), we obtain that the values $c_{x,y} + \sigma_y(\tau)$ should be weighted by the transition probabilities $p_{x,y}$. In the case of the non-stationary process, i.e., if the probabilities and the costs are changing in time, the expected total cost of the dynamical system is defined and calculated in a similar way; in formula (1.53) we change $p_{x,y}$ by $p_{x,y}(\tau)$ and $c_{x,y}$ by $c_{x,y}(\tau)$.

It is easy to observe that the formula for the expected total cost given above can be represented as follows

$$\sigma_x(\tau+1) = \sum_{y\in X} p_{x,y}c_{x,y} + \sum_{y\in X} p_{x,y}\sigma_y(\tau), \quad \tau = 0, 1, 2, \dots, t-1.$$

If for an arbitrary state $x_i \in X$ in this formula we denote

$$\mu_i = \sum_{y\in X} p_{x_i,y}c_{x_i,y}, \quad \sigma_i(\tau) = \sigma_{x_i}(\tau), \quad i = 1, 2, \dots, n$$

and if we regard μ_i and $\sigma_i(\tau)$ as the components of the corresponding vectors

$$\mu = \begin{pmatrix} \mu_1 \\ \mu_2 \\ \vdots \\ \mu_n \end{pmatrix}, \quad \sigma(\tau) = \begin{pmatrix} \sigma_1(\tau) \\ \sigma_2(\tau) \\ \vdots \\ \sigma_n(\tau) \end{pmatrix}$$

then the formula for calculating the expected total cost can be written in the following matrix form:

$$\sigma(\tau+1) = \mu + P\sigma(\tau), \quad \tau = 0, 1, 2, \dots, t-1. \tag{1.54}$$

The component μ_i of the vector μ may be interpreted as the cost to be expected in the next transition out of state x_i and, therefore, we call it the *expected immediate cost* in the state x_i. An arbitrary component $\sigma_i(\tau)$ of the vector $\sigma(\tau)$ expresses the expected total cost of the system during τ transitions if the system starts transitions in the state x_i.

Applying t times this formula and taking into account that $\sigma(0) = \mathbf{0}$, where $\mathbf{0}$ is a vector with zero components, we obtain

$$\sigma(t) = P^0\mu + P^1\mu + P^2\mu + \cdots + P^{t-1}\mu,$$

where P^0 is the identity matrix, i.e., $P^0 = I$.

For an arbitrary state $x_i \in X$ we denote

$$\omega_i(t) = \frac{1}{t}\sigma_i(t), \quad t = 1, 2, \dots.$$

This value expresses the *expected average cost per transition* of the system during t state transitions if the system starts transitions in x_i. We call the vector $\omega(t)$ with the components $\omega_i(t)$, $i = 1, 2, \dots, n$ the *vector of average costs* of the system if it starts transitions in x_i.

Let us consider an arbitrary discrete Markov process for which there exists the limit

$$\lim_{t\to\infty} P^t = Q.$$

Then there exists the limit

$$\lim_{t\to\infty} \omega(t) = \omega$$

and $\omega = Q\mu$. This fact can be proved by using the following property. Let $Q(0), Q(1), Q(2), \dots, Q(t), \dots$ be an arbitrary sequence of real matrices for which there exists the limit

$$\lim_{t\to\infty} Q(t) = Q.$$

Then there exists the limit

$$\lim_{t\to\infty} \frac{\sum_{k=0}^{t-1} Q(k)}{t}$$

and this limit is equal to Q, i.e.,

$$\lim_{t\to\infty} \frac{\sum_{k=0}^{t-1} Q(k)}{t} = Q.$$

So, if we put $Q(t) = P^t$ then we obtain

$$\lim_{t\to\infty} \frac{1}{t} \sum_{k=0}^{t-1} P^k = Q.$$

This means that

$$\lim_{t\to\infty} \omega(t) = \lim_{t\to\infty} \frac{1}{t} \sum_{k=0}^{t-1} P^k \mu = Q\mu.$$

Using the results described above we can determine the vector of expected average costs $\omega(t)$ if $t \to \infty$ in the following way: We find the limit matrix Q and the vector μ and then calculate $\omega = Q\mu$. This means that in the case of a large number of states' transitions t the vector of the expected total costs $\sigma(t)$ can be approximated with the vector $tQ\mu$, i.e., $\sigma(t) \approx tQ\mu$.

We have proved the property mentioned above in the case if there exists the limit $\lim_{t\to\infty} P^t = Q$. In the general case the existence of the limit $\lim_{t\to\infty} \omega(t) = \omega$ for an arbitrary matrix P can be proved by using the Cesaro limit

$$\lim_{t \to \infty} \frac{1}{t} \sum_{k=0}^{t-1} P^k = Q.$$

The result described above allows us to prove the following lemma:

Lemma 1.25 *For an arbitrary Markov process the vector of limiting average costs ω satisfies the following equations:*

$$\omega = P\omega \tag{1.55}$$
$$\omega = Q\omega. \tag{1.56}$$

Proof Using formula (1.54) we can write the following relation

$$\frac{\sigma(t+1)}{t} = \frac{\mu}{t} + P\frac{\sigma(t)}{t},$$

i.e.,

$$\frac{t+1}{t}\frac{\sigma(t+1)}{t+1} = \frac{\mu}{t} + P\frac{\sigma(t)}{t}.$$

If in the last equation we take the limit if $t \to \infty$ then we obtain the equality (1.55).
We prove formula (1.56) using (1.55). From (1.55) we obtain

$$\omega = P\omega, \quad \omega = P^2\omega, \quad \ldots, \quad \omega = P^t\omega.$$

This implies

$$\omega = \frac{1}{t}\sum_{\tau=0}^{t-1} P^{\tau}\omega.$$

If in this formula we take the limit for $t \to \infty$ then we obtain (1.56). □

In the following we study the asymptotic behavior of the expected total cost and of the average cost per transition in the Markov processes with transition costs using the z-transform. We can see that the z-transform allows us to formulate a more adequate asymptotic formula for determining the expected total cost and average cost per transition for the dynamical system in Markov chains.

1.7.2 An Asymptotic Behavior Analysis of the Expected Total Cost Based on the z-Transform

Now we show that the z-transform can be used for the estimation of the expected total cost of the dynamical system in the case of a large number of transitions. To obtain an asymptotic formula for the expected total cost in Markov processes we apply the z-transform to the equation

$$\sigma(t+1) = \mu + P\sigma(t).$$

Based on the properties of the z-transform from Sect. 1.1.4 we get

$$z^{-1}(F_\sigma(z) - F_\sigma(0)) = \frac{1}{1-z}\mu + PF_\sigma(z).$$

Through rearrangement we obtain

$$\begin{aligned} F_\sigma(z) - \sigma(0) &= \frac{z}{1-z}\mu + zPF_\sigma(z), \\ (I - zP)F_\sigma(z) &= \frac{z}{1-z}\mu + \sigma(0) \end{aligned}$$

or

$$F_\sigma(z) = \frac{z}{1-z}(I - zP)^{-1}\mu + (I - zP)^{-1}\sigma(0). \tag{1.57}$$

For many practical problems $\sigma(0)$ is identically zero and, therefore, the Eq. (1.57) reduces to

$$F_\sigma(z) = \frac{z}{1-z}(I - zP)^{-1}\mu.$$

In general $\sigma(0)$ may be different from zero.

In Sect. 1.1.4 we have shown that the inverse transformation of $(I - zP)^{-1}$ has the form $Q + T(t)$, where Q is the limit matrix and $T(t)$ is a sum of differential matrices with geometrically decreasing coefficients. This means that for $(I - zP)^{-1}$ the following relation can be written

$$(I - zP)^{-1} = \frac{1}{1-z}Q + \mathbb{T}(z),$$

where $\mathbb{T}(z)$ is the z-transform of $T(t)$.

If we substitute this formula in (1.57) then we obtain

$$F_\sigma(z) = \frac{z}{(1-z)^2}Q\mu + \frac{z}{1-z}\mathbb{T}(z)\mu + \frac{1}{1-z}Q\sigma(0) + \mathbb{T}(z)\sigma(0). \tag{1.58}$$

To identify the components of $\sigma(t)$ by using the inverse transformation for $F_\sigma(z)$ we have to analyze each component of $F_\sigma(z)$. First of all we observe that the term $zQ\mu/(1-z)^2$ represents the ramp of the magnitude $Q\mu$. A more detailed analysis of the term $z\mathbb{T}(z)\mu/(1-z)$ allows us to conclude that it represents a step of magnitude $\mathbb{T}(1)\mu$ plus geometric terms that tend to zero as t becomes very large. Furthermore, let us assume that all roots of the equation $\det(I - zP) = 0$ are different. Then $\mathbb{T}(z)$

can be expressed as follows

$$\mathbb{T}(z) = \sum_i \frac{D_i}{1 - \alpha_i z},$$

where D_i are the matrices that do not depend on z and α_i represent some constants each of which is less then 1. Therefore, $z\mathbb{T}(z)/(1 - z)$ can be represented in the following way

$$\frac{z}{1-z}\mathbb{T}(z) = \frac{1}{1-z}\sum_i \frac{D_i}{1-\alpha_i} - \sum_i \frac{D_i}{(1-\alpha_i)(1-\alpha_i z)}.$$

If after that we take the inverse transformation then we obtain

$$\sum_i \frac{D_i}{1-\alpha_i} - \sum_i \frac{D_i \alpha_i^t}{1-\alpha_i}.$$

Now it is evident that

$$\sum_i \frac{D_i}{1-\alpha_i} = \mathbb{T}(1).$$

In a similar way we can show that the inverse transformation of $z\mathbb{T}(z)/(1 - z)$ is $T(1)$ if the equation $\det(I - zP) = 0$ admits multiple roots.

So, the term $z\mathbb{T}(z)\mu/(1-z)$ in (1.58) expresses the ramp of the magnitude $\mathbb{T}(1)\mu$. The quantity $Q\sigma(0)/(1 - z)$ is a step of magnitude $Q\sigma(0)$ and $\mathbb{T}(z)\sigma(0)$ represents geometric components that vanish if t is large.

Finally, if we take the inverse transformation for $F_\sigma(z)$ then we obtain the asymptotic representation of the expected total cost

$$\sigma(t) = tQ\mu + \mathbb{T}(1)\mu + Q\sigma(0) + \epsilon(t),$$

where $\epsilon(t) \to 0$ if $t \to \infty$.

If we come back to the expression $\omega = Q\mu$ then our formula can be written as follows

$$\sigma(t) = t\omega + \varepsilon + \epsilon(t), \tag{1.59}$$

where

$$\varepsilon = \mathbb{T}(1)\mu + Q\sigma(0)$$

and ω represents the limiting vector of the average costs; a component ω_j of the vector ω expresses the average cost per transition of the system if it starts transitions in the state x_j. The component $\mathbb{T}(1)$ can be determined using $O(n^4)$ elementary operations based on the approach described in the Sects. 1.3 and 1.4.

Below we give an example which illustrates how to calculate the expected total cost in Markov processes based on the formula above.

Example 1 Consider the Markov process with the corresponding matrices of probability transitions and cost transitions

$$P = \begin{pmatrix} 0.5 & 0.5 \\ 0.4 & 0.6 \end{pmatrix}; \quad C = \begin{pmatrix} 9 & 3 \\ 3 & -7 \end{pmatrix}.$$

We are seeking for the estimation of the vector of the expected total costs $\sigma(t)$ and the vector of average costs per transition in this Markov process.

First of all we find the vector $\mu = \begin{pmatrix} \mu_1 \\ \mu_2 \end{pmatrix}$ where $\mu_1 = 0.5 \cdot 9 + 0.5 \cdot 3 = 6$, $\mu_2 = 0.4 \cdot 3 - 0.6 \cdot 7 = -3$, i.e., $\mu = \begin{pmatrix} 6 \\ -3 \end{pmatrix}$. If we take σ identically equal to zero then we have

$$\sigma(t) = tQ\mu + \mathbb{T}(1)\mu. \tag{1.60}$$

In Sect. 1.1.4 it is shown that for our example we have

$$(I - zP)^{-1} = \frac{1}{1-z}\begin{pmatrix} \frac{4}{9} & \frac{5}{9} \\ \frac{4}{9} & \frac{5}{9} \end{pmatrix} + \frac{1}{1 - \frac{1}{10^t}}\begin{pmatrix} \frac{5}{9} & -\frac{5}{9} \\ -\frac{4}{9} & \frac{4}{9} \end{pmatrix} = \frac{1}{1-z}Q + \mathbb{T}(z)$$

that implies

$$Q = \begin{pmatrix} \frac{4}{9} & \frac{5}{9} \\ \frac{4}{9} & \frac{5}{9} \end{pmatrix}; \quad \mathbb{T}(1) = \begin{pmatrix} \frac{50}{81} & -\frac{50}{81} \\ -\frac{40}{81} & \frac{40}{81} \end{pmatrix}.$$

If we introduce S, $\mathbb{T}(1)$ in (1.60) and take into account that $\mu = \begin{pmatrix} 6 \\ -3 \end{pmatrix}$ then we obtain

$$\sigma(t) = t\begin{pmatrix} 1 \\ 1 \end{pmatrix} + \begin{pmatrix} \frac{50}{9} \\ \frac{40}{9} \end{pmatrix}.$$

So,

$$\sigma_1(t) = t + \frac{50}{9}, \quad \sigma_2(t) = t - \frac{40}{9}.$$

The vector of average cost per transition ω is determined as follows:

$$\omega = Q\mu = \begin{pmatrix} 1 \\ 1 \end{pmatrix}.$$

A more detailed interpretation of this example can be found in [47].

1.7.3 Determining the Expected Total Cost for Non-stationary Markov Processes

Consider a Markov process with a given matrix of probabilities P and a cost matrix C where the elements of these matrices depend on time, i.e., $p_{x,y} = p_{x,y}(t)$, $c_{x,y} = c_{x,y}(t)$ and $\sum_{y \in X} p_{x,y}(t) = 1$ for every $t = 0, 1, 2. \ldots$. So, we consider a non-stationary Markov process determined by the dynamic matrices $P(t) = (p_{x,y}(t))$ and $C(t) = (c_{x,y}(t))$. For an arbitrary $x \in X$ the expected total cost $\sigma_x(t)$ during t transitions of the system if it starts transitions in x can be calculated on the bases of the following recursive formula

$$\sigma_x(\tau) = \sum_{y \in X} p_{x,y}(t-\tau)(c_{x,y}(t-\tau) + \sigma_y(\tau-1)), \quad \tau = 1, 2, \ldots, t$$

starting with $\sigma_x(0) = 0$ for every $x \in X$. After t steps we obtain $\sigma_x(t)$. Note that in the recursive formula above $\sigma_x(\tau)$ expresses the expected total cost during τ transitions of the system if it starts transitions in x at the moment of time $t - \tau$. Based on this calculation procedure we can show that the following formula for $\sigma(t)$ holds

$$\begin{aligned} \sigma(t) = {} & \mu(0) + P(0)\mu(1) + P(0)P(1)\mu(2) + \ldots \\ & + P(0)P(1)P(2)\cdots P(t-2)\mu(t-1), \end{aligned}$$

where

$$\mu(\tau) = \begin{pmatrix} \mu_1(\tau) \\ \mu_2(\tau) \\ \vdots \\ \mu_n(\tau) \end{pmatrix}, \quad \tau = 0, 1, 2, \ldots, t-1,$$

are the column vectors with the components

$$\mu_j(\tau) = \sum_{y\in X} p_{x_j,y}(\tau)c_{x_j,y}(\tau), \quad \tau = 0, 1, 2, \dots, t-1; \quad j = 1, 2, \dots, n.$$

This formula can easily be proved by using the induction principle on the number of transitions t. The value $\mu_j(\tau)$ in this formula expresses the immediate cost of the system in the state x at the moment of time τ.

1.7.4 Definition and Calculation of the Variance of the Total Cost

For a Markov process with probability matrix $P = (p_{x,y})$ and associated cost matrix $C = (c_{x,y})$ we can define the variance $D_{x_{i_0}}(t)$ of the total cost during t transitions for the dynamical system if it starts transitions in the state $x(0) = x_{i_0}$ at the moment of time $t = 0$ [34, 115]. For an arbitrary state $x \in X$ and an arbitrary t we define and calculate the variance $D_x(t)$ during t transitions using the following recursive formula

$$D_x(\tau+1) = \sum_{y\in X} p_{x,y}((c_{x,y} - \mu_x)^2 + D_y(\tau)), \quad \tau = 0, 1, 2, \dots, t-1$$

where

$$\mu_x = \sum_{w\in X} p_{x,w}c_{x,w}$$

and $D_x(0) = 0$ for every $x \in X$. This formula can be treated in the same way as the formula for the expected total cost. The variance for the non-stationary case is defined in a similar way.

If we denote the column vector by $D(t)$ with components $D_x(t)$ and the column vector by μ^v with components

$$\mu_x^v = \sum_{y\in X}(c_{x,y} - \mu_x)^2$$

then the formula above in the extended form can be expressed as follows

$$D(t) = P^0\mu^v + P^1\mu^v + P^2\mu^v + \dots + P^{t-1}\mu^v.$$

In a similar way as for the average cost it can be shown that for $D(t)$ there exists

$$\overline{D} = \lim_{t\to\infty} \frac{1}{t} D(t)$$

and $\overline{D} = Q\mu^{v}$. A component $\overline{D}_x$ of vector D can be treated as the limiting average variance for the dynamical system if it starts transitions in the state x [114]. So, for the case of a large number of transitions the variance vector $D(t)$ can be expressed as

$$D(t) = t\overline{D} + \varepsilon' + \epsilon'(t),$$

where

$$\varepsilon' = \mathbb{T}(1)\mu^{v} + QD(0)$$

and $\epsilon'(t)$ tends to zero if $t \rightarrow \infty$, i.e., $D(t)$ for a large t the value $D(t)$ can be approximated with $t\overline{D}$, i.e., $D(t) \approx tQ\mu^{v}$.

1.8 Markov Processes with Discounted Costs

Consider a Markov Process with a given stochastic matrix of transition probabilities $P = (p_{x,y})$ and a matrix of transition costs $C = (c_{x,y})$. Assume that the future costs of states' transitions of the dynamical system from one state to another in the considered process are discounted according to a given *discount factor* (discount rate) γ, where $0 < \gamma < 1$. This means that at the moment of time t the cost of the system's transition from a state $x \in X$ to a state $y \in X$ is $c_{x,y}(t) = \gamma^t c_{x,y}$. For such a process the expected total discounted cost during t transitions if the system starts transitions in a state $x \in X$ is defined and calculated on the basis of the following recursive formula

$$\sigma_x(\tau+1) = \sum_{y\in X} p_{x,y}(c_{x,y} + \gamma\sigma_y(\tau)), \quad \tau = 0, 1, 2, \dots t-1.$$

Using the vector notations from Sect. 1.7.1 this formula can be expressed in the matrix form

$$\sigma(\tau+1) = \mu + \gamma P\sigma(\tau), \quad \tau = 0, 1, 2, \dots t-1,$$

where $\sigma(0)$ is the vector with zero components. This implies that the expected total discounted cost during t transitions can be calculated by applying the formula

$$\sigma(t) = P^0\mu + \gamma P^1\mu + \gamma^2 P^2\mu + \cdots + \gamma^{t-1}P^{t-1}\mu, \tag{1.61}$$

where μ is the vector with the components $\mu_i = \sum_{y\in X} p_{x_i,y}c_{x_i,y}$, $i = 1, 2, \dots, n$.

In the following we can see that for Markov processes with discounted costs the limit $\lim_{t\rightarrow\infty}\sigma(t) = \sigma < \infty$ always exists (see [47, 135, 140]). Using this property we can take the limit in the relation $\sigma(t+1) = \mu + \gamma P\sigma(t)$ for $t \rightarrow \infty$ and we obtain the equation

$$\sigma = \mu + \gamma P\sigma.$$

For $0 \le \gamma < 1$ this equation has a unique solution with respect to σ. So, the vector of limiting expected total discounted costs σ can be found by solving the system of linear equations

$$(I - \gamma P)\sigma = \mu$$

where $\det(I - \gamma P) \neq 0$ for $0 \le \gamma < 1$.

To prove the properties mentioned above and to study the asymptotic behavior of the expected total discounted costs in Markov processes we will use again the z-transform.

If we apply the z-transform to the relation $\sigma(t+1) = \mu + \gamma P \sigma(t)$ then we have

$$z^{-1}(F_\sigma(z) - F_\sigma(0)) = \frac{1}{1-z}\mu + \gamma P F_\sigma(z).$$

After rearrangement we obtain

$$F_\sigma(z) = \frac{z}{1-z}(I - \gamma z P)^{-1}\mu + (I - \gamma z P)^{-1} F_\sigma(0).$$

In this case the inverse matrix of $(I - \gamma z P)$ can be represented as follows

$$(I - \gamma z P)^{-1} = \frac{1}{1 - \gamma z} Q + \mathbb{T}(\gamma z). \tag{1.62}$$

Thus,

$$F_\sigma(z) = \frac{z}{(1-z)(1-\gamma z)} Q\mu + \frac{z}{1-z}\mathbb{T}(\gamma z)\mu + \left(\frac{1}{1-\gamma z} Q + \mathbb{T}(\gamma z)\right) F_\sigma(0).$$

If we substitute here

$$\frac{z}{(1-z)(1-\gamma z)} = \frac{1}{1-\gamma}\left(\frac{1}{1-z} - \frac{1}{1-\gamma z}\right)$$

then we can find the inverse z-transform of the component

$$\frac{z}{(1-z)(1-\gamma z)} Q\mu;$$

the pre-image of this component corresponds to

$$\frac{1}{1-\gamma} Q\mu - \frac{\gamma^t}{1-\gamma} Q\mu.$$

It is easy to observe that the inverse transform of $z\mathbb{T}(z)/(1-z)$ is $T(\gamma)$. Indeed, if all roots of the characteristic polynomial $\det(I - zP) = 0$ are different then

$$\mathbb{T}(\gamma z) = \sum_i \frac{D_i}{1 - \alpha_i \gamma z},$$

where D_i represent matrices that do not depend on z and α_i are the corresponding proper values. Therefore,

$$\frac{z}{1-z}\mathbb{T}(z) = \frac{1}{1-z}\sum_i \frac{D_i}{1-\alpha_i\gamma z} - \sum_i \frac{D_i}{(1-\alpha_i\gamma)(1-\alpha_i\gamma z)}.$$

Based on this formula for $z\mathbb{T}(z)/(1-z)$ we can find the inverse z-transform which has the form

$$\sum_i \frac{D_i}{1-\alpha_i\gamma} - \sum_i \frac{D_i(\alpha_i\gamma)^t}{1-\alpha_i\gamma}.$$

This allows us to conclude that

$$\sum_i \frac{D_i}{1-\alpha_i\gamma} = T(\gamma).$$

In an analogues way we can show that the inverse z-transform of $z\mathbb{T}(z)/(1-z)$ is $T(\gamma)$ if the equation $\det(I - zP) = 0$ admits multiple roots. Finally, we can observe also that all coefficients in the representation of $F_\sigma(0)$ tend to zero for $t \to \infty$. Therefore, for large t the expected total discounted cost can be expressed by the following asymptotic formula

$$\sigma(t) = \frac{1}{1-\gamma} Q\mu + T(\gamma)\mu + \epsilon(t),$$

where $\epsilon(t) \to 0$ for $t \to \infty$. If we take the relation (1.62) into account then we obtain that for $\sigma(t)$ there exists $\lim_{t\to\infty}\sigma(t) = \sigma$ and

$$\sigma = (I - \gamma P)^{-1}\mu, \tag{1.63}$$

where

$$(I - \gamma P)^{-1} = \frac{1}{1-\gamma} Q + T(\gamma).$$

So, if $0 < \gamma < 1$ then there exists $(I - \gamma P)^{-1}$ and the vector of limiting discounted costs σ can be calculated by using formula (1.62).

In general formula (1.62) can be obtained from the recursive formula (1.61) for $t \to \infty$. Indeed for $0 < \gamma < 1$ we have

$$\lim_{t\to\infty} \sigma(t) = \sum_{\tau=0}^{\infty} (\gamma P)^{\tau} \mu.$$

The matrix P is stochastic and $0 < \gamma < 1$. Therefore, all proper values are less than 1 and we obtain

$$\sum_{t=0}^{\infty} (\gamma P)^{\tau} = (I - \gamma P)^{-1}.$$

This means that $\lim_{t\to\infty} \sigma(t) = \sigma = (I - \gamma)^{-1}\mu$, i.e., formula (1.63) holds.
Example Let a Markov process with discounted costs be given, where $\lambda = 0.8$,

$$P = \begin{pmatrix} 0.5 & 0.5 \\ 0.6 & 0.4 \end{pmatrix}; \quad C = \begin{pmatrix} 5 & 3 \\ 6 & -4 \end{pmatrix},$$

and consider the problem of determining $\sigma = \begin{pmatrix} \sigma_1 \\ \sigma_2 \end{pmatrix}$. This vector can be found according to formula (1.63) or by solving the system of linear equations $(I - \gamma P)\sigma = \mu$, where $\mu = \begin{pmatrix} \mu_1 \\ \mu_2 \end{pmatrix}$. So, if we calculate $\mu_1 = 5 \times 0.5 + 3 \times 0.5 = 4$; $\mu_2 = 6 \times 0.6 - 4 \times 0.4 = 2$ then we find σ_1 and σ_2 by solving the system of linear equations

$$\begin{cases} 0.6\sigma_1 - 0.4\sigma_2 = 4, \\ -0.48\sigma_1 + 0.58\sigma_2 = 2. \end{cases}$$

The solution of this system is $\sigma_1 = 20$, $\sigma_2 = 20$.

1.9 Semi-Markov Processes with Transition Costs

So far we have studied Markov processes in which the time between transitions is constant. Now we will consider a more general class of stochastic discrete processes where the transition time τ from one state $x \in X$ to another state $y \in Y$ is an integer random variable that takes values from a given interval $[1, t']$, i.e., $\tau \in \{1, 2, \dots, t'\}$. Such processes are related to Semi-Markov processes [97]. We consider a stochastic discrete process that is determined by a finite set of states X and a probability function

$$p : X \times X \times \{1, 2, \dots, t'\} \rightarrow [0, 1]$$

that satisfies the condition

$$\sum_{y \in X} \sum_{\tau=1}^{t} p_{x,y,\tau} = 1, \quad \forall x \in X,$$

where $p_{x,y,\tau}$ expresses the probability of the system to pass from the state x to the state y using τ units of time. Let $P_{x_{i_0}}(x, \bar{t})$ denote the probability of the system $\mathbb{L}$ to reach the state x at the moment of time $\bar{t}$ if it starts transitions in the state x_{i_0} at the moment of time $t = 0$. It is easy to observe that this probability for the considered process can be calculated by using the following recursive formula

$$P_{x_{i_0}}(x, t) = \sum_{y \in X} \sum_{\tau=1}^{t} p_{y,x,\tau} P_{x_{i_0}}(y, t - \tau), \quad t = 0, 1, 2, \dots, \bar{t},$$

where $P_{x_{i_0}}(x_{i_0}, 0) = 1$ and $P_{x_{i_0}}(x, 0) = 0$ for $x \in X \setminus \{x_{i_0}\}$. Here we set $p_{x,y,\tau} = 0$ if $\tau > t'$. If in the considered process for arbitrary $x, y \in X$ the probabilities $p_{x,y,\tau}$ satisfy the condition

$$p_{x,y,\tau} = 0, \quad \tau = 2, 3, \dots t', \tag{1.64}$$

then we obtain a Markov process and the formula above for calculating $P_{x_{i_0}}(x, t)$ is transformed into formula (1.1).
For a Semi-Markov process with transition costs along the cost function

$$c : X \times X \times \{1, 2, \dots, t'\} \rightarrow \mathbb{R}$$

is given that determines the values $c_{x,y,\tau}$ for every $x, y \in X$ and $\tau \in \{1, 2, \dots, t\}$. Here $c_{x,y,\tau}$ is the cost of the system $\mathbb{L}$ to pass from the state x to the state y by using τ units of time. The total expected cost $\sigma_x(\bar{t})$ at the moment of time $\bar{t}$ if the system starts transitions in a state x at the moment of time $t = 0$ is a random variable induced by the cost and probability transition functions.

We define and calculate this value using the following formula

$$\sigma_x(t) = \sum_{y \in X} \sum_{\tau=1}^{t} p_{x,y,\tau}(c_{x,y,\tau} + \sigma_y(t - \tau)), \quad t = 1, 2 \dots, \bar{t} - 1, \tag{1.65}$$

where $\sigma_x(0) = 0$, $\forall x \in X$. If the probabilities $p_{x,y,\tau}$ satisfy the condition (1.64) then formula (1.65) becomes formula (1.53) for the calculation of the expected total cost in Markov processes. Formula (1.65) can be written in the following form

$$\sigma_x(t) = \mu_x^t + \sum_{y \in X} \sum_{\tau=1}^{t} p_{x,y,\tau} \sigma_y(t-\tau), \quad t = 1, 2, \ldots, \bar{t},$$

where

$$\mu_x^t = \sum_{y \in X} \sum_{\tau=1}^{t} p_{x,y,\tau} c_{x,y,\tau}$$

represents the immediate cost in the state $x \in X$. In an analogous way as for Markov processes here we can introduce the discount factor γ, $0 < \gamma < 1$. If we assume that the cost at the next discrete moment of time is discounted with the rate γ then we can define the expected total discounted cost $\sigma_x^{\gamma}(\bar{t})$ by using the following recursive formula

$$\sigma_x^{\gamma}(t) = \sum_{y \in X} \sum_{\tau=1}^{t} p_{x,y,\tau}(c_{x,y,\tau} + \gamma^{\tau} \sigma_y(t-\tau)), \quad t = 0, 1, 2 \ldots, \bar{t}.$$

It is easy to observe that for an arbitrary Semi-Markov process with transition costs an auxiliary Markov process with transition costs can be constructed. We obtain this auxiliary Markov process if we represent a transition from the state x to the state y by using $\tau \in \{1, 2, \ldots, t'\}$ units of time in a Semi-Markov process as a sequence of τ unit time transitions via $\tau - 1$ fictive intermediate states $x_1^{\tau}, x_2^{\tau}, \ldots, x_{\tau-1}^{\tau}$. So, we regard a transition from x to y by using τ units of time in the Semi-Markov process as a sequence of transitions

$$x \to x_1^{\tau}, \quad x_1^{\tau} \to x_2^{\tau}, \quad \ldots, \quad x_{\tau-2}^{\tau} \to x_{\tau-1}^{\tau}, \quad x_{\tau-1}^{\tau} \to y,$$

for an auxiliary Markov process where the corresponding transition probabilities are defined as follows:

$$p_{x,x_1^{\tau}} = p_{x,y,\tau}; \quad p_{x_1^{\tau},x_2^{\tau}} = p_{x_2^{\tau},x_3} = \cdots = p_{x_{\tau-2}^{\tau},x_{\tau-1}^{\tau}} = p_{x_{\tau-1}^{\tau},y} = 1.$$

We define the transition costs for the auxiliary Markov process with new fictive states as follows:

$$c_{x,x_1^{\tau}} = c_{x,y,\tau}; \quad c_{x_1^{\tau},x_2^{\tau}} = c_{x_2^{\tau},x_3^{\tau}} = \cdots = c_{x_{\tau-2}^{\tau},x_{\tau-1}^{\tau}} = c_{x_{\tau-1}^{\tau},y} = 0.$$

After the construction above the problem of calculating the probabilities $P_x(y,t)$ and the expected total costs in Semi-Markov processes can be reduced to the problem of calculating the corresponding characteristics for the auxiliary Markov process. Moreover, the proposed approach allows us to determine the limiting state probabilities, the average cost per transition and the expected total cost for Semi-Markov processes with transition costs.

1.10 Determining the Expected Total Cost for Markov Processes with Stopping States

For Markov processes with transition costs the expected total cost may not exists if the system makes transitions indefinitely. In this section we consider a class of Markov processes with transition costs for which the expected total cost exists and can be efficiently calculated.

Let a Markov process determined by the matrix of transition probability $P = (p_{x,y})$ and the matrix of transition costs $C = (c_{x,y})$ be given. In addition, we assume that for the Markov process a stopping state $z \in X$ as it is defined in Sect. 1.5.5 is given. We consider the problem of determining the expected total cost for the dynamical system if it starts transitions in a state $x \in X$ and stops transitions in a given state z as soon as this state is reached. We show that if the stopping state $z \in X$ corresponds to a positive recurrent state of the unichain Markov process then the expected total cost exists for an arbitrary starting state and it can be efficiently calculated. We show also how to define and calculate the expected total cost in a process with a given stopping state for an arbitrary Markov chain with transition costs.

At first let us consider our problem in the case of unichain processes if the stopping state coincides with an absorbing state $z \in X$. In this case the expected total cost for the problem with a stopping state z can be calculated for an arbitrary starting state $x \in X$ considering $c_{z,z} = 0$, $\sigma_z = 0$ and applying the recursive formula (1.54). So, here we have only to fix $c_{z,z} = 0$ in the matrix C if $c_{z,z} \neq 0$ and then to apply formula (1.54). Based on the results from Sect. 1.7.2 we may conclude that in the considered case this iterative procedure is convergent. Thus, the expected total cost σ_x in the unichain Markov process exists for an arbitrary starting state $x \in X$. Therefore, if in the recursive formula (1.54) we take the limit when t tends to ∞ then we obtain the system of linear equations

$$\sigma = \mu + P\sigma, \quad \sigma_{z,z} = 0$$

which has a unique solution. The existence of a unique solution of this system of equations for the unichain Markov process with stopping state $z \in X$ where $c_{z,z} = 0$, can be proved if we represent $\sigma = \mu + P\sigma$ in the following form

$$(I - P)\sigma = \mu. \tag{1.66}$$

The rank of the matrix $(I - P)$ is equal to $n - 1$ because the column vectors of the matrix P that correspond to the states $x \in X \setminus \{z\}$ are linear independent. Therefore, if we add the condition $\sigma_z = 0$ to the system (1.66) then we obtain the system of linear equations

$$(I - P)\sigma = \mu, \quad \sigma_z = 0, \tag{1.67}$$

which has a unique solution because of $\mu_z = 0$. So, we can determine σ_x for every $x \in X$. The solution of the system (1.67) can be found using the well known classical numerical method as well as the iterative procedure from [119, 128]. After replacing $\sigma_{z,z} = 0$ in $\sigma = \mu + P\sigma$ we obtain the system for which the iterative procedure from [119, 128] can be efficiently used for an approximation.

Below an example for calculating the expected total cost in a Markov unichain with an absorbing stopping state is given.

Example 1 Let a Markov unichain with absorbing stopping state $z = 4$ be given, where the probability and the transition matrices are defined as follows:

$$P = \begin{pmatrix} 0.3 & 0.3 & 0.4 & 0 \\ 0.5 & 0 & 0.3 & 0.2 \\ 0 & 0.6 & 0 & 0.4 \\ 0 & 0 & 0 & 1.0 \end{pmatrix}, \quad C = \begin{pmatrix} 2 & 1 & 2 & 0 \\ -3 & 0 & -1 & 1 \\ 0 & 1 & 0 & 2 \\ 0 & 0 & 0 & 0 \end{pmatrix}.$$

So, $X = \{1, 2, 3, 4\}$ and $z = 4$.

Consider the problem of determining the expected total costs when the system starts transitions in the states $x \in X \setminus \{4\}$ and stops transitions in the state $z = 4$, i.e., we have to calculate σ_1, σ_2, σ_3, σ_4, where $\sigma_4 = 0$.
We find

$$\begin{aligned} \mu_1 &= 0, 3 \cdot 2 + 0.3 \cdot 1 + 0.4 \cdot 2 = 1.7, \\ \mu_2 &= 0.5 \cdot (-3) + 0.3 \cdot (-1) + 0.2 \cdot 1 = -1.6, \\ \mu_3 &= 0.6 \cdot 1 + 0.4 \cdot 2 = 1.4 \end{aligned}$$

and solve the system of linear equations (1.67), i.e., we determine the solution of the following system of linear equations

$$\begin{cases} 0.7\sigma_1 - 0.3\sigma_2 - 0.4\sigma_3 = 1.7, \\ -0.5\sigma_1 + \sigma_2 - 0.3\sigma_3 - 0.2\sigma_4 = -1.6, \\ -0.6\sigma_2 + \sigma_3 - 0.4\sigma_4 = 1.4, \\ \sigma_4 = 0. \end{cases}$$

The solution of this system is

$$\sigma_1 = 4, \quad \sigma_2 = 1, \quad \sigma_3 = 2, \quad \sigma_4 = 0.$$

If the stopping state $z \in X$ corresponds to a positive recurrent state in the unichain Markov process then the problem of determining the expected total cost can be calculated by modifying the state z into an absorbing state. This means that the matrix P is changed into a new matrix P' which is obtained from P where the elements $p_{z,y}$

that correspond to the row z are changed as follows: $p_{z,z} = 1$, $p_{z,y} = 0, \forall y\ X \setminus \{z\}$. In such a way we obtain a new unichain Markov process with a new probability transition matrix, where z is an absorbing state and we can calculate the expected total cost for an arbitrary starting state $x \in X$.

For defining and calculating the expected total cost of first hitting a given state $z \in X$ of an arbitrary Markov process it is necessary to introduce an additional hypothesis related to probability transitions because the first hitting probability $\pi'_{x,z}$ for some $x \in X$ may be less then one or equal to zero. It is evident that if $\pi'_{x,z} = 0$ then the expected total cost σ_x in the considered Markov process with stopping state z makes no sense. Therefore, we can consider the problem of calculating σ_x only for $x \in \overline{X} = X \setminus X^0$, where

$$X^0 = \{x \in X \mid \pi_{x,z} = 0\}.$$

This means that the expected total cost in the considered process with a given stopping state z can be calculated by using only the matrices $\overline{P} = (p_{x,y}), \overline{C} = (c_{x,y})$ generated by the elements $p_{x,y}$ and $c_{x,y}$ for $x, y \in \overline{X}$. Here $\overline{P}$ is a submatrix of P and the condition $\sum_{y \in \overline{X}} p_{x,y} = 1$ for some $x \in \overline{X}$ may fail to hold. We obtain the condition $\sum_{y \in \overline{X}} p_{x,y} = 1, \forall x \in \overline{X}$ if we change the matrix $\overline{P}$ into a new matrix $\overline{P}' = (p'_{x,y})$, where

$$p'_{x,y} = \frac{1}{\sum\limits_{v \in \overline{X}} p_{x,v}} p_{x,y}, \quad \forall x \in \overline{X} \setminus \{z\}; \quad p_{z,z} = 1; \quad p_{z,x} = 0, \quad \forall x \in \overline{X} \setminus \{z\}.$$

Here the probabilities $p'_{x,y}$ can be treated as the conditional probability of the system's transition from the state x to the state y in the case if the state z is reached, i.e., when $\pi'_{x,y} = 1$. Such a transformation of the probabilities allows us to estimate the expected total cost if the dynamical system stops transitions in the state z. Therefore, in this case for each state $x \in \overline{X}$ the corresponding probability transitions $p_{x,y}$ from x for $y \in \overline{X}$ and $p'_{x,y}$ from x for $y \in \overline{X}$ should be proportional. It is easy to observe that after the transformation mentioned above the immediate costs μ'_x and the probability of first hitting the absorbing state in the auxiliary problem satisfy the following properties:

- $\min\limits_{y \in \overline{X}} c'_{x,y} \leq \mu_x \leq \max\limits_{y \in \overline{X}} c'_{x,y}, \quad \forall x \in \overline{X}$;
- the probability of first hitting the state z from arbitrary $x \in \overline{X}$ is equal to 1.

Thus, to calculate the expected total cost in a Markov process with stopping state z we have to construct the auxiliary Markov process with probability matrix $\overline{P}$ and cost matrix $\overline{C}$, then to determine the matrices $\overline{P}'$ and $\overline{C}'$, where $\overline{C}'$ is obtained from $\overline{C}$ by fixing $c_{z,x} = 0, \forall x \in \overline{X}$ and after that to calculate the expected total cost of first hitting the state z from an arbitrary state $x \in \overline{X}$. These values express the corresponding expected total costs in the initial Markov process with stopping state z.

Example 2 Consider the problem of determining the expected total costs σ_x for the Markov process with the matrices $P,\ C$ from Example 1 if $z = 3$.

For this example we have $X^0 = \{4\}$ and $\overline{X} = \{1, 2, 3\}$ because $\pi'_{4,3} = 0$ and $\pi'_{x,3} \neq 0$ for $x = 1, 2, 3$. Thus, we obtain

$$\overline{P} = \begin{pmatrix} 0.3 & 0.3 & 0.4 \\ 0.5 & 0 & 0.3 \\ 0 & 0.6 & 0 \end{pmatrix}, \quad \overline{C} = \begin{pmatrix} 2 & 1 & 2 \\ -3 & 0 & -1 \\ 0 & 1 & 2 \end{pmatrix}.$$

Using the procedures mentioned above we determine

$$\overline{P}' = \begin{pmatrix} 0.3 & 0.3 & 0.4 \\ 0.625 & 0 & 0.375 \\ 0 & 0 & 1 \end{pmatrix}, \quad \overline{C}' = \begin{pmatrix} 2 & 1 & 2 \\ -3 & 0 & -1 \\ 0 & 0 & 0 \end{pmatrix}.$$

For the auxiliary Markov process with the matrices $\overline{P}'$ and $\overline{C}'$ we solve the system of linear equations $(I - \overline{P}')\sigma' = \mu', \sigma'_z = 0$.
So, we calculate

$$\begin{aligned} \mu'_1 &= 0, 3 \cdot 2 + 0.3 \cdot 1 + 0.4 \cdot 2 = 1.7, \\ \mu'_2 &= 0, 625 \cdot (-3) + 0.375 \cdot (-1) = 2.25, \\ \mu'_3 &= 0, \end{aligned}$$

and solve the following system of linear equations

$$\begin{cases} 0.7\sigma'_1 - 0.3\sigma'_2 - 0.4\sigma'_3 = 1.7, \\ -0.625\sigma'_1 + \sigma'_2 - 0.375\sigma'_3 = -2.25, \\ \sigma'_3 = 0. \end{cases}$$

The solution of this system is $\sigma'_1 = 1.041$, $\sigma'_2 = 1.5994$, $\sigma'_3 = 0$ and for the initial problem we can take $\sigma_1 = \sigma'_1, \sigma_2 = \sigma'_2, \sigma_3 = \sigma'_3$. These values express the expected total cost from $x \in \{1, 2, 3\}$ to $z = 3$ in the case if the system stops transitions in $z = 3$.

The expected total cost for the discounted Markov processes with stopping state z can be determined by using the following system of linear equations

$$(I - \gamma P)\sigma = \mu, \quad \sigma_z = 0.$$

If $z \in X$ is an absorbing state for a Markov unichain with $c_{z,z} = 0$ then this system has a unique solution for an arbitrary $\gamma \in (0, 1]$.

Chapter 2
Stochastic Optimal Control Problems and Markov Decision Processes with Infinite Time Horizon

The aim of this chapter is to develop methods and algorithms for determining the optimal solutions of stochastic discrete control problems and Markov decision problems with an infinite time horizon. We denote such methods and algorithms on the bases of the results from the previous chapter and classical optimization methods. The set of states of the system in the considered problems is finite and the starting state is fixed. We study the stochastic discrete processes that may be controlled in some dynamical states. The average and the expected total discounted costs optimization principles for such processes are applied and new classes of a stochastic control model are formulated. Based on such a concept we study a class of stochastic discrete control problems that emphasis Markov decision problems and deterministic optimal control problems with an infinite time horizon. We obtain the stochastic versions of classical discrete control problems assuming that the dynamical system in the control process may admit dynamical states in which the vector of control parameters is changing in a random way according to given distribution functions of the probabilities on given feasible sets. So, in the considered control problems we assume that the dynamics of the system may contain controllable states as well as uncontrollable states. These problems are formulated on networks and polynomial time algorithms for determining their optimal solutions are proposed. In the case that the dynamical system contains only controllable states the proposed algorithms become algorithms for determining the optimal stationary strategies of the classical deterministic control problems with an infinite time horizon. The proposed methods and algorithms are extended to Markov decision processes.

We develop a linear programming approach to Markov decision processes and show how to use the duality theory for determining solutions of the decision problems with average and expected total discounted optimization criteria. Based on such an approach we describe algorithms for solving new classes of stochastic discrete optimization problems. Polynomial time algorithms for Markov decision problems with average and expected total discounted costs optimization criteria are proposed and formulated.

Furthermore, some numerical examples are given and the computational complexity aspects of the described methods and algorithms are analyzed.

D. Lozovanu and S. Pickl, *Optimization of Stochastic Discrete Systems and Control on Complex Networks*, Advances in Computational Management Science 12,
DOI 10.1007/978-3-319-11833-8_2

2.1 Problem Formulation and the Main Concept of Optimal Control Models with Infinite Time Horizon

The infinite horizon decision problem can be regarded as approximation model for decision problems with an finite time horizon in the case of a large sequence of decisions. Often, it is easier to solve the infinite horizon problem and to use the solution of this to obtain a solution of the finite horizon problem with a large number of decisions. The Markov decision processes and the classical control problems with infinite time horizon are related to such kind of models that are widely used for studying and solving many practical finite horizon decision problems. Below we formulate a class of stochastic discrete optimal control problems with average and expected total discounted costs optimization criteria that combine the statements of deterministic optimal control problems with infinite time horizon and Markov decision processes [5, 114]. We start with a formulation of the stochastic optimal control problem that represents a generalization of the following deterministic control model.

Let a discrete dynamical system $\mathbb{L}$ with a finite set of states $X \subset \mathbb{R}^n$ be given where at every time-step $t = 0, 1, 2, \ldots$, the state of the system $\mathbb{L}$ is $x(t) \in X$. At the starting moment of time $t = 0$ the state of the dynamical system $\mathbb{L}$ is $x(0) = x_0$. Assume that the dynamics of the system $\mathbb{L}$ is described by the system of difference equations

$$x(t+1) = g_t(x(t), u(t)), \quad t = 0, 1, 2, \ldots \tag{2.1}$$

where

$$x(0) = x_0 \tag{2.2}$$

and

$$u(t) = (u_1(t), u_2(t), \ldots, u_m(t)) \in \mathbb{R}^m$$

represents the *vector of the control parameters* (see [6, 11, 132]). For any time step t and an arbitrary state $x(t) \in X$ the feasible set $U_t(x(t))$ of the vector $u(t)$ of control parameters is given, i.e.,

$$u(t) \in U_t(x(t)), \quad t = 0, 1, 2, \ldots. \tag{2.3}$$

We assume that in (2.1) the vector functions

$$g_t(x(t), u(t)) = (g_t^1(x(t), u(t)),\ g_t^2(x(t), u(t)), \ldots,\ g_t^n(x(t), u(t)))$$

are determined uniquely by $x(t)$ and $u(t)$ at every time step $t = 0, 1, 2, \ldots$. So, $x(t+1)$ is determined uniquely by $x(t)$ and $u(t)$.

Additionally, we assume that at each moment of time t the cost

$$c_t(x(t), x(t+1)) = c_t(x(t),\ g_t(x(t), u(t)))$$

of the system's transition from the state $x(t)$ to the state $x(t+1)$ is known.

Let

$$x_0 = x(0), x(1), x(2), \ldots, x(t), \ldots$$

be a trajectory generated by given vectors of the control parameters

$$u(0), u(1), \ldots, u(t-1), \ldots .$$

Then after a fixed number of transitions τ of the dynamical system we can calculate the *integral-time cost (total cost)* which we denote by $F^{\tau}_{x0}(u(t))$, i.e.,

$$F^{\tau}_{x0}(u(t)) = \sum_{t=0}^{\tau-1} c_t(x(t),\ g_t(x(t), u(t))). \tag{2.4}$$

In [6, 11] the following discrete optimal control problem with finite time horizon has been considered: Find for given τ the vectors of control parameters

$$u(0), u(1), u(2), \ldots, u(\tau-1)$$

which satisfy the conditions (2.1)–(2.3) and minimize the functional (2.4). The solution of this optimal control problem can be found by using *dynamic programming techniques* [6, 79].

Here we consider the *infinite horizon control model*. We assume that τ is not bounded, i.e., $\tau \to \infty$. It is evident that if $\tau \to \infty$ then the integral-time cost

$$\lim_{\tau\to\infty} \sum_{t=0}^{\tau-1} c_t(x(t),\ g_t(x(t), u(t)))$$

for a given control may not exist. Therefore, we study in this case the asymptotic behavior of the integral-time cost $F^{\tau}_{x0}(u(t))$ by a trajectory determined by a feasible or an optimal control. To estimate this value we apply the concept from [5, 6], i.e., for a fixed control u if τ is too large we estimate $F^{\tau}_{x0}(u(t))$ asymptotically using the function $\phi_u(\tau) = K\varphi(\tau)$ such that

$$\lim_{\tau\to\infty} \frac{1}{\varphi(\tau)} \sum_{t=0}^{\tau-1} c_t(x(t),\ g_t(x(t), u(t))) = K, \tag{2.5}$$

where K is a constant.

So, in control problems with an infinite time horizon we are seeking for a control u^* with a suitable limiting function $\phi_{u^*}(\tau)$.

Based on the asymptotic approach mentioned above we may conclude that for a given control, if τ is too large, the value $F^{\tau}_{x0}(u(t))$ can be approximated by $K\varphi(\tau)$.

Moreover, we can see that for the stationary case of the control model with the costs that do not depend on time the function $\phi_u(\tau)$ is linear. This means that $\varphi(\tau) = \tau$ and $F^{\tau}_{x0}(u(t))$ for a large τ can be approximated by $\phi_u(\tau) = K\tau$.

In the following we study only stationary control problems. For such problems the vector functions g_t and the feasible sets $U_t(x(t))$ do not depend on time, i.e., $g_t(x,u) = g(x,u)$ and $U_t(x) = U(x)$, $\forall x \in X, t = 0, 1, 2, \ldots$. Moreover, the control at every discrete moment of time depends only on the state $x \in X$ and the cost of the system's transition from the state $x \in X$ to the state $y \in Y$ does not depend on time, i.e., $c_t(x(t), x(t+1)) = c(x, y)$, $\forall x, y \in X$ and every $t = 0, 1, 2, \ldots$ if $x = x(t)$, $y = x(t+1)$.

Thus, for the considered stationary control problems the integral-time cost by a trajectory during τ transitions can be asymptotically expressed as $F_{x_0}^{\tau}(u(t)) = K\varphi(\tau)$, where $\varphi(\tau) = \tau$. In this case for the dynamical system $\mathbb{L}$ the constant K in (2.5) expresses the *average cost per transition* along a trajectory determined by the control $u(t)$. Therefore, for the infinite horizon optimal control problem the objective function which has to be minimized is defined as follows:

$$F_{x_0}(u(t)) = \lim_{\tau \to \infty} \frac{1}{\tau} \sum_{t=0}^{\tau-1} c(x(t),\ g(x(t), u(t))). \tag{2.6}$$

In [5] it is shown that for the stationary case of the problem the optimal control u^* does not depend on time or on the starting state and it can be found in the set of stationary controls.

Another class of control problems with an infinite time horizon which is widely used for practical problems is characterized by a discounting objective cost function [8]

$$\widehat{F}_{x_0}(u(t)) = \sum_{t=0}^{\infty} \gamma^t c_t(x(t),\ g_t(x(t), u(t))). \tag{2.7}$$

Here γ is a *discount factor* that satisfies the condition $0 < \gamma < 1$ and $\widehat{F}_{x_0}(u(t))$ is called the *total discounted cost*. In a control problem with such an optimization criterion we are seeking for the control which minimizes the functional (2.7).

In [28, 114, 129, 140] it is shown that if $0 < \gamma < 1$ and the costs $c_t(x(t), g_t(x(t), u(t)))$ are bounded then for the stationary case of the control problem with a discounted objective optimization criterion the optimal stationary control exists.

The problems formulated above correspond to deterministic models in which the decision maker is able to fix the vector of control parameters $u(t)$ from a given feasible set $U_t(x(t))$ in each dynamical state $x(t)$; the states $x(t) \in X$ in these models are called *controllable states*.

The main results we describe in the following are related to stochastic versions of the control problems formulated above. We consider the control models in which the dynamical system in the control process may admit dynamical states $x(t)$ where the corresponding vector of control parameters $u(t)$ is changed in a random way according to given distribution functions

$$p : U_t(x(t)) \to [0, 1], \quad \sum_{i=1}^{k(x(t))} p\left(u^i_{x(t)}\right) = 1 \tag{2.8}$$

on the corresponding dynamical feasible set $U_t(x(t))$. Here $k(x(t)) = |U_t(x(t))|$, i.e., we consider the control models with finite feasible sets.

We regard each dynamical state $x(t)$ of the system in the considered control problem as a position (x, t) and we assume that the set of positions

$$Z = \{(x, t) = x(t) \,|x(t) \in X, \quad t = 0, 1, 2, \ldots\}$$

is divided into two subsets

$$Z = Z^C \cup Z^N, \quad Z^C \cap Z^N = \emptyset$$

such that Z^C corresponds to the set of *controllable states* and Z^N corresponds to the set of *uncontrollable states*. This means that for the stochastic control problems we have the following behavior of the dynamics in the control process: If the starting state $x(0)$ belongs to the set of controllable states Z^C then the decision maker fixes the vector of control parameters $u(0)$ from the feasible set $U_0(x(0))$ and we obtain the next state $x(1)$; if the state $x(0)$ belongs to the set Z^N then the system passes to the next state $x(1)$ in a random way. If at the moment of time $t = 1$ the state $x(1)$ belongs to the set of controllable states Z^C then the decision maker fixes the vector of control parameters $u(1)$ from $U_1(x(1))$ and we obtain the next state $x(2)$; if $x(1)$ belongs to the set of uncontrollable states Z^N then the system passes to the next state $x(2)$ in a random way and so on indefinitely.

It is evident that for a fixed control the average cost per transition and the discounted total cost in this process represent the random variables induced by the distribution functions on feasible sets in the uncontrollable states and the control in the controllable states.

To define the *expected average cost per transition* and *expected discounted total cost* in the considered stochastic control problems for a fixed control we will apply the concept of Markov decision processes in the following way:

Let $u'(t) \in U_t(x(t))$ be the given feasible vectors in the controllable states $x(t) \in Z^C$. Then we may assume that we have the following distribution functions

$$p : U_t(x(t)) \to \{0, 1\} \quad \text{ for } x(t) \in Z^C$$

where $p(u'(t)) = 1$ and $p(u(t)) = 0, \forall u(t) \in U_t(x(t)) \backslash \{u'(t)\}$.

These distribution functions in the controllable states together with the distribution functions (2.8) in the uncontrollable states determine a Markov process. For this Markov process with transition probabilities $p_{z,v}$ and transition costs $c_{z,v}$ for $(z, v) \in Z \times Z$ we can determine the expected average and the expected discounted total costs which we denote, respectively, by $F_{x_0}(u(t))$ and $\widehat{F}_{x_0}(u(t))$. In such a way we obtain the corresponding optimization problems in which we are seeking for the controls that minimize the expected average and discounted total costs, respectively.

Thus, we shall use the combined concept of deterministic and stochastic control models from [36, 81–94, 96, 108, 109], and will develop algorithms for determining optimal strategies of the considered problems. Mainly, we will study the stationary versions of the control problems with a finite set of states for the dynamic system and will describe algorithms based on linear programming. In the general case, for non-stationary control problems, the optimal control may not exist. Some special classes of non-stationary problems may admit the solution and the optimal control can be found by using a special calculation procedure.

2.2 An Optimal Stationary Control with an Average Cost Criterion and Algorithms for Solving Stochastic Control Problems on Networks

In this section we consider the stationary stochastic discrete optimal control problem with average cost criterion. We formulate this problem on networks and describe polynomial time algorithms for determining the optimal control by using a linear programming approach.

2.2.1 Problem Formulation

Let a discrete dynamical system $\mathbb{L}$ with a finite set of states X be given, where $|X| = n$. At every discrete moment of time $t = 0, 1, 2, \ldots$ the state of $\mathbb{L}$ is $x(t) \in X$. The dynamics of the system is described by a directed *graph of states' transitions* $G = (X, E)$ where the set of vertices X corresponds to the set of states of the dynamical system and an arbitrary directed edge $e = (x, y) \in E$ expresses the possibility of the system $\mathbb{L}$ to pass from the state $x = x(t)$ to the state $y = x(t + 1)$ at every discrete moment of time t. So, a directed edge $e = (x, y)$ in G corresponds to a stationary control of the system in the state $x \in X$ which provides a transition from $x = x(t)$ to $y = x(t + 1)$ for every discrete moment of time t. We assume that graph G does not contain deadlock vertices, i.e., for each x there exists at least one leaving directed edge $e = (x, y) \in E$. In addition, we assume that to each edge $e = (x, y) \in E$ a quantity c_e is associated which expresses the cost (or the reward [47]) of the system $\mathbb{L}$ to pass from the state $x = x(t)$ to the state $y = x(t)$ for every $t = 0, 1, 2, \ldots$.

The cost c_e for an arbitrary edge $e = (x, y)$ is denoted by $c_{x,y}$. A sequence of directed edges $E' = \{e_0, e_1, e_2, \ldots, e_t, \ldots\}$ where $e_t = (x(t), x(t + 1))$, $t = 0, 1, 2, \ldots$ determines in G a control of the dynamical system with a fixed starting state $x_0 = x(0)$. An arbitrary control in G generates a trajectory $x_0 = x(0), x(1), x(2), \ldots$ for which the average cost per transition can be defined in the following way

$$f(E') = \lim_{t\to\infty} \frac{1}{t} \sum_{\tau=0}^{t-1} c_{e_\tau}.$$

In [5] it is shown that this value exists and $|f_{x_0}(E')| \leq \max_{e \in E'} |c_e|$. Moreover, in [5] it is shown that if G is strongly connected then for an arbitrary fixed starting state $x_0 = x(0)$ there exists the optimal control $E^* = \{e_0^*, e_1^*, e_2^* \ldots\}$ for which

$$f(E^*) = \min_{E'} \lim_{t \to \infty} \frac{1}{t} \sum_{\tau=0}^{t-1} c_{e_\tau}$$

and this optimal control does not depend either on the starting state or on time. Therefore, the optimal control for this problem can be found in the set of stationary strategies $\mathbb{S}$. A *stationary strategy* in G is defined as a map:

$$s : x \to y \in X(x) \quad \text{for } x \in X,$$

where $X(x) = \{y \in X \mid e = (x, y) \in E\}$.

Let s be a stationary strategy. Denote by $G_s = (X, E_s)$ the subgraph of G generated by edges of the form $e = (x, s(x))$ for $x \in X$. Then it is easy to observe that in G_s there exists a unique directed cycle C_s which can be reached from x_0 through the directed edges from E_s. Moreover, we can see that the mean cost of this cycle is equal to the average cost per transition of the dynamical system by the trajectory generated by the stationary strategy s. Thus, if G is a strongly connected directed graph then the problem of determining the optimal control on G is equivalent to the problem of finding in G the cycle C_G^* for which

$$\frac{\sum_{e \in E(C_G^*)} c_e}{n(C_G^*)} = \min_{C_G} \frac{\sum_{e \in E(C_G)} c_e}{n(C_G)},$$

where $E(C_G)$ is the set of directed edges of the directed cycle C_G in G that can be reached from a starting vertex and $n(C_G)$ is the number of its edges. If the cycle C_G^* is known then the optimal control for an given arbitrary starting state $x_0 = x(0)$ in G can be found in the following way: We fix the transitions through the directed edges of the graph in order to reach a vertex of the directed cycle C_G^* and then we preserve transitions through the directed edges of this cycle.

Polynomial and strongly polynomial time algorithms for determining the optimal average cost cycles in a weighted directed graph and the optimal stationary strategies for control problems on networks have already been proposed in [53, 65, 79, 117].

In the following we will consider the stochastic version of the problem formulated above. We assume that the set of states X of the dynamical system may admit states in which the system $\mathbb{L}$ makes transitions to the next state in a random way according to a given distribution function of probabilities on the set of possible transitions from these states. So, the set of states X is divided into two subsets X_C and X_N ($X = X_C \cup X_N$, $X_C \cap X_N = \emptyset$), where X_C represents the set of states $x \in X$ in which the transitions of the system to the next state y can be controlled by the decision maker at every discrete moment of time t and X_N represents the set of states $x \in X$ in which the decision maker is not able to control the transition because the system passes to the next state y randomly. Thus, for each $x \in X_N$ a probability distribution function $p_{x,y}$ on the set of possible transitions (x, y) from x to $y \in X(x)$ is given, i.e.,

$$\sum_{y \in X(x)} p_{x,y} = 1, \quad \forall x \in X_N; \quad p_{x,y} \geq 0, \quad \forall y \in X(x). \tag{2.9}$$

Here $p_{x,y}$ expresses the probability of the system's transition from the state x to the state y for every discrete moment of time t. Note, that the condition $p_{x,y} = 0$ for a directed edge $e = (x, y) \in E$ is equivalent with the condition that G does not contain this edge.

In the same way as for the deterministic problem here we assume that to each directed edge $e = (x, y) \in E$ a cost c_e is associated.

We call the graph G with the properties mentioned above *decision network* and denote it by (G, X_C, X_N, c, p, x_0). So, this network is determined by the directed graph G with a fixed starting state x_0, the subsets X_C, X_N, the cost function $c : E \rightarrow \mathbb{R}$ and the probability function $p : E_N \rightarrow [0, 1]$ on the subset of the edges $E_N = \{e = (x, y) \in E \mid x \in X_N, \; y \in X\}$ where p satisfies the condition (2.9). If the control problem is considered for an arbitrary starting state then we denote the network by (G, X_C, X_N, c, p).

We define a stationary strategy for the control problem on networks as a map:

$$s : x \rightarrow y \in X(x) \quad \text{for} \quad x \in X_C.$$

Let s be an arbitrary stationary strategy. Then we can determine the graph $G_s = (X, E_s \cup E_N)$, where $E_s = \{e = (x, y) \in E \mid x \in X_C, \; y = s(x)\}$, $E_N = \{e = (x, y) \mid x \in X_N, \; y \in X\}$. This graph corresponds to a Markov process with the probability matrix $P^s = (p^s_{x,y})$, where

$$p^s_{x,y} = \begin{cases} p_{x,y}, & \text{if } x \in X_N \text{ and } y \in X; \\ 1, & \text{if } x \in X_C \text{ and } y = s(x); \\ 0, & \text{if } x \in X_C \text{ and } y \neq s(x). \end{cases}$$

In the considered Markov process for an arbitrary state $x \in X_C$ the transition $(x, s(x))$ from the states $x \in X_C$ to the states $y = s(x) \in X$ is made with the probability $p_{x,s(x)} = 1$ if the strategy s is applied. For this Markov process we can determine the average cost per transition for an arbitrary fixed starting state $x_i \in X$ in such a way as we have defined it in Sect. 1.7.2. Thus, we can determine the vector of average costs ω^s which corresponds to the strategy s. As we have shown in Sect. 1.7.2 the vector ω^s can be calculated according to the formula $\omega^s = Q^s \mu^s$, where Q^s is the limit matrix of the Markov process generated by the stationary strategy s and μ^s is the corresponding vector of the immediate costs, i.e., $\mu^s_x = \sum_{y \in X(x)} p^s_{x,y} c^s_{x,y}$. A component ω^s_x of the vector ω^s represents the average cost per transition in our problem with a given starting state x and a fixed strategy s, i.e.,

$$f_x(s) = \omega^s_x.$$

In such a way we can define the value of the objective function $f_{x_0}(s)$ for the control problem on a network with a given starting state x_0 when the stationary strategy s is applied.

The control problem on the network (G, X_C, X_N, c, p, x_0) consists of finding a stationary strategy s^* for which

$$f_{x_0}(s^*) = \min_s f_{x_0}(s).$$

In the next section we can see that the optimal stationary strategy in the considered problem does not depend on the starting state. We show that a polynomial time algorithm for determining the optimal solution of this problem can be elaborated. Moreover, we show that the proposed algorithm can be extended to Markov decision processes.

2.2.2 A Linear Programming Approach for Determining Optimal Stationary Strategies on Perfect Networks

We consider the stochastic control problem on the network (G, X_C, X_N, c, p, x_0) with $X_C \neq \emptyset$, $X_N \neq \emptyset$ and assume that G is a strongly connected directed graph. Additionally, we assume that in G for an arbitrary stationary strategy $s \in \mathbb{S}$ the subgraph $G_s = (X, E_s \cup E_N)$ is strongly connected. This means that the Markov chain induced by the probability transition matrix P^s is irreducible for an arbitrary strategy s. We call the decision network with such a condition a *perfect network*. At first we describe an algorithm for determining the optimal stationary strategies for the control problem on perfect networks. Then we show that the proposed algorithm can be extended for the problem if an arbitrary strategy s generates a Markov unichain. For a unichain control problem the graph G^s induced by a stationary strategy may not be strongly connected but it contains a unique strongly connected component that is reachable from every $x \in X$.

So, in this section we consider the control problem that the average cost per transition is the same for an arbitrary starting state, i.e.,

$$f_x(s) = \omega^s, \quad \forall x \in X.$$

We will consider in the next section the case of a multichain control problem, i.e., the case that for different starting states the average cost per transition may be different.

Let $s \in \mathbb{S}$ be an arbitrary strategy. Taking into account that for every fixed $x \in X_C$ we have a unique $y = s(x) \in X(x)$ then we can identify the map s with the set of boolean values $s_{x,y}$ for $x \in X_C$ and $y \in X(x)$, where

$$s_{x,y} = \begin{cases} 1, & \text{if } y = s(x); \\ 0, & \text{if } y \neq s(x). \end{cases}$$

For the optimal stationary strategy s^* we denote the corresponding boolean values by $s^*_{x,y}$.

Assume that the network (G, X_C, X_N, c, p, x_0) is perfect. Then the following lemma holds.

Lemma 2.1 *A stationary strategy s^* is optimal if and only if it corresponds to an optimal solution q^*, s^* of the following mixed integer bilinear programming problem: Minimize*

$$\psi(s,q) = \sum_{x \in X_C} \sum_{y \in X(x)} c_{x,y} s_{x,y} q_x + \sum_{z \in X_N} \mu_z q_z \tag{2.10}$$

subject to

$$\begin{cases} \sum_{x \in X_C} s_{x,y} q_x + \sum_{z \in X_N} p_{z,y} q_z = q_y, \quad \forall y \in X; \\ \sum_{x \in X_C} q_x + \sum_{z \in X_N} q_z = 1; \\ \sum_{y \in X(x)} s_{x,y} = 1, \quad \forall x \in X_C; \\ s_{x,y} \in \{0,1\}, \ \forall x \in X_C, \ y \in X; \quad q_x \geq 0, \ \forall x \in X, \end{cases} \tag{2.11}$$

where

$$\mu_z = \sum_{y \in X(z)} p_{z,y} c_{z,y}, \quad \forall z \in X_N.$$

Proof Denote $\mu_x = \sum_{y \in X(x)} c_{x,y} s_{x,y}$ for $x \in X_C$. Then μ_x for $x \in X_C$ and μ_z for $z \in X_N$ represent, respectively, the immediate cost of the system in the states $x \in X_C$ and $z \in X_N$ if the strategy $s \in S$ is applied. Indeed, we can treat the values $s_{x,y}$ for $x \in X_C$ and $y \in X(x)$ as probability transitions from the state $x \in X_C$ to the state $y \in X(x)$.

Therefore, for fixed s the solution $q^s = (q^s_{x_{i_1}}, q^s_{x_{i_2}}, \dots, q^s_{x_{i_n}})$ of the system of linear equations

$$\begin{cases} \sum_{x \in X_C} s_{x,y} q_x + \sum_{z \in X_N} p_{z,y} q_z = q_y, \quad \forall y \in X; \\ \sum_{x \in X_C} q_x + \sum_{z \in X_N} q_z = 1; \end{cases} \tag{2.12}$$

corresponds to the vector of limit probabilities in the ergodic Markov chain determined by the graph $G_s = (X, E_s \cup E_N)$ with the probabilities $p_{x,y}$ for $(x,y) \in E_N$ and $p_{x,y} = s_{x,y}$ for $(x,y) \in E_C$ $(E_C = E \setminus E_N)$. Therefore, for given s the value

$$\psi(s, q^s) = \sum_{x \in X_C} \mu_x q_x + \sum_{z \in X_N} \mu_z q_z$$

expresses the average cost per transition for the dynamical system if the strategy s is applied, i.e.,

$$f_x(s) = \psi(s, q^s), \quad \forall x \in X.$$

So, if we solve the optimization problem (2.10), (2.11) on a perfect network then we find the optimal strategy s^*. □

Remark 2.2 In the case of a perfect network the objective function $\psi(s, q)$ on the feasible set of solutions of the system (2.11) depends only on s, because q_x for $x \in X$ can be uniquely expressed via $s_{x,y}$ $(x \in X_C,\ y \in X)$ according to (2.12). Moreover, for perfect networks the condition $q_x \geq 0$ for $x \in X$ in (2.11) holds if $s_{x,y} \geq 0$, $\forall x \in X_C,\ y \in X$. Therefore, the condition $q_x \geq 0$ for $x \in X$ in (2.11) is redundant and can be omitted. This condition is essential only for multichain control problems.

In the following for an arbitrary vertex $y \in X$ we will denote by $X_C^-(y)$ the set of vertices from X_C which contain directed leaving edges $e = (x, y) \in E$ that end in y, i.e., $X_C^-(y) = \{x \in X_C \mid (x, y) \in E\}$; in an analogues way we define the set $X^-(y) = \{x \in X \mid (x, y) \in E\}$.

Based on the lemma above we can prove the following result.

Theorem 2.3 *Let $\alpha^*_{x,y}$ $(x \in X_C,\ y \in X)$, q^*_x $(x \in X)$ be a basic optimal solution of the following linear programming problem:*
Minimize

$$\overline{\psi}(\alpha, q) = \sum_{x \in X_C} \sum_{y \in X(x)} c_{x,y} \alpha_{x,y} + \sum_{z \in X_N} \mu_z q_z \tag{2.13}$$

subject to

$$\begin{cases} \sum\limits_{x \in X_C^-(y)} \alpha_{x,y} + \sum\limits_{z \in X_N} p_{z,y} q_z = q_y, \quad \forall y \in X; \\ \sum\limits_{x \in X_C} q_x + \sum\limits_{z \in X_N} q_z = 1; \\ \sum\limits_{y \in X(x)} \alpha_{x,y} = q_x, \quad \forall x \in X_C; \\ \alpha_{x,y} \geq 0, \ \forall x \in X_C,\ y \in X; \ q_x \geq 0, \ \forall x \in X. \end{cases} \tag{2.14}$$

Then the optimal stationary strategy s^* on a perfect network can be found as follows:

$$s^*_{x,y} = \begin{cases} 1, & \textit{if } \alpha^*_{x,y} > 0; \\ 0, & \textit{if } \alpha^*_{x,y} = 0, \end{cases}$$

where $x \in X_C$, $y \in X(x)$. Moreover, for every starting state $x \in X$ the optimal average cost per transition is equal to $\overline{\psi}(\alpha^*, q^*)$, i.e.,

$$f_x(s^*) = \sum_{x \in X_C} \sum_{y \in X(x)} c_{x,y}\alpha^*_{x,y} + \sum_{z \in X_N} \mu_z q^*_z$$

for every $x \in X$.

Proof To prove the theorem it is sufficient to apply Lemma 2.1 and to show that the bilinear programming problem (2.10), (2.11) with boolean variables $s_{x,y}$ for $x \in X_C$, $y \in X$ can be reduced to the linear programming problem (2.13), (2.14). Indeed, we observe that the restrictions $s_{x,y} \in \{0, 1\}$ in the problems (2.10), (2.11) can be replaced by $s_{x,y} \geq 0$ because the optimal solutions after such a transformation of the problem are not changed. In addition, the restrictions

$$\sum_{y \in X(x)} s_{x,y} = 1, \quad \forall x \in X_C$$

can be changed by the restrictions

$$\sum_{y \in X(x)} s_{x,y} q_x = q_x, \quad \forall x \in X_C$$

because for the perfect network it holds $q_x > 0$, $\forall x \in X_C$.

Based on the properties mentioned above in the problem (2.10), (2.11) we may replace the system (2.11) by the following system

$$\begin{cases} \sum_{x \in X_C^-(y)} s_{x,y} q_x + \sum_{z \in X_N} p_{z,y} q_z = q_y, \quad \forall y \in X; \\ \sum_{x \in X_C} q_x + \sum_{z \in X_N} q_z = 1; \\ \sum_{y \in X(x)} s_{x,y} q_x = q_x, \quad \forall x \in X_C; \\ s_{x,y} \geq 0, \ \forall x \in X_C, \ y \in X; \quad q_x \geq 0, \ \forall x \in X. \end{cases} \tag{2.15}$$

Thus, we may conclude that problem (2.10), (2.11) and problem (2.10), (2.15) have the same optimal solutions. Taking into account that for the perfect network $q_x > 0$, $\forall x \in X$ we can introduce in problem (2.10), (2.15) the notations $\alpha_{x,y} = s_{x,y} q_x$ for $x \in X_C$, $y \in X(x)$. This leads to the problem (2.13), (2.14). It is evident that $\alpha_{x,y} \neq 0$ if and only if $s_{x,y} = 1$. Therefore, the optimal stationary strategy s^* can be found according to the rule given in the theorem. □

Remark 2.4 In Theorem 2.3 the linear programming problem (2.13), (2.14) can be changed by the following equivalent linear programming problem:

Minimize

$$\overline{\psi}(\alpha, q) = \sum_{x \in X_C} \sum_{y \in X(x)} c_{x,y} \alpha_{x,y} + \sum_{z \in X_N} \mu_z q_z \tag{2.16}$$

subject to

$$\begin{cases} \sum\limits_{x \in X_C^-(y)} \alpha_{x,y} - \sum\limits_{x \in X(y)} \alpha_{y,x} + \sum\limits_{x \in X_N} p_{x,y} q_x = 0, \quad \forall y \in X_C; \\ \sum\limits_{x \in X_C^-(y)} \alpha_{x,y} - q_y + \sum\limits_{x \in X_N} p_{x,y} q_x = 0, \quad \forall y \in X_N; \\ \sum\limits_{x \in X_C} \sum\limits_{y \in X(x)} \alpha_{x,y} + \sum\limits_{x \in X_N} q_x = 1; \\ \alpha_{x,y} \geq 0, \ \ \forall x \in X_C, \ y \in X; \ \ q_x \geq 0, \ \ \forall x \in X_N. \end{cases} \tag{2.17}$$

This problem is obtained from (2.13), (2.14) if we take into account Remark 2.2 and eliminate q_x for $x \in X_C$ from (2.14). If we solve this problem then we should take into account that $\alpha_{x,y} = s_{x,y} q_x, \ \forall x \in X_C, \ y \in X(x)$, where $q_x = \sum_{y \in X(x)} \alpha_{x,y}$, $\forall x \in X_C$.

So, if the network (G, X_C, X_N, c, p, x_0) is perfect then we can find the optimal stationary strategy s^* by using the following algorithm.

Algorithm 2.5 Determining the Optimal Stationary Strategy on Perfect Networks

(1) Formulate the linear programming problem (2.13), (2.14) and find a basic optimal solution $\alpha^*_{x,y}$ $(x \in X_C, \ y \in X)$, q^*_x $(x \in X)$.
(2) Fix a stationary strategy s^* where $s^*_{x,y} = 1$ for $x \in X_C, \ y \in X(x)$ if $\alpha^*_{x,y} > 0$; otherwise put $s^*_{x,y} = 0$.

Below an example for determining the optimal control problem on networks by using linear programming is given.

Example Consider a stochastic control problem for which the network is represented in Fig. 2.1, i.e.,

$$G = (X, E), \ \ X = \{1, 2, 3, 4\}, \ \ X_C = \{1, 2\}, \ \ X_N = \{3, 4\},$$
$$E = \{(1, 3), \ (1, 4), \ (2, 3), \ (2, 4), \ (3, 1), \ (3, 4), \ (4, 2), \ (4, 3)\}.$$

The transition cost for directed edges from E and the transition probabilities for directed edges originating in the vertices 3 and 4 are given by:

$$\begin{array}{llll} c_{1,3} = 1, & c_{2,3} = 3, & c_{3,1} = 2, & c_{4,2} = 1, \\ c_{1,4} = 2, & c_{2,4} = 1, & c_{3,4} = 4, & c_{4,3} = 3, \\ p_{3,1} = 0.5, & p_{3,4} = 0.5, & p_{4,2} = 0.5, & p_{4,3} = 0.5\,. \end{array}$$

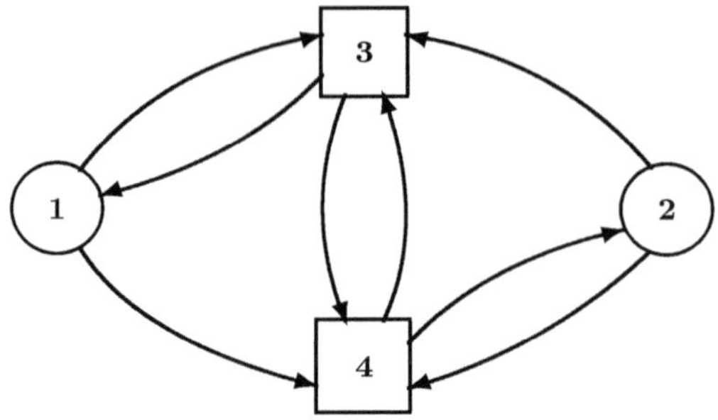

Fig. 2.1 The perfect network for the control problem

We are seeking for the optimal stationary strategy s^* which gives the solution of the problem for an arbitrary starting state $x \in X$.

It is easy to see that the network is perfect and, therefore, we can determine the optimal strategy by solving the linear programming problem (2.13), (2.14).

For this example we have

$$\overline{\psi}(\alpha, q) = c_{1,3}\alpha_{1,3} + c_{1,4}\alpha_{1,4} + c_{2,3}\alpha_{2,3} + c_{2,4}\alpha_{2,4} + \mu_3 q_3 + \mu_4 q_4,$$

where

$$\mu_3 = p_{3,1}c_{3,1} + p_{3,4}c_{3,4} = 0.5 \cdot 2 + 0.5 \cdot 4 = 3,$$
$$\mu_4 = p_{4,2}c_{4,2} + p_{4,3}c_{4,3} = 0.5 \cdot 1 + 0.5 \cdot 3 = 2.$$

So, to determine the optimal stationary strategy s^* we need to solve the linear programming problem:
Minimize

$$\overline{\psi}(\alpha, q) = \alpha_{1,3} + 2\alpha_{1,4} + 3\alpha_{2,3} + \alpha_{2,4} + 3q_3 + 2q_4$$

subject to

$$\begin{cases} 0.5q_3 = q_1, \\ 0.5q_4 = q_2, \\ \alpha_{1,3} + \alpha_{2,3} + 0.5q_4 = q_3, \\ \alpha_{1,4} + \alpha_{2,4} + 0.5q_3 = q_4, \\ \alpha_{1,3} + \alpha_{1,4} = q_1, \\ \alpha_{2,3} + \alpha_{2,4} = q_2, \\ q_1 + q_2 + q_3 + q_4 = 1, \\ q_i \ge 0,\ i = 1, 2, 3, 4; \quad \alpha_{i,j} \ge 0,\ i, j = 1, 2, 3, 4. \end{cases}$$

It is easy to check that the optimal solution of this problem is

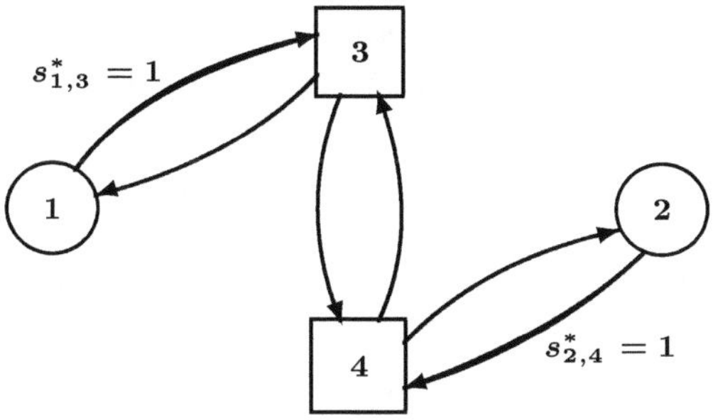

Fig. 2.2 The network induced by the optimal strategy

$$\alpha^*_{1,4} = 0, \quad \alpha^*_{2,3} = 0, \quad \alpha^*_{1,3} = \frac{1}{6}, \quad \alpha^*_{2,4} = \frac{1}{6},$$
$$q^*_1 = \frac{1}{6}, \quad q^*_2 = \frac{1}{6}, \quad q^*_3 = \frac{2}{6}, \quad q^*_4 = \frac{2}{6} \quad \text{and } \varphi(\alpha^*, q^*) = 2.$$

So, $s^*_{1,4} = 0,\ s^*_{2,3} = 0,\ s^*_{1,3} = 1,\ s^*_{2,4} = 1.$

In Fig. 2.2 a network is presented which corresponds to an optimal stationary strategy $s^*_{1,3} = 1, s^*_{2,4} = 1$.

2.2.3 Remark on the Application of the Unichain Linear Programming Model for an Arbitrary Network

The linear programming problem (2.13), (2.14) can be solved on an arbitrary decision network (G, X_C, X_N, c, p). A basic optimal solution α^*, q^* determines the strategy

$$s^*_{x,y} = \begin{cases} 1, & \text{if } \alpha^*_{x,y} > 0; \\ 0, & \text{if } \alpha^*_{x,y} = 0, \end{cases}$$

and a subset $X^* = \{x \in X \mid q_{x^*} > 0\}$, where s^* provides the optimal average cost per transition for the dynamical system $\mathbb{L}$ when it starts transitions in the states $x_0 \in X^*$.

This means that for an arbitrary network Algorithm 2.5 determines the optimal stationary strategy of the problem only in the case if the system starts transitions in the states $x \in X^*$. So, in the general case the algorithm finds a strategy s^* and a distinct positive recurrent class X^* in X with the minimal average cost per transition of the system $\mathbb{L}$ for an arbitrary starting state $x_0 \in X^*$.

For a unichain control problem Algorithm 2.5 determines the strategy s^* and the recurrent class X^*. In this case the remaining states $x \in X \backslash X^*$ in X correspond to transient states and the optimal stationary strategies in the states $x \in X \backslash X^*$ can be chosen in order to reach X^*. Therefore, the linear programming model (2.13), (2.14) can be used for determining the optimal stationary strategy for an arbitrary unichain control problem.

2.2.4 Determining the Solutions for an Arbitrary Unichain Control Problem and for the Deterministic Case

As we have noted the linear programming model (2.13), (2.14) can be used for studying the control problem on a network of arbitrary structure. Here we show how to use the linear programming model (2.13), (2.14) for determining the optimal stationary strategies of the control problem in the following two cases:

(1) the network is not perfect but for an arbitrary stationary strategy s the matrix P^s corresponds to a recurrent Markov chain;
(2) the network contains only controllable states, i.e., $X_N = \emptyset$.

First let us analyze the problem in the case (1). In this case an arbitrary strategy s in G generates a graph G_s with unique strongly connected components $G'_s = (X'_s, E'_s)$ that can be reached from any vertex $x \in X$. The optimal stationary strategy s^* in G can be found from a basic optimal solution by fixing $s^*_{x,y} = 1$ for the basic variables. This means that in G we can find the optimal stationary strategy as follows:

We solve the linear programming problem (2.13), (2.14) and find a basic optimal solution α^*, q^*. Then we find the subset of vertices $X^* = \{x \in X \mid q^*_x > 0\}$ which in G corresponds to a strongly connected subgraph $G^* = (X^*, E^*)$. On this subgraph we determine the optimal solution of the problem using the algorithm described in the previous section. It is evident that if $x_0 \in X^*$ then we obtain the solution of the problem with fixed starting state x_0. To determine the solution of the problem for an arbitrary starting state we may select successively vertices $x \in X \setminus X^*$ which contain outgoing directed edges that end in X^* and will add them at each time to X^* using the following rule:

- if $x \in X_C \cap (X \setminus X^*)$ then we fix an directed edge $e = (x, y)$, put $s^*_{x,y} = 1$ and change X^* by $X^* \cup \{x\}$;
- if $x \in X_N \cap (X \setminus X^*)$ then change X^* by $X^* \cup \{x\}$.

Thus, in the case (1) we can determine the optimal stationary strategy of the control problem on the network (G, X_C, X_N, c, p, x_0).

In the case (2) ($X_N = \emptyset$) we have a deterministic model and the linear programming problem (2.13), (2.14) becomes the linear programming problem from [65, 117]. Thus, the linear programming model generalizes the deterministic model from [65, 117] and from Theorem 2.3 we obtain the following result.

Lemma 2.6 *Let $G = (X, E)$ be a strongly connected directed graph with $X_N = \emptyset$ and let $\alpha^*_{x,y}$, $(x, y) \in E$ be the basic optimal solution of the linear programming problem:*
Minimize

$$\overline{\psi}(\alpha) = \sum_{x \in X_C} \sum_{y \in X(x)} c_{x,y} \alpha_{x,y} \tag{2.18}$$

subject to

$$\begin{cases} \sum_{x \in X^-(y)} \alpha_{x,y} - \sum_{z \in X(y)} \alpha_{y,z} = 0, \quad \forall y \in X; \\ \sum_{x \in X} \sum_{y \in X(x)} \alpha_{x,y} = 1; \\ \alpha_{x,y} \geq 0, \quad \forall (x, y) \in E. \end{cases} \tag{2.19}$$

Then the subgraph $G' = (X', E')$ generated by the directed edges $(x, y) \in E$ with $\alpha^*_{x,y} > 0$ has a structure of a directed cycle and an optimal stationary strategy s^* for the control problem on G with a given starting state x_0 can be found as follows:

- *fix a simple directed path which connects x_0 with the directed cycle G' and find the set of edges E'' of this directed path;*
- *fix the stationary strategy s^* where $s^*_{x,y} = 1$ if $(x, y) \in E' \cup E''$; otherwise put $s^*_{x,y} = 0$.*

Proof If $X_N = \emptyset$ then problem (2.13), (2.14) is transformed into the following problem:
Minimize (2.18) subject to

$$\begin{cases} \sum_{x \in X^-(y)} \alpha_{x,y} = q_y, \quad \forall y \in X; \\ \sum_{x \in X} q_y + \sum_{z \in X_N} q_z = 1; \\ \sum_{y \in X(x)} \alpha_{x,y} = q_x, \quad \forall x \in X; \\ \alpha_{x,y} \geq 0, \quad \forall x, y \in X; \quad q_x \geq 0, \quad \forall x \in X. \end{cases} \tag{2.20}$$

After the elimination of q_x and q_y from the system (2.20) we obtain the system (2.19). In such a way we obtain that (2.18), (2.19) becomes the mean cost cycle problem on G and the algorithm from the lemma above determines the optimal solution of the problem. □

Based on the lemma above we can propose the following algorithm for finding the solution of the problem in the case $X_N = \emptyset$.

Algorithm 2.7 Determining the Optimal Solution for the Deterministic Control Problem

1. Formulate the linear programming problem (2.18), (2.19) and find a basic optimal solution $\alpha^*_{x,y}$ and the corresponding directed graph $G' = (X', E')$ which has the structure of a directed cycle;
2. Fix a simple directed path which connects x_0 with the directed cycle G' and find the set of edges E'' of this directed path;

3. Fix a stationary strategy s^* where $s^*_{x,y} = 1$ if $(x, y) \in E' \cup E''$; otherwise put $s^*_{x,y} = 0$.

So, the deterministic control problem can be efficiently solved on an arbitrary network if $X_N = \emptyset$.

2.2.5 Dual Linear Programming for the Unichain Control Problem and an Algorithm for Determining the Optimal Strategies

For the linear programming model (2.16), (2.17) we consider the following dual problem:
Maximize

$$\overline{\psi}'(\varepsilon, \omega) = \omega \tag{2.21}$$

subject to

$$\begin{cases} \varepsilon_x - \varepsilon_y + \omega \le c_{x,y}, & \forall\, x \in X_C,\ y \in X(x); \\ \varepsilon_x - \sum\limits_{z \in X} p_{x,z}\varepsilon_z + \omega \le \mu_x, & \forall\, x \in X_N. \end{cases} \tag{2.22}$$

Remark 2.8 The conditions $q_y \ge 0$, $\forall x \in X_N$ in the unichain primal linear programming problem are redundant. Therefore, the constraints 2.22 in the problem (2.21), (2.22) can be replaced by the following constraints

$$\begin{cases} \varepsilon_x - \varepsilon_y + \omega \le c_{x,y}, & \forall\, x \in X_C,\ y \in X(x); \\ \varepsilon_x - \sum\limits_{z \in X} p_{x,z}\varepsilon_z + \omega = \mu_x, & \forall\, x \in X_N. \end{cases} \tag{2.23}$$

The optimal stationary strategies of the unichain control problem correspond to basic optimal solutions of this problem and can be found by using the following theorem.

Theorem 2.9 *An arbitrary optimal solution* ε^*_x $(x \in X)$, ω^* *of the problem* (2.21), (2.22) *for a unichain control model on the network* (G, X_C, X_N, c, p) *possesses the following property:*

(1) $\min\limits_{y \in X(x)} \{c_{x,y} + \varepsilon^*_y - \varepsilon^*_x - \omega^*\} = 0, \ \ \forall x \in X_C;$

(2) $\mu_x + \sum\limits_{z \in X(x)} p_{x,z}\varepsilon^*_z - \varepsilon^*_x - \omega^* = 0, \ \ \forall x \in X_N;$

(3) a stationary strategy $s^* : X_C \to X$ *is optimal if and only if* $(x, s^*(x)) \in E^*_C$, $\forall x \in X_C$, *where*

$$E_C^* = \{e = (x, y) \in E_C \mid c_{x,y} + \varepsilon_y^* - \varepsilon_x^* - \omega^* = 0\}.$$

The value ω^* is equal to the optimal average cost in the unichain control problem on the network (G, X_C, X_N, c, p).

Proof The properties (1) and (2) of the theorem represent the optimality conditions for the dual linear programming problem (2.21), (2.22). If $\alpha_{x,y}^*$, $(x \in X_C, y \in X(x))$, q_x^* $(x \in X)$ is a basic solution of the primal problem (2.16), (2.17), where $\alpha_{x,y}^* = s_{x,y}^* q_x^*$, $q^* = \sum_{y \in X(x)} \alpha_{x,y}$, then we can take $s_{x,y}^* = 1$ for $(x, y) \in E_C$ that satisfies the conditions (1), (2) and $s_{x,y} = 0$ in the other case. This means that an optimal stationary strategy in G is determined by the map $s^* : X_C \to X$ for which $(x, s^*(x)) \in E_C^*$, $\forall x \in X_C$. □

Corollary 2.10 *Each subset $E_{s^*} = \{e = (x, s^*(x)) \in E_C^* \mid x \in X_C\}$ in G generates a subgraph $G_{s^*} = (X, E_{s^*} \cup E_C)$ that corresponds to a Markov unichain, i.e., G_{s^*} contains a unique strongly connected component that is reachable from every $x \in X$. The values of the boolean variable $s_{x,y}^*$, $x \in X$, $y \in X(x)$ that correspond to an optimal solution of the problem can be found by fixing*

$$s_{x,y}^* = \begin{cases} 1, & \text{if } (x, y) \in E_{s^*}; \\ 0, & \text{if } (x, y) \notin E_{s^*}. \end{cases}$$

Corollary 2.11 *Let s be an arbitrary strategy for the control problem on the network (G, X_C, X_N, c, p) and $P^s = (p_{x,y}^s)$ be the transition probability matrix induced by this strategy,*

$$p_{x,y}^s = \begin{cases} p_{x,y}, & \text{if } x \in X_N \text{ and } y \in X; \\ 1, & \text{if } x \in X_C \text{ and } y = s(x); \\ 0, & \text{if } x \in X_C \text{ and } y \neq s(x). \end{cases}$$

Then in the Markov process induced by this transition probability matrix it holds

$$q_x^s \left(\mu_x^s + \varepsilon_x^s - \sum_{z \in X} p_{x,z}^s \varepsilon_z^s - \omega^s \right) = 0 \quad \forall x \in X,$$

where q_x^s is a limiting probability in the state $x \in X$ and $\mu_x^s = \sum_{y \in X(x)} p_{x,y}^s c_{x,y}$.

From Theorem 2.9 we can make the following conclusions. For an arbitrary unichain control problem there exist a function $\varepsilon^* : X \to \mathbb{R}$ and a value ω^* that satisfy the conditions

(1) $\overline{c}_{x,y} = c_{x,y} + \varepsilon_y^* - \varepsilon_x^* - \omega^* \geq 0, \quad \forall x \in X_C, \ \forall y \in X(x)$;

(2) $\min_{y \in X} \overline{c}_{x,y} = 0, \quad \forall x \in X_C$;

(3) $\overline{\mu}_x = \mu_x + \sum_{y \in X} p_{x,y} \varepsilon_x^* - \varepsilon_x^* - \omega_x^* = 0, \ \ \forall x \in X_N$.

If in the decision network (G, X_C, X_N, c, p) we change the cost function c by $\overline{c}$ then we obtain a new control problem on the network $(G, X_C, X_N, \overline{c}, p)$. Such a transformation of the cost function in the control problem does not change the optimal stationary strategies. In the new control problem the cost function $\overline{c}$ satisfies the conditions $\min_{y \in X(x)} \overline{c}_{x,y} = 0, \ \ \forall x \in X_C$ and $\overline{\mu}_x = 0, \ \ \forall x \in X_N$. For this problem the optimal average cost $\overline{\omega}_x^*$ for every $x \in X$ is equal to zero and an optimal stationary strategy can be found by fixing an arbitrary map s^* such that $(x, s^*(x)) \in E_C^*$, where $E_C^* = \{(x, y) \in E_C \,|\, \overline{c}_{x,y} = 0\}$.

We call the cost function $\overline{c}_{x,y} = c_{x,y} + \varepsilon_y^* - \varepsilon_x^* - \omega^*, \ (x, y) \in E$ a *potential transformation* induced by the *potential function* $\varepsilon^* : X \to \mathbb{R}$ and the values ω_x^* for $x \in X$. Furthermore, we call the new problem with the cost function $\overline{c}$ a *control problem in canonical form*.

2.2.6 The Potential Transformation and Optimality Conditions for Multichain Control Problems

The aim of this section is to formulate and prove the optimality conditions for an average multichain stochastic control problem. For this reason we extend the notions of the potential function and *potential transformation* for a multichain control problem and study their main properties. Based on these properties we prove the optimality conditions and show how to reduce the average multichain control problem to an auxiliary one in canonical form for which the optimal solutions can easily be found. We show that such a transformation of the control problem into an auxiliary problem in canonical form always exists. Finally, we show that the problem of determining optimal stationary strategies in a multichain control problem can be formulated as a linear programming problem.

We define the *decision network in canonical form* $(G, X_C, X_N, \overline{c}, p)$ for a multichain control problem on the network (G, X_C, X_N, c, p) by using the potential transformation

$$\overline{c}_{x,y} = c_{x,y} + \varepsilon_y - \varepsilon_x - h_x, \quad \forall x \in X, \ \forall y \in X(x), \tag{2.24}$$

where the function $\varepsilon : X \to \mathbb{R}$ and the values h_x for $x \in X$ satisfy the conditions:

(1) $\overline{c}_{x,y} = c_{x,y} + \varepsilon_y - \varepsilon_x - h_x \geq 0, \quad \forall x \in X_C, \ y \in X(x)$;

(2) $\min\limits_{y \in X} \overline{c}_{x,y} = 0, \quad \forall x \in X_C$;

(3) $\overline{\mu}_x = \mu_x + \sum\limits_{y \in X} p_{x,y}\varepsilon_y - \varepsilon_x - h_x = 0, \quad \forall x \in X_N$;

(4) $h_x = \min\limits_{y \in X(x)} h_y, \quad \forall x \in X_C, \ \forall y \in X(x)$;

(5) $h_x = \sum_{y \in X} p_{x,y} h_y, \quad \forall x \in X_N$

(6) $E_h(x) \cap E_{\overline{c}}(x) \neq \emptyset$, where

$$E_h(x) = \left\{ (x, y) \in E_C \mid y \in \underset{z \in X(x)}{\operatorname{argmin}} \{h_z\} \right\}, \quad x \in X_C$$

and

$$E_{\overline{c}}(x) = \left\{ (x, y) \in E_C \mid y \in \underset{z \in X(x)}{\operatorname{argmin}} \{\overline{c}_{x,z}\} \right\}, \quad x \in X_C.$$

In general, the potential transformation (2.24) can also be considered for an arbitrary network. However, the optimal stationary strategies in the control problem after such a potential transformation may differ from the optimal stationary strategies in the initial network. The potential transformation with the properties mentioned above preserves the optimal strategy of the multichain control problem.

If the decision network in canonical form is known then the optimal stationary strategy for the stochastic multichain control problem can be found in a similar way as for the unichain case of the problem, i.e., we fix a strategy $s^* : X_C \to X$ such that $(x, s^*(x)) \in E^*_{\overline{c}}$. Moreover, the potential transformation $\overline{c}$ that satisfies the conditions (1)–(6) gives the values of the optimal average costs $\omega^*_x = h_x$ in the states $x \in X$ for a multichain control problem on the network (G, X_C, X_N, c, p).

In the following we show that for an arbitrary network (G, X_C, X_N, c, p) that there exists a network in canonical form $(G, X_C, X_N, \overline{c}, p)$ that obtains the optimal stationary strategy s^* and the optimal average costs ω^*_x for $x \in X$. We ground all these results on the basis of the following optimality principle for a multichain control problem.

Theorem 2.12 *For an arbitrary decision network (G, X_C, X_N, c, p) there exists a potential transformation*

$$\overline{c}_{x,y} = c_{x,y} + \varepsilon^*_y - \varepsilon^*_x - h^*_x, \quad \forall x \in X, \ y \in X(x)$$

of the cost function c that satisfies the following conditions:

(1) $\overline{c}_{x,y} = c_{x,y} + \varepsilon^*_y - \varepsilon^*_x - h^*_x \geq 0, \quad \forall x \in X_C, \ y \in X(x);$

(2) $\min_{y \in X} \overline{c}_{x,y} = 0, \quad \forall x \in X_C;$

(3) $\overline{\mu}_x = \mu_x + \sum_{y \in X} p_{x,y} \varepsilon^*_y - \varepsilon^*_x - h^*_x = 0, \quad \forall x \in X_N;$

(4) $h^*_x = \min_{y \in X(x)} h^*_y, \quad \forall x \in X_C, \ \forall y \in X(x);$

(5) $h^*_x = \sum_{y\in X} p_{x,y}h^*_y, \quad \forall x \in X_N;$

(6) $E^*_{h^*}(x) \cap E^*_{\overline{c}}(x) \neq \emptyset, \ \forall x \in X_C,$ *where*

$$E^*_{h^*}(x) = \left\{ (x, y) \in E_C \mid y \in \underset{z\in X(x)}{\operatorname{argmin}}\{h_z\} \right\}, \ x \in X_C$$

and

$$E^*_{\overline{c}}(x) = \left\{ (x, y) \in E_C \mid y \in \underset{z\in X(x)}{\operatorname{argmin}}\{\overline{c}_{x,z}\} \right\}, \ x \in X_C.$$

The values ε^*_x *for* $x \in X$ *correspond to a basic solution of the system of linear equations*

$$\begin{cases} c_{x,y} + \varepsilon_y - \varepsilon_x - h^*_x = 0, \quad \forall x \in X_C, \ (x, y) \in E^*_{h^*}(x); \\ \mu_x + \sum\limits_{y\in X} p_{x,y}\varepsilon_y - \varepsilon_x - h^*_x = 0, \quad \forall x \in X_N \end{cases} \tag{2.25}$$

and determines the decision network in canonical form $(G, X_C, X_N, \overline{c}, p)$ *for the control problem on the network* (G, X_C, X_N, c, p)*, where* $\overline{c}_{x,y} = c_{x,y} + \varepsilon^*_y - \varepsilon^*_x - h^*_x$, $\forall x \in X, \ y \in X(x)$.

The values h^*_x *for* $x \in X$ *coincide with the corresponding optimal average costs* ω^*_x *for* $x \in X$ *and an optimal stationary strategy for the control problem on the network can be found by fixing an arbitrary map* $s^* : X_C \to X$ *such that* $(x, s^*(x)) \in E^*_{h^*}(x) \cap E^*_{\overline{c}}(x), \ \forall x \in X_C$.

This theorem is tightly connected with the existence of the solutions for the *bias equations in average Markov decision processes* (see [115, 140]). In the terms of bias equations this theorem can be formulated in the following way:

Theorem 2.13 *The system of equations*

$$\begin{cases} \varepsilon_x + h_x = \min\limits_{y\in X}\{c_{x,y} + \varepsilon_y\}, \quad \forall x \in X_C; \\ \varepsilon_x + h_x = \mu_x + \sum\limits_{y\in X} p_{x,y}\varepsilon_y, \quad \forall x \in X_N \end{cases} \tag{2.26}$$

has solutions with respect to ε_x *for* $x \in X$ *under the set of solutions of the following system of equations*

$$\begin{cases} h_x = \min\limits_{y\in X(x)} h_y, \qquad \forall x \in X_C; \\ h_x = \sum\limits_{y\in X(x)} p_{x,y}h_x, \quad \forall x \in X_N. \end{cases} \tag{2.27}$$

If ε_x^*, h_x^* $(x \in X)$ is the solution of these equations then h_x^* for $x \in X$ coincides with the optimal average costs ω_x^*.

To prove Theorem 2.12 we need some auxiliary results.

Let $s : X \to X$ be a feasible strategy for the control problem on the decision network (G, X_C, X_N, c, p) and $P^s = (p_{x,y}^s)$ be the transition probability matrix of the Markov chain induced by the strategy s, i.e.,

$$p_{x,y}^s = \begin{cases} p_{x,y}, & \text{if } x \in X_N \text{ and } y = X(x); \\ 1, & \text{if } x \in X_C \text{ and } y = s(x); \\ 0, & \text{if } x \in X_C \text{ and } y \neq s(x). \end{cases} \tag{2.28}$$

Denote by $Q^s = (q_{x,y}^s)$ the limit matrix in the Markov chain with probability transition matrix P^s and by $X_1^s, X_2^s, \dots, X_k^s$ the corresponding irreducible sets in this Markov chain.

Lemma 2.14 *Let ω_x^s be the average cost per transition of the system for a feasible strategy $s : X_C \to X$ of the control problem on the decision network (G, X_C, X_N, c, p). Then for an arbitrary potential function $\varepsilon : X \to \mathbb{R}$ and arbitrary real values h_x for $x \in X$ the average cost per transition $\overline{\omega}_x^s$ of the system on the potential transformed network $(G, X_C, X_N, p, \overline{c})$ satisfies the condition*

$$\overline{\omega}_x^s = \omega_x^s - \sum_{z \in X} q_{x,z}^s h_z, \quad \forall x \in X. \tag{2.29}$$

Proof Let s be a feasible stationary strategy of the control problem. Consider a potential transformation $\overline{c}_{x,y} = c_{x,y} + \varepsilon_y - \varepsilon_x - h_x, (x, y) \in E$ determined by an arbitrary function $\varepsilon : X \to \mathbb{R}$ and arbitrary real values h_z for $z \in X$. Then after the potential transformation the average cost $\overline{\omega}_x^s$ for an arbitrary $x \in X$ can be calculated as follows:

$$\begin{aligned}
\overline{\omega}_x^s &= \sum_{z \in X} \overline{\mu}_z^s q_{x,z}^s = \sum_{z \in X} \sum_{y \in X(z)} p_{z,y}^s \overline{c}_{z,y} q_{x,z}^s \\
&= \sum_{z \in X} \sum_{y \in X(z)} p_{z,y}^s (c_{z,y} + \varepsilon_y^s - \varepsilon_z^s - h_z) q_{x,z}^s = \sum_{z \in X} \sum_{y \in X(z)} p_{z,y}^s c_{z,y} q_{x,z}^s \\
&\quad + \sum_{z \in X} q_{x,z}^s \sum_{y \in X(z)} p_{z,y}^s \varepsilon_y^s - \sum_{z \in X} q_{x,z}^s \sum_{y \in X(z)} p_{z,y}^s \varepsilon_z^s - \sum_{z \in X} q_{x,z}^s h_z \sum_{y \in X(z)} p_{z,y}^s \\
&= \omega_x^s + \sum_{z \in X} q_{x,z}^s \Big(\sum_{y \in X(z)} p_{z,y}^s \varepsilon_y - \sum_{y \in X(z)} p_{z,y}^s \varepsilon_z \Big) - \sum_{z \in X} q_{x,z}^s h_z,
\end{aligned}$$

i.e., we have

$$\overline{\omega}_x^s = \omega_x^s + \sum_{z \in X} q_{x,z}^s \Big(\sum_{y \in X(x)} p_{z,y}^s \varepsilon_y^s - \varepsilon_z^s \Big) - \sum_{z \in X} q_{x,z}^s h_z, \quad \forall x \in X. \tag{2.30}$$

Now we show that for an arbitrary strategy s it holds

$$\sum_{z\in X} q^s_{x,z}\Big(\sum_{y\in X(z)} p^s_{z,y}\varepsilon^s_y - \varepsilon^s_z\Big) = 0, \quad \forall x \in X. \tag{2.31}$$

Let $X^s_1, X^s_2, \dots, X^s_k$ be the corresponding irreducible sets in the Markov chain induced by the strategy s. Then in the graph $G_s = (X, E_s \cup E_N)$ each subset X^s_i of X generates a strongly connected graph that corresponds to a distinct irreducible Markov chain and in each irreducible set the average costs for an arbitrary starting state is the same.

If we denote by $\omega^{s,i}$ the average cost for the corresponding states in the irreducible sets X^s_i then we have

$$\sum_{z\in X} q^s_{x,z}\Big(\sum_{y\in X(z)} p^s_{z,y}\varepsilon^s_y - \varepsilon^s_z\Big)$$

$$= \sum_{z\in X} q^s_{x,z}\Big(\Big(\mu^s_z + \sum_{y\in X(z)} p^s_{z,y}\varepsilon^s_y - \varepsilon^s_z - \omega^s_z\Big) + (\omega^s_z - \mu^s_z)\Big)$$

$$= \sum_{i=1}^{k}\sum_{z\in X^s_i} q^s_{x,z}\Big(\Big(\mu^s_z + \sum_{y\in X(z)} p^s_{z,y}\varepsilon^s_y - \varepsilon^s_z - \omega^{s,i}\Big) + (\omega^{s,i} - \mu^s_z)\Big)$$

$$= \sum_{i=1}^{k}\sum_{z\in X^s_i} q^s_{x,z}\Big(\mu^s_z + \sum_{y\in X(z)} p^s_{z,y}\varepsilon^s_y - \varepsilon^s_z - \omega^{s,i}\Big) + \sum_{i=1}^{k}\sum_{z\in X^s_i} q^s_{x,z}(\omega^{s,i} - \mu^s_z).$$

Here, according to Corollary 2.11 it holds

$$\mu^s_z + \sum_{y\in X(z)} p^s_{z,y}\varepsilon^s_y - \varepsilon^s_z - \omega^{s,i} = 0, \quad \forall z \in X^s_i,\ i = 1, 2, \dots, k.$$

Therefore, we obtain

$$\sum_{z\in X} q^s_{x,z}\Big(\sum_{y\in X(z)} p^s_{z,y}\varepsilon^s_y - \varepsilon^s_z\Big) = \sum_{i\in 1}^{k}\sum_{z\in X^s_i} q^s_{x,z}(\omega^{s,i} - \mu^s_z)$$

$$= \sum_{i=1}^{k}\omega^{s,i}\sum_{z\in X^s_i} q^s_{x,z} - \sum_{i=1}^{k}\sum_{z\in X^s_i} q^s_{x,z}\mu^s_z = \sum_{i=1}^{k}\omega^{s,i} - \sum_{i=1}^{k}\omega^{s,i} = 0.$$

So, condition (2.31) holds.
If we introduce (2.31) in (2.30) then we obtain (2.29). □

Corollary 2.15 *Let s be an arbitrary feasible strategy for the control problem on the network (G, X_C, X_N, c, p) and $Q^s = (q^s_{x,y})$ be the matrix of limiting probabilities in the Markov chain induced by the strategy s. Then for an arbitrary potential function $\varepsilon : X \to \mathbb{R}$ the following condition holds*

$$\sum_{z\in X} q^s_{x,z}\left(\sum_{y\in X(x)} p^s_{z,y}\varepsilon^s_y - \varepsilon^s_z\right) = 0, \quad \forall x \in X. \tag{2.32}$$

If in (2.24) we fix $h_x = h, \ \forall x \in X$ then we obtain the following potential transformation

$$\overline{c}_{x,y} = c_{x,y} + \varepsilon_y - \varepsilon_x - h, \quad \forall x \in X, \ \forall y \in X(x), \tag{2.33}$$

In this case from Lemma 2.14 we obtain the following result.

Corollary 2.16 *Let ω^s_x be the average cost per transition of the system for a feasible strategy $s : X_C \to X$ of the control problem on the decision network (G, X_C, X_N, c, p). Then for an arbitrary potential function $\varepsilon : X \to \mathbb{R}$ and $h \in \mathbb{R}$ the average cost per transition $\overline{\omega}^s_x$ of the system on the potential transformed network $(G, X_C, X_N, \overline{c}, p)$ satisfies the condition*

$$\overline{\omega}^s_x = \omega^s_x - h, \quad \forall x \in X, \ \forall s. \tag{2.34}$$

Corollary 2.16 shows that an arbitrary control problem with average cost criterion can be transformed into a similar one where the transition cost function $\overline{c}$ is nonnegative or positive. Indeed, if we take an arbitrary function $\varepsilon : X \to \mathbb{R}$ and $h = -M$, where $M \geq \max_{(x,y)\in E} |c_{x,y}|$, then the cost function $\overline{c}$ in the control problem becomes nonnegative or positive.

Lemma 2.17 *Assume that for a fixed strategy s the values $h^s_x, \ x \in X$ satisfy the condition*

$$h^s_x - \sum_{y\in X(x)} p^s_{x,y} h^s_y = 0, \quad \forall x \in X. \tag{2.35}$$

Then for an arbitrary potential function $\varepsilon : X \to \mathbb{R}$ the average cost $\overline{\omega}^s_x$ in the control problem on the network $(C, X_C, X_N, \overline{c}, p)$ with a transformed potential cost function

$$\overline{c}_{x,y} = c_{x,y} + \varepsilon_y - \varepsilon_x - h^s_x, \quad \forall x \in X, \ \forall y \in X(x)$$

can be calculated using the following formula

$$\overline{\omega}^s_x = \omega^s_x - h^s_x, \quad \forall x \in X. \tag{2.36}$$

If h_x^s for $x \in X$ satisfies the condition

$$h_x^s - \sum_{y \in X(x)} p_{x,y}^s h_y^s \le 0, \quad \forall x \in X, \tag{2.37}$$

then

$$\overline{\omega}_x^s \ge \omega_x^s - h_x^s, \quad \forall x \in X. \tag{2.38}$$

If h_x^s for $x \in X$ satisfies the condition

$$h_x^s - \sum_{y \in X(x)} p_{x,y}^s h_y^s \ge 0, \quad \forall x \in X, \tag{2.39}$$

then

$$\overline{\omega}_x^s \le \omega_x^s - h_x, \quad \forall x \in X, \ \forall s \in \mathbb{S}. \tag{2.40}$$

Proof According to Lemma 1.25 (see Eqs. (1.55), (1.56)) the condition (2.35) implies

$$h_x^s = \sum_{y \in X(x)} q_{x,y}^s h_y^s, \quad \forall x \in X. \tag{2.41}$$

If we introduce (2.41) in (2.29) then we obtain (2.35). In the case if h_x^s for $x \in X$ satisfies (2.37) we obtain $h_x^s \le \sum_{y \in X(x)} p_{x,y}^s h_y^s$. This implies (2.38). If h_x^s for $x \in X$ satisfies (2.39) then we obtain $h_x^s \ge \sum_{y \in X(x)} p_{x,y}^s h_y^s$. This implies (2.40). □

Lemma 2.18 *Let s be an arbitrary stationary strategy for the control problem on the network (G, X_C, X_N, c, p) and $P^s = (p_{x,y}^s)$ be the probability transition matrix induced by the strategy s, i.e., the elements $p_{x,y}^s$ of this matrix are defined according to (2.28). Then the system of linear equations*

$$\begin{cases} \mu_x^s + \sum\limits_{y \in X} p_{x,y}^s \varepsilon_y^s - \varepsilon_x^s - h_x^s = 0, & \forall x \in X; \\ h_x^s - \sum\limits_{y \in X(x)} p_{x,y}^s h_y^s = 0, & \forall x \in X; \end{cases} \tag{2.42}$$

has solutions. Moreover, if

$$h_x^s - \sum_{y \in X(x)} p_{x,y}^s h_y^s \le 0, \quad \forall x \in X \tag{2.43}$$

then

$$\mu_x^s + \sum_{y \in X} p_{x,y} \varepsilon_y^s - \varepsilon_x^s - h_x^s \ge 0, \quad \forall x \in X; \tag{2.44}$$

if

$$h_x^s - \sum_{y \in X(x)} p_{x,y}^s h_y^s \geq 0, \quad \forall x \in X \tag{2.45}$$

then

$$\mu_x^s + \sum_{y \in X} p_{x,y} \varepsilon_y^s - \varepsilon_x^s - h_x^s \leq 0, \quad \forall x \in X. \tag{2.46}$$

Proof We shall use the vector representation of the system (2.42). Denote by μ, h and ε the vectors with the corresponding components μ_x, h_x and ε_x for $x \in X$. Additionally, assume that the matrix P^s is represented in canonical form as it is defined in Sect. 1.1.3, i.e.,

$$P^s = \begin{pmatrix} P_1^s & 0 & \dots & 0 & 0 \\ 0 & P_2^s & \dots & 0 & 0 \\ \cdot & \cdot & \dots & \cdot & \cdot \\ \cdot & \cdot & \dots & \cdot & \cdot \\ \cdot & \cdot & \dots & \cdot & \cdot \\ 0 & 0 & \dots & P_k^s & 0 \\ W_1^s & W_2^s & \dots & W_k^s & W_{k+1}^s \end{pmatrix},$$

where P_r^s, $r = 1, 2, \dots, k$ represent the submatrices of P^s that corresponds to the ergodic classes X_r^s of the Markov multichain and W_r^s represent the submatrices of P^s that give the probability transitions from the states $x \in X \backslash (\bigcup_{r=1}^k X_r^s)$ to the states X_r^s; the elements of the matrix W_{k+1}^s represent the probability transitions $p_{x,y}$ between the states $x, y \in \bigcup_{r=1}^k X_r^s$. For each class X_r^s we shall use the vectors $\mu^{s,r}$, $h^{s,r}$ and $\varepsilon^{s,r}$ with the corresponding components $\mu_x^{s,r}$, $h_x^{s,r}$ and $\varepsilon_x^{s,r}$ for $x \in X_r^s$. Using these notations we can write the system (2.42) as follows

$$\begin{cases} \mu^{s,r} - (I^r - P_r^s)\varepsilon^{s,r} - h^{s,r} = 0, \quad r = 1, 2, \dots, k; \\ \mu^{s,k+1} - \sum_{r=1}^{k} (I^{k+1} - W_r^s)\varepsilon^{s,r} + (I^{k+1} - W_{k+1}^s)\varepsilon^{s,k+1} - h^{s,k+1} = 0; \\ (I^r - P_r^s)h^{s,r} = 0, \quad r = 1, 2, \dots, k; \\ \sum_{r=1}^{k} (I^r - W_r^s)h^{s,r} + (I^{k+1} - W_{k+1}^s)h^{s,r} = 0. \end{cases} \tag{2.47}$$

In this system each equation

$$\mu^{s,r} - (I^r - P_r^s)\varepsilon^{s,r} - h^{s,r} = 0$$

that corresponds to the class X_r^s, $r \in \{1, 2, \dots, k\}$ has a solution. This solution can be found on the bases of Theorem 2.9. According to this theorem we obtain

$h_x^{s,r} = \omega^{s,r}$, $\forall x \in X_r^s$, where $\omega^{s,r}$ is the average cost of the ergodic class X_r^s. In (2.47) each equation

$$(I^r - P_r^s)h^{s,r} = 0, \quad r \in \{1, 2, \ldots, k\}$$

is redundant and therefore can be deleted. Thus, from the last equation of (2.47) we can determine

$$h^{s,r} = -(I^{k+1} - W_{k+1})^{-1} \sum_{r=1}^{k} (I^r - W_r^s)h^{s,r}.$$

Note that for $(I^{k+1} - W_{k+1})$ there always exists the inverse matrix (see [7, 21, 115]). If we introduce this expression in the equation

$$\mu^{s,k+1} - \sum_{r=1}^{k} (I^{k+1} - W_r^s)\varepsilon^{s,r} + (I^{k+1} - W_{k+1}^s)\varepsilon^{s,k+1} - h^{s,k+1} = 0$$

of the system (2.47) then we can determine uniquely $\varepsilon^{s,k+1}$. So, the system (2.42) obtains solutions.

The second part of the lemma follows from the procedure given above to determine the solution of the system (2.42).

The condition (2.43) implies

$$\mu^{s,k+1} - \sum_{r=1}^{k} (I^{k+1} - W_r^s)\varepsilon^{s,r} + (I^{k+1} - W_{k+1}^s)\varepsilon^{s,k+1} - h^{s,k+1} \geq 0$$

and the condition (2.45) implies

$$\mu^{s,k+1} - \sum_{r=1}^{k} (I^{k+1} - W_r^s)\varepsilon^{s,r} + (I^{k+1} - W_{k+1}^s)\varepsilon^{s,k+1} - h^{s,k+1} \leq 0.$$

In (2.42) the solution of the system of equations

$$\mu^{s,r} - (I^r - P_r^s)\varepsilon^{s,r} - h^{s,r} = 0, \quad r = 1, 2, \ldots, k$$

does not depend on the conditions $(I^r - P_r^s)h^{s,r} \leq 0$ and $(I^r - P_r^s)h^{s,r} \geq 0$. So, the lemma holds. □

Corollary 2.19 *For an arbitrary stationary strategy s on the decision network (G, X_C, X_N, c, p) there exist ε_x^s and h_x^s for $x \in X$ that satisfy the conditions*

(1) $c_{x,y} + \varepsilon_y^s - \varepsilon_x^s - h_x^s = 0, \quad \forall x \in X_C, y = s(x);$

(2) $\mu_x + \sum_{y\in X(x)} p_{x,y}\varepsilon_y^s - \varepsilon_x^s - h_x = 0, \quad \forall x \in X_N;$

(3) $h_x^s = h_y^s, \quad \forall x \in X_C, y = s(x);$

(4) $h_x^s = \sum_{y\in X(x)} p_{x,y} h_y^s, \quad \forall x \in X_N.$

If h_x^s for $x \in X$ satisfies the conditions

$$h_x^s \le h_y^s \ \ for \ \ x \in X_C, y = s(x) \ \ and \ \ h_x^s \le \sum_{y\in X(x)} p_{x,y} h_y^s \ \ for \ \ x \in X_N$$

then

$$c_{x,y} + \varepsilon_y^s - \varepsilon_x^s - h_x^s \ge 0, \quad \forall x \in X_C, \ y = s(x);$$

$$\mu_x + \sum_{y\in X(x)} p_{x,y}\varepsilon_y^s - \varepsilon_x^s - h_x \ge 0, \quad \forall x \in X_N.$$

If h_x^s for $x \in X$ satisfies the conditions

$$h_x^s \ge h_y^s \ \ for \ \ x \in X_C, y = s(x) \ \ and \ \ h_x^s \ge \sum_{y\in X(x)} p_{x,y} h_z^s \ \ for \ \ x \in X_N$$

then

$$c_{x,y} + \varepsilon_y^s - \varepsilon_x^s - h_x^s \le 0, \quad \forall x \in X_C, \ y = s(x);$$
$$\mu_x + \sum_{y\in X(x)} p_{x,y}\varepsilon_y^s - \varepsilon_x^s - h_x \le 0, \quad \forall x \in X_N.$$

If in Lemma 2.18 we vary the strategy s then as a consequence from this lemma we obtain the following result.

Lemma 2.20 *Let (G, X_C, X_N, c, p) be an arbitrary decision network. Then there exist a function $\varepsilon^* : X \to \mathbb{R}$ and the values h_x^* for $x \in X$ such that for an arbitrary stationary strategy s of the control problem on the network it holds*

$$\begin{cases} \mu_x^s + \sum_{y\in X} p_{x,y}^s \varepsilon_y^* - \varepsilon_x^* - h_x^* \ge 0, & \forall x \in X; \\ h_x^* - \sum_{y\in X(x)} p_{x,y}^s h_y^* \le 0, & \forall x \in X; \end{cases} \tag{2.48}$$

Moreover, there exists a stationary strategy s^* such that

$$h_x^* - \sum_{y\in X(x)} p_{x,y}^{s^*} h_y^* = 0, \qquad \forall x \in X, \tag{2.49}$$

where ε_x^* for $x \in X$ represents a solution of the system of the equation

$$\mu_x^{s^*} + \sum_{y \in X} p_{x,y}^{s^*} \varepsilon_y - \varepsilon_x - h_x^* = 0, \quad \forall x \in X. \tag{2.50}$$

Corollary 2.21 *For an arbitrary decision network (G, X_C, X_N, c, p) there exist the values ε_x^*, h_x^* for $x \in X$ that represent the solution of the system of linear inequalities*

$$\begin{cases} c_{x,y} + \varepsilon_y^* - \varepsilon_x^* - h_x^* \geq 0, \quad \forall x \in X_C, & \forall y \in X(x); \\ \mu_x + \sum\limits_{y \in X(x)} p_{x,y} \varepsilon_y^* - \varepsilon_x^* - h_x^* \geq 0, & \forall x \in X_N; \\ h_x^* - h_y^* \leq 0, & \forall x \in X_C, \ \forall y \in X(x); \\ h_x^* - \sum\limits_{y \in X(x)} p_{x,y} h_y^* \leq 0, & \forall x \in X_N, \end{cases} \tag{2.51}$$

where h_x^* for $x \in X$ satisfy the condition

$$\begin{cases} \min\limits_{y \in X(x)} \{h_y^* - h_x^*\} = 0, \quad \forall x \in X_C; \\ h_x^* - \sum\limits_{y \in X(x)} p_{x,y} h_y^* = 0, \ \forall x \in X_N \end{cases} \tag{2.52}$$

and ε_x^* for $x \in X$ represents a solution of the system of linear equations

$$\begin{cases} c_{x,y} + \varepsilon_y - \varepsilon_x - h_x^* = 0, \quad \forall (x,y) \in E_{h^*}^*; \\ \mu_x + \sum\limits_{y \in X(x)} p_{x,y} \varepsilon_y - \varepsilon_x - h_x^* = 0, \ \forall x \in X_N, \end{cases} \tag{2.53}$$

where $E_{h^*}^* = \big\{(x,y) \in E_C \,|\, x \in X_C, \ y \in \mathrm{argmin}_{z \in X(x)} h_z^*\big\}$.

Proof of Theorem 2.12. According to Lemma 2.20 and Corollary 2.21 for the decision network (G, X_C, X_N, c, p) there exist the function $\varepsilon^* : X \to \mathbb{R}$ and the values h_x^* for $x \in X$ that satisfy the conditions (2.51)–(2.53). Thus, we can determine as a basic solution of the system (2.51) and the potential transformation

$$\overline{c}_{x,y} = c_{x,y} + \varepsilon_y^* - \varepsilon_x^* - h_x^*, \quad \forall x \in X_C, \ \forall y \in X(x)$$

that corresponds to the network $(G, X_C, X_N, \overline{c}, p)$ in canonical form. We obtain an optimal stationary strategy s^* for the problem on this network if for every $x \in X$ we fix $s^*(x) = y^*$, where y^* satisfies the condition $\overline{c}_{x,y^*}^{s^*} = 0$. Based on Lemma (2.17) we have

$$0 = \overline{\omega}_x^{s^*} = \omega_x^{s^*} - h_x^*, \quad \forall x \in X,$$

and for an arbitrary other strategy s it holds $\omega_x^s - h^* \geq 0$. So, $\omega_x^* = h_x^*, \ \forall x \in X$.

2.2.7 Linear Programming for Multichain Control Problems and an Algorithm for Determining Optimal Stationary Strategies

We develop the linear programming approach for a multichain control problem on the bases of the optimality criterion established in Theorem 2.12 and Corollary 2.21. If the vectors $\overline{h}^*$ and $\overline{\varepsilon}^*$ with the corresponding components h_x^* and ε_x^* satisfy the condition (2.52), then we obtain the vector of optimal average costs ω^*. Consequently we have to determine the "maximal" $\overline{h}^*$ that satisfies (2.51). This means that we have to maximize the positive linear combination of components of $\overline{h}^*$ that satisfy (2.51).

Thus, we can determine ε^* and ω^* if we solve the following linear programming problem:
Maximize

$$\overline{\psi}'(\varepsilon,\omega) = \sum_{x\in X} \theta_x \omega_x \tag{2.54}$$

subject to

$$\begin{cases} \varepsilon_x - \varepsilon_y + \omega_x \le c_{x,y}, \forall x \in X_C, & \forall y \in X(x); \\ \varepsilon_x - \sum\limits_{y\in X(x)} p_{x,y}\varepsilon_y + \omega_x \le \mu_x, & \forall x \in X_N; \\ \omega_x - \omega_y \le 0, & \forall x \in X_C,\ \forall y \in X(x); \\ \omega_x - \sum\limits_{y\in X(x)} p_{x,y}\omega_y \le 0, & \forall x \in X_N; \end{cases} \tag{2.55}$$

where $\theta_x > 0,\ \forall x \in X$ and $\sum_{x\in X} \theta_x = 1$.

Note that in this model in the case of the unichain control problem the restrictions $\omega_x - \omega_y \le 0$ for $x \in X_C,\ y \in X(x)$ and $\omega_x - \sum_{y\in X(x)} p_{x,y}\omega_y \le 0$ for $x \in X_N$ become redundant in (2.55), because here we can take $\omega_x = \omega_y,\ \forall x, y \in X$. Thus, this model generalizes the linear programming model (2.21), (2.22). Using this model we can propose the following algorithm for determining the solution of the multichain control problem.

Remark 2.22 In (2.55) the inequalities that correspond to the states $x \in X_N$ can be changed by equalities, i.e., the constraints (2.55) in the problem (2.54), (2.55) can be replaced by the constraints

$$\begin{cases} \varepsilon_x - \varepsilon_y + \omega_x \le c_{x,y}, & \forall x \in X_C,\ \forall y \in X(x); \\ \varepsilon_x - \sum\limits_{y\in X(x)} p_{x,y}\varepsilon_y + \omega_x = \mu_x, & \forall x \in X_N; \\ \omega_x - \omega_y \le 0, & \forall x \in X_C,\ \forall y \in X(x); \\ \omega_x - \sum\limits_{y\in X(x)} p_{x,y}\omega_y = 0, & \forall x \in X_N. \end{cases} \tag{2.56}$$

Thus, the optimal solutions of the problems (2.54), (2.55) and (2.54), (2.56) are the same.

Algorithm 2.23 Determining the Optimal Stationary Strategies for the Multichain Control Problem

(1) Formulate the linear programming problem (2.54), (2.55) and determine an optimal solution ε^*, ω^* that satisfies the conditions (2.52), (2.53).
(2) Formulate the potential transformation

$$\overline{c}_{x,y} = c_{x,y} + \varepsilon^*_y - \varepsilon^*_x - \omega^*_x, \quad \forall (x, y) \in E.$$

(3) Determine the set

$$E^*_{\overline{c}}(x) = \left\{ (x, y) \in E_C \;\middle|\; y \in \operatorname*{argmin}_{z \in X(x)} \overline{c}_{x,z} \right\}, \quad \forall x \in X_C;$$

$$E^*_{\omega^*}(x) = \left\{ (x, y) \in E_C \;\middle|\; y \in \operatorname*{argmin}_{z \in X(x)} \omega^*_z \right\}, \quad \forall x \in X_C;$$

(4) Fix a strategy $s^* : X_C \to X$ such that $s^*(x) = y$ for every $x \in X$, where $(x, y) \in E_{\overline{c}^*}(x) \cap E^*_{\omega^*}(x)$.

Below we illustrate Algorithm 2.23 based on the following example.

Example Consider the stochastic control problem on network (G, X_1, X_2, c, p) with the structure of the graph $G = (X, E)$ given in Fig. 2.3.

In this graph the vertices are represented by circles and squares. The vertices represented by circles correspond to the controllable states of the dynamical system and the vertices represented by squares correspond to uncontrollable states.

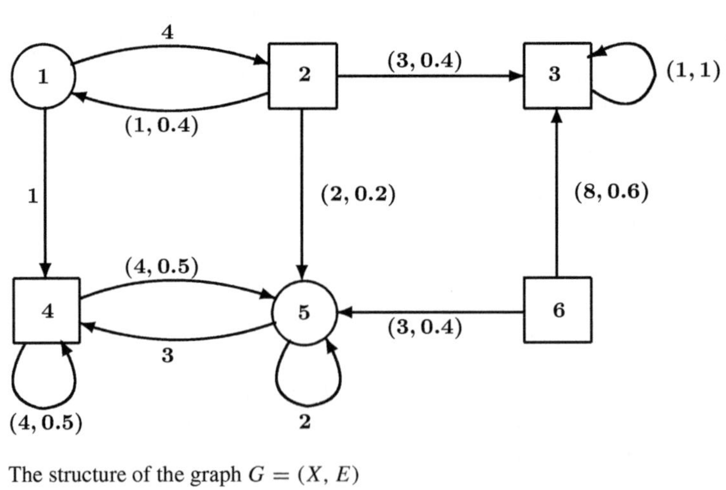

Fig. 2.3 The structure of the graph $G = (X, E)$

So,
$X = \{1, 2, 3, 4, 5, 6\}$; $X_C = \{1, 5\}$; $X_N = \{2, 3, 4, 6\}$;

$$\begin{aligned} E &= \{(1,2),\ (1,4),\ (2,1),\ (2,3),\ (2,5),\ (3,3),\ (4,4),\ (4,5),\\ &\qquad (5,5),\ (5,4),\ (6,3),\ (6,5)\};\\ E_C &= \{(1,2),\ (1,4),\ (5,5),\ (5,4)\};\\ E_N &= \{(2,1),\ (2,3),\ (2,5),\ (3,3),\ (4,4),\ (4,5),\ (6,3),\ (6,5)\}. \end{aligned}$$

The values of the cost function $c : E \to \mathbb{R}$ and of the transition probability function $p : E \to \mathbb{R}$ are written close to the edges in the picture. For the edges $e = (x, y) \in E_N$ these values are written in parentheses, where the first quantity expresses the cost and the second one represents the probability transition from the state x to the state y. For the edges $e = (x, y) \in E_C$ only the costs are given which are written also close to the edges. Thus, for this example we obtain:

$c_{1,2} = 4,\quad c_{1,4} = 1,\quad c_{2,1} = 1,\quad c_{2,3} = 3,\quad c_{2,5} = 2,\quad c_{3,3} = 1,$
$c_{4,4} = 4,\quad c_{4,5} = 4,\quad c_{5,5} = 2,\quad c_{5,4} = 3,\quad c_{6,3} = 8,\quad c_{6,5} = 3;$
$p_{2,1} = 0.4,\quad p_{2,3} = 0.4,\quad p_{2,5} = 0.2,\quad p_{3,3} = 1,\quad p_{4,4} = 0.5,$
$p_{4,5} = 0.5,\quad p_{6,3} = 0.6,\quad p_{6,5} = 0.4.$

We apply Algorithm 2.23. Afterwards, we solve the linear programming problem: Maximize

$$\overline{\Psi}'(\varepsilon, \omega) = \theta_1\omega_1 + \theta_2\omega_2 + \theta_3\omega_3 + \theta_4\omega_4 + \theta_5\omega_5 + \theta_6\omega_6$$

subject to

$$\begin{cases} \varepsilon_1 - \varepsilon_2 + \omega_1 \le c_{1,2};\\ \varepsilon_1 - \varepsilon_4 + \omega_1 \le c_{1,4};\\ \varepsilon_5 - \varepsilon_4 + \omega_5 \le c_{5,4};\\ \varepsilon_5 - \varepsilon_5 + \omega_5 \le c_{5,5};\\ \varepsilon_2 - (p_{2,1}\varepsilon_1 + p_{2,3}\varepsilon_3 + p_{2,5}\varepsilon_5) + \omega_2 \le \mu_2;\\ \varepsilon_3 - p_{3,3}\varepsilon_3 + \omega_3 \le \mu_3;\\ \varepsilon_4 - (p_{4,4}\varepsilon_4 + p_{4,5}\varepsilon_5) + \omega_4 \le \mu_4;\\ \varepsilon_6 - (p_{6,3}\varepsilon_3 + p_{6,5}\varepsilon_5) + \omega_6 \le \mu_6;\\ \omega_1 - \omega_2 \le 0, \quad \omega_1 - \omega_4 \le 0;\\ \omega_5 - \omega_5 \le 0, \quad \omega_5 - \omega_4 \le 0;\\ \omega_2 - (p_{2,1}\omega_1 + p_{2,3}\omega_3 + p_{2,5}\omega_5) \le 0;\\ \omega_3 - p_{3,3}\omega_3 \le 0;\\ \omega_4 - (p_{4,4}\omega_4 + p_{4,5}\omega_5) \le 0;\\ \omega_6 - (p_{6,3}\omega_3 + p_{6,5}\omega_5) \le 0. \end{cases}$$

Here

$$\theta_1 = \theta_2 = \theta_3 = \theta_4 = \theta_5 = \theta_6 = \frac{1}{6}$$

and

$$\mu_2 = 2,\ \mu_3 = 1,\ \mu_4 = 4,\ \mu_6 = 6.$$

If we introduce these data in the linear programming model above then we obtain the problem:
Maximize

$$\overline{\Psi}'(\varepsilon, \omega) = \frac{1}{6}\omega_1 + \frac{1}{6}\omega_2 + \frac{1}{6}\omega_3 + \frac{1}{6}\omega_4 + \frac{1}{6}\omega_5 + \frac{1}{6}\omega_6$$

subject to

$$\begin{cases} \varepsilon_1 - \varepsilon_2 + \omega_1 \le 4; \\ \varepsilon_1 - \varepsilon_4 + \omega_1 \le 1; \\ \varepsilon_5 - \varepsilon_4 + \omega_5 \le 3; \\ \qquad \omega_5 \le 2; \\ \varepsilon_2 - 0.4\varepsilon_1 - 0.4\varepsilon_3 - 0.2\varepsilon_5 + \omega_2 \le 2; \\ \qquad \omega_3 \le 1; \\ \varepsilon_4 - 0.5\varepsilon_4 - 0.5\varepsilon_5 + \omega_4 \le 4; \\ \varepsilon_6 - 0.6\varepsilon_3 - 0.4\varepsilon_5 + \omega_6 \le 6; \\ \omega_1 - \omega_2 \le 0, \quad \omega_1 - \omega_4 \le 0, \quad \omega_5 - \omega_4 \le 0; \\ \omega_2 - 0.4\omega_1 - 0.4\omega_3 - 0.2\omega_5 \le 0; \\ \omega_4 - 0.5\omega_4 - 0.5\omega_5 \le 0; \\ \omega_6 - 0.6\omega_3 - 0.4\omega_5 \le 0. \end{cases}$$

The optimal solution of this problem that satisfies the conditions (2.52), (2.53) is:

$$\varepsilon_1^* = 0,\ \varepsilon_2^* = -\frac{8}{3},\ \varepsilon_3^* = -\frac{25}{3},\ \varepsilon_4^* = 4,\ \varepsilon_5^* = 0,\ \varepsilon_6^* = -\frac{2}{5};$$
$$\omega_1^* = \frac{4}{3},\ \omega_2^* = \frac{4}{3},\ \omega_3^* = 1,\ \omega_4^* = 2,\ \omega_5^* = 2,\ \omega_6^* = \frac{7}{5}.$$

If we determine the potential transformation

$$\overline{c}_{x,y} = c_{x,y} + \varepsilon_y^* - \varepsilon_x^* - \omega_x^*, \quad \forall (x, y) \in E$$

then we obtain

$$\overline{c}_{1,2} = 0,\ \overline{c}_{1,4} = \frac{11}{3},\ \overline{c}_{2,1} = \frac{7}{3},\ \overline{c}_{2,3} = -4,\ \overline{c}_{2,5} = \frac{10}{3},$$
$$\overline{c}_{3,3} = 0,\ \overline{c}_{4,5} = -2,\ \overline{c}_{4,4} = 2,\ \overline{c}_{5,4} = 5,\ \overline{c}_{5,5} = 0,\ \overline{c}_{6,3} = -\frac{4}{3},\ \overline{c}_{6,5} = 2;$$

$$\overline{\mu}_2 = 0,\ \overline{\mu}_3 = 0,\ \overline{\mu}_4 = 0,\ \overline{\mu}_6 = 0.$$

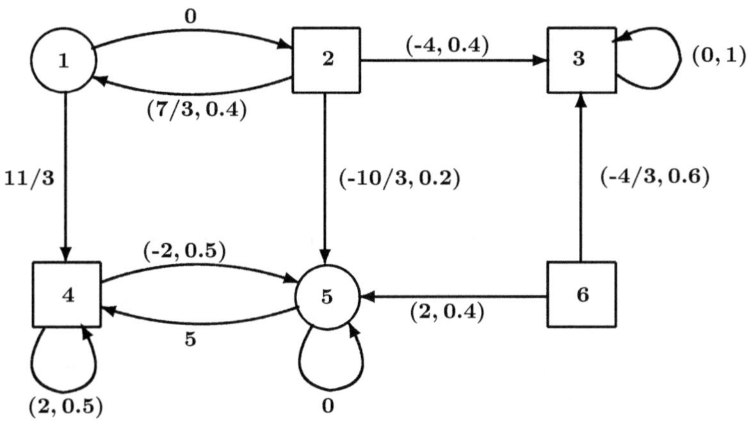

Fig. 2.4 The network $(G, X_C, X_N, \overline{c}, p)$ in canonical form

The network $(G, X_C, X_N, \overline{c}, p)$ in canonical form is represented by Fig. 2.4. This network satisfies the conditions:

(1) $\min\{\overline{c}_{1,1}, \overline{c}_{1,4}\} = 0, \ \min\{\overline{c}_{5,5}, \overline{c}_{5,4}\} = 0;$

(2) $\overline{\mu}_2 = 0, \ \overline{\mu}_3 = 0, \ \overline{\mu}_4 = 0, \ \overline{\mu}_6 = 0.$

For a given optimal solution ε_x^*, ω_x^* for $x \in X$ we have $E_{h^*}^*(1) = E_{\overline{c}}(1) = \{(1, 2)\}$ and $E_{h^*}^*(5) = E_{\overline{c}}(5) = \{(5, 5)\}$.

Therefore, if we fix $s^*(1) = 2$; $s^*(5) = 5$ then we obtain the optimal stationary strategy $s^* : 1 \to 2; \ 5 \to 5$. The corresponding network induced by the optimal stationary strategy s^* is represented by Fig. 2.5.

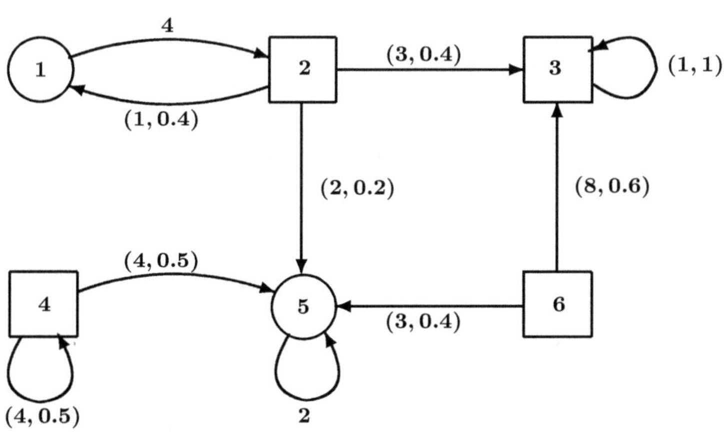

Fig. 2.5 The network induced by the optimal strategy

Another optimal solution of the linear programming problem for this example is:

$$\varepsilon_1^* = 0,\ \varepsilon_2^* = -\frac{8}{3},\ \ \varepsilon_3^* = -\frac{13}{2},\ \varepsilon_4^* = \frac{1}{3},\ \varepsilon_5^* = -\frac{11}{3},\ \varepsilon_6^* = -\frac{7}{15};$$
$$\omega_1^* = \frac{4}{3},\ \ \omega_2^* = \frac{4}{3},\ \ \omega_3^* = 1,\ \ \omega_4^* = 2,\ \ \omega_5^* = 2,\ \ \omega_6^* = \frac{7}{5}.$$

If we calculate $\overline{c}_{x,y}$ and $\overline{\mu}_x$ that correspond to this optimal solution then we obtain

$$\overline{c}_{1,2} = 0,\ \overline{c}_{1,4} = 0,\ \overline{c}_{5,4} = 3,\ \overline{c}_{5,5} = 0,\ \overline{\mu}_2 = 0,\ \overline{\mu}_3 = 0,\ \overline{\mu}_4 = 0,\ \overline{\mu}_6 = 0.$$

It is easy to observe that in this case $E_{\overline{c}}^*(x) \neq E_{h^*}^*(x)$ for $x = 1$. However, we can determine the optimal solution $s^*(1) = 2$, $s^*(5) = 5$ if we fix the strategy s^* such that $(x, s^*(x)) \in E_{\overline{c}}^*(x) \cap E_{h^*}^*(x)$ for $x = 1$ and $x = 2$, i.e., we obtain the same optimal stationary strategy as in the previous case.

Remark 2.24 If for a multichain control problem it is necessary to determine the optimal stationary strategy s^* only for a fixed starting state x_0 then it is sufficient to solve the linear programming problem:
Maximize

$$\overline{\psi}'(\varepsilon, \omega) = \omega_{x_0} \tag{2.57}$$

subject to (2.54). The optimal strategy for the considered problem can be found using Algorithm 2.23 if in the item 1 we exchange the problem (2.54), (2.55) by the problem (2.55), (2.57).

If in the example above we fix $x_0 = 1$ and solve the linear programming problem (2.55), (2.57) then we obtain the optimal solution ε^*, ω^*, where

$$\varepsilon_1^* = 0,\ \ \varepsilon_2^* = -\frac{8}{3},\ \ \varepsilon_3^* = -\frac{25}{3},\ \ \varepsilon_4^* = 4,\ \ \varepsilon_5^* = 0;$$
$$\omega_1^* = \frac{4}{3},\ \ \omega_2^* = \frac{4}{3},\ \ \omega_3^* = 1,\ \ \omega_4^* = 2,\ \ \omega_5^* = 2$$

and ε_6^*, ω_6^* are arbitrary values that satisfy the conditions

$$\varepsilon_3^* - 0.6\varepsilon_3^* - 0.4\varepsilon_5^* + \omega_6^* \leq 6,\ \ \omega_6^* - 0.6\omega_3^* - 0.4\omega_5^* \leq 0.$$

Here ε_6^* may differ from $-2/5$ and ω_6^* may differ from $7/5$. In this case we obtain the same optimal strategy $s^* : 1 \to 2;\ 5 \to 5$ but we do not obtain ε_6^* and ω_6^*. If we solve the problem (2.55), (2.57) for $x_0 = 6$ then we obtain

$$\varepsilon_6^* = -\frac{25}{3},\ \omega_6^* = \frac{7}{5},\ \varepsilon_3^* = 0,\ \omega_3^* = 1,\ \varepsilon_5^* = 0,\ \omega_5^* = 2.$$

The remaining variables may be arbitrary.

2.2.8 Primal and Dual Linear Programming Models for the Multichain Problem

The problem (2.54), (2.55) generalizes the unichain dual linear programming model (2.21), (2.22). Therefore, we can regard (2.54), (2.55) as the dual problem of a primal multichain linear programming model. If we dualize (2.54), (2.55) then we obtain a problem which generalizes the problem (2.16), (2.17). This problem can be formulated as follows:
Minimize

$$\overline{\psi}(\alpha, \beta, \lambda, q) = \sum_{x \in X_C} \sum_{y \in X(x)} c_{x,y} \alpha_{x,y} + \sum_{z \in X_N} \mu_z q_z \tag{2.58}$$

subject to

$$\begin{cases} \sum\limits_{x \in X_C^-(y)} \alpha_{x,y} - \sum\limits_{x \in X(y)} \alpha_{y,x} + \sum\limits_{x \in X_N} p_{x,y} q_x = 0, \quad \forall y \in X_C; \\ \sum\limits_{x \in X_C^-(y)} \alpha_{x,y} - q_y + \sum\limits_{x \in X_N} p_{x,y} q_x = 0, \quad \forall y \in X_N; \\ \sum\limits_{y \in X(x)} \alpha_{x,y} + \sum\limits_{y \in X} \beta_{x,y} - \sum\limits_{y \in X_C^-(x)} \beta_{y,x} - \sum\limits_{y \in X_N^-(x)} p_{y,x} \lambda_y = \theta_x, \forall x \in X_C; \\ q_x + \lambda_x - \sum\limits_{y \in X_N^-(x)} p_{y,x} \lambda_x = \theta_x, \ \forall x \in X_N; \\ \alpha_{x,y}, \beta_{x,y} \geq 0, \ \forall x \in X_C, \ y \in X(x); \ q_x, \lambda_x \geq 0, \ \forall x \in X_N. \end{cases} \tag{2.59}$$

It is easy to see that this linear programming model generalizes the unichain linear programming model (2.16), (2.17). The last two restrictions (equalities) in (2.59) generalize the constraint

$$\sum_{x \in X_C} \sum_{y \in X(x)} \alpha_{x,y} + \sum_{x \in X_N} q_x = 1.$$

In the following we shall regard the linear programming problem (2.54), (2.55).

2.2.9 An Algorithm for Solving the Multichain Control Problem Using a Dual Unichain Model

For multichain control problems the optimal average costs in different states may be different. Therefore, the set of states X can be divided into several subsets $X_1, X_2, \ldots, X_k$ such that each subset X_i, $i \in \{1, 2, \ldots, k\}$ contains the states with the same optimal average costs and there are no states from different subsets with the same optimal average costs.

Let ω^i be the corresponding optimal average cost of the states $x \in X_i$, $i = 1, 2, \ldots, k$ and assume that $\omega^1 < \omega^2 < \cdots < \omega^k$. In this section we show that the average costs ω^i and the corresponding subsets X_i can be found successively by solving k unichain linear programming problems (2.21), (2.22).

At the first step of the algorithm we solve the linear programming problem:
Maximize

$$\overline{\psi}'(\varepsilon, h) = h \tag{2.60}$$

subject to

$$\begin{cases} \varepsilon_x - \varepsilon_y + h \leq c_{x,y}, & \forall x \in X_C,\ y \in X(x); \\ \varepsilon_x - \sum\limits_{z \in X} p_{x,z}\varepsilon_z + h \leq \mu_x, & \forall x \in X_N. \end{cases} \tag{2.61}$$

Let ε_x^1 $(x \in X)$, h^1 be an optimal solution of this problem on the network (G, X_C, X_N, c, p). Then this solution satisfies the conditions:

(1) $c_{x,y}^1 = c_{x,y} + \varepsilon_y^1 - \varepsilon_x^1 - h^1 \geq 0, \qquad \forall x \in X_C,\ y \in X(x);$

(2) $\mu_x^1 = \mu_x + \sum\limits_{y \in X(x)} p_{x,y}\varepsilon_y^1 - \varepsilon_x^1 - h^1 \geq 0, \qquad \forall x \in X_N;$

(3) There exists a nonempty subset X_1 from X where
$\min\limits_{y \in X(x)} c_{x,y}^1 = \min\limits_{y \in X_1(x)} c_{x,y}^1 = 0,\ \forall x \in X_1 \cap X_C;$
$\mu_x^1 = 0,\ \forall x \in X_1 \cap X_N,$
and X_1 is a maximal subset in X with such a property.

If in the network (G, X_C, X_N, c, p) we make the potential transformation

$$c_{x,y}^1 = c_{x,y} + \varepsilon_y^1 - \varepsilon_x^1 - h^1, \qquad \forall x \in X,\ y \in X(x)$$

then we obtain the network (G, X_C, X_N, c^1, p) with a new cost function c^1 on E. According to Lemma 2.14 and Corollary 2.16 the optimal stationary strategies of the control problem on this network are the same as the optimal stationary strategies on the network (G, X_C, X_N, c, p). Moreover, we have here

$$\overline{\omega}_x^1 = \omega_x - h^1, \quad \forall x \in X,$$

where ω_x for $x \in X$ represents the corresponding optimal average costs of the states $x \in X$ in the primal problem and $\overline{\omega}_x^1$ are the optimal average costs of the states in the control problem on the network with transformation potential function c^1.

Thus, after the first step of the algorithm we obtain the subset X_1, the value of the optimal average cost $\omega^1 = h^1$ for the states $x \in X_1$, the function $\varepsilon^1 : X \to \mathbb{R}$

and the network (G, X_C, X_N, c^1, p) with a new cost function c^1, where the optimal average costs $\overline{\omega}_x^1$ in the problem with the new network satisfy the condition:

$$\overline{\omega}_x^1 = 0, \;\; \forall x \in X_1; \;\; \overline{\omega}_x^1 = \omega_x - h^1 > 0, \;\; \forall x \in X \setminus X_1.$$

At the second step of the algorithm we solve the linear programming problem: Minimize the objective function (2.60) subject to

$$\begin{cases} \varepsilon_x - \varepsilon_y + h \leq c^1_{x,y}, & \forall x \in X_C \setminus X_1, \; y \in X(x); \\ \varepsilon_x - \sum\limits_{z \in X} p_{x,z}\varepsilon_z + h \leq \mu^1_x, & \forall x \in X_N \setminus X_1; \\ \varepsilon_x - \varepsilon_y \leq c^1_{x,y}, & \forall x \in X_1 \cap X_C, \; y \in X(x); \\ \varepsilon_x - \sum\limits_{z \in X)} p_{x,z}\varepsilon_z \leq \mu^1_x, & \forall x \in X_1 \cap X_N. \end{cases} \tag{2.62}$$

This system is obtained from (2.61) by changing $c_{x,y}$ and μ_x by $c_{x,y}$ and μ_x, and setting $h = 0$ in the inequalities that correspond to the states $x \in X_1$.

Let ε_x^2 $(x \in X)$, h^2 be an optimal solution of this problem on the network (G, X_C, X_N, c^1, p). Then this solution satisfies the conditions:

(1) $c^2_{x,y} = c_{x,y} + \varepsilon^2_y - \varepsilon^2_x - h^2 \geq 0, \qquad \forall x \in X_C, \; y \in X(x);$

(2) $\mu^2_x = \mu_x + \sum\limits_{y \in X(x)} p_{x,y}\varepsilon^2_y - \varepsilon^2_x - h^2 \geq 0, \qquad \forall x \in X_N;$

(3) There exists a nonempty subset X_2 from X where

$$\min_{y \in X(x)} c^2_{x,y} = \min_{y \in X_2(x)} c^2_{x,y} = 0, \; \forall x \in X_2 \cap X_C;$$

$$\mu^2_x = 0, \; \forall x \in X^2 \cap X_N,$$

and X_2 is a maximal subset in X with such a property.

After that we make the potential transformation

$$c^2_{x,y} = c^1_{x,y} + \varepsilon^2_y - \varepsilon^2_x - h^2, \qquad \forall x \in X, \; y \in X(x)$$

in the network (G, X_C, X_N, c^1, p) and we obtain the network (G, X_C, X_N, c^2, p) with a new cost function c^2 on E. According to Lemma 2.14 and Corollary 2.16 the optimal stationary strategies of the control problem on this network are the same as the optimal stationary strategies on the network (G, X_C, X_N, c^1, p). Moreover, here we have

$$\overline{\omega}^2_x = \overline{\omega}^1_x - h^2, \;\; \forall x \in X \setminus X_1,$$

where $\overline{\omega}^1_x$ for $x \in X \setminus X_1$ represent the corresponding optimal average costs of the states in the problem before the potential transformation is made and $\overline{\omega}^2_x$ are the optimal average costs of the states $x \in X \backslash X_1$ in the control problem after the potential transformation is made.

Thus, after the second step of the algorithm we obtain the subset X_2, the value of the optimal average cost h^2 for the states $x \in X_2$, the function $\varepsilon^2 : X \to \mathbb{R}$ and the network (G, X_C, X_N, c^2, p) with a new cost function c^2, where for the optimal average costs $\overline{\omega}_x^2$ in the problem we may set:

$$\overline{\omega}_x^2 = 0, \ \forall x \in X_1 \cup X_2; \quad \overline{\omega}_x^2 = \overline{\omega}_x^1 - h^2 > 0, \ \forall x \in X \backslash (X_1 \cup X_2).$$

At the next step of the algorithm we solve the linear programming problem: Minimize the objective function (2.60) subject to

$$\begin{cases} \varepsilon_x - \varepsilon_y + h \le c_{x,y}^2, & \forall x \in X_C \backslash (X_1 \cup X_2), \ y \in X(x); \\ \varepsilon_x - \sum\limits_{z \in X} p_{x,z}\varepsilon_z + h \le \mu_x^2, & \forall x \in X_C \backslash (X_1 \cup X_2); \\ \varepsilon_x - \varepsilon_y \le c_{x,y}^2, & \forall x \in (X_1 \cup X_2) \cap X_C; \\ \varepsilon_x - \sum\limits_{z \in X} p_{x,z}\varepsilon_z \le \mu_x^2, & \forall x \in (X_1 \cup X_2) \cap X_N. \end{cases} \tag{2.63}$$

This system is obtained from (2.62) by exchanging $c_{x,y}^1$ and μ_x^1 by $c_{x,y}^2$ and μ_x^2, and setting $h = 0$ in the inequalities that corresponds to the states $x \in X_2$.

After a finite number of steps we obtain the subsets

$$X_1, X_2, \ldots, X_k \ \ (X = X_1 \cup X_2 \cup \cdots \cup X_k),$$

the potential functions $\varepsilon^i : X \to \mathbb{R}, \quad i = 1, 2, \ldots, k$ and the values $h^1, h^2, \ldots, h^k$, where

$$\omega^i = \sum_{j=1}^{i} h^j, \quad j = 1, 2, \ldots, k.$$

If we find $\varepsilon_x^* = \sum_{i=1}^k \varepsilon_x^i$ and fix $\omega_x^* = \omega^{i^*}$ for $x \in X_i$ then we determine the potential transformation

$$\overline{c}_{x,y} = c_{x,y} + \varepsilon_y^* - \varepsilon_x^* - \omega_x^*, \ \forall x \in X, y \in X(x),$$

that satisfies the conditions (1) – (6) of Theorem 2.12. This means that we determine the network $(G, X_C, X_N, \overline{c}, p)$ and the optimal stationary strategy s^*.

Example Consider the stochastic control problem on the network with the data from the example given in the previous section. The network is represented by Fig. 2.3, where $X = X_C \cup X_N, \ X_C = \{1, 5\}, \ X_N = \{2, 3, 4, 6\}$, and the costs and transition probabilities are written again along the edges.

We apply the algorithm described above. At the first step of the algorithm we solve the linear programming problem:
Minimize

$$\overline{\psi}'(\varepsilon, h) = h$$

subject to

$$\begin{cases} \varepsilon_1 - \varepsilon_2 + h \leq 4; \\ \varepsilon_1 - \varepsilon_4 + h \leq 1; \\ \varepsilon_5 - \varepsilon_4 + h \leq 3; \\ \varepsilon_5 - \varepsilon_5 + h \leq 2; \\ \varepsilon_2 - 0.4\varepsilon_1 - 0.4\varepsilon_3 - 0.2\varepsilon_5 + h \leq 2; \\ \varepsilon_3 - \varepsilon_3 + h \leq 1; \\ \varepsilon_4 - 0.5\varepsilon_4 - 0.5\varepsilon_5 + h \leq 4; \\ \varepsilon_6 - 0.6\varepsilon_3 - 0.4\varepsilon_5 + h \leq 6. \end{cases}$$

An optimal solution of this problem is $h^1 = 1,\ \varepsilon_1^1 = 0,\ \varepsilon_2^1 = 0,\ \varepsilon_3^1 = 0,\ \varepsilon_4^1 = 0,\ \varepsilon_5^1 = 0,\ \varepsilon_6^1 = 0 = 1$. We calculate $c_{x,y}^1$ and μ_x^1 using the formula

$$c_{x,y}^1 = c_{x,y} + \varepsilon_y^1 - \varepsilon_x^1 - h^1, \ \forall x \in X_1,\ y \in X(x);$$
$$\mu_x^1 = \mu_x + \sum_{y \in X(x)} p_{x,y} c_{x,y} - \varepsilon_x^1 - h^1, \ x \in X_2$$

and determine $c_{1,2}^1 = 3,\ c_{1,4}^1 = 0,\ c_{5,4}^1 = 2,\ c_{5,5}^1 = 1;\ \mu_2^1 = 1,\ \mu_3^1 = 0,\ \mu_4^1 = 3,\ \mu_6^1 = 5$. After the first step of the algorithm we obtain:

$$X_1 = \{3\};\ h^1 = 1;\ \varepsilon_1^1 = 0,\ \varepsilon_2^1 = 0,\ \varepsilon_3^1 = 0,\ \varepsilon_4^1 = 0,\ \varepsilon_5^1 = 0,\ \varepsilon_6^1 = 0.$$

At the second step of the algorithm we solve the linear programming problem:
Minimize

$$\overline{\psi}'(\varepsilon, h) = h$$

subject to

$$\begin{cases} \varepsilon_1 - \varepsilon_2 + h \leq 3; \\ \varepsilon_1 - \varepsilon_4 + h \leq 0; \\ \varepsilon_5 - \varepsilon_4 + h \leq 2; \\ \varepsilon_5 - \varepsilon_5 + h \leq 1; \\ \varepsilon_3 - \varepsilon_3 \leq 0; \\ \varepsilon_2 - 0.4\varepsilon_1 - 0.4\varepsilon_3 - 0.2\varepsilon_5 + h \leq 1; \\ \varepsilon_4 - 0.5\varepsilon_4 - 0.5\varepsilon_5 + h \leq 3; \\ \varepsilon_6 - 0.6\varepsilon_3 - 0.4\varepsilon_5 + \omega \leq 5. \end{cases}$$

An optimal solution of this problem is

$$h^2 = \frac{1}{3},\ \varepsilon_1^2 = 0,\ \varepsilon_2^2 = -\frac{8}{3},\ \varepsilon_3^2 = -\frac{25}{3},\ \varepsilon_4^2 = 4,\ \varepsilon_5^2 = 0,\ \varepsilon_6^2 = -\frac{2}{5}.$$

We calculate $c_{x,y}^2$ and μ_x^2 using formula

$$c_{x,y}^2 = c_{x,y}^1 + \varepsilon_y^2 - \varepsilon_x^2 - h^2;\ \ \mu_x^2 = \mu_x^1 + \sum_{z\in X} p_{x,y}\varepsilon_z - h^2$$

and find

$$c_{1,2}^2 = 0,\ c_{1,4}^2 = \frac{2}{3},\ c_{5,4}^2 = \frac{17}{3},\ c_{5,5}^2 = \frac{2}{3},\ \mu_2^2 = 0,\ \mu_3^2 = 0,\ \mu_4^2 = \frac{2}{3},\ \mu_6^2 = \frac{1}{15}.$$

After the second step of the algorithm we obtain: $X_2 = \{1, 2\}$;

$$h^2 = \frac{1}{3},\ \ \varepsilon_1^2 = 0,\ \ \varepsilon_2^2 = -\frac{8}{3},\ \ \varepsilon_3^2 = -\frac{25}{3},\ \ \varepsilon_4^2 = 4,\ \ \varepsilon_5^2 = 0,\ \ \varepsilon_6^2 = -\frac{2}{5}.$$

At the third step of the algorithm we solve the linear programming problem: Minimize

$$\overline{\psi}'(\varepsilon, h) = h$$

subject to

$$\begin{cases} \varepsilon_1 - \varepsilon_2 \le 0; \\ \varepsilon_1 - \varepsilon_4 \le \frac{11}{3}; \\ \varepsilon_5 - \varepsilon_4 + h \le \frac{17}{3}; \\ \varepsilon_3 - \varepsilon_3 \le 0; \\ \varepsilon_5 - \varepsilon_5 + h \le \frac{2}{3}; \\ \varepsilon_2 - 0.4\varepsilon_1 - 0.4\varepsilon_3 - 0.2\varepsilon_5 \le \frac{2}{3}; \\ \varepsilon_4 - 0.5\varepsilon_4 - 0.5\varepsilon_5 + h \le \frac{2}{3}; \\ \varepsilon_6 - 0.6\varepsilon_3 - 0.4\varepsilon_5 + h \le \frac{1}{15}. \end{cases}$$

An optimal solution of this problem is

$$h^3 = \frac{1}{15},\ \varepsilon_1^3 = 0,\ \varepsilon_2^3 = 0,\ \varepsilon_3^3 = 0,\ \varepsilon_4^3 = 0,\ \varepsilon_5^3 = 0,\ \varepsilon_6 = 0.$$

Using this solution we find

$$c^3_{1,2} = 0,\ c^4_{1,4} = \frac{11}{3},\ c^3_{5,5} = \frac{3}{5},\ c^3_{5,4} = \frac{26}{5},\ \mu^3_2 = 0,\ \ \mu^3_3 = 0,\ \mu^3_4 = \frac{3}{5},\ \mu^3_6 = 0.$$

After this step we obtain:

$$X_3 = \{6\};\ h^3 = \frac{1}{15},\ \varepsilon^3_1 = 0,\ \varepsilon^3_2 = 0,\ \varepsilon^3_3 = 0,\ \varepsilon^3_4 = 0,\ \varepsilon^3_5 = 0,\ \varepsilon_6 = 0.$$

At the fourth step of the algorithm we solve the linear programming problem: Minimize

$$\overline{\psi}\,'(\varepsilon, h) = h$$

subject to

$$\begin{cases} \varepsilon_1 - \varepsilon_2 \leq 0; \\ \varepsilon_1 - \varepsilon_4 \leq \dfrac{11}{3}; \\ \varepsilon_5 - \varepsilon_4 + h \leq \dfrac{28}{5}; \\ \varepsilon_3 - \varepsilon_3 \leq 0; \\ \varepsilon_5 - \varepsilon_5 + h \leq \dfrac{3}{5}; \\ \varepsilon_2 - 0.4\varepsilon_1 - 0.4\varepsilon_3 - 0.2\varepsilon_5 \leq \dfrac{2}{3}; \\ \varepsilon_4 - 0.5\varepsilon_4 - 0.5\varepsilon_5 + h \leq \dfrac{3}{5}; \\ \varepsilon_6 - 0.6\varepsilon_3 - 0.4\varepsilon_5 \leq 0. \end{cases}$$

An optimal solution of this system is $h^4 = 3/5$, $\varepsilon^3_1 = 0$, $\varepsilon^3_2 = 0$, $\varepsilon^3_3 = 0$, $\varepsilon^3_4 = 0$, $\varepsilon^3_5 = 0$, $\varepsilon_6 = 0$. Using this solution we find $c^4_{1,2} = 0$, $c^4_{2,4} = 11/3$, $c^4_{5,4} = 5$, $c^4_{5,5} = 0$, $\mu^4_2 = 0$, $\mu^4_3 = 0$, $\mu^4_4 = 0$, $\mu^4_6 = 0$. After this step we obtain $X_4 = \{4, 5\}$ and $h^4 = 3/5$.

Thus, finally we have $X = X_1 \cup X_2 \cup X_3 \cup X_4$, where

$$X_1 = \{3\},\ X_2 = \{1, 2\},\ X_3 = \{6\},\ X_4 = \{4, 5\},$$

and

$$\omega^1 = h^1,\ \omega^2 = h^1 + h^2,\ \omega 3 = h^1 + h^2 + h^3,\ \omega^4 = h^1 + h^2 + h^3 + h^4,$$

i.e.,

$$\omega^1 = 1,\ \omega^2 = \frac{4}{3},\ \omega^3 = \frac{7}{5},\ \omega^4 = 2.$$

In addition we can find

$$\varepsilon_1^* = \varepsilon_1^1 + \varepsilon_1^2 + \varepsilon_1^3 + \varepsilon_1^4 = 0; \quad \varepsilon_2^* = \varepsilon_2^1 + \varepsilon_2^2 + \varepsilon_2^3 + \varepsilon_2^4 = -\frac{8}{3};$$
$$\varepsilon_3^* = \varepsilon_3^1 + \varepsilon_3^2 + \varepsilon_3^3 + \varepsilon_3^4 = -\frac{25}{3}; \quad \varepsilon_4^* = \varepsilon_4^1 + \varepsilon_4^2 + \varepsilon_4^3 + \varepsilon_4^4 = 4;$$
$$\varepsilon_5^* = \varepsilon_5^1 + \varepsilon_5^2 + \varepsilon_5^3 + \varepsilon_5^4 = 0; \quad \varepsilon_6^* = \varepsilon_6^1 + \varepsilon_6^2 + \varepsilon_6^3 + \varepsilon_6^4 = -\frac{2}{5}.$$

If we make the potential transformation of the cost function c for ω^* and ε^* found above then we obtain the network in canonical form $(G, X_C, X_N, \overline{c}, p)$ represented by Fig. 2.4 that gives the optimal stationary strategies.

2.2.10 An Approach for Solving the Multichain Control Problem Using a Reduction Procedure to a Unichain Problem

We consider the stochastic control problem on the network (G, X_C, X_N, c, p, x_0) with fixed starting state x_0 and describe an approximation algorithm for determining the optimal solutions which is based on a reduction procedure of the multichain problem to the unichain case.

We describe the reduction procedure in the case if the graph G satisfies the condition that for an arbitrary vertex $x \in X_C$ each outgoing directed edge $e = (x, y)$ ends in X_N, i.e., we assume that

$$E_C = \{e = (x, y) \in E \mid x \in X_C, \ y \in X_N\}.$$

If the graph G does not satisfy this condition then the considered control problem can be reduced to a similar control problem on an auxiliary network $(G', X'_C, X'_N, c', p', x_0)$, where the graph G' satisfies the condition mentioned above. Graph $G' = (X', E')$ is obtained from $G = (X, E)$, where each directed edge $e = (x, y) \in E_C$ is changed by the following two directed edges $e^1 = (x, x_e)$ and $e^2 = (x_e, y)$.

We include each vertex x_e in X'_N and to each edge $e' = (x_e, y)$ we associate the cost $c'_{x_e,y} = c_{x,y}$ and the transition probability $p'_{x_e,y} = 1$. To the edges $e' = (x, x_e)$ we associate the cost $c'_{x,x_e} = c_{(x,y)}$, where $e = (x, y)$. For the edges $e \in E_N$ in the new network we preserve the same costs and transition probabilities as in the initial network, i.e., the cost function c' on E_N and on the set of edges (x, x_e) for $x \in X_C, \ e \in E_C$ is induced by the cost function c. Thus, in the auxiliary network the graph G' is determined by the set of vertices $X' = X'_C \cup X'_N$ and the set of edges $E' = E'_C \cup E'_N$, where $X'_C = X_C$; $X'_N = X_N \cup \{x_e, \ e \in E_C\}$; $E'_C = \{e' = (x, x_e) \mid x \in X_C, e = (x, y) \in E_C\}$; $E'_N = E_N \cup \{e' = (x_e, y) \mid e = (x, y) \in E_C, y \in X\}$. It is evident that there exists a bijective mapping between the set of strategies in the states $x \in X_C$ of the network (G, X_C, X_N, c, p, x_0) and the set of strategies in the states $x \in X_C$ of the network $(G', X'_C, X'_N, c', p', x_0)$ that preserves the average costs of the problems on the corresponding networks.

Thus, without loss of generality we may consider that G possesses the property that for an arbitrary vertex $x \in X_C$ each outgoing directed edge $e = (x, y)$ ends in X_N. Additionally, let us assume that the vertex x_0 in G is reachable from every vertex $x \in X_N$. Then an arbitrary strategy s in the considered problem induces a transition probability matrix $P^s = (p^s_{x,y})$ that corresponds to a Markov unichain with a positive recurrent class X^+ that contains the vertex x_0.

Therefore, if we solve the control problem on the network then we obtain the solution of the problem with fixed starting state x_0. So, we obtain such a solution if the network satisfies the condition that for an arbitrary strategy s the vertex x_0 in G_s is attainable for every $x \in X_N$. Now let us assume that this property does not take place. In this case we can reduce our problem to a similar problem on a new auxiliary network $(G'', X''_C, X''_N, p'', c'', x_0)$ for which the property mentioned above holds. This network is obtained from the initial one by the following way: We construct the graph $G'' = (X, E'')$ which is obtained from $G = (X, E)$ by adding new directed edges $e''_{x_0} = (x, x_0)$ from $x \in X_N \backslash \{x_0\}$ to x_0, if for some vertices $x \in X_N \backslash \{x_0\}$ in G there are no directed edges $e = (x, x_0)$ from x to x_0. We define the costs of directed edges $(x, y) \in E''$ in G'' as follows: If $e'' = (x, y) \in E$ then the cost $c''_{e''}$ of this edge in G'' is the same as in G, i.e., $c''_{e''} = c_{e''}$ for $e'' \in E$; if $e'' = (x, x_0) \in E'' \backslash E$ then we put $c''_{e''} = 0$. The probabilities $p''_{x,y}$ for $(x, y) \in E''$ where $x \in X_N$ we define by using the following rule: We fix a small positive value ϵ and put $p''_{x,y} = p_{x,y} - \epsilon p_{x,y}$ if $(x, y) \in E'' \backslash E, \quad y \neq x_0$ and in G there is no directed edge $e = (x, x_0)$ from x to x_0; if in G for a vertex $x \in X \backslash \{x_0\}$ there exists a leaving directed edge $e = (x, x_0)$ then for an arbitrary outgoing directed edge $e = (x, y), \ y \in X(x)$ we put $p''_{x,y} = p_{x,y}$; for the directed edges $(x, x_0) \in E' \backslash E$ we put $p''_{x,x_0} = \epsilon$.

Let us assume that the probabilities $p_{x,y}$ for $(x, y) \in E$ are given in the form of irreducible decimal fractions $p_{x,y} = a_{x,y}/b_{x,y}$.

Additionally, assume that $\epsilon \le 2^{-2L-2}$, where

$$L = \sum_{(x,y)\in E} \log(a_{x,y} + 1) + \sum_{(x,y)\in E} \log(b_{x,y} + 1) + \sum_{e\in E} \log(|c_e| + 1) + 2\log(n) + 1.$$

Here L is the length of the binary-coded data of the matrix P and of the cost vector c with integer components; each probability $p_{x,y}$ is given by the integer couple $a_{x,y}$, $b_{x,y}$. Then, based on the results from [57, 58] for our auxiliary optimization problem (with approximated data) we can conclude that the solution of this problem will correspond to the solution of our initial problem.

If we consider the control problem on the auxiliary network $(G', X'_C, X'_N, c', p', x_0)$ then we can observe that an arbitrary optimal basic solution of the linear programming problem (2.13), (2.14) satisfies the condition $q^*_{x_0} > 0$ and therefore we can determine the optimal stationary strategy s'^* for the auxiliary problem using Algorithm 2.5. In addition we can observe that if for our stochastic control problem on the network there exists the optimal stationary strategy s^* then it coincides with an optimal stationary strategy s'^* of the stochastic control problem on the auxiliary network, i.e., $s^* = s'^*$. Moreover, the optimal values of the objective functions $f_{x_0}(s^*)$ can be obtained from the optimal value of the objective function $f'_{x_0}(s'^*)$ in the auxiliary problem using the approximation procedure. So, to find the optimal solution of the problem on the network (G, X_C, X_N, c, p, x_0) it is necessary to construct the auxiliary network $(G', X'_C, X'_N, c', p', x_0)$ where for each vertex $x \in X'_N$ an arbitrary directed edge $e' = (x, y)$ ends in X_N. Then we construct the network $(G'', X''_C, X''_N, c'', p'', x_0)$ and the auxiliary stochastic optimal control problem on this network. If the optimal stationary strategy s'^* in the auxiliary problem is found then we fix $s^* = s'^*$ on X_C.

Example Consider the multichain control problem on the network (G, X_C, X_N, c, p, x_0) represented by Fig. 2.6. In this network the vertices represented by circles correspond to the controllable states of the dynamical system and the vertices represented by squares correspond to uncontrollable states. To each edge that originates in the vertices which correspond to the controllable states the associated cost is written along the edge. To each edge that originates in the vertices that correspond to uncontrollable states the associated cost and the transition probability are written in parentheses. The starting state x_0 is fixed and it corresponds to vertex 2, i.e., $x_0 = 2$.

For this network we have $X_C = \{1, 4, 6\}$, $X_N = \{2, 3, 5\}$ and there exist two edges $(1, 1)$, $(1, 4)$ which start in X_C and end in X_C. The corresponding network $(G', X'_C, X'_N, c', p', x_0)$ is represented on Fig. 2.7. This network is obtained from the network on Fig. 2.7 by adding two new vertices $1'$ and $1''$ on the edges $(1, 1)$ and $(1, 4)$.

The network $(G'', X''_C, X''_N, c'', p'', x_0)$ is represented in Fig. 2.8. This network is obtained from the network in Fig. 2.7 by adding the directed edges (x, x_0) that start in the vertices $x \in X''_N = \{1', 1'', 5, 3\}$ and end in $x_0 = 2$, where $c_{x,x_0} = 0$

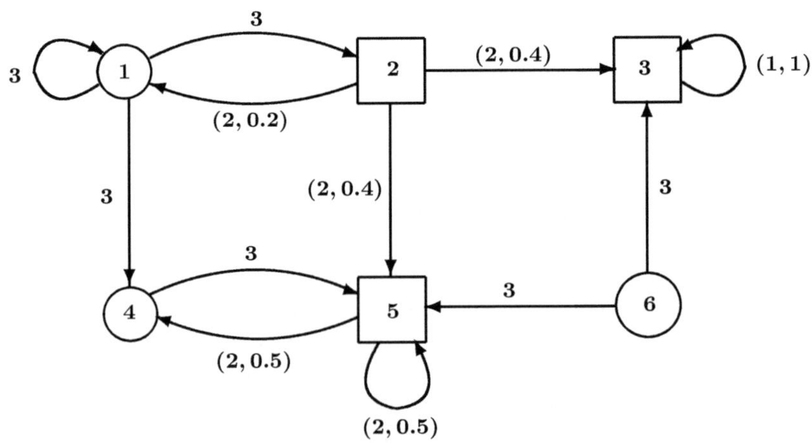

Fig. 2.6 The network (G, X_C, X_N, c, p, x_0)

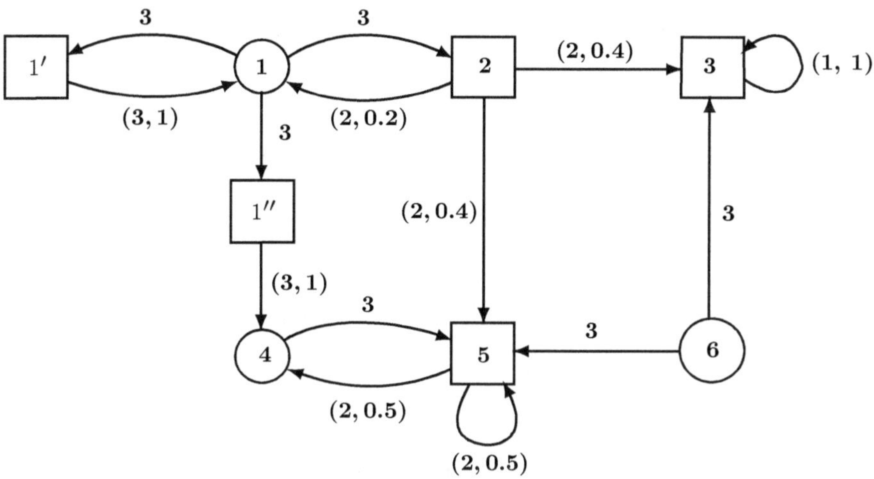

Fig. 2.7 The network $(G', X'_C, X'_N, c', p', x_0)$

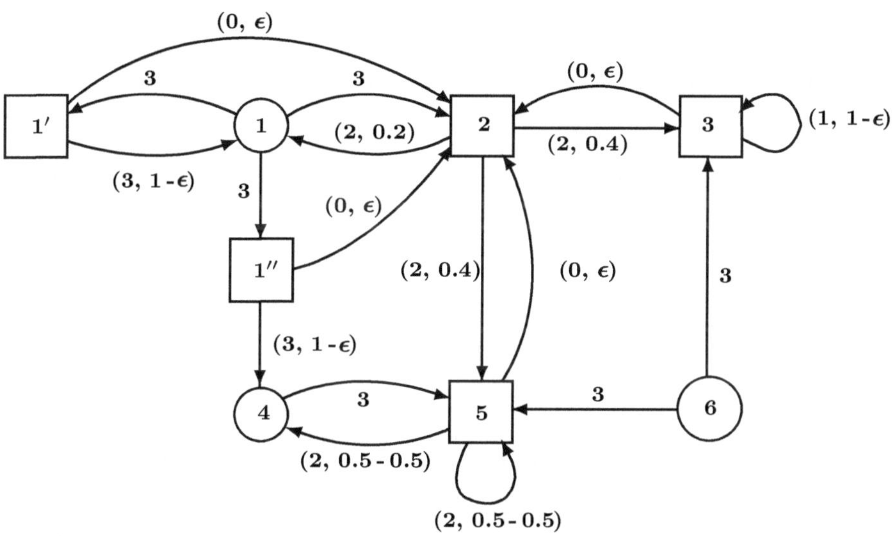

Fig. 2.8 The network $(G'', X''_C, X''_N, c'', p'', x_0)$

and $p_{x,x_0} = \epsilon$. The corresponding probabilities $p_{x,y}$ for the directed edges (x, y) for $x \in X''_N$ are defined as follows: $p''_{x,y} = p_{x,y} - p_{x,y}\epsilon$. The control problem on the auxiliary network possesses the property that an arbitrary strategy s'' generates a Markov unichain. Therefore, for this problem we can use the linear programming model (2.13), (2.14) or the linear programming model (2.21), (2.22) with $\epsilon = 10^{-4}$.

In both cases we determine the same optimal stationary strategy

$$s^{*''} : 1 \to 2; \quad 4 \to 5; \quad 6 \to 3.$$

This means that the optimal solution for the initial problem is

$$s^{*} : 1 \to 2; \quad 4 \to 5; \quad 6 \to 3.$$

In the following we show that the linear programming models for the stochastic control problem can be extended for Markov decision processes which lead to the linear programming models from [25, 45, 46, 51, 115].

2.3 A Linear Programming Approach for Markov Decision Problems with an Average Cost Optimization Criterion

We extend now the linear programming approach and algorithms from the previous section for the Markov decision problem with an average cost optimization criterion. We show that an arbitrary Markov decision problem can be transformed into a stochastic control problem on a network and vice versa, an arbitrary stochastic control problem on a network can be formulated as a Markov decision problem. Thus, the considered problems are equivalent and therefore the linear programming approach can be developed and specified for Markov decision problems.

2.3.1 Problem Formulation

A *Markov decision process* [4, 115] is determined by a tuple (X, A, p, c), where X is a finite state space, A is a finite *set of actions*, p is a nonnegative real function $p : A \times X \times X \to R^{+}$ that satisfies the condition $\sum_{y \in X} p_{x,y}^{a} = 1, \ \ \forall a \in A$ and $c : A \times X \times X \to \mathbb{R}$ is a real function. The function p for a fixed action $a \in A$ and arbitrary $x, y \in X$ determines the probability $p_{x,y}^{a}$ of the system's transition from the state $x \in X$ at the moment of time t to state y at the moment of time $t + 1$ for every $t = 0, 1, 2, \ldots$. For a fixed action $a \in A$ and arbitrary $x, y \in X$ the function c determines the cost $c_{x,y}^{a}$ of the system's transition from the state $x = x(t)$ to the state $y = x(t + 1)$ for $t = 0, 1, 2, \ldots$. In the considered Markov process the functions p and c do not depend on time, i.e., we have a stationary Markov decision process. If in each state $x \in X$ we fix an action from $a \in A$ then we obtain a Markov process induced by these actions. The problem with an average cost optimization criterion for the Markov decision process (X, A, p, c) with given starting state x_0 consists in determining the actions in the states of the system that provide the minimal (or maximal) average cost per transition for the Markov process induced by the chosen actions. In the following we will study this problem in terms of stationary strategies.

We define a stationary strategy s for Markov decision process as a map

$$s : x \rightarrow a \in A(x) \quad \text{for} \quad x \in X,$$

where $A(x)$ represents the set of actions in the state $x \in X$. An arbitrary stationary strategy s induces a simple Markov process with the transition probability matrix $P^s = (p^s_{x,y})$ and the transition cost matrix $C^s = (c^s_{x,y})$. For this Markov process with probability and cost matrices P^s, C^s we can determine the expected average cost per transition $\omega^s_{x_0}$ if the dynamical system starts transitions in the state x_0 at the moment of time $t = 0$. We denote this quantity by $f_{x_0}(s)$, i.e.,

$$f_{x_0}(s) = \omega^s_{x_0}.$$

We consider the Markov decision problem with an average cost criterion, i.e., we are seeking for a strategy s^* for which

$$f_{x_0}(s^*) = \min_s f_{x_0}(s).$$

For an arbitrary Markov decision problem we may assume that the action sets in different states are different, i.e., $A(x) \neq A(y)$. However, it is easy to observe that an arbitrary problem can be reduced to the case $|A(x)| = |A(y)| = |A|, \ \forall x, y \in X$ introducing some copies of the actions in the states $y \in X$ if for two different states $x, y \in X$ it holds $|A(y)| < |A(x)|$.

In the case $|A(x)| = |A(y)| = |A|, \ \forall x, y \in X$ a Markov decision process can be given by $2|A|$ matrices $P^{a_k} = (p^{a_k}_{x,y}), \ C^{a_k} = (c^{a_k}_{x,y}), \quad k = 1, 2, \ldots, |A|$, where $\sum_{y \in X} p^{a_k}_{x,y} = 1, \ \forall a_k \in A, \forall x \in X$.

A fixed strategy $s : x \rightarrow a_k \in A(x)$ for $x \in X$ generates a Markov process with the probability transition matrix P^s and the transition cost matrix C^s induced by the rows of the corresponding matrices P^{a_k} and $C^{a_k}, \quad k = 1, 2, \ldots, |A|$, respectively.

Using the matrix representation of the Markov decision processes we can show that the stochastic control problem with average cost criterion can be represented as a Markov decision problem. Indeed, the matrix representation of the control problem corresponds to the case if $X = X_C \cup X_N, \ X_C \cap X_N = \emptyset$, where for an arbitrary state $x_i \in X_C$ the probabilities $p^{a_k}_{x_i,y}$ are equal to 0 or 1 and for an arbitrary state $x_i \in X_N$ the corresponding $i - th$ rows in the matrices $P^{a_1}, P^{a_2}, \ldots, P^{a_{|A|}}$ and $C^{a_1}, C^{a_2}, \ldots, C^{a_{|A|}}$ are the same. This means that an arbitrary stochastic control problem can be transformed into a Markov decision problem.

In the next section we show that an arbitrary Markov decision problem with average cost criterion can be reduced to a stochastic control problem on an auxiliary network. We can observe that the mentioned reduction procedure can be realized in polynomial time. Thus, the considered problems are equivalent from a computational point of view. Using the reduction procedure of the Markov decision problem to a stochastic control problem we can extend the algorithms from the previous section for determining the optimal solution for Markov decision problems.

2.3.2 Reduction of Markov Decision Problems to Stochastic Control Problems

Let us show that the problem of determining the optimal stationary strategies s^* in a Markov decision process (X, A, p, c) with average cost criterion can be reduced to the problem of determining the optimal stationary strategy in the control problem on a network $(G', X'_C, X'_N, p', c', x'_0)$, where $G' = (X', E')$, X'_C, X'_N, p', c' and x_0 are defined in the following way: The set of vertices $X' = X'_C \cup X'_N$ contains $(|A|+1)|X|$ vertices, where $|X'_C| = |X|$ and $|X'_N| = |A||X|$. So, the set of controllable states in the control problem consists of a copy of the set of states X and the set of uncontrollable states X'_N consists of $|A|$ copies of the set of states X. Therefore, we define X'_C and X_N as follows:

$$X'_C = \{x' = x \mid x \in X\}; \quad X'_N = \bigcup_{a \in A} X^a,$$

where

$$X^a = \{x^a = (x, a) \mid x \in X\} \text{ for } a \in A.$$

We also represent the set of directed edges E' as a couple of two disjoint subsets $E' = E'_C \cup E'_N$, where E'_C is the set of outgoing edges from $x' \in X'_C$ and E'_N is the set of outgoing edges from $x^a \in X'_N$. The states E'_C and E'_N are defined as follows:

$$E'_C = \{(x, (x, a)) \mid x \in X'_C; \ (x, a) \in X'_N, \ a \in A\};$$
$$E'_N = \{((x, a), y) \mid (x, a) \in X'_N, \ y \in X'_C, \ p^a_{x,y} > 0, \ a \in A\}.$$

On the set of directed edges E' we define the cost function $c' : E' \to \mathbb{R}$, where

$$c'_{e'} = 0, \ \ \forall e' = (x, (x, a)) \in E'_C;$$

$$c'_{e'} = 2\, c^a_{x,y} \ \text{ for } \ e' = ((x, a), y) \in E'_N \ \ (x, y \in X, \ a \in A).$$

On E'_2 we define the transition probability function $p' : E'_N \to [0, 1]$, where $p'_{e'} = p^a_{x,y}$ for $e' = ((x, a), y) \in E'_N$.

It is easy to observe that between the set of stationary strategies $\mathbb{S}$ in the Markov decision process and the set of strategies $\mathbb{S}'$ in the control problem on the network $(G', X'_C, X'_N, c', p', x_0)$ there exists a bijective mapping that preserves the average cost per transition. Therefore, if we find the optimal stationary strategy for the control problem on the network then we can determine the optimal stationary strategy in the Markov decision process.

The network constructed above gives a graphical interpretation of the Markov decision process via the structure of the graph G, where the actions and all possible transitions for an arbitrary fixed action are represented by arcs and nodes. A more simple graphical interpretation of the Markov decision process may be given using the graph of probability transitions $G_p = (X, E_p)$, which is induced by the probability function $p : X \times X \times A \to [0, 1]$. This graph may contain parallel directed edges where each directed edge corresponds to an action. The set of vertices X corresponds to the set of states and the set of edges E_p consists of $|A|$ subsets $E_p^1, E_p^2, \ldots, E_p^{|A|}$ $\left(E_p = \bigcup_{i=1}^{|A|} E_p^i\right)$, where $E_p^i = \{e^{a_i} = (x, y)^{a_i} \mid p_{x,y}^{a_i} > 0\}$, $i = 1, 2, \ldots, |A(x)|$.

An example how to construct the graph $G_p = (X, E_p)$ and how to determine the solution of the Markov decision problem using the reduction procedure to an auxiliary control problem on the network is given below.

Example Consider a Markov decision process (X, A, p, c) where $X = \{1, 2\}$, $A = 1, 2$ and the possible values of the corresponding probability and cost functions $p : X \times X \times A \to [0, 1]$, $c : X \times X \times A \to \mathbb{R}$ are defined as follows:

$$
\begin{aligned}
p_{1,1}^{a_1} &= 0.7,\ p_{1,2}^{a_1} = 0.3,\ p_{2,1}^{a_1} = 0.6,\ p_{2,2}^{a_1} = 0.4,\\
p_{1,1}^{a_2} &= 0.4,\ p_{1,2}^{a_2} = 0.6,\ p_{2,1}^{a_2} = 0.5,\ p_{2,2}^{a_2} = 0.5;\\
c_{1,1}^{a_1} &= 1,\ c_{1,2}^{a_1} = 0,\ c_{2,1}^{a_1} = -2,\ c_{2,2}^{a_1} = 5,\\
c_{1,1}^{a_2} &= 0,\ c_{1,2}^{a_2} = 4,\ c_{2,1}^{a_2} = 2,\ c_{2,2}^{a_2} = -3.
\end{aligned}
$$

We consider the problem of finding the optimal stationary strategy for the corresponding Markov decision problem with minimal average costs and an arbitrary fixed starting state.

The data concerned with the actions in the considered Markov decision problem can be represented in a suitable form using the probability matrices

$$
P^{a_1} = \begin{pmatrix} 0.7 & 0.3 \\ 0.6 & 0.4 \end{pmatrix}, \quad P^{a_2} = \begin{pmatrix} 0.4 & 0.6 \\ 0.5 & 0.5 \end{pmatrix}
$$

and the matrices of transition cost

$$
C^{a_1} = \begin{pmatrix} 1 & 0 \\ -2 & 5 \end{pmatrix}, \quad C^{a_2} = \begin{pmatrix} 0 & 4 \\ 2 & -3 \end{pmatrix}.
$$

In Fig. 2.9 this Markov process is represented by the multigraph $G_p = (X, E_p)$ with the set of vertices $X = \{1, 2\}$.

The set of directed edges E_p contains parallel directed edges that correspond to probability transitions from one state to another for different actions. We call this graph *multigraph of the Markov decision process*.

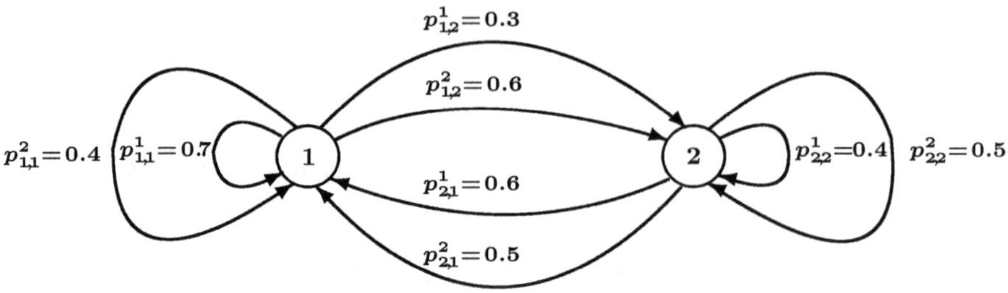

Fig. 2.9 The graph of the Markov decision process

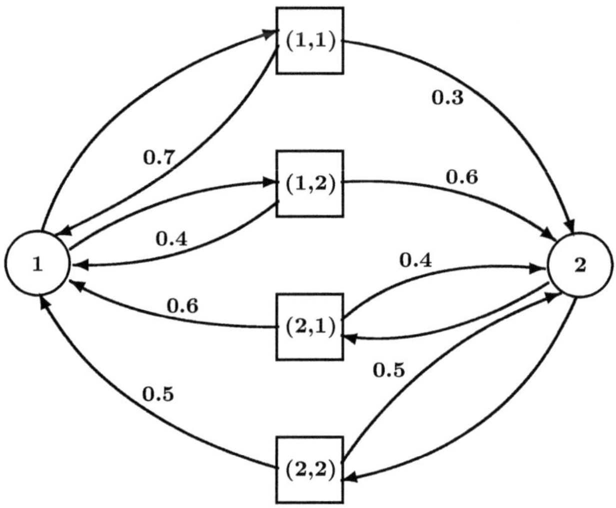

Fig. 2.10 The graph G' for the control problem

In Fig. 2.10 the graph $G' = (X', E')$ is represented. In G' the sets X'_C, X'_N, E'_C, E'_N are defined as follows:

$$X'_C = \{1, 2\}, \;\; X'_N = X^1 \cup X^2 = \{(1, 1), (1, 2), (2, 1), (2, 2)\}$$

where

$$X^1 = \{(1, 1), (1, 2)\}, \;\; X^2 = \{(2, 1), (2, 2)\}$$

and

$$\begin{aligned} E'_C &= \{(1, (1, 1)), \;\; (1, (1, 2)), \;\; (2, (2, 1)), \;\; (2, (2, 2))\}, \\ E'_N &= \{((1, 1), 1), \;\; ((1, 1), 2), \;\; ((2, 1), 1), \;\; ((2, 2), 1), \\ &\qquad ((1, 2), 1), \;\; ((1, 2), 2), \;\; ((2, 1), 2), \;\; ((2, 2), 2)\}. \end{aligned}$$

The probabilities $p'_e = p'_{(x,a),y} = p^a_{x,y}$ for directed edges $((x,a),y) \in E'_N$ are written along the edges in Fig. 2.10 and the costs of the directed edges from E' are defined in the following way:

$$c'_{1,(1,1)} = c'_{1,(1,2)} = 0, \qquad c'_{2,(2,1)} = c'_{2,(2,2)} = 0,$$

$$\begin{array}{l} c'_{(1,1),1} = 2,\ c'_{(1,1),2} = 0,\ c'_{(2,1),1} = -4,\ c'_{(2,2),1} = \ \ 4, \\ c'_{(1,2),1} = 0,\ c'_{(1,2),2} = 8,\ c'_{(2,1),2} = \ 10,\ c'_{(2,2),2} = -6. \end{array}$$

The set of possible stationary strategies for this Markov decision process consists of four strategies, i.e., $\mathbb{S} = \{s^1, s^2, s^3, s^4\}$ where

$$\begin{array}{l} s^1 : 1 \to a_1, \quad 2 \to a_1; \\ s^2 : 1 \to a_1, \quad 2 \to a_2; \\ s^3 : 1 \to a_2, \quad 2 \to a_1; \\ s^4 : 1 \to a_2, \quad 2 \to a_2. \end{array}$$

A fixed strategy s in the Markov decision process generates a simple Markov process with transition costs, where the corresponding matrices P^s, C^s are formed from the rows of the matrices P^{a_i} and C^{a_i}, $i = 1, 2$. As an example, if we fix the strategy s_2 then we obtain a simple Markov process with transition costs generated by the following matrices P^{s_2} and C^{s_2}:

$$P^{s_2} = \begin{pmatrix} 0.7 & 0.3 \\ 0.5 & 0.5 \end{pmatrix}, \quad C^{s_2} = \begin{pmatrix} 1 & 0 \\ 2 & -3 \end{pmatrix}.$$

It is easy to check that this Markov process is ergodic and the limit matrix of this process is

$$Q^{s_2} = \begin{pmatrix} \frac{5}{8} & \frac{3}{8} \\ \frac{5}{8} & \frac{3}{8} \end{pmatrix}.$$

We can determine the components of the vector of immediate costs $\mu^{s_2} = \begin{pmatrix} \mu_1^{s_2} \\ \mu_2^{s_2} \end{pmatrix}$ using formula $\mu_i^{s_2} = p_{i,1}^{s_2} c_{i,1}^{s_2} + p_{i,2}^{s_2} c_{i,2}^{s_2}$, $i = 1, 2$, i.e., $\mu_1^{s_2} = 0.7$ and $\mu_2^{s_2} = 0.5$. In such a way we determine $f_1(s_2) = f_2(s_2) = 1/4$. Analogously, it can be calculated by $f_1(s_1) = f_2(s_1) = 22/30$, $f_1(s_3) = f_2(s_3) = 16/10$ and $f_1(s_4) = f_2(4) = 9/11$. We can see that the optimal stationary strategy for the Markov decision problem with minimal average cost criterion is s^2. This strategy can be found by solving the following linear programming problem on the auxiliary network (G', X'_C, X'_N, p', c'):

Minimize

$$\overline{\psi}(\alpha, q) = 1.4q_{1,1} + 4.8q_{1,2} + 1.6q_{2,1} - q_{2,2}$$

subject to

$$\begin{cases} 0.7q_{1,1} + 0.4q_{1,2} + 0.6q_{2,1} + 0.5q_{2,2} = q_1, \\ 0.3q_{1,1} + 0.6q_{1,2} + 0.4q_{2,1} + 0.5q_{2,2} = q_2, \\ \alpha_{1,(1,1)} = q_{1,1}, \\ \alpha_{1,(1,2)} = q_{1,2}, \\ \alpha_{2,(2,1)} = q_{2,1}, \\ \alpha_{2,(2,2)} = q_{2,2}, \\ \alpha_{1,(1,1)} + \alpha_{1,(1,2)} = q_1, \\ \alpha_{2,(2,1)} + \alpha_{2,(2,2)} = q_2, \\ q_{1,1} + q_{1,2} + q_{2,1} + q_{2,2} + q_1 + q_2 = 1, \\ \alpha_{1,(1,1)},\ \alpha_{1,(1,2)},\ \alpha_{2,(2,1)},\ \alpha_{2,(2,2)} \geq 0, \\ q_{1,1},\ q_{1,2},\ q_{2,1},\ q_{2,2},\ q_1,\ q_2 \geq 0. \end{cases}$$

The optimal solution of this problem is

$$q_1^* = \frac{5}{16}, \quad q_2^* = \frac{3}{16}, \quad q_{1,1}^* = \frac{5}{16}, \quad q_{2,2}^* = \frac{3}{16}, \quad q_{1,2}^* = 0, \quad q_{2,1}^* = 0,$$
$$\alpha_{1,(1,1)}^* = \frac{5}{16}, \quad \alpha_{2,(2,2)}^* = \frac{3}{16}, \quad \alpha_{1,(1,2)}^* = 0, \quad \alpha_{2,(2,1)}^* = 0$$

and the optimal value of the objective function is $\overline{\psi}(\alpha^*, q^*) = 1/4$.

The optimal strategy s^* on G' we can find using Theorem 2.3, i.e., we fix

$$s_{1,(1,1)}^* = 1, \quad s_{1,(1,2)}^* = 0, \quad s_{2,(2,1)}^* = 0, \quad s_{2,(2,2)}^* = 1.$$

This means that the optimal stationary strategy for the Markov decision problem is

$$s^* : 1 \to a_1, \quad 2 \to a_2$$

and the average cost per transaction is $f_1(s^*) = f_2(s^*) = 1/4$.

The auxiliary graph with distinguished optimal strategies in the controllable states $x_1 = 1$ and $x_2 = 2$ is represented in Fig. 2.11. The unique outgoing directed edge $(1, (1, 1))$ from vertex 1 that ends in vertex $(1, 1)$ corresponds to the optimal strategy $1 \to a_1$ in the state $x = 1$ and the unique outgoing directed edge $(2, (2, 2))$ from vertex 2 that ends in vertex $(2, 2)$ corresponds to the optimal strategy $2 \to a_2$ in the state $x = 2$.

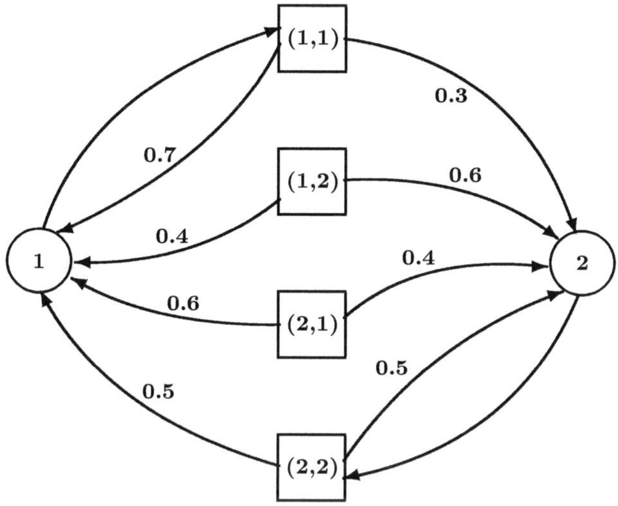

Fig. 2.11 The graph induced by the optimal strategy in the control problem

2.3.3 A Linear Programming Approach for the Average Markov Decision Problem and an Algorithm for Determining the Optimal Strategies

In the previous sections we have shown that the optimal stationary strategies for Markov decision processes can be found by constructing an auxiliary stochastic control problem and applying the linear programming algorithm for the control problem on an auxiliary network. Below we show how to apply a linear programming algorithm directly to the Markov decision problem with an average cost optimization criterion without constructing the auxiliary stochastic control problem.

At first we describe the linear programming algorithm for a special class of Markov decision processes.

We consider Markov decision processes with the property that an arbitrary stationary strategy $s : X \to A$ generates an ergodic Markov chain, i.e., we assume that the graph $G_p^s = (X, E_p^s)$ of the matrix of probability transitions $P^s = (p_{x,y}^s)$ is strongly connected. In general, we can see that the linear programming approach can be used for an arbitrary Markov decision problem where an arbitrary stationary strategy generates a unichain. We call such Markov decision processes *perfect Markov decision processes*. It is easy to observe that if for an arbitrary strategy $s : A \to X$ in the Markov decision process each row of the matrix $P^s = (p_{x,y}^s)$ contains at least $[(|X| + 1)/2] + 1$ nonzero elements then the corresponding graph $G_p^s = \left(X, E_p^s\right)$ contains a unique strongly connected component that can be reached from every $x \in X$ [19], i.e., in this case the matrix P^s corresponds to a Markov unichain.

Let $s : X \to A$ be an arbitrary strategy ($s \in S$) for a Markov decision process. Then for every fixed $x \in X$ we have a unique action $a = s(x) \in A(x)$ and therefore we can identify the map s with the set of boolean values $s_{x,a}$ for $x \in X$ and $a \in A(x)$, where

$$s_{x,a} = \begin{cases} 1, & \text{if } a = s(x); \\ 0, & \text{if } a \neq s(x). \end{cases}$$

In a similar way for the optimal stationary strategy s^* we shall proceed with the boolean values $s^*_{x,a}$.

Assume that the Markov decision process is perfect. Then the following lemma holds.

Lemma 2.25 *A stationary strategy s^* is optimal if and only if it corresponds to an optimal solution of the following mixed integer bilinear programming problem: Minimize*

$$\psi(s, q) = \sum_{x \in X} \sum_{a \in A(x)} \mu_{x,a} s_{x,a}\, q_x \tag{2.64}$$

subject to

$$\begin{cases} \sum\limits_{x \in X} \sum\limits_{a \in A(x)} p^a_{x,y} s_{x,a} q_x = q_y, \quad \forall y \in X; \\ \sum\limits_{x \in X} q_x = 1; \\ \sum\limits_{a \in A(x)} s_{x,a} = 1, \quad \forall x \in X; \\ \qquad s_{x,a} \in \{0, 1\}, \ \forall x \in X, \ a \in A(x); \quad q_x \geq 0, \ \forall x \in X, \end{cases} \tag{2.65}$$

where

$$\mu_{x,a} = \sum_{y \in X} c^a_{x,y}\, p^a_{x,y}$$

is the immediate cost in the state $x \in X$ for a fixed action $a \in A(x)$.

Proof For a fixed strategy s the system (2.65) has a unique solution with respect to q_x, $x \in X$ which represents the limiting probabilities of the recurrent Markov chains with the matrix of probability transition P^s. The value of the objective function (2.64) for this solution expresses the average cost per transition for an arbitrary fixed starting state. Therefore, for a fixed strategy s we have $f_x(s) = \psi(s, q^s)$, $\forall x \in X$. This means that if we solve the optimization problem (2.64), (2.65) for the perfect Markov decision process then we obtain the optimal stationary strategy s^*. □

Remark 2.26 For a perfect Markov decision processes the objective function $\psi(s,q)$ on the set of feasible solutions depends only on $s_{x,a}$ for $x \in X, a \in A(x)$. Moreover, the conditions $q_x \geq 0$ for $x \in X$ in (2.65) hold if $s_{x,a} \geq 0, \forall x \in X, a \in A(x)$ and therefore in the case of perfect Markov processes can be omitted. The conditions $q_x \geq 0, \ \forall x \in X$ in (2.65) are essential for non perfect Markov processes.

Based on Lemma 2.25 we can prove the following result.

Theorem 2.27 *Let $\alpha^*_{x,a}$ $(x \in X,\ a \in A(x))$, q^*_x $(x \in X)$ be a basic optimal solution of the following linear programming problem:*
Minimize

$$\overline{\psi}(\alpha,q) = \sum_{x\in X}\sum_{a\in A(x)} \mu_{x,a}\alpha_{x,a} \tag{2.66}$$

subject to

$$\begin{cases} \sum\limits_{x\in X}\sum\limits_{a\in A(x)} p^a_{x,y}\,\alpha_{x,a} = q_y, \quad \forall y \in X; \\ \sum\limits_{x\in X} q_x = 1; \\ \sum\limits_{a\in A(x)} \alpha_{x,a} = q_x, \quad \forall x \in X; \\ \quad \alpha_{x,a} \geq 0, \ \forall x \in X,\ a \in A(x);\ q_x \geq 0, \ \forall x \in X, \end{cases} \tag{2.67}$$

where

$$\mu_{x,a} = \sum_{y\in X} c^a_{x,y}\, p^a_{x,y} \ \textit{for} \ x \in X.$$

Then the optimal stationary strategy s^* for a perfect Markov decision process can be found as follows:

$$s^*_{x,a} = \begin{cases} 1, & \textit{if } \alpha^*_{x,a} > 0; \\ 0, & \textit{if } \alpha^*_{x,a} = 0, \end{cases}$$

where $x \in X,\ a \in A(x)$.

Moreover, for every starting state $x \in X$ the optimal average cost per transition is equal to $\overline{\psi}(\alpha^*, q^*)$, i.e.,

$$f_x(s^*) = \sum_{x\in X}\sum_{a\in A(x)} \mu_{x,a}\alpha^*_{x,a}$$

for every $x \in X$.

Proof The proof of this theorem is similar to the proof of Theorem 2.3. Applying Lemma 2.25 we obtain that the bilinear programming problem (2.64), (2.65) with boolean variables $s_{x,a}$ for $x \in X,\ a \in A(x)$ can be reduced to the linear programming problem (2.66), (2.67). We observe that the restriction $s_{x,a} \in \{0, 1\}$ in the problem (2.64), (2.65) can be replaced by $s_{x,a} \geq 0$ because the optimal basic solutions after such a transformation of the problem are not changed. In addition the restrictions

$$\sum_{a \in A(x)} s_{x,a} = 1, \quad \forall x \in X$$

can be changed by the restrictions

$$\sum_{a \in A(x)} s_{x,a}\, q_x = q_x, \quad \forall x \in X$$

because the condition $q_x > 0,\ \ \forall x \in X$ for the perfect Markov process holds. This means that the system (2.65) in the problem (2.64), (2.65) can be replaced by the following system

$$\begin{cases} \displaystyle\sum_{x \in X} \sum_{a \in A(x)} p_{x,y}^{a}\, s_{x,a}\, q_x = q_y, \quad \forall y \in X; \\ \displaystyle\sum_{x \in X} q_x = 1; \\ \displaystyle\sum_{a \in A(x)} s_{x,a}\, q_x = q_x, \quad \forall x \in X; \\ \qquad s_{x,a} \geq 0, \ \forall x \in X,\ a \in A(x); \quad q_x \geq 0, \ \forall x \in X. \end{cases} \tag{2.68}$$

In such a way we may conclude that problem (2.64), (2.65) and problem (2.64), (2.68) have the same optimal solutions. Taking into account that for the perfect network we have $q_x > 0,\ \ \forall x \in X$ then in problem (2.64), (2.68) we can introduce the notations $\alpha_{x,a} = s_{x,a}\, q_x$ for $x \in X,\ a \in A(x)$, i.e., we obtain the problem (2.66), (2.67). It is evident that $\alpha_{x,a} \neq 0$ if and only if $s_{x,y} = 1$. Therefore, the optimal stationary strategy s^* can be found according to the rule formulated in the theorem. □

It is easy to observe that q_x in the system (2.67) can be eliminated if we take into account that

$$\sum_{a \in A(x)} \alpha_{x,a} = q_x, \quad \forall x \in X.$$

Then theorem 2.27 can be formulated in the following way.

Theorem 2.28 *Let* $\alpha^*_{x,a}$ $(x \in X,\ a \in A(x))$, *be a basic optimal solution of the following linear programming problem:*

Minimize

$$\overline{\psi}(\alpha) = \sum_{x\in X}\sum_{a\in A(x)} \mu_{x,a}\alpha_{x,a} \tag{2.69}$$

subject to

$$\begin{cases} \sum_{a\in A(y)} \alpha_{y,a} - \sum_{x\in X}\sum_{a\in A(x)} p_{x,y}^{a}\,\alpha_{x,a} = 0, \;\; \forall\, y \in X; \\ \sum_{x\in X}\sum_{a\in A(x)} \alpha_{x,a} = 1; \\ \qquad \alpha_{x,a} \geq 0, \;\; \forall x \in X, \; a \in A(x). \end{cases} \tag{2.70}$$

Then the optimal stationary strategy s^* for the perfect Markov decision process can be found as follows:

$$s^*_{x,a} = \begin{cases} 1, & \text{if } \alpha^*_{x,a} > 0; \\ 0, & \text{if } \alpha^*_{x,a} = 0, \end{cases}$$

where $x \in X,\ a \in A(x)$. Moreover, for every starting state $x \in X$ the optimal average cost per transition is equal to $\overline{\psi}(\alpha^*, q^*)$, i.e.,

$$f_x(s^*) = \sum_{x\in X}\sum_{a\in A(x)} \mu_{x,a}\,\alpha^*_{x,a}$$

for every $x \in X$.

Thus, based on theorems proven above the optimal stationary strategy for the Markov decision problem can be found using the following algorithm.

Algorithm 2.29 Determining the Optimal Stationary Strategies for the Perfect Markov Decision Problem

(1) Formulate the linear programming problem (2.66), (2.67) and find a basic optimal solution $\alpha^*_{x,y}$, q^*_x, q^*_z of this problem;
(2) Fix $s^*_{x,a} = 1$ for (x, a) that corresponds to the basic components of the optimal solution and set $s^*_{x,a} = 0$ for the remaining components.

Example Consider the Markov decision problem with an average cost criterion from Sect. 2.3.2. The corresponding multigraph of the Markov decision process is represented in Fig. 2.9.

The optimal stationary strategy s^* of this problem can be found by solving the linear programming problem (2.66), (2.67), i.e.:
Minimize

$$\overline{\psi}(\alpha, q) = 0.7\alpha_{1,1} + 2.4\alpha_{1,2} + 0.8\alpha_{2,1} - 0.5\alpha_{2,2}$$

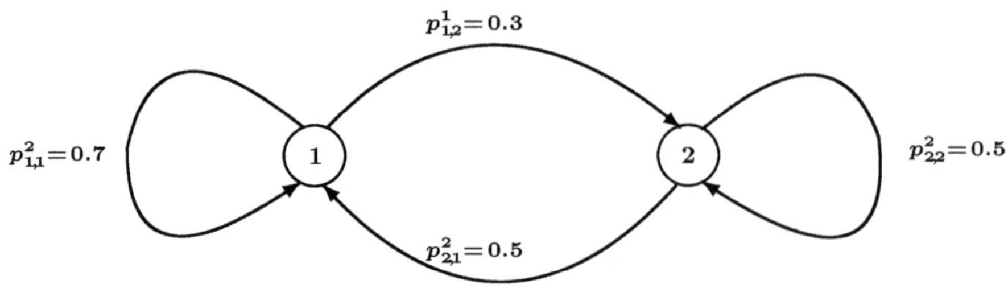

Fig. 2.12 The graph induced by the optimal strategy in Markov decision problem

subject to

$$\begin{cases} 0.7\alpha_{1,1} + 0.6\alpha_{2,1} + 0.4\alpha_{1,2} + 0.5\alpha_{2,2} = q_1, \\ 0.3\alpha_{1,1} + 0.4\alpha_{2,1} + 0.6\alpha_{1,2} + 0.5\alpha_{2,2} = q_2, \\ q_1 + q_2 = 1, \\ \alpha_{1,1} + \alpha_{1,2} = q_1, \\ \alpha_{2,1} + \alpha_{2,2} = q_2, \\ \alpha_{1,1},\ \alpha_{1,2},\ \alpha_{2,1},\ \alpha_{2,2} \geq 0, \quad q_1,\ q_2 \geq 0. \end{cases}$$

The optimal solution of this problem is

$$q_1^* = \frac{5}{8}, \quad q_2^* = \frac{3}{8}, \quad \alpha_{1,1}^* = \frac{5}{8}, \quad \alpha_{2,2}^* = \frac{3}{8}, \quad \alpha_{1,2}^* = 0, \quad \alpha_{2,1}^* = 0$$

and the corresponding average cost is equal to $1/4$, i.e., $\psi(\alpha^*, q^*) = 1/4$.

The optimal solution of the problem corresponds to the optimal stationary strategy $s_{1,1}^* = 1,\ s_{1,2}^* = 0,\ s_{2,1}^* = 0,\ s_{2,2}^* = 1$ i.e. $s^* : 1 \to a_1,\ 2 \to a_2$.

So, the optimal stationary strategy s^* determines the Markov process with the following probability and cost matrices

$$P^{s^*} = \begin{pmatrix} 0.7 & 0.3 \\ 0.5 & 0.5 \end{pmatrix}, \qquad C^{s^*} = \begin{pmatrix} 1 & 0 \\ 2 & -3 \end{pmatrix}.$$

The graph of transition probabilities of this Markov process is represented in Fig. 2.12.

The result described above shows that the Markov decision problem with an average cost criterion can be transformed into a stochastic optimal control problem on the auxiliary network $(G', X_C, X_N, p', c', x_0)$. This means that the linear programming algorithm proposed in the previous sections can be developed and specified for Markov decision problems with an average and discounted costs optimization criteria.

2.3.4 A Dual Linear Programming Model for an Average Markov Decision Problem

Consider the linear programming problem (2.69), (2.70) for an arbitrary unichain Markov decision process. As we have shown the solution of this problem always exists. If we dualize (2.69), (2.70) then we obtain the following problem:
Maximize

$$\psi'(\varepsilon, \omega) = \omega \tag{2.71}$$

subject to

$$\varepsilon_x - \sum_{y \in X} p_{x,y}^a \varepsilon_y + \omega \leq \mu_{x,a}, \quad \forall x \in X, \ \forall a \in A. \tag{2.72}$$

Based on duality theory of linear programming we obtain the following result.

Theorem 2.30 *The linear programming problem* (2.71), (2.72) *has solutions and an arbitrary optimal solution* ε^*, ω^* *of the problem possesses the following property: For each* $x \in X$ *there exists an action* $a^* \in A(x)$ *that satisfies the condition*

$$\min_{a \in A(x)} \left\{ \mu_{x,a^*} + \sum_{y \in X} p_{x,y}^{a^*} \varepsilon_y^* - \varepsilon_x^* - \omega^* \right\} = 0, \quad \forall x \in X. \tag{2.73}$$

The action a^* *in each state* $x \in X$ *determines the optimal stationary strategy* $s^*(x) = a^*$ *and* ω^* *is equal to the optimal value of the average cost in the Markov decision process.*

This theorem represents the optimization criterion for unichain Markov decision problems with average expected cost. Based on this criterion we can determine the optimal stationary strategies of the problem in the unichain case using the following algorithm.

Algorithm 2.31 Determining the Optimal Solution of a Unichain Markov Decision Problem Using a Dual Linear Programming Model

(1) Formulate the linear programming problem (2.71), (2.72) and find an optimal solution ε^*, ω^* of this problem;
(2) For each $x \in X$ fix $s^*(x) = a^*$, where $a^* \in A(x)$ satisfies condition (2.73).

2.3.5 Optimality Conditions for Multichain Decision Problems and a Linear Programming Approach

The optimality conditions for a *multichain Markov decision problem* with an average optimization cost criterion can be derived from the optimality conditions for an average multichain control problem if we take into account the mentioned relationship between Markov decision processes and stochastic control models. Based on Theorem 2.13 and the results from Sect. 2.3.2 we can formulate the following optimality principle for an average multichain decision problem.

Theorem 2.32 *Let a Markov decision process* (X, A, p, c) *be given. Then the system of equations*

$$\varepsilon_x + \omega_x = \min_{a \in A(x)} \left\{ \mu_{x,a} + \sum_{y \in X} p^a_{x,y} \varepsilon_y \right\}, \quad \forall x \in X; \tag{2.74}$$

has a solution under the set of solutions of the system of equations

$$\omega_x = \min_{a \in A(x)} \left\{ \sum_{y \in X} p^a_{x,y} \omega_x \right\}, \quad \forall x \in X, \tag{2.75}$$

i.e., the system of equations (2.75) *has such a solution* $\omega^*_x,\ x \in X$ *for which there exists a solution* $\varepsilon^*_x,\ x \in X$ *of the system of equations*

$$\varepsilon_x + \omega^*_x = \min_{a \in A(x)} \left\{ \mu_{x,a} + \sum_{y \in X} p^a_{x,y} \varepsilon_y \right\}, \quad \forall x \in X. \tag{2.76}$$

The values ω^*_x *for* $x \in X$ *coincide with the optimal average costs* $\omega_x,\ x \in X$ *for the Markov decision problem and an optimal stationary strategy*

$$s^* : x \to a \in A(x) \text{ for } x \in X$$

for an average Markov decision problem can be found by fixing a map $s^*(x) = a \in A(x)$ *such that*

$$a \in \underset{a \in A(x)}{\operatorname{argmin}} \left\{ \sum_{y \in X} p^a_{x,y} \omega^*_x \right\}$$

and

$$a \in \underset{a \in A(x)}{\operatorname{argmin}} \left\{ \mu_{x,a} + \sum_{y \in X} p^a_{x,y} \varepsilon^*_y \right\}.$$

Note that the Lemmas 2.14, 2.20 are valid also for average Markov decision problems if the strategies s and s^* we treat as strategies $s : X \to A$; $s^* : X \to A$ for the Markov decision problem.

Then from these lemmas we obtain the proof of Theorem 2.32. The proof of this theorem also follows from the Theorems 2.12, 2.13 and the reduction procedure from Markov decision problem to stochastic control problem described in Sect. 2.3.2.

From Theorem 2.32 we can make the following conclusion. To determine a solution of the Markov decision problem it is necessary to determine ω_x for $x \in X$ that satisfies (2.75) and for which there exists ε_x for $x \in X$ that satisfies (2.74). This is equivalent with the problem of determining the "maximal" vector ω with the components ω_x for $x \in X$ that satisfies the conditions

$$\varepsilon_x + \omega_x \le \mu_{x,a} + \sum_{y \in X} p^a_{x,y} \varepsilon_y, \quad \forall x \in X, \ \forall a \in A(x);$$

$$\omega_x \le \sum_{y \in X} p^a_{x,y} \omega_x, \quad \forall x \in X, \ \forall a \in A(x).$$

Thus, we have to maximize a positive linear combination of components of ω under the restrictions given above, i.e., we obtain the following linear programming problem:
Maximize

$$\psi'(\varepsilon, \omega) = \sum_{x \in X} \theta_x \omega_x \tag{2.77}$$

subject to

$$\begin{cases} \varepsilon_x + \omega_x \le \mu_{x,a} + \sum\limits_{y \in X} p^a_{x,y} \varepsilon_y, & \forall x \in X, \ \forall a \in A(x); \\ \omega_x \le \sum\limits_{y \in X} p^a_{x,y} \omega_x, & \forall x \in X, \ \forall a \in A(x) \end{cases} \tag{2.78}$$

where $\theta > 0$, $\forall x \in X$ and $\sum_{x \in X} \theta_x = 1$.

From Theorem 2.32 we obtain the following result.

Corollary 2.33 *For an arbitrary strategy $s : X \to A$ the following system of linear equations*

$$\begin{cases} \varepsilon_x + \omega_x = \mu_{x,s(x)} + \sum\limits_{y \in X} p^{s(x)}_{x,y} \varepsilon_y, & \forall x \in X; \\ \omega_x = \sum\limits_{y \in X} p^{s(x)}_{x,y} \omega_x, & \forall x \in X \end{cases} \tag{2.79}$$

has a solution.

2.3.6 Primal and Dual Linear Programming Models for a Multichain Markov Decision Problem

We can regard the linear programming model (2.77), (2.78) as a dual model for a *primal multichain linear programming problem*. So, if we consider the dual model or (2.77), (2.78) then we obtain the following linear programming problem:
Minimize

$$\overline{\psi}(\alpha) = \sum_{x \in X} \sum_{a \in A(x)} \mu_{x,a} \alpha_{x,a} \tag{2.80}$$

subject to

$$\begin{cases} \sum_{a \in A(y)} \alpha_{y,a} - \sum_{x \in X} \sum_{a \in A(x)} p^a_{x,y}\, \alpha_{x,a} = 0, \ \ \forall y \in X; \\ \sum_{a \in A(y)} \alpha_{y,a} + \sum_{a \in A(y)} \beta_{y,a} - \sum_{x \in X} \sum_{a \in A(x)} p^a_{x,y} \beta_{x,a} = \theta_y, \ \forall y \in X; \\ \qquad \alpha_{x,a} \geq 0, \ \beta_{y,a} \geq 0, \ \ \forall x \in X, \ a \in A(x), \end{cases} \tag{2.81}$$

where $\theta > 0, \ \forall y \in X$ and $\sum_{y \in X} \theta_y 1$.
This problem generalizes the unichain linear programming problem (2.69), (2.70) from Sect. 2.3.3. In (2.81) the restrictions

$$\sum_{a \in A(y)} \alpha_{y,a} + \sum_{a \in A(y)} \beta_{y,a} - \sum_{x \in X} \sum_{a \in A(x)} p^a_{x,y} \beta_{x,a} = \theta_y, \ \ \forall y \in X \tag{2.82}$$

with the condition $\sum_{y \in X} \theta_y = 1$ generalize the constrain

$$\sum_{x \in X} \sum_{a \in A(y)} \alpha_{y,a} = 1 \tag{2.83}$$

in the unichain model. It is easy to check that by summing (2.82) over y, we obtain the equality (2.83).

2.4 Iterative Algorithms for Markov Decision Processes and Control Problems with an Average Cost Criterion

As we have shown the Markov decision problem and optimal control problems with average cost criterion can be solved using the linear programming approach. Here we show that these problems can be solved using iterative algorithms. These algorithms are based on the optimization criteria proved in previous sections. We can observe that the optimization criterion for a stochastic control problem in the case $X_C = \emptyset$ leads to the equation which can be derived directly from formula (1.59).

Indeed, using formula (1.59) we can write the following two equivalent equations

$$\sigma(t) = t\omega + \varepsilon + \epsilon(t),$$
$$\sigma(t-1) = (t-1)\omega + \varepsilon + \epsilon(t-1),$$

where $\epsilon(t)$ and $\epsilon(t-1)$ tend to zero if t tends to infinity. If we introduce the expression of $\sigma(t)$ and $\sigma(t-1)$ in the recursive formula

$$\sigma(t) = \mu + P\sigma(t-1)$$

then we obtain

$$t\omega + \varepsilon + \epsilon(t) = \mu + P((t-1)\omega + \varepsilon + \epsilon(t-1).$$

Through rearrangement we get

$$\varepsilon + t\omega - (t-1)P\omega = \mu + P\varepsilon + P\epsilon(t-1) - \epsilon(t).$$

Here $\omega = P\omega$. In addition for a Markov unichain all components of the vector ω are the same, i.e., $\omega_1 = \omega_2 = \cdots = \omega_n = \omega$ (here $\omega_i = \omega_{x_i}$). So, if $t \to \infty$ then $\epsilon(t),\ \epsilon(t-1) \to 0$ and we obtain

$$\varepsilon_i + \omega = \mu_{x_i} + [P\varepsilon]_i, \quad i = 1, 2, \ldots, n. \tag{2.84}$$

This is the system of equations for a unichain Markov process. It is well known that in the case of unichain processes the rank of the matrix $(I - P)$ is equal to $n - 1$ (see [98]). Based on this fact in [98] it has been shown that the system of equations (2.84) has a unique solution once it is setting $\varepsilon_i = 0$ for some i. This means that two different vectors ε' and ε'' which represent the solutions of this equation differ only by some constant for each component. Therefore, the system of equations (2.84) allows us to determine the average cost per transition in unichain Markov processes with transition costs. The existence of the solution of this system of equations (2.84) also follows from Theorem 2.30.

The system of equations for the decision problem in the case of unichain processes is the following

$$\varepsilon_i + \omega = \min_{a \in A(x_i)} \left(\mu_{x_i,a} + [P^a\varepsilon]_i\right), \quad i = 1, 2, \ldots, n. \tag{2.85}$$

According to Theorem 2.30 the system of equations (2.85) has solutions. The solution of this system of equations and the optimal stationary strategy for unichain Markov decision problems can be found using the following iterative algorithm.

Algorithm 2.34 Determining the Solution of a Unichain Markov Decision Problem

Preliminary step (Step 0): Fix an arbitrary stationary strategy

$$s^0 : x_i \to a \in A(x_i) \ \text{ for } \ x_i \in X.$$

General step (Step k, $k > 0$): Calculate

$$\mu_{x_i,a^{k-1}} = \sum_{y \in X(x_i)} p_{x_i,y}^{s^{k-1}(x_i)} c_{x_i,y}^{s^{k-1}(x_i)}$$

for every $x_i \in X$. Then solve the system of linear equations

$$\varepsilon_i^{k-1} + \omega^{k-1} = \mu_{x_i,s^{k-1}(x_i)} + [P^{s^{k-1}} \varepsilon^{k-1}]_i, \quad i = 1, 2, \ldots, n,$$
$$\varepsilon_n^{k-1} = 0,$$

and find $\varepsilon_1^{k-1}, \varepsilon_2^{k-1}, \ldots, \varepsilon_{n-1}^{k-1}$ and ω^{k-1}. After that determine a new strategy

$$s^k : x_i \to a \in A(x_i) \ \text{ for } \ x_i \in X,$$

where

$$s^k(x_i) = \underset{a \in A(x_i)}{\operatorname{argmin}} \left(\mu_{x_i,a} + [P^a \varepsilon^{k-1}]_i\right), \quad i = 1, 2, \ldots, n.$$

Check if the following condition holds

$$s^k(x_i) = s^{k-1}(x_i), \quad \forall x_i \in X. \tag{2.86}$$

If the condition (2.86) holds then fix

$$s^* = s^k, \quad \omega^* = \omega^k$$

as the optimal solution of the problem; otherwise go to the next step $k + 1$.

The correctness and the convergence of this algorithm follow from the results described above and the results from [115, 118–120, 140].

The algorithm described above can be specified for determining the optimal stationary strategies in the stochastic control problem with an average cost optimization criterion.

Algorithm 2.35 Determining the Solution for a Stochastic Control Problem

Let the average control problem on a perfect network determined by the graph of state's transition $G = (X, E)$ with the set of controllable states X_C, the set of uncontrollable states X_N, the probability function $p : E_N \to [0, 1]$ which satisfies the condition from Sect. 2.3 and the cost function $c : E \to \mathbb{R}$ be given.

Preliminary step (Step 0): Fix an arbitrary stationary strategy

$$s^0 : x_i \to x_j \in X(x_i) \text{ for } x_i \in X_C.$$

General step (Step k, k > 0): Determine the probability matrix $P^{s^{k-1}} = \left(p^{s^{k-1}}_{x_i,x_j}\right)$, where

$$p^{s^{k-1}}_{x_i,x_j} = \begin{cases} p_{x_i,x_j}, & \text{if } x_i \in X_N \text{ and } (x_i, x_j) \in E_2; \\ 1, & \text{if } x_i \in X_C \text{ and } x_j = s^{k-1}(x_i); \\ 0, & \text{if } x_i \in X_C \text{ and } x_j \neq s^{k-1}(x_i). \end{cases}$$

Then calculate

$$\mu_{x_i,s^{k-1}(x_i)} = \sum_{x_j \in X(x_i)} p^{s^{k-1}(x_i)}_{x_i,x_j} c_{x_i,x_j}$$

for every $x_i \in X$. After that solve the system of linear equations

$$\varepsilon_i^{k-1} + \omega^{k-1} = \mu_{x_i,s^{k-1}(x_i)} + [P^{s^{k-1}} \varepsilon^{k-1}]_i, \quad i = 1, 2, \ldots, n,$$

$$\varepsilon_n^{k-1} = 0,$$

and find $\varepsilon_1^{k-1}, \varepsilon_2^{k-1}, \ldots, \varepsilon_{n-1}^{k-1}$ and ω^{k-1}. Then determine a new strategy

$$s^k : x_i \to x_j \in X(x_i) \quad \text{for} \quad x_i \in X_C,$$

where

$$s^k(x_i) = \underset{x_j \in X(x_i)}{\operatorname{argmin}} \left(c_{x_i,x_j} + \varepsilon_j^{k-1}\right), \quad \forall x_i \in X_C.$$

Check if the following condition holds

$$s^k(x_i) = s^{k-1}(x_i), \quad \forall x_i \in X_C. \tag{2.87}$$

If the condition (2.87) holds then fix

$$s^* = s^k, \quad \omega^* = \omega^k$$

as the optimal solution of the problem; otherwise go to the next step $k + 1$.

In the case $X_N = \emptyset$ this algorithm is transformed into the algorithm for solving a deterministic control problem. In this case the algorithm correctly finds the solution of the problem if each stationary strategy in G generates a subgraph G_s which contains a unique directed cycle.

The algorithms described above determine the optimal stationary strategies for a Markov decision problem and a stochastic optimal control problem if an arbitrary strategy in these problems generates a unichain process.

For the multichain case of the problem the algorithm uses the multichain bias equations (2.74)–(2.76).

Algorithm 2.36 Determining the Solution of a Multichain Markov Decision Problem

Preliminary step (Step 0): Fix an arbitrary stationary strategy

$$s^0 : x_i \to a \in A(x_i) \text{ for } x_i \in X.$$

General step (Step k, $k \geq 1$): Determine the matrix $P^{s^{k-1}}$ and $\mu^{s^{k-1}}$ that corresponds to the strategy s^{k-1}. Find $\omega^{s^{k-1}}$ and $\varepsilon^{s^{k-1}}$ which satisfy the conditions

$$\begin{cases} (P^{s^{k-1}} - I)\omega^{s^{k-1}} = 0; \\ \mu^{s^{k-1}} + (P^{s^{k-1}} - I)\varepsilon^{s^{k-1}} - \omega^{s^{k-1}} = 0. \end{cases}$$

Then find a strategy s^k such that

$$s^k \in \operatorname*{argmin}_s \left\{ P^s \omega^{s^{k-1}} \right\}$$

and set $s^k = s^{k-1}$ if

$$s^{k-1} \in \operatorname*{argmin}_s \left\{ P^s \omega^{s^{k-1}} \right\}.$$

After that check if $s^k = s^{k-1}$? If $s^k = s^{k-1}$ then go to next step $k+1$; otherwise choose the strategy s^k such that

$$s^k \in \operatorname*{argmin}_s \left\{ \mu^s + P^s \varepsilon^{s^{k-1}} \right\}$$

and set $s^k = s^{k-1}$ if

$$s^{k-1} \in \operatorname*{argmin}_s \left\{ \mu^s + P^s \varepsilon^{s^{k-1}} \right\}.$$

After that check if $s^k = s^{k-1}$? If $s^k = s^{k-1}$ then STOP and set $s^* = s^{k-1}$; otherwise go to the next step $k+1$.

The convergence of the algorithms based on iterative procedures are proved in [115, 121–123]. In a similar way as for the unichain case of the problem the algorithm described above can be specified for a multichain stochastic control problem on networks. The computational complexity of the Markov decision problems in the general case is studied in [106].

2.5 A Discounted Stochastic Control Problem and Algorithms for Determining the Optimal Strategies on Networks

Now we consider the infinite horizon discounted stochastic control problem. Following the concept from the previous sections we formulate the discounted stochastic control problem on networks and describe algorithms for determining the optimal stationary strategies using a linear programming approach. Then we extend this approach for Markov decision problems with an expected total discounted cost optimization criterion.

2.5.1 Problem Formulation

Let a time-discrete system $\mathbb{L}$ with finite set of states X be given and assume that the dynamics of the system is described by a directed graph of states' transitions $G = (X, E)$ with the vertex set X and edge set E. Thus, an arbitrary directed edge $e = (x, y) \in E$ expresses the possibility of the system to pass from the state $x = x(t)$ to the state $y = x(t)$ at every discrete moment of time $t = 0, 1, 2, \dots$. On an edge set E a cost function $c : E \to \mathbb{R}$ is defined that indicates a cost c_e to each directed edge $e = (x, y) \in E$ if the system makes a transition from the state $x = x(t)$ to the state $y = x(t+1)$ for every $t = 0, 1, 2, \dots$. We define the stationary control for the system $\mathbb{L}$ in G as a map

$$s : x \to y \in X(x) \quad \text{for} \quad x \in X,$$

where $X(x) = \{y \in X \mid (x, y) \in E\}$.

Let s be an arbitrary stationary control. Then the set of edges of the form $(x, s(x))$ in G generates a subgraph $G_s = (X, E_s)$ where each vertex $x \in X$ contains one leaving directed edge. So, if the starting state $x_0 = x(0)$ is fixed then the system makes transitions from one state to another through the corresponding directed edges $e_0^s, e_1^s, e_2^s, \dots, e_t^s, \dots$, where $e_t^s = (x(t), x(t+1)), \quad t = 0, 1, 2, \dots$. This sequence of directed edges generates a trajectory $x_0 = x(0), x(1), x(2), \dots$ which leads to a unique directed cycle. For an arbitrary stationary strategy s and a fixed starting state x_0 the discounted expected total cost $\sigma_{x_0}^{\gamma}(s)$ is defined as follows

$$\sigma_{x_0}^{\gamma}(s) = \sum_{t=0}^{\infty} \gamma^t c_{e_t^s},$$

where $\gamma, \quad 0 < \gamma < 1$, is a given discount factor.

Based on the results from [47, 114] it is easy to show that for an arbitrary stationary strategy s there exists $\sigma^{\gamma}_{x0}(s)$. If we denote by $\sigma^{\gamma}(s)$ the column vector with components $\sigma^{\gamma}_{x}(s)$ for $x \in X$ then $\sigma^{\gamma}_{x0}(s)$ can be found by solving the system of linear equations

$$(I - \gamma P^{s})\sigma^{\gamma}(s) = c^{s}, \tag{2.88}$$

where c^{s} is the vector with corresponding components $c_{x,s(x)}$ for $x \in X$, I is the identity matrix and P^{s} the matrix with elements $p^{s}_{x,y}$ for $x, y \in X$ defined as follows:

$$p^{s}_{x,y} = \begin{cases} 1, & \text{if } y = s(x); \\ 0, & \text{if } y \neq s(x). \end{cases}$$

It is well known that for $0 < \gamma < 1$ the rank of the matrix $I - \gamma P^{s}$ is equal to $|X|$ and the system (2.88) has solutions for arbitrary c^{s} (see [114, 140]). Thus, we can determine $\sigma^{\gamma}_{x0}(s^{*})$ for an arbitrary starting state x_0.

In the considered deterministic discounted control problem on G we are seeking for a stationary control s^{*} such that

$$\sigma^{\gamma}_{x0}(s^{*}) = \min_{s} \sigma^{\gamma}_{x0}(s).$$

We formulate and study this problem in a more general case considering its *stochastic version*. We assume that the dynamical system may admit states in which the vector of control parameters is changed in a random way. So, the set of states X is divided into two subsets $X = X_C \cup X_N$, $X_C \cap X_N = \emptyset$, where X_C represents the set of states in which the decision maker is able to control the dynamical system and where X_N represents the set of states in which the dynamical system makes transitions to the next state in a random way. This means that for every $x \in X$ on the set of feasible transitions $E(x)$ the distribution function $p : E(x) \to \mathbb{R}$ is defined such that $\sum_{e \in E(x)} p_e = 1$, $p_e \geq 0$, $\forall e \in E(x)$ and the transitions from the states $x \in X_N$ to the next states are made randomly according to these distribution functions. Here, in a similar way as for the deterministic problem we assume that to each directed edge $e = (x, y) \in E$ a cost c_e of system's transition from the state $x = x(t)$ to the state $y = x(t + 1)$ for $t = 0, 1, 2, \ldots$ is associated. In addition we assume that the discount factor γ, $0 < \gamma < 1$, and the starting state x_0 are given. We define a stationary control on G as a map

$$s : x \to y \in X(x) \quad \text{for} \quad x \in X_C.$$

Let s be an arbitrary stationary strategy. We define the graph $G_s = (X, E_s \cup E_N)$, where $E_s = \{e = (x, y) \in E \,|\, x \in X_C, \; y = s(x)\}$, $E_N = \{e = (x, y) \,|\, x \in X_N, \; y \in X\}$. This graph corresponds to a Markov process with the probability matrix $P^{s} = (p^{s}_{x,y})$, where

$$p^s_{x,y} = \begin{cases} p_{x,y}, & \text{if } x \in X_N \text{ and } y \in X; \\ 1, & \text{if } x \in X_C \text{ and } y = s(x); \\ 0, & \text{if } x \in X_C \text{ and } y \neq s(x). \end{cases}$$

For this Markov process with associated costs c_e, $e \in E$ we can define the expected total discounted cost $\sigma^{\gamma}_{x_0}(s)$ as we have introduced in Chap. 1. We consider the problem of determining the strategy s^* for which

$$\sigma^{\gamma}_{x_0}(s^*) = \min_s \sigma^{\gamma}_{x_0}(s).$$

Without loss of generality we may consider that G has the property that an arbitrary vertex in G is reachable from x_0; otherwise we can delete all vertices that could not be reached from x_0.

2.5.2 A Linear Programming Approach for a Discounted Control Problem on Networks

We develop a linear programming approach for the discounted stochastic control problem on the network (G, X, E, c, p, x_0) with a given discount factor γ using the same logical scheme as in Sect. 2.2. We identify an arbitrary stationary strategy s in G with the set of boolean variables $s_{x,y}$ for $x \in X_C$ and $y \in X(x)$, where

$$s_{x,y} = \begin{cases} 1, & \text{if } y = s(x); \\ 0, & \text{if } y \neq s(x). \end{cases} \tag{2.89}$$

In the following we will simplify the notations and instead σ^{γ}_x we shall use σ_x.

Lemma 2.37 *For a fixed strategy s the values σ^{γ}_x, $x \in X$ determine the unique optimal basic solution of the following linear programming problem:*
Maximize

$$\varphi^s_{x_0}(\sigma) = \sigma_{x_0} \tag{2.90}$$

subject to

$$\begin{cases} \sigma_x - \gamma \sum\limits_{y \in X(x)} s_{x,y}\, \sigma_y \ \leq \ \sum\limits_{y \in X(x)} c_{x,y}\, s_{x,y}\,, & \forall x \in X_C; \\ \sigma_x - \gamma \sum\limits_{y \in X(x)} p_{x,y}\, \sigma_y \ \leq \ \mu_x, & \forall x \in X_N; \end{cases} \tag{2.91}$$

where

$$\mu_x = \sum_{y \in X(x)} c_{x,y} \, p_{x,y}, \quad \forall x \in X_N$$

and

$$\sum_{y \in X(x)} s_{x,y} = 1, \quad \forall x \in X_C; \quad s_{x,y} \in \{0, 1\}, \quad \forall x \in X_C, \ y \in X.$$

Proof If for a fixed strategy s we treat the values $s_{x,y}$ as the transition probabilities from the states $x \in X$ to the states $y \in X$ then the condition (2.88) in the extended form can be written as follows:

$$\begin{cases} \sigma_x - \gamma \sum\limits_{y \in X(x)} s_{x,y} \, \sigma_y = \sum\limits_{y \in X(x)} c_{x,y} \, s_{x,y} \,, \quad \forall x \in X_C; \\ \sigma_x - \gamma \sum\limits_{y \in X(x)} p_{x,y} \, \sigma_y = \mu_x, \quad \forall x \in X_N. \end{cases} \tag{2.92}$$

This system determines uniquely the values σ_x for $x \in X$. Therefore, for fixed s the linear programming problem (2.90), (2.92) has a solution and the optimal value of the objective function is equal to σ_{x_0}.

It is evident that if in this system we change the costs $c_{x,y}$ by new costs $c'_{x,y}$ such that $c'_{x,y} \leq c_{x,y}$ then we obtain a new linear programming problem for which the corresponding optimal value σ'_{x0} of the objective function is less or equal to σ_{x_0}. Thus, if we change the system of linear equations (2.92) by the following system of linear inequalities

$$\begin{cases} \sigma_x - \gamma \sum\limits_{y \in X(x)} s_{x,y} \, \sigma_y \leq \sum\limits_{y \in X(x)} c_{x,y} \, s_{x,y} \,, \quad \forall x \in X_C; \\ \sigma_x - \gamma \sum\limits_{y \in X(x)} p_{x,y} \, \sigma_y \leq \mu_x, \quad \forall x \in X_N, \end{cases} \tag{2.93}$$

then for a fixed strategy s we obtain a new linear programming problem (2.90), (2.93) with the optimal solution σ_{x_0}. So, the lemma holds. □

Now we consider the optimization problem (2.90), (2.93) in the case if $s_{x,y}$ are arbitrary boolean variables and correspond to the possible stationary strategies. So, if we add the condition $s_{x,y} \in \{0, 1\}, \ \forall x \in X_C, \ y \in X$ to (2.93) then we obtain the mixed integer bilinear programming problem in which we have to maximize with respect to σ_x and minimize with respect to s.

Based on Lemma 2.37 we can prove the following result.

Theorem 2.38 *Let $\alpha^*_{x,y}$ $(x \in X_C,\ y \in X)$, β^*_x $(x \in X)$ be an optimal solution of the following linear programming problem:*
Minimize

$$\phi_{x_0}(\alpha, \beta) = \sum_{x \in X_C} \sum_{y \in X(x)} c_{x,y}\, \alpha_{x,y} + \sum_{x \in X_N} \mu_x \beta_x \tag{2.94}$$

subject to

$$\begin{cases} \beta_y - \gamma \sum\limits_{x \in X^-_C(y)} \alpha_{x,y} - \gamma \sum\limits_{x \in X^-_N(y)} p_{x,y}\, \beta_x = 1, \quad y = x_0; \\ \beta_y - \gamma \sum\limits_{x \in X^-_C(y)} \alpha_{x,y} - \gamma \sum\limits_{x \in X^-_N(y)} p_{x,y}\, \beta_x = 0, \quad \forall y \in X \backslash \{x_0\}; \\ \qquad \sum\limits_{y \in X(x)} \alpha_{x,y} = \beta_x, \quad \forall x \in X_C; \\ \beta_x \geq 0,\ \forall x \in X; \quad \alpha_{x,y} \geq 0, \quad \forall x \in X_C,\ y \in X(x), \end{cases} \tag{2.95}$$

where

$$\mu_x = \sum_{y \in X(x)} c_{x,y}\, p_{x,y}, \quad \forall x \in X_N.$$

Then

$$\frac{\alpha^*_{x,y}}{\beta^*_x} \in \{0, 1\},\ \forall y \in X(x), \forall y \in X^*_C,$$

where $X^+_C = \{x \in X_C \mid \beta_x > 0\}$ and an optimal stationary strategy for the discounted stochastic control problem on the network can be found as follows:

- *if $x \in X^+_C$ then fix*

$$s^*_{x,y} = \frac{\alpha^*_{x,y}}{\beta^*_x}, \forall x \in X(x);$$

- *if $x \in X \backslash X^+_C$ then fix an arbitrary $s_{x,y} \in \{0, 1\}$ for every $y \in X(x)$ such that*

$$\sum_{y \in X(x)} s_{x,y} = 1.$$

Proof According to Lemma 2.37 for a fixed strategy s the values $\sigma_x,\ x \in X$ can be found by solving the linear programming problem (2.90), (2.91).

Considering the dual problem for (2.90), (2.91) with respect to σ_x for a fixed strategy s we obtain the following optimization problem:
Minimize

$$\phi_{x_0}^s(\beta) = \sum_{x \in X_C} \sum_{y \in X(x)} c_{x,y}\, s_{x,y}\, \beta_x + \sum_{x \in X_N} \mu_x \beta_x \tag{2.96}$$

subject to

$$\begin{cases} \beta_y - \gamma \sum\limits_{x \in X_C^-(y)} s_{x,y}\, \beta_x - \gamma \sum\limits_{x \in X_N^-(y)} p_{x,y}\, \beta_x = 1, \quad y = x_0; \\ \beta_y - \gamma \sum\limits_{x \in X_C^-(y)} s_{x,y}\, \beta_x - \gamma \sum\limits_{x \in X_N^-(y)} p_{x,y}\, \beta_x = 0, \quad \forall y \in X \setminus \{x_0\}; \\ \beta_x \geq 0, \quad \forall x \in X. \end{cases} \tag{2.97}$$

In this system $s_{x,y}$ for $x \in X_C,\ y \in X(x)$ the following condition is satisfied:

$$\begin{cases} \sum\limits_{y \in X(x)} s_{x,y} = 1, \quad \forall x \in X_C; \\ s_{x,y} \geq 0, \quad \forall x \in X,\ y \in X(x). \end{cases} \tag{2.98}$$

Then, an optimal strategy s^* of the control problem on G corresponds to an extreme point of the set of solutions of system (2.98). It is easy to observe that system (2.97) is consistent for an arbitrary feasible solution of system (2.97), and therefore, an optimal stationary strategy s^* can be determined by minimizing (2.96) with respect to $s_{x,y}$ and β_x subject to (2.97), (2.98). Thus, if we add condition (2.98) to condition (2.97) (and after that we minimize (2.96) with respect to $s_{x,y}$ and β_x), then we obtain the following nonlinear programming problem:
Minimize

$$\phi_{x_0}(s, \beta) = \sum_{x \in X_C} \sum_{y \in X(x)} c_{x,y} s_{x,y} \beta_x + \sum_{x \in X_N} \mu_x \beta_x \tag{2.99}$$

subject to

$$\begin{cases} \beta_y - \gamma \sum\limits_{x \in X_C^-(y)} s_{x,y}\, \beta_x - \gamma \sum\limits_{x \in X_N^-(y)} p_{x,y}\, \beta_x = 1, \quad y = x_0; \\ \beta_y - \gamma \sum\limits_{x \in X_C^-(y)} s_{x,y}\, \beta_x - \gamma \sum\limits_{x \in X_N^-(y)} p_{x,y}\, \beta_x = 0, \quad \forall y \in X \setminus \{x_0\}; \\ \qquad \sum\limits_{y \in X(x)} s_{x,y} = 1, \quad \forall x \in X_C; \\ \quad \beta_x \geq 0, \quad \forall x \in X; \quad s_{x,y} \geq 0, \quad \forall x \in X,\ y \in X(x). \end{cases} \tag{2.100}$$

This is a bilinear programming problem however it can be easily reduced to a linear programming problem (2.96), (2.97) using the following elementary transformations:

We change in (2.100) the restrictions $\sum_{y\in X(x)} s_{x,y} = 1, \ \forall x \in X_C$ by $\sum_{y\in X(x)} s_{x,y}\beta_y = \beta_y, \ \forall x \in X_C$ and then we introduce the notations

$$\alpha_{x,y} = s_{x,y}\beta_y, \quad \forall x \in X, \ y \in X(x). \tag{2.101}$$

It is easy to observe that if $\alpha^*_{x,y}$ $(x \in X_C, \ y \in X)$, β^*_y $(y \in X)$ is a basic optimal solution of problem (2.94), (2.95) then for each $x \in X_C^+$ among $\alpha^*_{x,y}, \ y \in X(x)$ only one it is different from zero and it is equal to β^*_x. Moreover, if $\alpha^*_{x,y}$ $(x \in X_C, \ y \in X)$, β^*_y $(y \in X)$ is a basic optimal solution of the problem (2.94), (2.95) then $\alpha^*_{x,y} = 0, \ \forall x \in X\backslash X_C^+, \forall y \in X)$ and therefore for $x \in X\backslash X_C^+, \ y \in X(x)$ in the optimal solution of problem (2.99), (2.100) we can fix arbitrary $s^*_{x,y} \in \{0,1\}$ for $y \in X(x)$ such that $\sum_{y\in X(x)} s_{x,y} = 1$. So, we can determine the optimal stationary strategy for the control problem on network according to the rule formulated in the theorem. □

Note that in the considered control problem with fixed stating state x_0 the vertices $x \in X_c^+$ of the graph G correspond to the states in which the decision person makes the optimal control. The vertices $x \in X\backslash X_C^+$ of graph G correspond to the states of the dynamical system that couldn't be reached in the process of the optimal control made by the decision person. Therefore for the optimal solution of the control problem on G with fixed starting state x_0 we can set $s^*_{x,y} = 0, \forall x \in X\backslash X_C^+, \forall y \in X(x)$. This does not affect the sense of the control problem on networks.

From Theorem 2.38 in the case $X = X_C$ (i.e. $X_N = \emptyset$) we obtain conditions for determining the optimal stationary strategies of the deterministic discounted control problem.

Corollary 2.39 *Let $X = X_C$ and $\alpha^*_{x,y}$ $(x \in X, \ y \in X)$, β^*_x $(x \in X)$ be an optimal solution of the following linear programming problem:*
Minimize

$$\phi_{x_0}(\alpha,\beta) = \sum_{x\in X}\sum_{y\in X(x)} c_{x,y}\alpha_{x,y} \tag{2.102}$$

subject to

$$\begin{cases} \beta_y - \gamma \displaystyle\sum_{x\in X^-(y)} \alpha_{x,y} = 1, \quad y = x_0; \\ \beta_y - \gamma \displaystyle\sum_{x\in X^-(y)} \alpha_{x,y} = 0, \quad \forall y \in X\backslash\{x_0\}; \\ \displaystyle\sum_{y\in X(x)} \alpha_{x,y} = \beta_x, \quad \forall x \in X; \\ \beta_x \ge 0, \ \forall x \in X; \quad \alpha_{x,y} \ge 0, \quad \forall x \in X, \ y \in X(x). \end{cases} \tag{2.103}$$

Then the optimal stationary strategy s^* of the discounted stochastic control problem on network can be found by fixing

$$s^*_{x,y} = \frac{\alpha^*_{x,y}}{\beta^*_x}, \quad \forall x \in X_C^+,\ y \in X(x)$$

and arbitrary $s^*_{x,y} \in \{0, 1\}$ for $x \in X \setminus X_C^+$ such that $\sum_{y \in X(x)} s_{x,y} = 1$;

Based on Theorem 2.38 we can propose the following algorithm for determining the optimal solution of the discounted control problem on the network.

Algorithm 2.40 Determining the Optimal Stationary Strategy for the Discounted Stochastic Control Problem

(1) Formulate the linear programming problem (2.96), (2.97);
(2) Determine an optimal solution $\alpha^*_{x,y}$ $(x \in X_C,\ y \in X)$, β^*_y $(y \in X)$ of the problem (2.96), (2.97) and fix

$$s^*_{x,y} = \frac{\alpha^*_{x,y}}{\beta^*_x}, \quad \forall x \in X_C,\ y \in X(x)$$

and an arbitrary $s^*_{x,y} \in \{0, 1\}$ for every $x \in X \setminus X_C^+$ such that $\sum_{y \in X(x)} s_{x,y} = 1$.

The results described above allow us to determine the stationary strategy for the problem with a fixed starting state x_0. In the general case, if it is necessary to find the optimal stationary strategy for an arbitrary starting state $x \in X$ then we can use the following results.

Lemma 2.41 *For a fixed strategy s the values σ_x^γ, $x \in X$ determine the unique optimal basic solution of the following linear programming problem:*
Maximize

$$\varphi^s(\sigma) = \sum_{x \in X} \sigma_x \tag{2.104}$$

subject to (2.91).

The proof of this lemma is identical to the proof of Lemma 2.37. Based on this lemma we can prove the following theorem.

Theorem 2.42 *Let $\alpha^*_{x,y}$ $(x \in X_C,\ y \in X)$, β^*_x $(x \in X)$ be a basic optimal solution of the following linear programming problem:*
Minimize

$$\phi(\alpha, \beta) = \sum_{x \in X_C} \sum_{y \in X(x)} c_{x,y}\, \alpha_{x,y} + \sum_{x \in X_N} \mu_x \beta_x \tag{2.105}$$

subject to

$$\begin{cases} \beta_y - \gamma \sum_{x \in X_C^-(y)} \alpha_{x,y} - \gamma \sum_{x \in X_N^-(y)} p_{x,y}\, \beta_x = 1, \quad \forall y \in X; \\ \sum_{y \in X(x)} \alpha_{x,y} = \beta_x, \quad \forall x \in X_C; \\ \beta_x \ge 0, \quad \forall x \in X; \quad \alpha_{x,y} \ge 0, \quad \forall x \in X,\ y \in X(x). \end{cases} \tag{2.106}$$

If in the graph $G = (X, E)$ *each vertex* $x \in X$ *contains at least one leaving directed edge then* $\beta_x^* > 0,\ \forall x \in X_C$ *and*

$$\frac{\alpha_{x,y}^*}{\beta_x^*} \in \{0, 1\}, \quad \forall x \in X_C,\ y \in X(x).$$

The optimal stationary strategy s^* *of the discounted stochastic control problem on the network can be found by fixing*

$$s_{x,y}^* = \frac{\alpha_{x,y}^*}{\beta_x^*}, \quad \forall x \in X_C,\ y \in X(x).$$

The proof of this theorem is similar to the proof of Theorem 2.38 and the solution of the problem can be found by using the linear programming problem (2.105), (2.106). Based on this theorem we determine the optimal stationary strategies for an arbitrary starting state $x \in X$.

In the problem (2.105), (2.105) we can eliminate β_y from those restrictions that correspond to vertices $y \in X_C$ if we take into account the relation $\beta_y = \sum_{x \in X(y)} \alpha_{y,x}$ for $y \in X_C$. After that from Theorem 2.42 we obtain the following corollary:

Corollary 2.43 *Let* $\alpha_{x,y}^*$ $(x \in X_C,\ y \in X)$, β_x^* $(x \in X)$ *be a basic optimal solution of the following linear programming problem:*
Minimize

$$\phi(\alpha, \beta) = \sum_{x \in X_C} \sum_{y \in X(x)} c_{x,y}\, \alpha_{x,y} + \sum_{x \in X_N} \mu_x \beta_x \tag{2.107}$$

subject to

$$\begin{cases} \sum_{x \in X(y)} \alpha_{y,x} - \gamma \sum_{x \in X_C^-(y)} \alpha_{x,y} - \gamma \sum_{x \in X_N^-(y)} p_{x,y}\, \beta_x = 1, \quad y \in X_C; \\ \beta_y - \gamma \sum_{x \in X_C^-(y)} \alpha_{x,y} - \gamma \sum_{x \in X_N^-(y)} p_{x,y}\, \beta_x = 1, y \in X_N; \\ \beta_x \ge 0, \ \forall x \in X_N; \quad \alpha_{x,y} \ge 0, \quad \forall x \in X_C,\ y \in X(x), \end{cases} \tag{2.108}$$

If in the graph $G = (X, E)$ each vertex $x \in X$ contains at least one leaving directed edge then $\sum_{y \in X(x)} \alpha^*_{x,y} > 0,\ \forall x \in X_C$ and

$$\frac{\alpha^*_{x,y}}{\sum_{y \in X} \alpha^*_{x,y}} \in \{0, 1\}, \quad \forall x \in X_C,\ y \in X(x).$$

The optimal stationary strategy s^* of the discounted stochastic control problem on the network can be found by fixing

$$s^*_{x,y} = \frac{\alpha^*_{x,y}}{\sum_{y \in X(x)} \alpha_{x,y}}, \qquad \forall x \in X_C,\ y \in X(x).$$

2.5.3 Dual Linear Programming Models for a Discounted Control Problem

If we dualize the linear programming problem (2.94), (2.95) then on the basis of duality theory we obtain the following result.

Theorem 2.44 *Let w^*_x $(x \in X_C)$, σ^*_x $(x \in X)$ be the optimal solution of the linear programming problem:*
Maximize

$$\varphi_{x_0}(\sigma, w) = \sigma_{x_0} \tag{2.109}$$

subject to

$$\begin{cases} w_x - \gamma\sigma_y \le c_{x,y}, & \forall x \in X_C,\ y \in X(x); \\ -w_x + \sigma_x \le 0, & \forall x \in X_C; \\ \sigma_x - \gamma \displaystyle\sum_{y \in X(x)} p_{x,y}\sigma_y \le \mu_x, & \forall x \in X_N. \end{cases} \tag{2.110}$$

Then $w^*_x = \sigma^*_x,\ \forall x \in X_C$ and $\sigma^*_{x_0}$ is the optimal discounted expected total cost for the problem on the network with the starting state x_0. An optimal stationary strategy can be found by fixing $s^* : X_C \to X$ such that $(x, s^*(x)) \in E^*(x),\ \forall x \in X_C$, where $E^*(x) = \{(x, y) \mid y \in X(x),\ \sigma^*_x - \gamma\sigma^*_y - c_{x,y} = 0\}$.

As a consequence from this theorem we obtain the following result.

Corollary 2.45 *For an arbitrary discounted control problem on the network (G, X_C, X_N, c, p) with a given discount factor γ there exist the values σ^*_x for $x \in X$ that satisfy the following conditions:*

(1) $\bar{c}_{x,y} = c_{x,y} + \gamma\sigma^*_y - \sigma^*_x \ge 0,\ \forall x \in X_C, y \in X(x)$;
(2) $\min_{y \in X(x)} \{\bar{c}_{x,y}\} = 0,\ \forall x \in X_C$;

(3) $\overline{\mu}_x = \mu_x + \gamma \sum_{y \in X(z)} p_{x,y}\sigma_y^* - \sigma_x^* = 0, \ \forall x \in X_N.$

An arbitrary stationary strategy $s^* : X_C \to X$ *such that* $(x, s^*(x)) \in E^*(x)$, $\forall x \in X_C$, *where* $E^*(x) = \{(x, y) \mid y \in X(x), \overline{c}_{x,y} = 0\}$, *represents an optimal stationary strategy for the discounted control problem.*

In the case $X_N = \emptyset$, from this corollary we obtain the optimality condition for the deterministic discounted control problem. The conditions for determining the optimal strategy and the value of the optimal cost for the problem with $\gamma = 1$ in the case if this value exists can also be derived from Theorem 2.44 and Corollary 2.45.

The results formulated above can be extended to the problem of determining the optimal stationary strategy with an arbitrary starting state $x \in X$. For the problem (2.105), (2.106) we can construct the dual problem in a similar way and we then obtain the following result.

Theorem 2.46 *Let* σ_x^*, w_x^* $(x \in X)$ *be the optimal solution of the linear programming problem:*
Maximize

$$\varphi(\sigma, w) = \sum_{x \in X} \sigma_x \tag{2.111}$$

subject to (2.110). *Then* σ_x^* *for* $x \in X$ *represents the optimal discounted expected total costs for the problem on the network with starting states* $x \in X$. *An optimal stationary strategy can be found by fixing* $s^* : X_C \to X$ *such that* $(x, s^*(x)) \in E^*(x)$, $\forall x \in X_C$, *where* $E^*(x) = \{(x, y) \mid y \in X(x), \ \sigma_x^* - \gamma\sigma_y^* - c_{x,y} = 0\}$.

2.6 A Linear Programming Approach for a Discounted Markov Decision Problem

Consider a Markov decision process (X, A, p, c) with a finite set of states X, a finite set of actions A, the probability function $p : A \times X \times X \to [0, 1]$ that satisfies the condition $\sum_{y \in X} p_{x,y}^a = 1, \ \forall a \in A$ and the cost function $c : A \times X \times X \to \mathbb{R}$. In addition we assume that the discount factor γ, $0 \le \gamma < 1$, and the starting state x_0 are given.

Let us fix a stationary strategy

$$s : x \to a \in A(x) \quad \text{for} \quad x \in X,$$

that induces a simple Markov process with a transition probability matrix $P^s = (p_{x,y}^s)$ and a transition cost matrix $C^s = (c_{x,y})$. Then we can determine the discounted expected total costs $\sigma_{x_0}^{\gamma}(s)$ (in order to simplify the notation in the following we shall use $\sigma_{x_0}(s)$ instead of $\sigma_{x_0}^{\gamma}(s)$).

We consider the problem of determining the strategy s^* such that

$$\sigma_{x_0}(s^*) = \min_s \sigma_{x_0}(s).$$

In a similar way as for the control problem, here we identify an arbitrary strategy $s : X \to A$ with the set of the boolean variables $s_{x,a}$ for $x \in X$ and $a \in A$, i.e.,

$$s_{x,a} = \begin{cases} 1, & \text{if } a = s(x); \\ 0, & \text{if } a \neq s(x). \end{cases} \tag{2.112}$$

Lemma 2.47 *For a fixed strategy s the values σ_x^γ, $x \in X$ determine the unique optimal basic solution of the following linear programming problem:*
Maximize

$$\varphi_{x_0}^s(\sigma) = \sigma_{x_0} \tag{2.113}$$

subject to

$$\sigma_x - \gamma \sum_{y \in X} \sum_{a \in A(x)} s_{x,a}\, p_{x,y}^a\, \sigma_y \leq \sum_{a \in A(x)} s_{x,a}\, \mu_{x,a}, \quad \forall x \in X; \tag{2.114}$$

where

$$\mu_{x,a} = \sum_{a \in A(x)} c_{x,y}^a\, p_{x,y}^a, \quad \forall x \in X$$

and

$$\sum_{a \in A(x)} s_{x,a} = 1, \quad \forall x \in X; \quad s_{x,a} \in \{0, 1\}, \quad \forall x \in X,\ a \in A(x).$$

Proof For a fixed strategy s the solution of the system of linear equations

$$\sigma_x - \gamma \sum_{y \in X} \sum_{a \in A(x)} s_{x,a}\, p_{x,y}^a\, \sigma_y = \sum_{a \in A(x)} s_{x,a}\, \mu_{x,a}, \quad \forall x \in X \tag{2.115}$$

uniquely determines σ_x, $\forall x \in X$. Thus, the problem of maximization of the objective function (2.113) subject to (2.115) for a fixed strategy s has a unique feasible solution which is an optimal one. This implies that if for fixed s we consider the problem: Maximize (2.113) subject to

$$\sigma_x - \gamma \sum_{y \in X} \sum_{a \in A(x)} s_{x,a}\, p_{x,y}^a\, \sigma_y \leq \sum_{a \in A(x)} s_{x,a}\, \mu_{x,a}, \quad \forall x \in X \tag{2.116}$$

then it has the same optimal solution as the problem (2.113), (2.115). Moreover, if in the problem (2.113), (2.116) we vary the boolean variables $s_{x,y}$ and take the maximum with respect to σ and the minimum with respect to s then we obtain the optimal strategy for the control problem. □

Using the lemma above we can prove the following theorem.

Theorem 2.48 *Let $\alpha^*_{x,a}$, β^*_y $(x \in X,\ a \in A)$ be a basic optimal solution of the following linear programming problem:*
Minimize

$$\phi_{x_0}(\alpha, \beta) = \sum_{x \in X} \sum_{a \in A(x)} \mu_{x,a}\, \alpha_{x,a} \tag{2.117}$$

subject to

$$\begin{cases} \beta_y - \gamma \sum\limits_{x \in X} \sum\limits_{a \in A(x)} p^a_{x,y}\, \alpha_{x,a} = 1, \quad y = x_0; \\ \beta_y - \gamma \sum\limits_{x \in X} \sum\limits_{a \in A(x)} p^a_{x,y}\, \alpha_{x,a} = 0, \quad \forall y \in X \backslash \{x_0\}; \\ \qquad \sum\limits_{a \in A(x)} \alpha_{x,a} = \beta_x, \quad \forall x \in X; \\ \beta_y \geq 0, \quad \forall y \in X; \quad \alpha_{x,a} \geq 0, \quad \forall x \in X,\ a \in A(x). \end{cases} \tag{2.118}$$

Then the optimal stationary strategy s^* for the discounted Markov decision problem is determined as follows:

$$s^*_{x,a} = \begin{cases} 1, & \text{if } \alpha^*_{x,a} \neq 0; \\ 0, & \text{if } \alpha^*_{x,a} = 0. \end{cases} \tag{2.119}$$

Proof According to Lemma 2.47 the optimal stationary strategy s^* corresponds to the optimal solution of the problem (2.113), (2.114).

In a similar way as in the proof of Lemma 2.37 here we have that a stationary strategy s corresponds to an extreme point of the set of solutions of the following system

$$\begin{cases} \sum\limits_{a \in A(x)} s_{x,a} = 1, \quad \forall x \in X; \\ s_{x,a} \geq 0, \quad \forall x \in X,\ a \in A. \end{cases} \tag{2.120}$$

Therefore, if we dualize (2.113), (2.120) with respect to σ_x for a fixed strategy s then we obtain the following optimization problem:
Minimize

$$\phi_{x_0}(s, \beta) = \sum_{x \in X} \sum_{a \in A(x)} \mu_{x,a} s_{x,a}\, \beta_x \tag{2.121}$$

subject to

$$\begin{cases} \beta_y - \gamma \sum_{x \in X} \sum_{a \in A(x)} p_{x,y}^a \, s_{x,a} \, \beta_x = 1, \quad y = x_0; \\ \beta_y - \gamma \sum_{x \in X} \sum_{a \in A(x)} p_{x,y}^a \, s_{x,a} \, \beta_x = 0, \quad \forall y \in X \setminus \{x_0\}; \\ \beta_y \geq 0, \quad \forall y \in X. \end{cases} \tag{2.122}$$

Now if we minimize (2.113) with respect to $s_{x,a}$ and β_x and in (2.120) we take into account the following restriction

$$\sum_{a \in A(x)} s_{x,a} = 1, \quad \forall x \in X; \quad s_{x,a} \geq 0, \quad \forall x \in X, \, a \in A(x)$$

then we obtain the problem:
Minimize

$$\phi_{x_0}(s, \beta) = \sum_{x \in X} \sum_{a \in A(x)} \mu_{x,a} s_{x,a} \, \beta_x \tag{2.123}$$

subject to

$$\begin{cases} \beta_y - \gamma \sum_{x \in X} \sum_{a \in A(x)} p_{x,y}^a \, s_{x,a} \, \beta_x = 1, \quad y = x_0; \\ \beta_y - \gamma \sum_{x \in X} \sum_{a \in A(x)} p_{x,y}^a \, s_{x,a} \, \beta_x = 0, \quad \forall y \in X \setminus \{x_0\}; \\ \sum_{a \in A(x)} s_{x,a} = 1, \quad \forall x \in X; \\ \beta_y \geq 0, \quad \forall y \in X; \quad s_{x,a} \geq 0, \quad \forall x \in X, \, a \in A(x). \end{cases} \tag{2.124}$$

This bilinear programming problem can be easily reduced to a linear programming problem (2.117), (2.118) using the following elementary transformations: We change in (2.124) the restrictions $\sum_{a \in A(x)} s_{x,a} = 1, \ \forall x \in X$ by $\sum_{a \in A(x)} s_{x,a} \, \beta_x = \beta_x, \ \forall x \in X$ and then we introduce the notations

$$\alpha_{x,a} = s_{x,a} \, \beta_x, \quad \forall x \in X, \, a \in A(x). \tag{2.125}$$

If ${\alpha_{x,y}^a}^*$, β_y^* $(x, y \in X, \, a \in A)$ is a basic optimal solution of the linear programming problem (2.117), (2.118) then by using (2.125) we obtain $s_{x,a}^*$ according to (2.119). □

Based on Theorem 2.48 we can propose the following algorithm for determining the solution of the Markov decision problem.

Algorithm 2.49 Determining the Optimal Stationary Strategy for the Discounted Markov Decision Problem

(1) Formulate the linear programming problem (2.117), (2.118);
(2) Determine a basic optimal solution $\alpha^*_{x,a}$ $(x \in X,\ a \in A)$, β^*_y $(y \in X)$ of the problem (2.117), (2.118) and determine $s^*_{x,x}$ according to (2.119).

The results described above allow us to determine the stationary strategy for the discounted Markov decision problem in the case if the starting state x_0 is fixed. In the general case, if it is necessary to find the optimal stationary strategy for an arbitrary starting state $x \in X$ then we can use the following results.

Lemma 2.50 *For a fixed strategy s the values* σ^γ_x, $x \in X$ *determine the unique optimal basic solution of the following linear programming problem:*
Maximize

$$\varphi^s(\sigma) = \sum_{x \in X} \sigma_x$$

subject to

$$\sigma_x - \gamma \sum_{y \in X} \sum_{a \in A(x)} s_{x,a}\, p^a_{x,y}\, \sigma_y \ \le \sum_{a \in A(x)} s_{x,a}\, \mu_{x,a}, \quad \forall x \in X,$$

where

$$\mu_{x,a} = \sum_{a \in A(x)} c^a_{x,y}\, p^a_{x,y}, \quad \forall x \in X$$

and

$$\sum_{a \in A(x)} s_{x,a} = 1, \quad \forall x \in X; \quad s_{x,a} \in \{0, 1\}, \quad \forall x \in X,\ a \in A(x).$$

The proof of this lemma is similar to the proof of Lemma 2.47.
Using this lemma we can prove the following theorem:

Theorem 2.51 *Let* $\alpha^*_{x,a}$, β^*_y $(x \in X,\ y \in X,\ a \in A)$ *be a basic optimal solution of the following linear programming problem:*
Minimize

$$\phi(\alpha, \beta) = \sum_{x \in X} \sum_{a \in A(x)} \mu_{x,a} \alpha_{x,a} \tag{2.126}$$

subject to

$$\begin{cases} \beta_y - \gamma \sum_{x \in X} \sum_{a \in A(x)} p_{x,y}^a \, \alpha_{x,a} = 1, \;\; \forall y \in X; \\ \qquad\qquad \sum_{a \in A(x)} \alpha_{x,a} = \beta_x, \;\; \forall x \in X; \\ \beta_y \geq 0, \;\; \forall y \in X; \quad \alpha_{x,a} \geq 0, \;\; \forall x \in X, \; a \in A(x). \end{cases} \tag{2.127}$$

Then the optimal stationary strategy s^* for the discounted Markov decision problem is determined according to (2.119).

The proof of this theorem is identical to the prove of Theorem 2.48; here we have to apply Lemma 2.50 instead of Lemma 2.47.

It is easy to observe that the constraints $\beta_y \geq 0, \; \forall y \in X$, in (2.127) are redundant. Therefore, we can eliminate $\beta_x, \forall x \in X$, from (2.127) introducing the expressions $\sum_{a \in A(x)} \alpha_{x,a} = \beta_x$ for $x \in X$ in the first group of the constraints. After that from Theorem 2.51 we obtain the following corollary.

Corollary 2.52 *Let $\alpha^*_{x,a}$ $(x \in X, \; y \in X, \; a \in A)$ be a basic optimal solution of the following linear programming problem:*
Minimize

$$\phi(\alpha) = \sum_{x \in X} \sum_{a \in A(x)} \mu_{x,a} \alpha_{x,a} \tag{2.128}$$

subject to

$$\begin{cases} \sum_{a \in A(x)} \alpha_{x,a} - \gamma \sum_{x \in X} \sum_{a \in A(x)} p_{x,y}^a \, \alpha_{x,a} = 1, \;\; \forall y \in X; \\ \alpha_{x,a} \geq 0, \;\; \forall x \in X, \; a \in A(x). \end{cases} \tag{2.129}$$

Then the optimal stationary strategy s^* for the discounted Markov decision problem is determined as follows:

$$s_{x,a} = \begin{cases} 1, & \text{if } a = s(x); \\ 0, & \text{if } a \neq s(x). \end{cases}$$

2.6.1 A Dual Linear Programming Model for the Discounted Markov Decision Problem

We formulate the dual linear programming model for the discounted Markov decision problem using the problem (2.128), (2.129). Applying the duality linear programming theorems to this problem we obtain the following result:

Theorem 2.53 *Let σ_x^* $(x \in X)$ be the optimal solution of the linear programming problem:*
Maximize

$$\varphi(\sigma) = \sum_{x \in X} \sigma_x \tag{2.130}$$

subject to

$$\sigma_x - \gamma \sum_{y \in X} p_{x,y}^a \sigma_y \leq \mu_{x,a}, \quad \forall x \in X,\ a \in A(x). \tag{2.131}$$

Then σ_x^* for $x \in X$ represents the optimal discounted expected total costs for the problem on the network with starting states $x \in X$. An optimal stationary strategy can be found by fixing $s^* : X \to A$ such that $s^*(x) = a \in A^*(x)$, $\forall x \in X$, where $A^*(x) = \{a \in A(x) \mid \sigma_x - \gamma \sum_{y \in X} p_{x,y}^a \sigma_y = 0\}$.

Thus, the solution of the discounted Markov decision problem can be found by solving the dual linear programming problem (2.130), (2.131).

2.7 An Iterative Algorithm for Discounted Markov Decision Processes and Stochastic Control Problems

To determine the optimal discounted costs and the corresponding optimal strategy in the Markov processes with discounted costs we shall use the following system of equations with respect to $\sigma_{x_1}, \sigma_{x_2}, \ldots, \sigma_{x_n}$:

$$\sigma_{x_i} = \min_{a \in A(x_i)} \left[\mu_{x_i,a} + \gamma \sum_{x_j \in X} p_{x_i,x_j}^a \sigma_{x_j} \right], \quad i = 1, 2, \ldots, n.$$

According to Theorem 2.53 this system of equations has a solution. Below we describe an iterative algorithm for determining the solution of this system of equations and finding the optimal stationary strategies of the discounted Markov decision problem.

Algorithm 2.54 Determining the Optimal Stationary Strategies for the Discounted Markov Decision Problem

Preliminary step (Step 0): Fix an arbitrary stationary strategy

$$s^0 : x_i \to a \in A(x_i) \text{ for } x_i \in X.$$

General step (Step k, $k > 0$): Calculate

$$\mu_{x_i,s^{k-1}(x_i)} = \sum_{y \in X(x_i)} p_{x_i,y}^{s^{k-1}(x_i)} c_{x_i,y}^{s^{k-1}(x_i)}$$

for every $x_i \in X$. Then solve the system of linear equations

$$\sigma_{x_i} = \mu_{x_i,s^{k-1}(x_i)} + \gamma \sum_{x_j \in X} p_{x_i,x_j}^{s^{k-1}(x_i)} \sigma_{x_j}, \quad i = 1, 2, \ldots, n$$

and find the solution $\sigma_{x_1}^{k-1}, \sigma_{x_2}^{k-1}, \ldots, \sigma_{x_n}^{k-1}$. After that determine a new strategy

$$s^k : x_i \to a \in A(x_i) \text{ for } x_i \in X,$$

where

$$s^k(x_i) = \underset{a \in A(x_i)}{\operatorname{argmin}} \left[\mu_{x_i,a} + \gamma \sum_{x_j \in X} p_{x_i,x_j}^a \sigma_{x_i}^{k-1} \right], \quad i = 1, 2, \ldots, n.$$

Check if the following condition holds

$$s^k(x_i) = s^{k-1}(x_i), \quad \forall x_i \in X. \tag{2.132}$$

If the condition (2.132) holds then fix

$$s^* = s^k; \quad \sigma_{x_i}^* = \sigma_{x_i}^k, \quad \forall x_i \in X$$

as the optimal solution of the problem; otherwise go to the next step $k + 1$.

This algorithm can be specified for a stochastic control problem with a discounted cost criterion. The correctness and the convergence of this iterative algorithm can be derived from the results described above and the results from [32, 112, 128, 136].

Algorithm 2.55 Determining the Optimal Stationary Strategies for the Discounted Stochastic Control Problem

We consider the discounted control problem on the network (G, X_1, X_2, c, p) with a given discount factor γ. The dynamics of the system is described by a directed graph $G = (X, E)$ with the set of controllable states X_C and the set of uncontrollable states X_N. In addition we assume that the probability function $p : E_2 \to [0, 1]$ and the cost function $c : E \to \mathbb{R}$ are given.

Preliminary step (Step 0): Fix an arbitrary stationary strategy

$$s^0 : x_i \to x_j \in X(x_i) \text{ for } x_i \in X_C.$$

General step (Step k, $k > 0$): Determine the probability matrix $P^{s^{k-1}} = (p^{s^{k-1}}_{x_i,x_j})$, where

$$p^{s^{k-1}}_{x_i,x_j} = \begin{cases} p_{x_i,x_j}, & \text{if } x_i \in X_N \text{ and } (x_i, x_i) \in E_N; \\ 1, & \text{if } x_i \in X_C \text{ and } x_j = s^{k-1}(x_i); \\ 0, & \text{if } x_i \in X_C \text{ and } x_j \neq s^{k-1}(x_i). \end{cases}$$

Then calculate

$$\mu_{x_i,s^{k-1}(x_i)} = \sum_{y \in X(x_i)} p^{s^{k-1}(x_i)}_{x_i,y} c^{s^{k-1}(x_i)}_{x_i,y}$$

for every $x_i \in X$ and solve the system of linear equations

$$\sigma_{x_i} = \mu_{x_i,s^{k-1}(x_i)} + \gamma \sum_{x_j \in X} p^{s^{k-1}(x_i)}_{x_i,x_j} \sigma_{x_j}, \quad i = 1, 2, \ldots, n$$

and find the solution $\sigma^{k-1}_{x_1}, \sigma^{k-1}_{x_2}, \ldots, \sigma^{k-1}_{x_n}$. After that determine a new strategy

$$s^k : x_i \to a \in A(x_i) \text{ for } x_i \in X_C,$$

where

$$s^k(x_i) = \operatorname*{argmin}_{a \in A(x_i)} \left[\mu_{x_i,a} + \gamma \sum_{x_j \in X} p^a_{x_i,x_j} \sigma^{k-1}_{x_i} \right], \quad \forall x_i \in X_C.$$

Check if the following condition holds

$$s^k(x_i) = s^{k-1}(x_i), \quad \forall x_i \in X_C. \tag{2.133}$$

If the condition (2.133) holds then fix

$$s^* = s^k; \quad \sigma^*_{x_i} = \sigma^k_{x_i}, \quad \forall x_i \in X$$

as the optimal solution of the problem; otherwise go to the next step $k + 1$.

This algorithm finds the optimal stationary strategy for an arbitrary stochastic control problem. In the case when $X = X_C$ ($X_N = \emptyset$) we obtain an iterative algorithm for deterministic discounted control problems.

2.8 Determining the Optimal Expected Total Cost for Markov Decision Problems with a Stopping State

The algorithms proposed in the previous sections determine the optimal stationary strategies for discounted Markov decision problems in the case if the discount factor γ satisfies the condition $0 < \gamma < 1$. If $\gamma = 1$ then the expected total cost in these problems may not exist. Here we study a class of unichain decision problems for which γ may be equal to 1 and the expected total cost exists. Moreover, we can see that for some problems γ may be an arbitrary positive value. For the considered problems we show how to determine the optimal expected total cost using a linear programming approach and iterative procedures. To ensure the existence of the expected total cost in these problems we assume that for the dynamical system there exists a state in which transitions stop as soon as this state is reached [89]. Furthermore, we describe algorithms for determining optimal strategies in such problems.

2.8.1 Problem Formulation and a Linear Programming Approach

Let (X, A, p, c) be a Markov decision process with a finite set of states X, a finite set of actions A, the probability function $p : A \times X \times X \to \mathbb{R}^+$ that satisfies the condition $\sum_{y \in X} p_{x,y}^a = 1, \ \forall a \in A$ and the transition cost function $c : A \times X \times X \to \mathbb{R}$. In addition a discount factor γ for the Markov decision process is given, where $0 < \gamma \leq 1$. We consider the problem of determining the stationary strategy with minimal expected total cost for unichain Markov processes in the case if the dynamical system stops transitions in a given state $z \in X$. At first we assume that the Markov process is perfect. Moreover, we assume that for an arbitrary fixed action in this decision process the state $z \in X$ is an absorbing state. Obviously, in this case for $0 < \gamma < 1$ the optimal expected total costs σ_x and the optimal stationary strategy for an arbitrary starting state $x \in X \backslash \{z\}$ can be found using the linear programming models (2.128), (2.129) and (2.130), (2.131) considering $c_{z,z}^a = 0, \ \forall a \in A(z)$. If the optimal strategy s^* is found then we have only to fix $s^*(x)$ for $x \in X \backslash \{z\}$ because z is the stopping state. In this case the expected total cost for a given starting state σ_{x_0} can be found by solving the linear programming problem (2.117), (2.118). Now we can see that the considered linear programming models can be used for determining the solution of the decision problem with an absorbing stopping state $z \in X$ in the case $\gamma = 1$ if $c_{z,z}^a = 0, \ \forall a \in A(x)$. Indeed, for a fixed strategy s the rank of the matrix $(I - P^s)$ for the unichain process is equal to $|X| - 1$ and the system of equations $(I - P^s)\sigma = \mu^s$ has a unique solution if we put $\sigma_z = 0$. Thus, for unichain processes with absorbing state $z \in X$ the system of equations

$$\begin{cases} (I - P^s)\sigma = \mu^s; \\ \sigma_z = 0 \end{cases}$$

has a unique solution if $c_{z,z}^a = 0, \ \forall a \in A(z)$.

The properties mentioned above allow us to conclude that for a unichain decision problem with $0 < \gamma \le 1$ the following lemma holds.

Lemma 2.56 *A stationary strategy s^* is optimal if and only if it corresponds to an optimal solution σ^*, s^* of the following mixed integer bilinear programming problem:*
Maximize

$$\varphi_{x_0}(\sigma, s) = \sigma_{x_0} \tag{2.134}$$

subject to

$$\begin{cases} \sigma_x - \gamma \sum\limits_{y\in X} \sum\limits_{a\in A(x)} s_{x,a}\, p^a_{x,y}\, \sigma_y \le \sum\limits_{a\in A(x)} s_{x,a}\, \mu_{x,a}, \quad \forall x \in X\backslash\{z\}; \\ \sigma_z = 0; \\ \sum\limits_{a\in A(x)} s_{x,a} = 1, \quad \forall x \in X\backslash\{z\}; \\ s_{x,a} \in \{0,1\}, \quad \forall x \in X\backslash\{z\},\ a \in A(x), \end{cases} \tag{2.135}$$

where

$$\mu_{x,a} = \sum_{y\in X} p^a_{x,y}\, c^a_{x,y}.$$

Note that in (2.134), (2.135) the boolean variables $s_{x,a}$ for $x \in X\backslash\{z\}$, $a \in A(x)$ correspond to a strategy $s : X\backslash\{z\} \to X$, where $s_{x,a} = 1$ if $s(x) = a$ and $s_{x,a} = 0$ if $s(x) \ne a$. Based on this lemma, we can prove the following theorem.

Theorem 2.57 *Let ${\alpha_{x,a}}^*$, β^*_y ($x \in X\backslash\{z\}$, $y \in X\backslash\{z\}$, $a \in A$) be a basic optimal solution of the following linear programming problem:*
Minimize

$$\phi_{x_0}(\alpha, \beta) = \sum_{x\in X\backslash\{z\}} \sum_{a\in A(x)} \mu_{x,a}\, \alpha_{x,a} \tag{2.136}$$

subject to

$$\begin{cases} \beta_y - \gamma \sum\limits_{x\in X\backslash\{z\}} \sum\limits_{a\in A(x)} p^a_{x,y}\, \alpha_{x,a} \ge 1, \quad y = x_0; \\ \beta_y - \gamma \sum\limits_{x\in X\backslash\{z\}} \sum\limits_{a\in A(x)} p^a_{x,y}\, \alpha_{x,a} \ge 0, \quad \forall y \in X\backslash\{x_0, z\}; \\ \sum\limits_{a\in A(x)} \alpha_{x,a} = \beta_x, \quad \forall x \in X\backslash\{z\}; \\ \beta_y \ge 0, \ \forall y \in X\backslash\{z\}; \quad \alpha_{x,a} \ge 0, \ \forall x \in X\backslash\{z\},\ a \in A(x). \end{cases} \tag{2.137}$$

Then the optimal stationary strategy s^* for the discounted unichain decision problem with absorbing state $z \in X$ is determined as follows:

$$s^*_{x,a} = \begin{cases} 1, & \textit{if } \alpha^*_{x,a} \neq 0; \\ 0, & \textit{if } \alpha^*_{x,a} = 0. \end{cases} \tag{2.138}$$

The proof of Theorem 2.57 is obtained in the same way as Theorem 2.48. Lemma 2.56 and Theorem 2.57 differ from Lemma 2.47 and Theorem 2.48, respectively, only in a single restriction in the systems (2.135) and (2.137). These systems are obtained from (2.114) and (2.114), respectively, by deleting the constraints that correspond to the absorbing state z. In the proof of Theorem 2.57 we have only to assume that the expected total cost for the problem with an absorbing stopping state exists.

Remark 2.58 The values σ_x, $\forall x \in X$ for a unichain Markov decision problem with stopping state z with $c^a_{z,z} = 0$, $\forall a \in A(z)$ and $\gamma = 1$ coincide with the values ε_x, $\forall x \in X$ for an zero average cost Markov decision problem.

Based on Theorem 2.57 the optimal stationary strategy of the problem with stopping state can be found by using the following algorithm.

Algorithm 2.59 Determining the Optimal Stationary Strategy for a Markov Decision Problem with Stopping State

(1) Formulate the linear programming problem (2.136), (2.137);
(2) Determine a basic optimal solution $\alpha^*_{x,a}$ $(x \in X\backslash\{z\},\ a \in A)$, β^*_y $(y \in X\backslash\{z\})$ of the problem (2.136), (2.137) and determine $s^*_{x,x}$ according to (2.138).

Remark 2.60 Theorem 2.57 and Algorithm 2.59 are also valid for an arbitrary Markov decision problem with a stopping state z in the case $\gamma \geq 1$ if the cost function $c : X \times X \times A \rightarrow \mathbb{R}$ is strict positive and there exists a strategy s that induces a unichain process with a stopping absorbing state z. Thus, Theorem 2.57 in the case $\gamma \geq 1$ gives necessary and sufficient conditions for determining the optimal stationary strategies in the discounted decision problem with positive costs and given stopping state z.

If in the considered decision problem it is necessary to determine the optimal stationary strategies for an arbitrary starting state $x \in X\backslash\{z\}$ then we can use the following result.

Theorem 2.61 *Let $\alpha^*_{x,a}$ $(x \in X\backslash\{z\},\ a \in A)$ be a basic optimal solution of the following linear programming problem:*
Minimize

$$\phi(\alpha) = \sum_{x \in X\backslash\{z\}} \sum_{a \in A(x)} \mu_{x,a}\alpha_{x,a} \tag{2.139}$$

subject to

$$\begin{cases} \sum\limits_{a\in A(y)} \alpha_{y,a} - \gamma \sum\limits_{x\in X\setminus\{z\}} \sum\limits_{a\in A(x)} p_{x,y}^{a}\, \alpha_{x,a} \geq 1, \quad \forall y \in X\setminus\{z\}; \\ \alpha_{x,a} \geq 0, \quad \forall x \in X\setminus\{z\},\ a \in A(x). \end{cases} \tag{2.140}$$

Then the optimal stationary strategy s^* for the discounted Markov decision problem with an arbitrary starting state $x \in X\setminus\{z\}$ and given stopping state z is determined as follows:

$$s_{x,a}^{*} = \begin{cases} 1, & \text{if } \alpha_{x,a}^{*} \neq 0; \\ 0, & \text{if } \alpha_{x,a}^{*} = 0. \end{cases}$$

The proof of this theorem can be obtained by using the following lemma.

Lemma 2.62 *A stationary strategy s^* is optimal if and only if it corresponds to an optimal solution σ^*, s^* of the following mixed integer bilinear programming problem:*
Maximize

$$\varphi(\sigma, s) = \sum_{x\in X} \sigma_x$$

subject to (2.135).

If for the linear programming problem (2.139), (2.140) we construct the dual model in the same way as for the previous problems then we obtain the following result.

Theorem 2.63 *Let σ_x^* $(x \in X)$ be the optimal solution of the linear programming problem:*
Maximize

$$\varphi(\sigma) = \sum_{x\in X} \sigma_x \tag{2.141}$$

subject to

$$\sigma_x - \gamma \sum_{y\in X} p_{x,y}^{a} \sigma_y \leq \mu_{x,a}, \qquad \forall x \in X\setminus\{z\},\ a \in A(x), \tag{2.142}$$

where $0 < \gamma \leq 1$. Then σ_x^* for $x \in X$ represents the optimal discounted expected total cost for the problem with starting states $x \in X$. An optimal stationary strategy can be found by fixing $s^* : X\setminus\{z\} \to A$ such that $s^*(x) = a \in A^*(x)$, $\forall x \in X\setminus\{z\}$, where $A^*(x) = \{a \in A(x) \mid \sigma_x - \gamma \sum_{y\in X} p_{x,y}^{a} \sigma_y = \mu_{x,a}\}$.

We can obtain an iterative algorithm for the problems with a stopping state from the algorithm for a discounted Markov decision problem from Sect. 2.7 if at each iteration of the algorithm we solve the system of linear equations

$$\begin{cases} \sigma_z = 0; \\ \sigma_{x_i} = \mu_{x_i, s^{k-1}(x_i)} + \gamma \sum\limits_{x_j \in X \setminus \{z\}} p_{x_i, x_j}^{s^{k-1}(x_i)} \sigma_{x_j}, \quad \forall x_i \in X \setminus \{z\} \end{cases}$$

instead of the system of linear equations

$$\sigma_{x_i} = \mu_{x_i, s^{k-1}(x_i)} + \gamma \sum_{x_j \in X} p_{x_i, x_j}^{s^{k-1}(x_i)} \sigma_{x_j}, \quad \forall x_i \in X.$$

Thus, if in the general step of the iterative algorithm from the previous section we replace this system of the equations by the system of equations written above we obtain the iterative algorithm for the problem with an absorbing state. Now let us show how to solve the unichain Markov decision problem with a given stopping state $z \in x$ if z is not an absorbing state but is a positive recurrent state of the Markov process induced by an arbitrary stationary strategy. In this case the problem can be reduced to the case with an absorbing stopping state if we make the following minor transformations in the unichain Markov decision process: For an arbitrary action $a \in A(z)$ we set $p_{z,y}^a = 0, \ \forall y \in X \setminus \{z\}$; $p_{z,z}^a = 1$. Obviously, after such a transformation of the unichain decision process we obtain the optimal stationary strategies of the problem if the state z is reached.

2.8.2 *Optimality Conditions for the Control Problem on Network with a Stopping State*

Consider a discounted control problem for the decision network (G, X_C, X_N, c, p) with a given stopping state $z \in X$ and a given discount factor γ, $0 < \gamma \le 1$. Then on the basis of Theorem 2.42 the following result can be proved.

Theorem 2.64 *Assume that in G an arbitrary stationary strategy $s : x \to X(x)$ for $x \in X_C$ generates a subgraph $G_s = (X, E_s \cup E_N)$ where the vertex z can be reached from arbitrary $x \in X \setminus \{z\}$. Then the linear programming problem:*
Minimize

$$\phi(\alpha, \beta) = \sum_{x \in X_C} \sum_{y \in X(x)} c_{x,y} \alpha_{x,y} + \sum_{x \in X_N} \mu_x \beta_x \tag{2.143}$$

subject to

$$\begin{cases} \sum_{x \in X(y)} \alpha_{y,x} - \gamma \sum_{x \in X_C^-(y)} \alpha_{x,y} - \gamma \sum_{x \in X_N^-(y)} p_{x,y} \beta_x = 1, \;\; y \in X_C \backslash \{z\}; \\ \beta_y - \gamma \sum_{x \in X_C^-(y)} \alpha_{x,y} - \gamma \sum_{x \in X_N^-(y)} p_{x,y} \beta_x = 1, \;\; y \in X_N \backslash \{z\}; \\ \beta_x \geq 0, \; \forall x \in X_N; \;\; \alpha_{x,y} \geq 0, \;\; \forall x \in X_C, \; y \in X(x), \end{cases} \tag{2.144}$$

has a solution. If α^*, β^* *is an arbitrary basic solution of the problem* (2.143), (2.144) *then the optimal stationary strategy* s^* *for the discounted control problem with a stopping state* z *can be found by fixing* $s^*_{x,y} = 1$ *for* $x \in X_C, \; y \in X(x)$ *if* $\alpha^*_{x,y} > 0$, *and* $s_{x,y} = 0^*$ *in the other case.*

It is easy to observe that the problem (2.143), (2.144) is obtained from the problem (2.107), (2.108) by deleting the restriction that corresponds to a stopping state z. Thus, if the conditions of the theorem hold then the problem has a solution for arbitrary $\gamma \in (0, 1]$.

The optimality conditions for control problems on networks with a stopping state can be derived if we consider the dual model for the problem (2.143), (2.144) or from Theorem 2.63. If we specify this theorem for the problem on networks then we obtain the following result.

Theorem 2.65 *Let* (G, X_C, X_N, c, p) *be a perfect decision network with a given stopping state* z *and a given discount factor* $\gamma\,(0 < \gamma \leq 1)$, *where the function* $c : E \rightarrow \mathbb{R}$ *is strictly positive. Then the optimal expected discounted total cost* σ^*_x *of the control problem on the decision network exists for an arbitrary fixed starting state* $x \in X$. *The values* σ^*_x, *for* $x \in X \backslash \{z\}$ *can be found by solving the following linear programming problem:*
Maximize

$$\varphi(\sigma) = \sum_{x \in X} \sigma_x \tag{2.145}$$

subject to

$$\begin{cases} \sigma_x - \gamma \sigma_y \leq c_{x,y}, \;\; \forall x \in X_C \backslash \{x\}, \;\; y \in X(x); \\ \sigma_x - \gamma \sum_{y \in X} p_{x,y} \sigma_y \leq \mu_x, \qquad \forall x \in X_N \backslash \{z\} \end{cases} \tag{2.146}$$

and the optimal stationary strategy can be determined by fixing $s^* : X \backslash \{z\} \rightarrow A$ *such that* $s^*(x) = y \in X^*(x), \;\; \forall x \in X_C \backslash \{z\}$, *where* $X^*(x) = \{y \in X(x) \,|\, \sigma_x - \gamma \sigma_y = c_{x,y}\}$.

Corollary 2.66 *Let (G, X_C, X_N, c, p) be a perfect decision network that satisfies the conditions of* Theorem 2.65. *Then for an arbitrary $\gamma \in (0, 1]$ there exist the values σ_x^* for $x \in X$ that satisfy the conditions:*

(1) $c_{x,y} + \gamma\sigma_y^* - \sigma_x^* \geq 0, \ \forall x \in X_C \setminus \{x\}, \ y \in X(x);$

(2) $\min_{y \in X(x)} (c_{x,y} + \gamma\sigma_y^* - \sigma_x^*) = 0, \ \forall x \in X_C;$

(3) $\mu_x + \gamma \sum_{y \in X} p_{x,y}\sigma_y^* - \sigma_x^* = 0, \quad \forall x \in X_N \setminus \{z\};$

An optimal stationary strategy of the optimal control problem on the network (G, X_C, X_N, c, p) with stopping state z can be found by fixing $s^* : X \setminus \{z\} \to A$ such that $s^*(x) = y \in X^*(x), \ \forall x \in X_C \setminus \{z\}$, where $X^*(x) = \{y \in X(x) \,|\, c_{x,y} + \gamma\sigma_y^* - \sigma_x^* = 0\}$.

Remark 2.67 The control problem on the network with a given stopping state z in the case $X_N = 0, \ \gamma = 1$ becomes the problem of determining in G the minimum cost paths from $x \in X$ to z. If G is an acyclic graph with sink vertex then the problem has a solution for an arbitrary $\gamma > 0$.

2.8.3 A Dynamic Programming Algorithm for Solving Deterministic Non-stationary Control Problems on Networks

As we have noted the deterministic stationary control problem on networks with fixed stopping state z corresponds to the case $X_N = \emptyset$ and the solution can be found by using linear programming models (2.143), (2.144) and (2.145), (2.146). In this section we show that this problem can be solved for the non-stationary case using a dynamic programming method. We describe an algorithm for finding the solution of the deterministic control problem on the network when the costs on the edges may depend on time. So, we assume that $X_N = \emptyset$ and in the network to each directed edge $e = (x, y) \in E$ the cost function $c_e(t)$ that depends on t is associated. This means that if the system makes a transition from the state $x = x(t)$ to the state $y = x(t+1)$ then the cost is $c_{x,y}(t)$. Thus, the problem in this case is formulated in the following way:

For a given time-moment $\bar{t}$ and fixed starting and stopping states $x_0, x_f \in X$ it is necessary to determine in G a sequence of the system's transitions $(x(0), x(1))$, $(x(1), x(2))$, ..., $(x(\bar{t}-1), x(\bar{t}))$, which transfers the system $\mathbb{L}$ from a starting state $x_0 = x(0)$ to a stopping state $x_f = x(\bar{t})$ such that the total cost

$$F_{x_0x_f}(\bar{t}) = \sum_{t=0}^{\bar{t}-1} c_{(x(t),x(t+1))}(t)$$

of the system's transitions by a trajectory

$$x_0 = x(0), x(1), x(2), \dots, x(\bar{t}) = x_f$$

is minimal, where $(x(t), x(t+1)) \in E, \quad t = 0, 1, 2, \dots, \bar{t} - 1$.

We describe the dynamic programming algorithm for solving this problem. Denote by

$$F^*_{x_0,x_f}(\bar{t}) = \min_{x_0=x(0),x(1),\dots,x(\bar{t})=x_f} \sum_{t=0}^{\bar{t}-1} c_{(x(t),x(t+1))}(t)$$

the minimal total cost of the system's transition from x_0 to x_f with $\bar{t}$ stages, where $F^*_{x_0,x_f}(0) = 0$ in the case $x_0 = x_f$ and $F^*_{x_0 x_f}(\bar{t}) = \infty$ if x_f cannot be reached from x_0 by using $\bar{t}$ transitions.

If we introduce the values $F^*_{x_0 x(t)}(t)$ for $t = 0, 1, 2, \dots, \bar{t} - 1$ then it is easy to observe that for $F^*_{x_0 x(t)}(t)$ the following recursive formula can be gained:

$$F^*_{x_0 x(t)}(t) = \min_{x(t-1) \in X^-_G(x(t))} \left\{ F^*_{x_0,x(t-1)}(t-1) + c_{(x(t-1),x(t))}(t-1) \right\},$$

where

$$F^*_{x_0,x(0)}(0) = 0$$

and

$$X^-_G(y) = \{x \in X \mid e = (x, y) \in E\}.$$

Based on this recursive formula we can tabulate the values $F^*_{x_0 x(t)}(t)$, $t = 1, 2, \dots, \bar{t}$ for every $x(t) \in X$. These values and the solution of the problem can be found using $O(|X|^2 \bar{t})$ elementary operations (here we do not take into account the number of operations for calculating the values of the functions $c_e(t)$ for a given t).

The tabulation process should be organized in such a way that for every vertex $x = x(t)$ at a given moment in time t it is determined not only the cost $F^*_{x_0 x(t)}(t)$ but also the state $x^*(t-1)$ at the previous time-moments for which

$$\begin{aligned} F^*_{x_0,x(t)}(t) &= F^*_{x_0 x^*(t-1)} + c_{(x^*(t-1),x(t))}(t-1) \\ &= \min_{x(t-1) \in X^-_G(x(t))} \{F^*_{x_0,x(t-1)} + c_{(x(t-1),x(t))}(t-1)\}. \end{aligned}$$

So, if to each x at the time-moments $t = 0, 1, 2, \dots, \bar{t}$ we associate the labels $(t, x(t), F^*_{x_0 x(t)}, x^*(t-1))$, then the corresponding table allows us to find the optimal trajectory successively starting from the final position, $x_f = x^*(\bar{t}), x^*(\bar{t} - 1), \dots, x^*(1), x^*(0) = x_0$. In the example given below all possible labels for every x and every t are represented in Table 2.1.

Table 2.1 The values $F^*_{x_0,x(t)}$ and $x^*(t-1)$

t	x, F^*	0	1	2	3	4	5
0	$F^*_{x_0,x(0)}$	0	∞	∞	∞	∞	∞
	$x^*(0-1)$	–	–	–	–	–	–
1	$F^*_{x_0,x(1)}$	∞	1	∞	1	∞	∞
	$x^*(0)$	–	0*	–	0	–	–
2	$F^*_{x_0,x(2)}$	∞	3	2	∞	5	2
	$x^*(1)$	–	3	1*	–	3	1
3	$F^*_{x_0,x(3)}$	∞	∞	5	6	4	3
	$x^*(2)$	–	–	1	2*	2	2
4	$F^*_{x_0,x(4)}$	∞	12	∞	11	8	6
	$x^*(3)$	–	3*	–	2	2	2
5	$F^*_{x_0,x(5)}$	∞	19	16	∞	21	16
	$x^*(4)$	–	3	1	–	3	1*

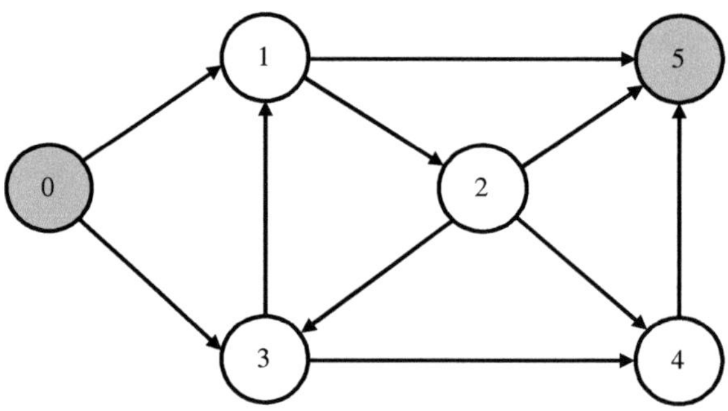

Fig. 2.13 The structure of the dynamic network

This problem can be extended to the case if the final state x_f should be reached at the moment of time $t(x_f)$ from a given interval $[\overline{t}_1, \overline{t}_2]$. If $\overline{t}_1 \neq \overline{t}_2$ then the problem can be reduced to $\overline{t}_2 - \overline{t}_1 + 1$ problems with $\overline{t} = \overline{t}_1$, $\overline{t} = \overline{t}_1 + 1$, $\overline{t} = \overline{t}_1 + 2, \ldots, \overline{t} = \overline{t}_2$, respectively; by comparing the minimal total costs of these problems we find the best one and $t(x_f)$.

An important case of the considered problem is if $\overline{t}_1 = 0$ and $\overline{t}_2 = \infty$. The solution of the problem with such a condition if the network may contain directed cycles has sense only for positive and non-decreasing cost functions $c_e(t)$ on the edges $e \in E$. Obviously, for this case we obtain $0 \leq t(x_f) \leq |X|$ and the problem can be solved in time $O(|X|^3)$ (the case with a free number of stages).

Example Let the dynamic network determined by the graph $G = (X, E)$ represented in Fig. 2.13 be given. The cost functions are the following:

$$c_{(0,1)}(t) = c_{(0,3)}(t) = c_{(2,5)}(t) = 1;$$
$$c_{(2,3)}(t) = c_{(3,1)}(t) = 2t; \quad c_{(3,4)}(t) = 2t + 2;$$
$$c_{(1,2)}(t) = c_{(2,4)}(t) = c_{(1,5)}(t) = t; \quad c_{(4,5)}(t) = 2t + 1.$$

We consider the problem of finding a trajectory in G from $x(0) = x_0 = 0$ to $x_f = 5$, where $T = 5$.

Using the recursive formula described above we get Table 2.1 with values $F^*_{x_0 x(t)}(t)$ and $x^*(t-1)$.

Starting from the final state $x_f = 5$ we find the optimal trajectory

$$5^* \leftarrow 1^* \leftarrow 3^* \leftarrow 2^* \leftarrow 1^* \leftarrow 0^*$$

with total cost $F_{x_0,x(5)}(5) = 16$.

The considered non-stationary control problem has been extended and generalized in [71, 79] as non linear minimum cost flow problems on dynamic networks. Algorithms based on *time-expanded network methods* for such a class of problems are described in [70, 71, 79, 93, 94].

2.9 Discrete Decision Problems with Varying Time of State's Transitions and Special Solution Algorithms

So far, in the control problems with average and discounted optimization cost criteria we have considered that the time between transitions in the control process is constant and it is equal to 1. We extend these problems and generalize these problems by assuming that the time of system's transition from one state to another in the decision process vary and it may be different from 1. Such a problem statement may be useful for studying and solving the decision models for the case of Semi-Markov processes. In this section we show that the deterministic problem with varying time of states' transitions can be reduced to the problem with a fixed unit time of system transitions from one state to another.

2.9.1 Problem Formulation

At first we formulate the control problem with an average cost optimization criterion when the transition time between the states is not constant.

Let the dynamical system $\mathbb{L}$ with a finite set of states $X \subseteq \mathbb{R}^n$ be given, where at every discrete moment of time $t = 0, 1, 2, \ldots$ the state of $\mathbb{L}$ is $x(t) \in X$. Assume, that the control of the system $\mathbb{L}$ at each time-moment $t = 0, 1, 2, \ldots$ for an arbitrary state $x(t)$ is realized by using the vector of control parameters $u(t) \in \mathbb{R}^m$ for which a feasible set $U_t(x(t))$ is given, i.e., $u(t) \in U_t(x(t))$. For arbitrary t and $x(t)$ on

$U_t\,(x(t))$ it is defined an integer function

$$\tau_{x(t)}\,:\; U_t\,(x(t)) \to \mathbb{N}$$

which represents to each control $u(t) \in U_t\,(x(t))$ an integer value $\tau_{x(t)}\,(u(t))$. This value expresses the time of system's transition from the state $x(t)$ to the state $x\big(t + \tau_{x(t)}\,(u(t))\big)$ if the control $u(t) \in U_t\,(x(t))$ has been applied at the moment t for a given state $x(t)$.

The dynamics of the system $\mathbb{L}$ is described by the following system of difference equations

$$\begin{cases} t_{j+1} &= t_j + \tau_{x(t_j)}\,\big(u(t_j)\big)\,; \\ x(t_{j+1}) &= g_{t_j}\,\big(x(t_j), u(t_j)\big)\,; \\ & u(t_j) \in U_{t_j}\,\big(x(t_j)\big)\,; \\ & j = 0, 1, 2, \ldots, \end{cases}$$

where

$$x(t_0) = 0, \quad t_0 = 0$$

is a given starting state of the dynamical system $\mathbb{L}$. Here we suppose that the functions g_t and $\tau_{x(t)}$ are known and t_{j+1} and $x(t_{j+1})$ are determined uniquely by $x(t_j)$ and $u(t_j)$ at each step j.

Let $u(t_j)$, $j = 0, 1, 2, \ldots$, be a control, which generates the trajectory $x(0)$, $x(t_1), x(t_2), \ldots, x(t_k), \ldots$. For this control we define the mean integral-time cost by a trajectory

$$F_{x_0}\,(u(t)) = \lim_{k\to\infty} \frac{\sum_{j=1}^{k-1} c_{t_j}\big(x(t_j),\; g_{t_j}\,\big(x(t_j),\; u(t_j)\big)\big)}{\sum_{j=0}^{k-1} \tau_{x(t_j)}(u(t_j))}$$

where $c_{t_j}\big(x(t_j),\; g_{t_j}\,\big(x(t_j),\; u(t_j)\big)\big) = c_{t_j}\,\big(x(t_j),\; x(t_{j+1})\big)$ represents the cost of the system $\mathbb{L}$ to pass from the state $x(t_j)$ to the state $x(t_{j+1})$ at the stage $[j,\, j+1]$.

We consider the problem of finding the time-moments $t = 0, t_1, t_2, \ldots, t_{k-1}$, ... and the vectors of control parameters $u(0), u(t_1), u(t_2), \ldots, u(t_{k-1}), \ldots$ which satisfy the conditions mentioned above and minimize the functional $F_{x_0}\,(u(t))$.

In the case of $\tau_{x(t)}(u(t)) \equiv 1$ for every t and $x(t)$ this problem becomes the control problem with unit time of states' transitions. The problem of determining the stationary control with unit time of states' transitions has been studied in [5, 53, 65, 73, 117]. In the mentioned papers it is assumed that $U_t(x(t))$, g_t and c_t do not depend on t, i.e., $g_t = g$, $c_t = c$ and $U_t(x) = U(x)$ for $t = 0, 1, 2, \ldots$. Richard Bellman showed in [5] that for the stationary case of the problem with unit time of states' transitions there exists an optimal stationary control $u^*(0), u^*(1), \ldots, u^*(t), \ldots$

such that

$$\lim_{k\to\infty} \frac{\sum_{t=0}^{k-1} c\big(x(t),\, g\,(x(t), u^*(t))\big)}{k}$$

$$= \inf_{u(t)} \lim_{k\to\infty} \frac{\sum_{t=0}^{k-1} c\big(x(t),\, g\,(x(t), u(t))\big)}{k} = \lambda < \infty.$$

Furthermore in [65, 117] it is shown that the stationary case of the problem can be reduced to the problem of finding the optimal mean cost cycle in a graph of states' transitions of a dynamical system. Based on these results in [18, 53, 73, 117] polynomial-time algorithms for finding the optimal stationary control are proposed. This variant of the problem can be solved by using the linear programming problem (2.18), (2.19) from Sect. 2.2.4.

Below we extend the results mentioned above to the general stationary case of the problem with arbitrary transit-time functions τ_x. We show that this problem can be formulated as the problem of determining the optimal mean cost cycles in the graph of states' transitions of the dynamical system for an arbitrary transition-time function on the edges.

For the discounted control problem with varying time of states' transitions the dynamics is determined in the same way as for the problem above; but the objective function which has to be minimized is defined as follows:

$$\widehat{F}_{x_0}(u(t)) = \sum_{j=0}^{\infty} \gamma^{t_j} c\big(x(t_j),\, g\,(x(t_j), u(t_j))\big),$$

where $\gamma,\ 0 < \gamma < 1$, is a given discounted factor.

2.9.2 A Linear Programming Approach for the Problem with Arbitrary Transition Costs

We consider the stationary case of the deterministic transition control problem, i.e., when g_t, c_t, $U_t(x(t))$, $u(t)$ do not depend on t and the transition function $\tau_{x(t)}$ depends only on the state x and on the control u_x in the state x. So, $g_t = g$, $c_t = c$, $U_t(x) = U(x)$, $\tau_{x(t)} = \tau(x, u_x)$ for $u(t) = u_x \in U(x)$, $\forall x \in X$, $t = 0, 1, 2, \ldots$.

In this case it is convenient to study the problem on a network where the dynamics of the system is described by the graph of states' transitions $G = (X, E)$. An arbitrary vertex x of G corresponds to a state $x \in X$ and an arbitrary directed edge $e = (x, y) \in E$ expresses the possibility of the system $\mathbb{L}$ to pass from the state $x(t)$ to the state $x(t + \tau_e)$, where τ_e is the time of the system's transition from the state x to the state y through the edge $e = (x, y)$. So, on the edge set E it is defined the function

$\tau : E \to \mathbb{R}^+$ which associates to each edge a positive number τ_e which means that if the system $\mathbb{L}$ at the moment of time t is in the state $x = x(t)$ then the system can reach the state y at the moment of time $t + \tau_e$ if it passes through the edge $e = (x, y)$, i.e., $y = x(t + \tau_e)$. In addition, on the edge set E it is defined the cost function $c : E \to \mathbb{R}$, which associates to each edge the cost c_e of the system's transition from the state $x = x(t)$ to the state $y = x(t + \tau_e)$ for an arbitrary discrete moment of time t. So, finally we have that to each edge $e = (x, y) \in E$ the cost c_e and the transition time τ_e from x to y are associated.

In G an arbitrary edge $e = (x, y)$ corresponds to a control in the initial problem and the set of edges $E(x) = \{e = (x, y) \mid (x, y) \in E\}$ originating in the vertex x corresponds to the feasible set $U(x)$ of the vectors of control parameters in the state x. The transition time function τ in G is induced by the transition time function τ_x for the stationary control problem.

It is easy to observe that the infinite horizon control problem with a varying time of states' transitions of the system on G can be regarded as the problem of finding in G the minimal mean cost cycle C_G^* that can be reached from the vertex x_0 where the vertex x_0 corresponds to the starting state $x_0 = x(0)$ of the dynamical system $\mathbb{L}$. Indeed, a stationary control in G corresponds to a fixed transition from a vertex $x \in X$ to another vertex $y \in X$ through a directed edge $e = (x, y)$ in G. Such a strategy of states' transitions of the dynamical system in G generates a trajectory which leads to a directed cycle C_G with the set of edges $E(C_G)$. Therefore, the considered stationary control problem on G is reduced to the problem of finding the minimal mean cost cycle that can be reached from x_0, where in G to each directed edge $e = (x, y) \in E$ the cost c_e and the transition time τ_e of the system's transition from the state $x = x(t)$ to the state $y = x(t + \tau_e)$ are associated.

If the minimal mean cost cycle C_G^* in G is known then the stationary optimal control for our problem can be found by the following way: In G we fix an arbitrary simple directed path $P(x_0, x_k)$ with the set of edges $E(P(x_0, x_k))$ which connects the vertex x_0 with the cycle C_G^*. After that for an arbitrary state $x \in X$ we choose a stationary control which corresponds to a unique directed edge $e = (x, y) \in E(P(x_0, x_k)) \cup E(C^*)$. For such a stationary control the following equality holds:

$$\inf_{u(t)} \lim_{k\to\infty} \frac{\sum_{j=0}^{k-1} c\big(x(t_0),\ g\left(x(t_j), u(t_j)\right)\big)}{\sum_{j=0}^{k-1} \tau_x(u(t_j))} = \frac{\sum_{e\in E(\mathbb{C}^*)} c_e}{\sum_{e\in E(\mathbb{C}^*)} \tau_e}.$$

Note that the condition $U(x) \neq \emptyset,\ \forall x \in X$, for the stationary case of the control problem means that in G each vertex x contains at least one leaving directed edge $e = (x, y)$. We will assume that in G every vertex $x \in X$ is attainable from x_0; otherwise we can delete vertices from X for which there are no directed paths $P(x_0, x)$ from x_0 to x. Moreover, without loss of generality, we may consider that G is a strongly connected graph. Then the problem of finding the optimal stationary control for the problem from Sect. 2.2.4 can be formulated as combinatorial optimization problem on G in which it is necessary to find a directed cycle C_G^* such that

$$\frac{\sum\limits_{e\in E(C_G^*)} c_e}{\sum\limits_{e\in E(C_G^*)} \tau_e} = \min_{C_G} \frac{\sum\limits_{e\in E(C_G)} c_e}{\sum\limits_{e\in E(C_G)} \tau_e}.$$

The problem of determining the minimal mean cost cycle in a double weighted directed graph has been studied in [19, 53, 63, 117]. In the cited works algorithms based on linear programming and parametrical methods are proposed. For the problem with a unit time of states' transitions in [53] a strongly polynomial time algorithm is proposed.

In the following we describe an approach which is based on linear programming. We can see that such an approach may be used for solving a more general class of problems, as example, for the multi-criterion version of minimal mean cost cycle problems [82].

We consider the following linear programming problem:
Minimize

$$z = \sum_{e\in E} c_e \alpha_e \tag{2.147}$$

subject to

$$\begin{cases} \sum\limits_{e\in E^+(x)} \alpha_e - \sum\limits_{e\in E^-(x)} \alpha_e = 0, \quad \forall x \in X; \\ \sum\limits_{e\in E} \tau_e \alpha_e = 1; \\ \alpha_e \geq 0, \quad \forall e \in E. \end{cases} \tag{2.148}$$

where $E^+(x) = \{e = (x, y) \in E \mid y \in X\}$, $E^-(x) = \{e = (y, x) \in E \mid y \in X\}$.

The following lemma holds.

Lemma 2.68 *Let* $\alpha = \left(\alpha_{e_1}, \alpha_{e_2}, \ldots, \alpha_{e_m}\right)$ *be a feasible solution of the system* (2.148) and $G_\alpha = (X_\alpha, E_\alpha)$ be the subgraph of G, generated by the set of edges $E_\alpha = \{e_i \in E \mid \alpha_{e_i} > 0\}$. Then an arbitrary extreme point $\alpha^0 = \left(\alpha^0_{e_1}, \alpha^0_{e_2}, \ldots, \alpha^0_{e_m}\right)$ of the polyhedron set determined by (2.148) corresponds to a subgraph $G_{\alpha^0} = (X_{\alpha^0}, E_{\alpha^0})$ which has the structure of a simple directed cycle and vice versa, i.e., if $G_{\alpha^0} = (X_{\alpha^0}, E_{\alpha^0})$ is a simple directed cycle in G then the solution $\alpha^0 = \left(\alpha^0_{e_1}, \alpha^0_{e_2}, \ldots, \alpha^0_{e_m}\right)$ with

$$\alpha^o_{e_i} = \begin{cases} \dfrac{1}{\sum_{e\in E_{\alpha^0}} \tau_e}, & \text{if } e_i \in E_{\alpha^0}; \\ 0, & \text{if } e_i \notin E_{\alpha^0} \end{cases}$$

corresponds to an extreme point of the set of solutions (2.148).

Proof Let $\alpha = \left(\alpha_{e_1}, \alpha_{e_2}, \ldots, \alpha_{e_m}\right)$ be an arbitrary feasible solution of the system (2.148). Then it is easy to observe that $G_\alpha = (X_\alpha, E_\alpha)$ contains at least one directed cycle. Indeed, for an arbitrary $x \in X_\alpha$ there exist at least one leaving edge $e' = (x, y) \in E_\alpha$ and at least one entering edge $e'' = (z, x) \in E_\alpha$; otherwise α does not satisfy condition (2.148).

Let us show that if G_α is not a simple directed cycle then α does not represent an extreme point of the set of solutions of the system (2.148). If G_α has not the structure of a simple directed cycle then it contains a simple directed cycle C with the set of edges $E(C_G) \subset E_\alpha$, i.e., $m' = |E(C_G)| < m$. Without loss of generality we may consider that $E(\mathbb{C}) = \{e_1, e_2, \ldots, e_{m'}\}$. Fix an arbitrary value θ such that $0 < \theta < \min_{e_i \in E(\mathbb{C})} \alpha_{e_i}$ and consider the following two solutions:

$$\alpha^1 = \frac{1}{1 - \theta \sum_{i=1}^{m} \tau_{e_i}} \left(\alpha_{e_1} - \theta,\ \alpha_{e_2} - \theta,\ \ldots,\ \alpha_{e_{m'}} - \theta,\ \alpha_{e_{m'+1}},\ \ldots,\ \alpha_{e_m}\right);$$

$$\alpha^2 = \frac{1}{\theta \sum_{i=1}^{m} \tau_{e_i}} \big(\underbrace{\theta, \theta, \ldots, \theta}_{m'}, 0, 0, \ldots, 0\big).$$

It is easy to check that α^1 and α^2 satisfy the condition (2.148), i.e., α^1 and α^2 are feasible solutions of the problem (2.147), (2.148). If we chose θ such that $0 < \theta \sum_{i=1}^{m'} \tau_{e_i} < 1$ then we obtain that α can be represented as a convex combination of feasible solutions α^1 and α^2, i.e.,

$$\alpha = \left(1 - \theta \sum_{i=1}^{m'} \tau_e\right)\alpha^1 + \left(\theta \sum_{i=1}^{m'} \tau_e\right)\alpha^2. \tag{2.149}$$

So, α is not an extreme point of the set of solutions (2.148). If G_α represents a simple directed cycle then the representation (2.149) is not possible, i.e., the second part of Lemma 2.68 holds. □

Using Lemma 2.68 we can prove the following result.

Theorem 2.69 *The optimal basic solution* $\alpha^* = \left(\alpha^*_{e_1}, \alpha^*_{e_2}, \ldots, \alpha^*_{e_m}\right)$ *of problem* (2.147), (2.148) *corresponds to a minimal mean cycle* $C^*_G = G_{\alpha^*}$ *in* G, *i.e.,*

$$\alpha^*(e_i) = \begin{cases} \dfrac{1}{\sum_{e \in E(\mathbb{C}^*)} \tau_e}, & \text{if } e \in E(\mathbb{C}^*); \\ 0, & \text{if } e \notin E(\mathbb{C}^*), \end{cases}$$

where $E(C^*_G)$ *is the set of edges of a directed cycle* C^*_G.

Proof According to Lemma 2.68 an arbitrary extreme point α^0 of the set of solutions of system (2.148) corresponds in G to the subgraph $G_{\alpha^0} = (X^0_\alpha, E^0_\alpha)$ which has the structure of a directed cycle. Taking into account that the optimal solution of problem (2.147), (2.148) is attained in an extreme point we obtain the proof of the theorem.□

The linear programming problem (2.147), (2.148) allows us to find the minimal mean cycle in the graph G with positive values $\tau_e = \tau_{x,y}$, for $e = (x, y) \in E$. More efficient algorithms for solving the problem can be obtained using the dual problem (2.147), (2.148).

Theorem 2.70 *If G is a strongly connected directed graph then there exists a function $\varepsilon : X \to \mathbb{R}$ and the value λ such that:*

(a) $\varepsilon_y - \varepsilon_x + c_{x,y} \geq \tau_{x,y} \cdot \lambda, \quad \forall (x, y) \in E;$
(b) $\min\limits_{y \in O^-(x)} \{\varepsilon_y - \varepsilon_x + c_{x,y} - \tau_{x,y}\lambda\} = 0, \quad \forall x \in X;$
(c) *an arbitrary cycle $\mathbb{C}^*$ of the subgraph $G^0 = (X, E^0)$ of G, generated by edges $(x, y) \in E$ for which $\varepsilon_y - \varepsilon_x + c_{x,y} - \tau_{x,y} \cdot \lambda = 0$ determines a minimal mean cycle in G.*

Proof We consider the dual problem for (2.147), (2.148):
Maximize

$$W = \lambda$$

subject to

$$\varepsilon_x - \varepsilon_y + \tau_{x,y}\lambda \leq c_{x,y}, \quad \forall (x, y) \in E.$$

If p is the optimal value of the problem then by using duality properties of the solution of the problem we obtain (a), (b) and (c). □

Based on results described above we can make the following conclusions.

1. If $\lambda = 0$ then the values $\varepsilon_x,\ x \in X$ can be treated as the cost of minimal paths from vertices $x \in X$ to a vertex x_f which belongs to the minimal mean cycle C_G^* (with $\lambda = 0$) in the graph G with given costs c_e of edges $e \in E$. So, if x_f is known then the cycle C_G^* can be found in the following way. We construct the tree of minimum cost directed paths from $x \in X$ to x_f and determine the values $\varepsilon_x,\ \forall x \in X$. Then in G we make a transformation of the costs $c'_{x,y} = \varepsilon_y - \varepsilon_x + c_{x,y}$ for $(x, y) \in E$ and find the subgraph $G^0 = (X, E^0)$ generated by edges (x, y) with $c'_{x,y} = 0$. After that we fix in G^0 a cycle C^* with zero cost of the edges. If the vertex x_f is not known then we have to construct the tree of minimal cost paths which respect to each $x_f \in X$. So, in this case with respect to each tree we find the subgraph $G^0 = (X, E^0)$. Then at least for one of such a subgraph we find a cycle C_G^* with zero cost ($c'_{x,y} = 0$) of the edges.
2. If $\lambda \neq 0$ and λ is known then the minimal mean cost cycle C^* can be found by the following way. In G we change the costs $c_{x,y}$ of edges $(x, y) \in E$ by $c_{x,y} - \tau_{x,y}\lambda$ and after that solve the problem with the new costs according to point 1.

3. If $\lambda \neq 0$ and it is not known then we find it using the bisection method on the segment $[h_0^1, h_0^2]$ where $h^1 = \min_{e\in E} c_e$, $h^2 = \max_{e\in E} c_e$. At each step k of the method we find the midpoint $\lambda_k = \left(h_k^1 + h_k^2\right)/2$ of the segment $[h_k^1, h_k^2]$ and check if in G with the cost $c_{x,y}^k - \tau_{x,y}\lambda_k$ there exists the cycle with negative cost. If at a given step there exists the cycle with negative cost then we fix $h_{k+1}^1 = h_k^1$, $h_{k+1}^2 = \lambda_k$; otherwise we put $h_{k+1}^1 = \lambda_k$, $h_{k+1}^1 = h_k^1$. In such a way we find λ with a given precision. After that the exact value of λ can be found from λ_k using a special roundoff procedure from [58].

The algorithm described above allows us to determine the solution of the problem in the case if $\tau_e \geq 0$, $\forall e \in E$. In general this problem can be considered for arbitrary τ_e and c_e. In this case we may use the following fractional linear programming problem: Minimize

$$z = \frac{\sum_{e\in E} c_e\alpha_e}{\sum_{e\in E} \tau_e\alpha_e} \tag{2.150}$$

subject to

$$\begin{cases} \sum\limits_{e\in E^+(x)} \alpha_e - \sum\limits_{e\in E^-(x)} \alpha_e = 0, \ \forall x \in X; \\ \sum\limits_{e\in E} \alpha_e = 1; \\ \qquad \alpha_e \geq 0, \ e \in E, \end{cases} \tag{2.151}$$

where $E^-(x) = \{e = (y, x) \in E \mid y \in X\}$; $E^+(x) = \{e = (x, y) \in E \mid y \in X\}$.

Of course, this model is valid if on the set of solutions of system (2.151) it holds $\sum_{e\in E} \tau_e\alpha_e \neq 0$. In a similar way as for the linear programming problem here we can show that an arbitrary optimal basic solution of the problem (2.150), (2.151) corresponds to an optimal mean directed cycle in G.

Let $\alpha = \left(\alpha_{e_1}, \alpha_{e_2}, \ldots, \alpha_{e_{|E|}}\right)$ be an arbitrary feasible solution of system (2.151) and denote by $G_\alpha = (X_\alpha, E_\alpha)$ the subgraph of G generated by the set of edges $E_\alpha = \{e \in E \mid \alpha_e > 0\}$. In [73] it is shown that an arbitrary extreme point $\alpha^0 = \left(\alpha_{e_1}^0, \alpha_{e_2}^0, \ldots, \alpha_{e_{|E|}}^0\right)$ of the set of solutions of system (2.151) corresponds to a subgraph $G_{\alpha^0} = (X_{\alpha^0}, E_{\alpha^0})$ which has the structure of an elementary directed cycle. Taking into account that for the problem (2.150), (2.151) there exists an optimal solution $\alpha^* = \left(\alpha_{e_1}^*, \alpha_{e_2}^*, \ldots, \alpha_{e_{|E|}}^*\right)$ which corresponds to an extreme point of the set of solutions (2.151) we obtain that

$$\max z = \frac{\sum_{e\in E_{\alpha^*}} c_e\alpha_e^*}{\sum_{e\in E_{\alpha^*}} \tau_e\alpha_e^*}$$

and the set of edges E_{α^*} generates a directed cycle G_{α^*} for which $\alpha_e^* = 1/|E_{\alpha^*}|$, $\forall e \in E_{\alpha^*}$. Therefore,

$$\max z = \frac{\sum_{e \in E_{\alpha^*}} c_e}{\sum_{e \in E_{\alpha^*}} \tau_e}.$$

So, an optimal solution of problem (2.150), (2.151) corresponds to the minimal mean cost cycle in the directed graph of states' transitions of the dynamical system.

This means that the fractional linear programming problem (2.150), (2.151) can be used for determining the optimal solution of the problem in the general case.

2.9.3 Reduction of the Problem to the Case with Unit Time of States' Transitions

As we have shown the deterministic control problem with an average cost criterion on the network can be solved for an arbitrary transition-time function using a linear programming problem (2.147), (2.148) or a linear fractional programming problem (2.150), (2.151). For the discounted control problem with varying time of state transitions a similar linear programming model could not be derived. However, both problems can be reduced to the corresponding cases of the problems with unit time of states' transitions of the system.

Below we describe a general scheme how to reduce the control problems with varying time of states' transitions to the case with unit time of states' transition of the system. We show that our problems can be reduced to the case with unit time of states' transitions on an auxiliary graph $G' = (X', E')$ which is obtained from $G = (X, E)$ using a special construction. This means that after such a reduction we can apply the linear programming approach described in Sect. 2.2.

Graph $G' = (X', E')$ with unit transitions on directed edges $e' \in E'$ is obtained from G where each directed edge $e = (x, y) \in E$ with corresponding transition time τ_e is changed by a sequence of directed edges

$$e_1' = (x, x_1^e),\ e_2' = (x_1^e, x_2^e),\ \ldots,\ e_{\tau_e}' = (x_{\tau_e - 1}^e, y).$$

This means that we represent a transition from a state $x = x(t)$ at the moment of time t to the state $y = x(t + \tau_e)$ at the moment of time $t + \tau_e$ in G in G' as the transition of a dynamical system from the state $x = x(t)$ at the time-moment t to $y = x(t + \tau_e)$ if the system makes transitions through a new fictive intermediate set of states $x_1', x_2', \ldots, x_{\tau-1}'$ at the corresponding discrete moments of time

$$t + 1,\ t + 2,\ \ldots,\ t + \tau_e - 1.$$

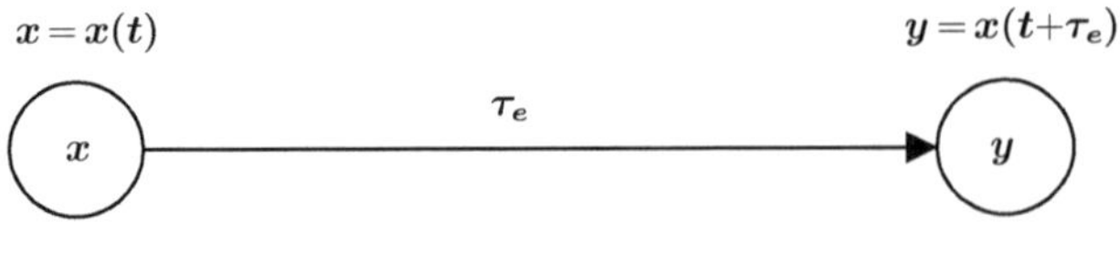

Fig. 2.14 The edge $e = (x, y)$ with the associated transition time τ_e

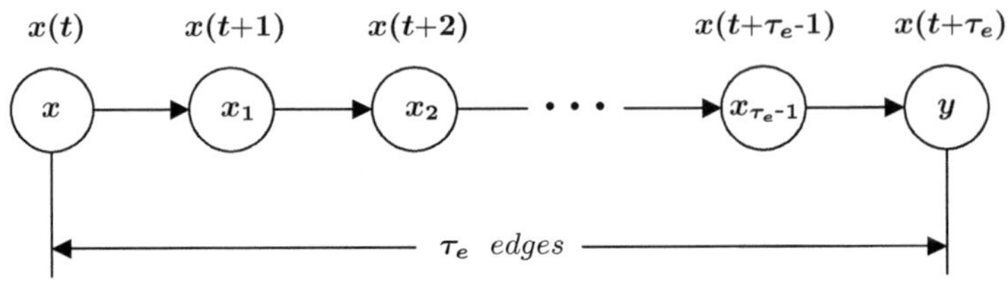

Fig. 2.15 The intermediate states for the edge $e = (x, y)$ in G'

The graphical interpretation of this construction is represented in Figs. 2.14 and 2.15. In Fig. 2.14 it is represented an arbitrary directed edge $e = (x, y)$ with the corresponding transition time τ_e in G. In Fig. 2.15 it is represented the sequence of directed edges e'_i and the intermediate states $x_1, x_2, \ldots, x_{\tau_e - 1}$ in G' that correspond to a directed edge $e = (x, y)$ in G. So, the set of of vertices X' of the graph G' consists of the set of states X and the set of intermediate states $XE = \{x_i^e \,|\, e \in E,\ i = 1, 2, \ldots, \tau_e\}$, i.e., $X' = X \cup XE$. Then the set of edges E' is defined as follows:

$$E' = \bigcup_{e \in E} \mathcal{E}^e, \quad \mathcal{E}^e = \{(x, x_1^e),\ (x_1^e, x_2^e),\ \ldots,\ (x_{\tau_e - 1}^e, y) \mid e = (x, y) \in E\}.$$

We define the cost function $c' : E' \to \mathbb{R}$ in the following way:

$$c'_{x, x^e} = c_{x,y}, \quad \text{if} \quad e = (x, y) \in E;$$
$$c'_{x_1^e, x_2^e} = c_{x_2^e, x_3^e} = \cdots = c_{x_{\tau_e - 1}, y} = 0.$$

It is evident that between the set of stationary strategies

$$s : x \to y \in X + (x) \quad \text{for} \quad x \in X$$

and the set of stationary strategies

$$s' : x' \to y' \in X'^{+}(x') \quad \text{for} \quad x' \in X'$$

there exists a bijective mapping such that the corresponding average and discounted costs on G and on C' are the same. So, if s'^* is the optimal stationary strategy of the problem with unit transitions on G' then the optimal stationary strategy s^* on

G is determined by fixing $s^*(x) = y$ if $s'^*(x) = x_1^e$, where $e = (x, y)$. For the stochastic versions of the control problem on $G = (X, E)$ the construction of the auxiliary graph is similar. Here we should take into account that the set of vertices (states) X are divided into two disjoint subsets X_C and X_N where X_C correspond to the set of controllable states and X_N corresponds to the set of uncontrollable states. Moreover, the probability function $p : E_N \to [0, 1]$ on the set $E_N = \{e = (x, y) \in E \mid x \in X_N\}$ is defined such that $\sum_{y \in X^+(x)} p_{x,y} = 1$. The graph $G' = (X', E')$ in the case of stochastic control problems is constructed in the same way as above. Here we have only to precise how to define the sets X'_C, X'_N and the probability function p' on the set $E'_N = \{e' = (x', y') \in E' \mid x' \in X'_N\}$ in G'. To obtain a bijective mapping between the stationary strategies of the problems in the initial graph G and the stationary strategies of the problem in the auxiliary graph it is necessary to take $X'_C = X_C$, $X'_N = X' \backslash X_C$ and to define the probability function $p' : E' \to [0, 1]$ as follows:

$$p'_{x',y'} = \begin{cases} p_{x,y}, & \text{if } x' = x,\ x' \in X_N \subset X'_N \text{ and } y' = x'_1; \\ 0, & \text{if } x' \in X'_N \backslash X_N. \end{cases}$$

The cost function on G' for the corresponding auxiliary stochastic control problems is defined in the same way as for deterministic problems.

In the following we extend the approach described above to Semi-Markov decision problems, which is valid for the stochastic control problem in its general form.

2.10 Determining the Optimal Strategies for Semi-Markov Decision Problems

The average and discounted Markov decision problems can be extended to Semi-Markov Decision Processes [113–115, 134, 140, 141]. A Semi-Markov decision process is determined by a finite state space X, a finite set of actions A, a nonnegative real function

$$p : A \times X \times X \times \{1, 2, \ldots, \overline{t}\} \to [0, 1]$$

that satisfies the condition

$$\sum_{y \in X} \sum_{\tau=1}^{\overline{t}} p^a_{x,y,\tau} = 1, \quad \forall a \in A$$

and the cost function

$$c : A \times X \times X \times \{1, 2, \ldots, \overline{t}\} \to \mathbb{R}.$$

Here the function p for a fixed action $a \in A$, arbitrary $x, y \in X$ and a fixed $\tau \in \{1, 2, \ldots, \bar{t}\}$ determines the probability $p^a_{x,y,\tau}$ of the system to pass from the state $x \in X$ to state y by using τ units of time. The function c for a fixed action a in the state $x \in X$, a given $y \in X$ and a fixed τ determines the cost $c^a_{x,y,\tau}$ of the system to pass from the state x to the state y using τ units of time. We define a stationary strategy s in the Semi-Markov decision process as a map

$$s : x \to a \in A(x) \quad \text{for} \quad x \in X,$$

where $A(x)$ represents the set of actions in the state $x \in X$. An arbitrary stationary strategy s induces a Semi-Markov process with the transition probabilitis $p^s_{x,y,\tau}$ and the transition costs $c^s_{x,y,\tau}$. For this Semi-Markov process with given transition costs we can define the average cost per transition $\omega_{x0}(s)$ and the expected total discounted cost $\sigma^{\gamma}_{x0}(s)$ if the system starts transitions in the state x_0 at the moment of time $t = 0$. The problems of determining stationary strategies with minimal average and expected total discounted cost for Semi-Markov decision processes can be formulated and studied in a similar way as for Markov decision processes.

Using the results from Sect. 1.9 we can reduce the considered decision problems to the corresponding problems for an auxiliary Markov decision process. Indeed, for an arbitrary action $a \in A$ in a state $x \in X$, a given $y \in X$ and fixed $\tau \in \{1, 2, \ldots, \bar{t}\}$ the transition from x to y in Semi-Markov decision process we represent as a sequence of τ transitions with unit time via τ fictive intermediate states

$$\begin{aligned} & x \to x_1^{a,\tau}, \\ & x_1^{a,\tau} \to x_2^{a,\tau}, \; \ldots, \; x_{\tau-2}^{a,\tau} \to x_{\tau-1}^{a,\tau}, \\ & x_{\tau-1}^{a,\tau} \to y, \end{aligned}$$

where the corresponding probabilities and transition costs are defined as follows:

$$\begin{aligned} p_{x,x_1^{a,\tau}} &= p^a_{x,y,\tau}; \\ p_{x_1^{a,\tau},x_2^{a,\tau}} &= p_{x_2^{a,\tau},x_3^{a,\tau}} = \cdots = p_{x_{\tau-2}^{a,\tau},x_{\tau-1}^{a,\tau}} = p_{x_{\tau-1}^{a,\tau},y} = 1; \\ c^a_{x,x_1^{a,\tau}} &= c^a_{x,y,\tau}; \\ c^a_{x_1^{a,\tau},x_2^{a,\tau}} &= c^a_{x_2^{a,\tau},x_3^{a,\tau}} = \cdots = c^a_{x_{\tau-2}^{a,\tau},x_{\tau-1}^{a,\tau}} = c^a_{x_{\tau-1}^{a,\tau},y} = 0. \end{aligned}$$

After that we consider a new Markov decision problem with a new set of states $\overline{X}' = X \cup X'$ obtained from X by adding the set of fictive intermediate states X' and new probability and cost functions defined above; the set of actions in the state $x \in X$ in the new problem are the same as for Semi-Markov decision problems and in each added fictive state there is a unique action determined by a unique transition to the next state with a probability equal to 1.

It is evident that if $\overline{s}^*$ is a optimal stationary strategy for the auxiliary decision problem then an optimal stationary strategy s^* of the Semi-Markov decision problem (with average or discounted optimization criterion) can be found in the following way:

$$s^*(x) = \overline{s}^*(x) \quad \text{for} \quad x \in X.$$

In such a way we can reduce the Semi-Markov decision problem to the corresponding auxiliary Markov decision problem.

Chapter 3
A Game-Theoretical Approach to Markov Decision Processes, Stochastic Positional Games and Multicriteria Control Models

In this chapter we formulate and study a class of stochastic positional games applying the game-theoretical concept to Markov decision problems with average and expected total discounted costs optimization criteria. We consider Markov decision processes that may be controlled by several actors (players). The set of states of the system in such processes is divided into several disjoint subsets which represent the corresponding position sets of the players. Each player has to determine which action should be taken in each state of his position set in order to minimize his own average cost per transition or the expected total discounted cost. The cost of system's transition from one state to another in the Markov process is given for each player separately. In addition the set of actions, the transition probability functions and the starting state are known. We assume that players use only stationary strategies, i.e., each player in his arbitrary position uses the same action for an arbitrary discrete moment of time. In the considered stochastic positional games we are seeking for a Nash equilibrium. The proposed approach is developed for stochastic control problems on networks. Furthermore, we show that the considered class of stochastic positional games can be used for studying cyclic games and Shapley stochastic games with average payoff functions of the players. The main results we describe in this chapter are concerned with the existence of Nash equilibria in the considered games and with the elaboration of algorithms for determining the optimal stationary strategies of the players. We show that Nash equilibria for the game model with average cost payoff functions of the players exists if an arbitrary situation generated by the stationary strategies of the players corresponds to a Markov unichain. For the model with a discounted payoff function we show that Nash equilibria always exist. Furthermore, the antagonistic game models for Markov decision problems with average and expected total discounted costs are considered and new results concerning with the existence of saddle points are derived. Algorithms based on classical numerical methods for determining optimal strategies of the players are analyzed and new combinatorial algorithms for solving positional games on networks are proposed.

Additionally, we formulate and study multi-objective dynamic decision problems with Stackelberg and Pareto optimization principles and describe some general schemes for determining the solutions of these problems. Some extensions and

D. Lozovanu and S. Pickl, *Optimization of Stochastic Discrete Systems and Control on Complex Networks*, Advances in Computational Management Science 12,
DOI 10.1007/978-3-319-11833-8_3

generalizations of multicriteria dynamic control problem are suggested and new algorithms based on dynamic programming are proposed.

3.1 Stochastic Positional Games with Average Payoff Functions of the Players

We introduce a class of stochastic positional games applying the concept of noncooperative games to Markov decision problems with an average cost optimization criterion. We call this new class of games *stochastic positional games with average payoff functions* or *average stochastic positional games* [69, 91]. The considered class of games extends and generalizes cyclic games [43, 75, 79], Markov decision problems with average cost optimization criteria [115, 135, 140] and the stochastic control problem from Sect. 2.2. Nash equilibria conditions for such a class of games are derived and some approaches for determining optimal strategies of the players are described.

3.1.1 Problem Formulation

We formulate the *stochastic positional game with average payoff functions* using the framework of a Markov decision process (X, A, p, c) with a finite set of states X, a finite set of actions A, a transition probability function $p : X \times X \times A \to [0, 1]$ that satisfies the condition

$$\sum_{y \in X} p^a_{x,y} = 1, \quad \forall x \in X, \quad \forall a \in A$$

and a transition cost function $c : X \times X \to \mathbb{R}$ which represents the costs $c_{x,y}$ of states' transitions for the dynamical system if it makes a transition from the state $x \in X$ to another state $y \in X$.

We consider the noncooperative model with m players in which m transition cost functions are given

$$c^i : X \times X \to \mathbb{R}, \quad i = 1, 2, \dots, m,$$

where $c^i_{x,y}$ expresses the cost of the system's transition from the state $x \in X$ to the state $y \in X$ for the player $i \in \{1, 2, \dots, m\}$. In addition we assume that the set of states X is divided into m disjoint subsets $X_1, X_2, \dots, X_m$

$$X = X_1 \cup X_2 \cup \dots \cup X_m \quad (X_i \cap X_j = \emptyset, \quad \forall i \neq j),$$

where X_i represents the positions set of the player $i \in \{1, 2, \dots, m\}$. So, the Markov process is controlled by m players, where each player $i \in \{1, 2, \dots, m\}$ fixes actions in his positions $x \in X_i$.

We consider the stationary game model, i.e., we assume that each player fixes actions in the states from his positions set using stationary strategies. We define the stationary strategies of the players as m maps:

$$\begin{array}{ll} s^1 : x \to a \in A^1(x) & \text{for } x \in X_1; \\ s^2 : x \to a \in A^2(x) & \text{for } x \in X_2; \\ \vdots & \vdots \\ s^m : x \to a \in A^m(x) & \text{for } x \in X_m, \end{array}$$

where $A^i(x)$ is the set of actions of the player i in the state $x \in X_i$. Without loss of generality we may consider $|A^i(x)| = |A^i| = |A|$, $\forall x \in X_i$, $i = 1, 2, \dots, m$. In order to simplify the notation we denote the set of possible actions in a state $x \in X$ for an arbitrary player by $A(x)$.

A stationary strategy s^i, $i \in \{1, 2, \dots, m\}$ in the state $x \in X_i$ means that at every discrete moment of time $t = 0, 1, 2, \dots$ the player i uses the action $a = s^i(x)$. Players fix their strategy independently and do not inform each other which strategies they use in the decision process.

If the players $1, 2, \dots, m$ fix their stationary strategies $s^1, s^2, \dots, s^m$, respectively, then we obtain a *situation* $s = (s^1, s^2, \dots, s^m)$. This situation corresponds to a simple Markov process determined by the probability distributions $p_{x,y}^{s^i(x)}$ in the states $x \in X_i$ for $i = 1, 2, \dots, m$. We denote by $P^s = (p_{x,y}^s)$ the matrix of probability transitions of this Markov process. If the starting state x_0 is given, then for the Markov process with the matrix of probability transitions P^s we can determine the average cost per transition $\omega_{x_0}^i(s^1, s^2, \dots, s^m)$ with respect to each player $i \in \{1, 2, \dots, m\}$ taking into account the corresponding matrix of transition costs $C^i = (c_{x,y}^i)$. So, on the set of situations we can define the payoff functions of the players as follows:

$$F_{x_0}^i(s^1, s^2, \dots, s^m) = \omega_{x_0}^i(s^1, s^2, \dots, s^m), \quad i = 1, 2, \dots, m.$$

In such a way we obtain a discrete noncooperative game in normal form which is determined by a finite set of strategies $\mathbb{S}^1, \mathbb{S}^1, \dots, \mathbb{S}^m$ of m players and the payoff functions defined above. In this game we are seeking for a *Nash equilibrium* [102], i.e., we consider the problem of determining the stationary strategies

$$s^{1*}, s^{2*}, \dots, s^{i-1*}, s^{i*}, s^{i+1*}, \dots, s^{m*}$$

such that

$$\begin{aligned} & F_{x_0}^i(s^{1*}, s^{2*}, \dots, s^{i-1*}, s^{i*}, s^{i+1*}, \dots, s^{m*}) \\ & \quad \le F_{x_0}^i(s^{1*}, s^{2*}, \dots, s^{i-1*}, s^{i}, s^{i+1*}, \dots, s^{m*}), \ \ \forall s^i \in \mathbb{S}^i, \ \ i = 1, 2, \dots, m. \end{aligned}$$

The game defined above is determined uniquely by the set of states X, the position sets $X_1, X_2, \ldots, X_m$, the set of actions A, the cost functions $c^i : X \times X \to \mathbb{R}$, $i = 1, 2, \ldots, m$, the probability function $p : X \times X \times A \to [0, 1]$ and the starting position x_0. Therefore, we denote it by

$$(X,\ A,\ \{X_i\}_{i=\overline{1,m}},\ \{c^i\}_{i=\overline{1,m}},\ p,\ x_0).$$

We call this game *stochastic positional game with average payoff functions.*

In the case $p^a_{x,y} = 0 \vee 1,\ \forall x, y \in X,\ \forall a \in A$ the stochastic positional game is transformed into the cyclic game studied in [43, 75, 79].

3.1.2 Determining Nash Equilibria for Stochastic Positional Games with Average Payoff Functions

To provide the existence of Nash equilibria for the considered stochastic positional game we shall use the following condition. We assume that an arbitrary situation $s = (s^1, s^2, \ldots, s^m)$ of the game generates a Markov unichain with the corresponding matrix of probability transitions $P^s = (p^s_{x,y})$. We call the Markov process with such a property with respect to the situations $s = (s^1, s^2, \ldots, s^m) \in S$ of the game *perfect Markov decision process*. We show that in this case the problem of determining Nash equilibria for a stochastic positional game can be formulated as continuous model that represents the game-theoretic variant of the following optimization problem:
Minimize

$$\psi(s, q) = \sum_{x\in X} \sum_{a\in A(x)} \mu_{x,a} s_{x,a}\, q_x \tag{3.1}$$

subject to

$$\begin{cases} \sum\limits_{x\in X} \sum\limits_{a\in A(x)} p^a_{x,y} s_{x,a} q_x = q_y, \quad \forall y \in X; \\ \sum\limits_{x\in X} q_x = 1; \\ \sum\limits_{a\in A(x)} s_{x,a} = 1, \quad \forall x \in X; \\ \qquad s_{x,a} \ge 0,\ \ \forall x \in X,\ a \in A(x), \end{cases} \tag{3.2}$$

where

$$\mu_{x,a} = \sum_{y\in X^+(x)} c_{x,y}\, p^a_{x,y}$$

is the immediate cost in the state $x \in X$ for a fixed action $a \in A(x)$.

It is easy to observe that the problem (3.1), (3.2) represents the continuous model for a Markov decision problem with an average cost criterion. Indeed, an arbitrary stationary strategy $s : X \to A$ can be identified with the set of boolean variables $s_{x,a} \in \{0, 1\}$, $x \in X$, $a \in A(x)$ that satisfies the conditions

$$\sum_{a \in A(x)} s_{x,a} = 1, \quad \forall x \in X; \quad s_{x,a} \geq 0, \ \forall x \in X, \ a \in A.$$

These conditions determine all feasible solutions of the system (3.2). The remaining restrictions in (3.2) correspond to the system of linear equations with respect to q_x for $x \in X$. This system of linear equations reflects the ergodicity condition for the limiting probability q_x, $x \in X$ in the Markov unichain, where q_x, $x \in X$, are determined uniquely for given $s_{x,a}$, $\forall x \in X, a \in A(x)$. Thus, the value of the objective function (3.1) expresses the average cost per transition in this Markov unichain and an arbitrary optimal solution $s^*_{x,a}$, q^*_x $(x \in X,\ a \in A)$ of the problem (3.1), (3.2) with $s^*_{x,a} \in \{0, 1\}$ represents an optimal stationary strategy for a Markov decision problem with an average cost criterion. If such an optimal solution is known, then an optimal action for the Markov decision problem can be found by fixing $a^* = s^*(x)$ for $x \in X$ if $s^*_{x,a} = 1$. The problem (3.1), (3.2) can be transformed into a linear programming problem using the notations $\alpha_{x,a} = s_{x,a}q_x$, $\forall x \in X, a \in A(x)$ (see [68]). Based on such a transformation of the problem we will describe some additionally properties of the optimal stationary strategies in Markov decision processes.

Lemma 3.1 *Let a Markov decision process (X, A, p, c) be given and consider the function*

$$\psi(s) = \sum_{x \in X} \sum_{a \in A(x)} \mu_{x,a} s_{x,a}\, q_x,$$

where q_x for $x \in X$ satisfies the condition

$$\begin{cases} \sum\limits_{x \in X} \sum\limits_{a \in A(x)} p^a_{x,y} s_{x,a} q_x = q_y, & \forall y \in X; \\ \sum\limits_{x \in X} q_x = 1. \end{cases} \tag{3.3}$$

Assume that an arbitrary stationary strategy s in the Markov decision process generates a Markov unichain, i.e., we have a perfect Markov decision process. Then the function $\psi(s)$ depends only on $s_{x,a}$ for $x \in X$, $a \in A(x)$, and on the set S of solutions of the system

$$\begin{cases} \sum\limits_{a \in A(x)} s_{x,a} = 1, & \forall x \in X; \\ s_{x,a} \geq 0, & \forall x \in X,\ a \in A(x), \end{cases} \tag{3.4}$$

the function $\psi(s)$ is monotone.

Proof According to Lemma 2.25 for perfect Markov decision processes an arbitrary basic solution of the system (3.4) corresponds to a stationary strategy. Moreover, for each such strategy the rank of the matrix of the system (3.3) is equal to $|X|$ and the system (3.3) has a unique solution with respect to q_x $(x \in X)$ (see [115, 140]). In the case of an arbitrary feasible solution of the system (3.4) the rank of the matrix of system (3.3) is also equal to $|X|$ and for each such solution the system (3.3) has a unique solution with respect to q_x $(x \in X)$ because the matrix of system (3.3) can be represented via the matrices that correspond to basic solutions of the system (3.4).

This means that the system (3.3) has a unique solution with respect to q_x $(x \in X)$ for an arbitrary solution of the system (3.4) and, therefore, for perfect Markov decision processes the function $\psi(s)$ depends only on s.

Now let us prove the second part of the lemma. We show that on the set of solutions of the system (3.4) the function $\psi(s)$ is monotone. For this reason it is sufficient to show that for arbitrary $s', s'' \in S$ with $\psi(s') \neq \psi(s'')$ the following relation holds

$$\min\{\psi(s'), \psi(s'')\} < \psi(\overline{s}) < \max\{\psi(s'), \psi(s'')\} \tag{3.5}$$

if

$$\overline{s} = \theta s' + (1-\theta)s'', \quad 0 < \theta < 1.$$

We can see that the relation (3.5) holds for an arbitrary $\overline{s} \in S(s', s'')$, where

$$S(s', s'') = \{\, s \mid \min\{s'_{x,a}, s''_{x,a}\} < s_{x,a} < \max\{s'_{x,a}, s''_{x,a}\}, \ \ \forall x \in X, a \in A(x)\}$$

and the equations

$$\psi(s) = \psi(s'), \quad \psi(s) = \psi(s'')$$

on the set

$$\overline{S}(s', s'') = \{\, s \mid \min\{s'_{x,a}, s''_{x,a}\} \leq s_{x,a} \leq \max\{s'_{x,a}, s''_{x,a}\}, \ \ \forall x \in X, a \in A(x)\}$$

have the unique solutions $s = s'$ and $s = s''$, respectively.

We prove the correctness of this property using the relationship of the problem (3.1), (3.2) with the following linear programming problem:

Minimize

$$\overline{\psi}(\alpha) = \sum_{x \in X} \sum_{a \in A(x)} \mu_x^a \, \alpha_{x,a} \tag{3.6}$$

subject to

$$\begin{cases} \sum\limits_{x\in X}\ \sum\limits_{a\in A(x)} p_{x,y}^{a}\,\alpha_{x,a} = q_y, & \forall y \in X; \\ \sum\limits_{x\in X} q_x = 1; & \\ \sum\limits_{a\in A(x)} \alpha_{x,a} = q_x, & \forall x \in X; \\ \alpha_{x,a} \geq 0, & \forall x \in X,\ a \in A(x). \end{cases} \tag{3.7}$$

The problem (3.6), (3.7) is obtained from (3.1), (3.2) introducing the substitutions $\alpha_{x,a} = s_{x,a}q_x$ for $x \in X,\ a \in A(x)$. These substitutions allow us to establish a bijective mapping between the set of feasible solutions of the problem (3.1), (3.2) and the set of feasible solutions of the linear programming problem (3.6), (3.7). So, if $\alpha_{x,a}$ for $x \in X,\ a \in A(x)$ and $\overline{\psi}(\alpha)$ are known then we can uniquely determine

$$s_{x,a} = \frac{\alpha_{x,a}}{q_x}, \quad \forall x \in X,\ a \in A(x) \tag{3.8}$$

for which $\psi(s) = \overline{\psi}(\alpha)$.

In particular, if an optimal basic solution α^*, q^* of the linear programming problem (3.6), (3.7) is found, then the optimal stationary strategy for a Markov decision problem can be found fixing

$$s_{x,a}^{*} = \begin{cases} 1, \text{ if } \alpha_{x,a}^{*} > 0; \\ 0, \text{ if } \alpha_{x,a}^{*} = 0. \end{cases}$$

Let s', s'' be arbitrary solutions of the system (3.4) where $\psi(s') < \psi(s'')$. Then there exist the corresponding feasible solutions α', α'' of the linear programming problem (3.6), (3.7) for which

$$\psi(s') = \overline{\psi}(\alpha'), \quad \psi(s'') = \overline{\psi}(\alpha''),$$
$$\alpha'_{x,a} = s'_{x,a}q'_x, \quad \alpha''_{x,y} = s''_{x,a}q''_x \ \forall x \in X,\ a \in A(x),$$

where q'_x, q''_x are determined uniquely by the system of linear equations (3.3) for $s = s'$ and $s = s''$, respectively. The function $\overline{\psi}(\alpha)$ is linear and therefore for an arbitrary $\overline{\alpha} = \theta\alpha' + (1-\theta)\alpha'',\ 0 \leq \theta \leq 1$ the following equality holds

$$\overline{\psi}(\overline{\alpha}) = \theta\overline{\psi}(\alpha') + (1-\theta)\overline{\psi}(\alpha''),$$

where $\overline{\alpha}$ is a feasible solution of the problem (3.6), (3.7), that in the initial problem (3.1), (3.2) corresponds to a feasible solution $\overline{s}$ for which

$$\psi(\overline{s}) = \overline{\psi}(\overline{\alpha}); \quad \overline{q}_x = \theta q'_x + (1-\theta)q''_x, \quad \forall x \in X.$$

Using (3.8) we have

$$\overline{s}_{x,a} = \frac{\overline{\alpha}_{x,a}}{\overline{q}_x}, \quad \forall x \in X, \ a \in A(x),$$

i.e.

$$\begin{aligned}\overline{s}_{x,a} &= \frac{\theta\alpha'_{x,a} + (1-\theta)\alpha''_{x,a}}{\theta q'_x + (1-\theta)q''_x} = \frac{\theta s'_{x,a}q'_x + (1-\theta)s''_{x,a}q''_x}{\theta q'_x + (1-\theta)q''_x} \\ &= \frac{\theta q'_x}{\theta q'_x + (1-\theta)q''_x} s'_{x,a} + \frac{(1-\theta)q''_x}{\theta q'_x + (1-\theta)q''_x} s''_{x,a}.\end{aligned}$$

So, we obtain

$$\overline{s}_{x,a} = \overline{\theta}_x s'_{x,a} + (1-\overline{\theta}_x)s''_{x,a},$$

where

$$\overline{\theta}_x = \frac{\theta q'_x}{\theta q'_x + (1-\theta)q''_x}, \quad 0 \le \theta \le 1.$$

It is easy to observe that $0 \le \overline{\theta}_x \le 1$, where $\overline{\theta}_x = 0, \ \forall x \in X$ if and only if $\theta = 0$ and $\overline{\theta}_x = 1, \ \forall x \in X$ if and only if $\theta = 1$. This means that for an arbitrary $\overline{s} \in S(s', s'')$ the condition (3.5) holds and the equations

$$\psi(s) = \psi(s'), \quad \psi(s) = \psi(s'')$$

on the set $\overline{S}(s', s'')$ have the unique solutions $s = s'$ and $s = s''$, respectively. Thus, the function $\psi(s)$ is monotone on the set of solutions of the system (3.4). □

Remark 3.2 The monotonicity property of the function $\psi(s)$ is induced by the monotonicity property of the linear function $\overline{\psi}(\alpha)$ that implicitly depends only on s.

Now we extend the results described above for the continuous model of the stochastic positional game with average payoff functions. We consider the model for perfect Markov decision processes.

Let us denote by S^i, $i \in \{1, 2, \dots, m\}$ the set of solutions of the system

$$\begin{cases} \sum\limits_{a \in A(x)} s^i_{x,a} = 1, & \forall x \in X_i; \\ s^i_{x,a} \ge 0, & \forall x \in X_i, \ a \in A(x). \end{cases} \tag{3.9}$$

So, S^i is a convex compact set and its arbitrary extreme point corresponds to a basic solution s' of the system (3.9), where $s'_{x,a} \in \{0, 1\}, \ \forall x \in X_1, a \in A(x)$. Thus, if s' is an arbitrary basic solution of the system (3.9), then $s' \in \mathbb{S}^i$.

On the set $S = S^1 \times S^2 \times \cdots \times S^m$ we define m payoff functions

$$\psi^i(s^1, s^2, \dots, s^m) = \sum_{j=1}^{m} \sum_{x \in X_j} \sum_{a \in A(x)} \mu^i_{x,a} s^j_{x,a} q_x, \quad i = 1, 2, \dots, m, \tag{3.10}$$

where

$$\mu^i_{x,a} = \sum_{y \in X} c^i_{x,y} p^a_{x,y}$$

is the immediate cost of player $i \in \{1, 2, \dots, m\}$ in the state $x \in X$ for a fixed action $a \in A(x)$; q_x for $x \in X$ are determined uniquely by the following system of linear equations

$$\begin{cases} \sum\limits_{i=1}^{m} \sum\limits_{x \in X_i} \sum\limits_{a \in A(x)} p^a_{x,y} s^i_{x,a} q_x = q_y, \quad \forall y \in X; \\ \sum\limits_{x \in X} q_x = 1 \end{cases} \tag{3.11}$$

when $s^1, s^2, \dots, s^m$ are given.

Below we show that for our game models the following properties hold:

- The set of Nash equilibria solutions of the continuous model is non empty if and only if the set of Nash equilibria solutions of the game in positional form is not empty;
- If $(s^1, s^2, \dots, s^m)$ is an extreme point of S then

  $$F^i_x(s^1, s^2, \dots, s^m) = \psi(s^1, s^2, \dots, s^m), \quad \forall x \in X, \quad i = 1, 2, \dots, m$$

  and all Nash equilibria solutions for the continuous game model that correspond to extreme points in S represent Nash equilibria solutions for the game in positional form.

As a corollary from Lemma 3.1 we obtain the following result:

Lemma 3.3 *For perfect Markov processes each payoff function*

$$\psi^i(s^1, s^2, \dots, s^m), \quad i \in \{1, 2, \dots, m\}$$

possesses the property that

$$\psi^i(\overline{s}^1, \overline{s}^2, \dots, \overline{s}^{i-1}, s^i, \overline{s}^{i+1}, \dots, \overline{s}^m)$$

is monotone with respect to $s^i \in S^i$ for arbitrary fixed $\overline{s}^k \in S^k$, $k = 1, 2, \dots, i-1, i+1, \dots, m$.

Using this lemma we can prove the following theorem.

Theorem 3.4 *Let* $(X, A, \{X_i\}_{i=\overline{1,m}}, \{c^i\}_{i=\overline{1,m}}, p, x)$ *be a stochastic positional game with a given starting position* $x \in X$ *and average payoff functions*

$$F_x^1(s^1, s^2, \dots, s^m),\ F_x^2(s^1, s^2, \dots, s^m),\ \dots,\ F_x^m(s^1, s^2, \dots, s^m)$$

of the players $1, 2, \dots, m$, *respectively. If for an arbitrary situation* $s = (s^1, s^2, \dots, s^m)$ *of the game the transition probability matrix* $P^s = (p^s_{x,y})$ *corresponds to a Markov unichain, then for the stochastic positional game with payoff functions* $F_x^i(s^1, s^2, \dots, s^m)$, $i = 1, 2, \dots, m$ *there exists a Nash equilibrium* $s^* = (s^{1*}, s^{2*}, \dots, s^{m*})$. *Moreover, for this game there exists a situation* $s^* = (s^{1*}, s^{2*}, \dots, s^{m*})$ *which is a Nash equilibrium for an arbitrary starting position* $x \in X$.

Proof According to Lemma 3.3 each payoff function $\psi^i(s^1, s^2, \dots, s^m)$, $i \in \{1, 2, \dots, m\}$ satisfies the condition that

$$\psi^i(\overline{s}^1, \overline{s}^2, \dots, \overline{s}^{i-1}, s^i, \overline{s}^{i+1}, \dots, \overline{s}^m)$$

is monotone with respect to $s^i \in S^i$ for arbitrary fixed $\overline{s}^k \in S^k$, $k = 1, 2, \dots, i-1, i+1, \dots, m$. In the considered game each subset S^i, $i \in \{1, 2, \dots, m\}$ is convex and compact.

Therefore, these conditions (see [23, 24, 116, 125]) provide the existence of a Nash equilibrium for the functions $\psi^i(s^1, s^2, \dots, s^m)$, $i \in \{i, 2, \dots, m\}$ on $S^1 \times S^2 \times \dots \times S^m$. Taking into account that S is a polyhedron set and the functions $\psi^i(\overline{s}^1, \overline{s}^2, \dots, \overline{s}^{i-1}, s^i, \overline{s}^{i+1}, \dots, \overline{s}^m)$ are monotone we obtain that the Nash equilibrium $s^{1*}, s^{2*}, \dots, s^{m*}$ corresponds to a basic solution of the system (3.9). This means that $(s^{1*}, s^{2*}, \dots, s^{m*})$ is a Nash equilibrium for the functions

$$F_x^1(s^1, s^2, \dots, s^m),\ F_x^2(s^1, s^2, \dots, s^m),\ \dots,\ F_x^m(s^1, s^2, \dots, s^m)$$

on the set of situations $\mathbb{S} = \mathbb{S}^1 \times \mathbb{S}^1 \times \dots \times \mathbb{S}^m$. □

Remark 3.5 The values

$$\omega_x^i = F_x^i(s^{1*}, s^{2*}, \dots, s^{m*}), \quad i = 1, 2, \dots, m$$

of the game for perfect Markov decision processes do not depend on the starting position x, i.e., $\omega_x^i = \omega_y^i$, $i = 1, 2, \dots, m$ for arbitrary $x, y \in X$.

Using the results described above we may conclude that in the case of perfect Markov decision processes a Nash equilibrium for stochastic positional games can be determined by using classical iterative methods for the continuous game models with payoff functions $\psi^i(s^1, s^2, \dots, s^m)$, $i \in \{i, 2, \dots, m\}$ on the set $S^1 \times S^2 \times \dots \times S^m$. If we refer these iterative methods to a discrete game model with payoff functions on $\mathbb{S}^1 \times \mathbb{S}^1 \times \dots \times \mathbb{S}^m$, then we obtain the iterative procedures where players fix successively their strategies in order to minimize their payoff functions, respectively, and finally in order to reach a Nash equilibrium.

In general, for stochastic positional games with average payoff functions of the players, a Nash equilibrium may not exists if the stationary strategies do not generate a Markov unichain. Moreover, a Nash equilibrium may not exists even for deterministic positional games (see [43,79]). So, Theorem 3.4 in the case $p_{x,y}^a \in \{0, 1\}$, gives conditions for the existence of Nash equilibria in *cyclic games with average payoff functions*.

3.1.3 Determining Nash Equilibria for Average Stochastic Positional Games Using the Potential Transformation

We extend the *potential transformation for stochastic positional games* using Theorem 2.71 and Eq. (2.85) in the case of unichain Markov decision processes. For such games Theorem 3.4 can be formulated in the terms of a potential transformation that gives new conditions for determining Nash equilibria solutions.

Theorem 3.6 *Let* $(X, A, \{X_i\}_{i=\overline{1,m}}, \{c^i\}_{i=\overline{1,m}}, p, \overline{x})$ *be a stochastic positional game with a given starting position* $\overline{x} \in X$ *and average payoff functions*

$$F_{\overline{x}}^1(s^1, s^2, \dots, s^m),\ F_{\overline{x}}^2(s^1, s^2, \dots, s^m),\ \dots,\ F_{\overline{x}}^m(s^1, s^2, \dots, s^m)$$

of the players $1, 2, \dots, m$, *respectively. Assume that for an arbitrary situation* $s = (s^1, s^2, \dots, s^m)$ *of the game the transition probability matrix* $P^s = (p_{x,y}^s)$ *corresponds to a Markov unichain. Then there exist the functions*

$$\varepsilon^i : X \to \mathbb{R}, \quad i = 1, 2, \dots, m$$

and the values $\omega^1, \omega^2, \dots, \omega^m$ *that satisfy the following conditions:*

(1) $\mu_{x,a}^i + \sum\limits_{y\in X} p_{x,y}^a \varepsilon_y^i - \varepsilon_x^i - \omega^i \geq 0, \ \ \forall x \in X_i, \ \ \forall a \in A(x),\ i = 1, 2, \dots, m;$

(2) $\min\limits_{a\in A(x)} \{\mu_{x,a}^i + \sum\limits_{y\in X} p_{x,y}^a \varepsilon_y^i - \varepsilon_x^i - \omega^i\} = 0, \ \ \forall x \in X_i,\ i = 1, 2, \dots, m;$

(3) *on each position set* X_i, $i \in \{1, 2, \dots, m\}$ *there exists a map* $s^{i^*} : X_i \to A$ *such that*

$$s^{i^*}(x) = a^* \in \operatorname*{argmin}_{a\in A(x)} \left\{\mu_{x,a}^i + \sum_{y\in X} p_{x,y}^a \varepsilon_y^i - \varepsilon_x^i - \omega^i\right\}$$

and

$$\mu_{x,a^*}^j + \sum_{y\in X} p_{x,y}^{a^*} \varepsilon_y^j - \varepsilon_x^j - \omega^j = 0, \ \ \forall x \in X_i, \ \ j = 1, 2, \dots, m.$$

The set of maps $s^{1*}, s^{2*}, \ldots, s^{m*}$ *determines a Nash equilibrium situation* $s^* = (s^{1*}, s^{2*}, \ldots, s^{m*})$ *for the stochastic positional game* $(X, A, \{X_i\}_{i=\overline{1,m}}, \{c^i\}_{i=\overline{1,m}}, p, \overline{x})$ *and*

$$F^i_{\overline{x}}(s^{1*}, s^{2*}, \ldots, s^{m*}) = \omega^i, \quad \forall \overline{x} \in X, \ i = 1, 2, \ldots, m.$$

Moreover, the situation $s^* = (s^{1*}, s^{2*}, \ldots, s^{m*})$ *is a Nash equilibrium for an arbitrary starting position* $\overline{x} \in X$.

Proof According to Theorem 3.4 for average stochastic positional games there exists a Nash equilibrium $s^* = (s^{1*}, s^{2*}, \ldots, s^{m*})$, and

$$\omega^i = F^i_{\overline{x}}(s^{1*}, s^{2*}, \ldots, s^{m*}), \quad \forall \overline{x} \in X, \ i = 1, 2, \ldots, m.$$

Let us fix the strategies $s^{1*}, s^{2*}, \ldots, s^{i-1*}, s^{i+1*}, \ldots, s^{m*}$ of the players $1, 2, \ldots, i-1, i+1, \ldots, m$ and consider the problem of determining the minimal average cost per transition with respect to the player i. Obviously, if we solve this decision problem then we obtain the strategy s^{i*}. We can determine the solution of this decision problem using the dual linear programming model (2.71), (2.72). According to Theorem 2.30 for this problem there exist the functions $\varepsilon^i : X \to \mathbb{R}$ and the values ω^i, $i = 1, 2, \ldots, m$ that satisfy the conditions:

(1) $\mu^i_{x,a} + \sum\limits_{y \in X} p^a_{x,y}\varepsilon^i_y - \varepsilon^i_x - \omega^i \geq 0, \quad \forall x \in X_i, \ \forall a \in A(x);$

(2) $\min\limits_{a \in A(x)} \left\{\mu^i_{x,a} + \sum\limits_{y \in X} p^a_{x,y}\varepsilon^i_y - \varepsilon^i_x - \omega^i\right\} = 0, \quad \forall x \in X_i.$

Moreover, for fixed strategies $s^{1*}, s^{2*}, \ldots, s^{i-1*}, s^{i+1*}, \ldots, s^{m*}$ of the corresponding players $1, 2, \ldots, i-1, i+1, \ldots, m$ we can select the strategy s^{i*} of the player i where

$$s^{i*}(x) \in \underset{a \in A(x)}{\operatorname{argmin}} \left\{\mu^i_{x,a} + \sum_{y \in X} p^a_{x,y}\varepsilon^i_y - \varepsilon^i_x - \omega^i\right\}$$

and $\omega^i = F^i_x(s^{1*}, s^{2*}, \ldots, s^{m*})$, $\forall x \in X$, $i = 1, 2, \ldots, m$. This means that the conditions (1)–(3) of the theorem hold. □

Corollary 3.7 *Let* $(X, A, \{X_i\}_{i=\overline{1,m}}, \{c^i\}_{i=\overline{1,m}}, p, \overline{x})$ *be a stochastic positional game for which there exist the functions* $\varepsilon^i : X \to \mathbb{R}$, $i = 1, 2, \ldots, m$ *and the values* $\omega^1, \omega^2, \ldots, \omega^m$ *that satisfy the conditions* (1)–(3) *of* Theorem 3.8. *Then for this game there exists a Nash equilibrium and the optimal stationary strategies of the players can be found according to the rule from the theorem.*

Theorem 3.6 generalizes condition (2.85) for the unichain average Markov decision problem. Based on this theorem we can determine the optimal stationary strategies of the players if ε^i and ω^i are known. The functions ε^i and the values ω^i for

$i \in \{1, 2, \dots, m\}$ can be found using a similar iterative algorithm as for the decision problem from the previous chapter. However, the problem of determining ε^i and ω^i seems to be too difficult from a computational point of view. The question concerned with the existence of polynomial time algorithms for determining the optimal strategies of the players in the considered games is an open problem. Moreover, the existence of the polynomial time algorithm for average stochastic positional games is an open problem even for a special deterministic antagonistic game. We will discuss the computational complexity of the problem of determining the optimal stationary strategies of the players in such games in the following sections.

3.1.4 Necessary and Sufficient Conditions for the Existence of Nash Equilibria in Average Stochastic Positional Games

Let $(X, A, \{X_i\}_{i=\overline{1,m}}, \{c^i\}_{i=\overline{1,m}}, p, x)$ be an average stochastic positional game in which a Nash equilibrium exists for an arbitrary starting position $x \in X$. Denote $\omega_x^i = F_x^i(s^{1*}, s^{2*}, \dots, s^{m*})$, $\forall x \in X$, $i = 1, 2, \dots, m$, where $s^* = (s^{1*}, s^{2*}, \dots, s^{m*})$ is a Nash equilibrium solution of the game. For this game in a similar way as for the average stochastic control problem we can define the potential transformation

$$\overline{c}^i_{x,y} = c^i_{x,y} + \varepsilon^i_y - \varepsilon^i_x - \omega^i_x, \quad \forall x, y \in X$$

of the costs with respect to each player $i \in \{1, 2, \dots, m\}$, where $\varepsilon^i : X \to \mathbb{R}$ is an arbitrary real function. It is easy to show that the considered potential transformations of the costs in the game do not change the optimal stationary strategies of the players. Therefore, based on this property and on results from the previous section we obtain the following result.

Theorem 3.8 *Let $(X, A, \{X_i\}_{i=\overline{1,m}}, \{c^i\}_{i=\overline{1,m}}, p, x)$ be an average stochastic positional game. Then in this game there exists a Nash equilibrium for an arbitrary starting position $x \in X$ if and only if there exist the functions*

$$\varepsilon^i : X \to \mathbb{R}, \ i = 1, 2, \dots, m$$

and the values $\omega_x^1, \omega_x^2, \dots, \omega_x^m$ for $x \in X$ that satisfy the following conditions:

(1) $\mu^i_{x,a} + \sum\limits_{y \in X} p^a_{x,y}\varepsilon^i_y - \varepsilon^i_x - \omega^i_x \geq 0, \quad \forall x \in X_i, \ \forall a \in A(x), \ i = 1, 2, \dots, m;$

(2) $\min\limits_{a \in A(x)} \{\mu^i_{x,a} + \sum\limits_{y \in X} p^a_{x,y}\varepsilon^i_y - \varepsilon^i_x - \omega^i_x\} = 0, \quad \forall x \in X_i, \ i = 1, 2, \dots, m;$

(3) *on each position set X_i, $i \in \{1, 2, \dots, m\}$ there exists a map $s^{i*} : X_i \to A$ such that*

$$s^{i*}(x) = a^* \in \underset{a\in A(x)}{\operatorname{argmin}} \left\{ \mu^i_{x,a} + \sum_{y\in X} p^a_{x,y}\varepsilon^i_y - \varepsilon^i_x - \omega^i \right\}$$

and

$$\mu^j_{x,a^*} + \sum_{y\in X} p^{a^*}_{x,y}\varepsilon^j_y - \varepsilon^j_x - \omega^j = 0, \quad \forall x \in X_i, \quad j = 1, 2, \ldots, m.$$

If such conditions hold then the set of maps $s^{1*}, s^{2*}, \ldots, s^{m*}$ *determines a Nash equilibrium of the game for an arbitrary starting position* $x \in X$ *and* $F^i_x(s^{1*}, s^{2*}, \ldots, s^{m*}) = \omega^i_x$, $i = 1, 2, \ldots, m$.

3.1.5 Nash Equilibria Conditions for Cyclic Games

In [43, 75, 79] the following positional game has been studied. Let a dynamical system $\mathbb{L}$ with a finite set of states X be given. The dynamics of the system is described by a directed graph of the state's transitions $G = (X, E)$, where the set of vertices X corresponds to the set of states of the system $\mathbb{L}$ and an arbitrary directed edge $e = (x, y) \in X$ expresses the possibility of the system's transition from the state $x = x(t)$ to the state $y = x(t+1)$ at every discrete moment of time $t = 0, 1, 2, \ldots$. We assume that the graph G possesses the property that each vertex $x \in X$ contains at least one outgoing directed edge. The starting state of the system is given, i.e., $x_0 = x(0)$, and the dynamics of the system is controlled by m players. For each player $i \in \{1, 2, \ldots, m\}$ a cost function $c^i : E \to \mathbb{R}$ is defined, where for $(x, y) = e \in E$ the value $c^i_{x,y}$ expresses the cost of the system's transition from the state $x = x(t)$ to the state $y = x(t+1)$, $t = 0, 1, 2, \ldots$. The set of states X is divided into m disjoint subsets $X_1, X_2, \ldots, X_m$

$$X = X_1 \cup X_2 \cup \cdots \cup X_m \quad (X_i \cap X_j = \emptyset, \quad \forall i \neq j),$$

where X_i represents the positions set of the player $i \in \{1, 2, \ldots, m\}$. Each player may control the dynamical system only in his positions. So, the control process on G is made by the players as follows:

If the starting state $x_0 = x(0)$ belongs to the set of positions X_i then player i on the set of possible transitions $E(x_0) = \{(x_0, y) \in E \mid y \in X\}$ selects a transition (x_0, x_1) from the state x_0 to the state $x_1 = x(1)$; in general, if at the moment of time t the dynamical system is in the state $x_t = x(t)$ and $x(t) \in X_i$ then the system is controlled by player i, i.e., player i selects the transition from the state x_t to a state $x_{t+1} \in X(x_t)$, where $X(t) = \{y \in X \mid (x_t, y) \in E\}$. In this dynamical process each player intends to minimize his average cost per transition by a trajectory $x_0, x_1, x_2, \ldots$, i.e.,

$$F^i = \lim_{t\to\infty} \frac{1}{t} \sum_{\tau=0}^{t-1} c^i_{x_t,x_{t+1}} \quad \to \min, \ \ i = 1, 2, \ldots, m.$$

In a more strict way this dynamical game can be formulated in terms of stationary strategies. We define the stationary strategies of the player as m maps

$$s^i : x \ \to \ y \in X(x) \ \ for \ \ x \in X_i; \ \ i = 1, 2, \ldots, m.$$

Let $s^1, s^2, \ldots, s^m$ be an arbitrary set of strategies of the players. We denote by $G_s = (X, E_s)$ the subgraph generated by edges $e = (x, s^i(x))$ for $x \in X_i$ and $i = 1, 2, \ldots, m$. Obviously, for fixed $s^1, s^2, \ldots, s^m$ the graph G_s possesses the property that a unique directed cycle C_s with the set of edges $E(C_s)$ can be reached from x_0. For fixed strategies $s^1, s^2, \ldots, s^m$ and a fixed starting state we define the quantities

$$F^i_{x_0}(s^1, s^2, \ldots, s^m) = \frac{1}{n(C_s)} \sum_{e\in E(C_s)} c^i_e, i = 1, 2, \ldots, m,$$

where $n(C_s)$ is the number of directed edges of the cycle C_s. The graph G is finite and therefore the set of strategies $\mathbb{S}^i$ of each player $i \in \{1, 2, \ldots, m\}$ is a finite set. In such a way on the set of situations $\mathbb{S} = \mathbb{S}^1 \times \mathbb{S}^2 \times \cdots \times \mathbb{S}^m$ the functions $F^1_{x_0}(s^1, s^2, \ldots, s^m)$, $F^2_{x_0}(s^1, s^2, \ldots, s^m)$, $\ldots$, $F^m_{x_0}(s^1, s^2, \ldots, s^m)$ determine a game in normal form which in [43, 75, 79] is called *cyclic game* and it is denoted by $(G, \{X_i\}_{i=\overline{1,m}}, \{c^i\}_{i=\overline{1,m}}, x_0)$.

It is easy to see that this game represents a particular case of a stochastic positional game from Sect. 3.1.1. Indeed, we can regard the cyclic game as a stochastic positional game induced by the set of states X with the partition $X = X_1 \cup X_2 \cup \cdots \cup X_m$, the cost functions c^i, $i = 1, 2, \ldots m$ and the set of action A with the corresponding probabilities p^a_{x,y_k}, where for a given state $x \in X$ we have $r(x) = |E(x)|$ actions that correspond to possible transitions through the edges $e_1 = (x, y_1)$, $e_2 = (x, y_2)$, $\ldots$, $e_{r(x)} = (x, y_{r(x)})$:

$$p^1_{x,y_1} = 1, \ \ p^1_{x,y_2} = 0, \ \ p^1_{x,y_3} = 0, \ \ \ldots, \ \ p^1_{x,y_{r(x)}} = 0;$$

$$p^2_{x,y_1} = 0, \ \ p^2_{x,y_2} = 1, \ \ p^2_{x,y_3} = 0, \ \ \ldots, \ \ p^2_{x,y_{r(x)}} = 0;$$

....

$$p^{r(x)}_{x,y_1} = 0, \ \ p^{r(x)}_{x,y_2} = 0, \ \ p^{r(x)}_{x,y_3} = 0, \ \ \ldots, \ \ p^{r(x)}_{x,y_{r(x)}} = 1;$$

So, cyclic games can be regarded as stochastic positional games with average payoff functions where $p^a_{x,y} = 0 \vee 1, \ \forall x, y \in X, \ \forall a \in A$.

As we have noted for cyclic games a Nash equilibrium may not exist. An example of a cyclic game for which no Nash equilibrium does exist is given in [43, 79]. We can obtain a Nash equilibrium existence criterion from Theorem 3.4.

Theorem 3.9 *If for an arbitrary situation of the game* $s = (s^1, s^2, \dots, s^m)$ *the corresponding graph* $G_s = (X, E_s)$ *contains a unique directed cycle such that it can be reached from an arbitrary* $x \in X$ *then in the cyclic game there exists a Nash equilibrium* $s^{1*}, s^{2*}, \dots, s^{m*}$. *Moreover* $s^{1*}, s^{2*}, \dots, s^{m*}$ *is a Nash equilibrium for an arbitrary starting position* $x \in X$ *of the game.*

We obtain this theorem as a corollary from Theorem 3.4 if we regard cyclic games as a special case of unichain stochastic positional games.

In the following we formulate a necessary and sufficient condition for the existence of Nash equilibria in so-called ergodic cyclic games with m players. To formulate this result we need the following definition.

Definition 3.10 Let $s^{1*}, s^{2*}, \dots, s^{m*}$ be a solution in the sense of Nash for a cyclic game determined by a network $(G, \{X_i\}_{i=\overline{1,m}}, \{c^i\}_{i=\overline{1,m}}, x_0)$, where $G = (X, E)$ is a strongly connected directed graph. We call this game an ergodic cyclic game if $s^{1*}, s^{2*}, \dots, s^{m*}$ represents the solution in the sense of Nash for a cyclic game on the network $(G, \{X_i\}_{i=\overline{1,m}}, \{c^i\}_{i=\overline{1,m}}, x)$ with an arbitrary starting position $x \in X$ and

$$F_x^i(s^{1*}, s^{2*}, \dots, s^{m*}) = F_y^i(s^{1*}, s^{2*}, \dots, s^{m*}), \quad \forall x, y \in X,\ i = 1, 2, \dots, m.$$

Theorem 3.11 *The dynamic c-game determined by the network* $(G,[4]\{X_i\}_{i=\overline{1,m}}, \{c^i\}_{i=\overline{1,m}}, x_0)$, *where* $G = (X, E)$ *is a strongly connected directed graph, is an ergodic one if and only if there exist on* X m *real functions*

$$\varepsilon^1 : X \to \mathbb{R}^1,\ \varepsilon^2 : X \to \mathbb{R}^1, \dots,\ \varepsilon^m : X \to \mathbb{R}^1,$$

and m *values* $\omega^1, \omega^2, \dots, \omega^m$ *such that the following conditions are satisfied:*

(a) $\varepsilon_y^i - \varepsilon_x^i + c_{x,y}^i - \omega^i \geq 0, \quad \forall (x, y) \in E_i,$
where $E_i = \{e = (x, y) \in E \mid x \in X_i\},\ i = 1, 2, \dots, m;$

(b) $\min_{y \in X_G(x)} \{\varepsilon_y^i - \varepsilon_x^i + c_{x,y}^i - \omega^i\} = 0, \quad \forall x \in X_i,\ i = 1, 2, \dots, m;$

(c) *the subgraph* $G' = (X, E')$ *generated by edge set* $E' = E_1^0 \cup E_2^0 \cup \dots \cup E_m^0$, $E_i^0 = \{e = (x, y) \in E_i \mid \varepsilon_y^i - \varepsilon_x^i + c_{x,y}^i - \omega^i = 0\}$, $i = 1, 2, \dots, m$, *has the property that it contains a connected subgraph* $\overline{G}^0 = (X, \overline{E}^0)$, *for which every vertex* $x \in X$ *has only one leaving edge* $e = (x, y) \in \overline{E}^0$ *and besides that*

$$\varepsilon_y^i - \varepsilon_x^i + c_{x,y}^i - \omega^i = 0, \quad \forall (x, y) \in \overline{E}^0,\ i = 1, 2, \dots, m.$$

The optimal solution of the problem can be determined by fixing the maps:

$$s^{1*} : x \to y \in X_{\overline{G}^0}(x) \quad for \quad x \in X_1;$$
$$s^{2*} : x \to y \in X_{\overline{G}^0}(x) \quad for \quad x \in X_2;$$
$$\vdots$$
$$s^{m*} : x \to y \in X_{\overline{G}^0}(x) \quad for \quad x \in X_m,$$

where $X_{\overline{G}^0}(x) = \{y \mid (x, y) \in \overline{E}^0\}$, *and*

$$F^i_{\overline{x}}(s^{1*}, s^{2*}, \dots, s^{m*}) = \omega^i, \quad \forall \overline{x} \in X, \ i = 1, 2, \dots, m.$$

Proof $\Longrightarrow$ Let $(G, \{X_i\}_{i=\overline{1,m}}, \{c^i\}_{i=\overline{1,m}}, x_0)$ be a network which determines an *ergodic cyclic game* with m players, i.e., in this game there exists a Nash equilibrium $s^{1*}, s^{2*}, \dots, s^{m*}$. Define

$$\omega^i = F^i_{x_0}(s^{1*}, s^{2*}, \dots, s^{m*}), \quad i = 1, 2, \dots, m. \tag{3.12}$$

It is easy to verify that if we change the cost function c^i by $\overline{c}^i = c^i - \omega^i$, $i = 1, 2, \dots, m$, then the obtained network $(G, \{X_i\}_{i=\overline{1,m}}, \{\overline{c}^i\}_{i=\overline{1,m}}, x_0)$ determines a new ergodic cyclic game which is equivalent to the initial one. For the new game $s^{1*}, s^{2*}, \dots, s^{m*}$ is a Nash equilibrium and

$$F^i_{x_0}(s^{1*}, s^{2*}, \dots, s^{m*}) = 0, \quad i = 1, 2, \dots, m.$$

This game can be regarded as the dynamic c-game from [9, 76] on the network $(G, \{X_i\}_{i=\overline{1,m}}, \{\overline{c}^i\}_{i=\overline{1,m}}, x_0)$ with a given starting position x_0 and a final position $x_0 \in C_{s^*}$, where C_{s^*} is a directed cycle generated by strategies $s^{1*}, s^{2*}, \dots, s^{m*}$ such that

$$\sum_{e \in E(C_{s^*})} \overline{c}^i_e = 0, \quad i = 1, 2, \dots, m.$$

Taking into account that our game is ergodic we may state, without loss of generality, that x_0 belongs to a directed cycle C_{s^*} generated by strategies $s^{1*}, s^{2*}, \dots, s^{m*}$. Therefore, the ergodic game with a network $(G, \{X_i\}_{i=\overline{1,m}}, \{\overline{c}^i\}_{i=\overline{1,m}}, x_0)$ can be regarded as the dynamic c-game from [9, 76] on the network $(G, \{X_i\}_{i=\overline{1,m}}, \{\overline{c}^i\}_{i=\overline{1,m}}, x_0)$ with a starting position x_0 and a final position x_0. So, according to Theorem 2 from [9] there exist m real functions

$$\varepsilon^1 : X \to \mathbb{R}^1, \ \varepsilon^2 : X \to \mathbb{R}^1, \dots, \ \varepsilon^m : X \to \mathbb{R}^1,$$

such that the following conditions are satisfied:

(1) $\varepsilon_y^i - \varepsilon_x^i + \overline{c}_{x,y}^i \geq 0, \quad \forall (x,y) \in E_i, \ i = 1, 2, \ldots, m;$

(2) $\min\limits_{x \in X_G(x)} \{\varepsilon_y^i - \varepsilon_x^i + \overline{c}_{x,y}^i\} = 0, \quad \forall x \in X_i, \ i = 1, 2, \ldots, m;$

(3) the subgraph $G' = (X, E')$, generated by the edge set $E' = E_1^0 \cup E_2^0 \cup \cdots \cup E_m^0$, $E_i^0 = \{e = (x,y) \in E_i \mid \varepsilon_y^i - \varepsilon_x^i + \overline{c}_{x,y}^i = 0\}, i = 1, 2, \ldots, m$, has the property that it contains a connected subgraph $\overline{G}^0 = (X, \overline{E}^0)$, for which every vertex $x \in X$ has only one leaving edge $e = (x,y) \in \overline{E}^0$ and besides that

$$\varepsilon_y^i - \varepsilon_x^i + \overline{c}_{x,y}^i = 0, \quad \forall (x,y) \in \overline{E}^0, \ i = 1, 2, \ldots, m.$$

If in the conditions (1), (2) and (3) mentioned above we take into account that $\overline{c}_{x,y}^i = \overline{c}_{x,y}^i - \omega^i, \ \forall (x,y) \in E, \ i = 1, 2, \ldots, m$, then we obtain the conditions (a), (b) and (c) from Theorem 3.11.

$\Longleftarrow$ Assume that the conditions (a), (b) and (c) of Theorem 3.11 hold. Then for the network $(G, \{X_i\}_{i=\overline{1,m}}, \{\overline{c}^i\}_{i=\overline{1,m}}, x_0)$ the conditions (1), (2) and (3) are satisfied. It is easy to check that an arbitrary set of strategies $s^{1*}, s^{2*}, \ldots, s^{m*}$, where

$$s^{i*} : x \to y \in X_{\overline{G}^0}(x), \quad i = 1, 2, \ldots, m,$$

is a Nash equilibrium for an ergodic cyclic game on the network $(G, \{X_i\}_{i=\overline{1,m}}, \{\overline{c}^i\}_{i=\overline{1,m}}, x_0)$ and

$$F_x^i(s^{1*}, s^{2*}, \ldots, s^{m*}) = 0, \quad i = 1, 2, \ldots, m.$$

This implies that $s^{1*}, s^{2*}, \ldots, s^{m*}$ determines a Nash equilibrium for the ergodic cyclic game on the network $(G, \{X_i\}_{i=\overline{1,m}}, \{c^i\}_{i=\overline{1,m}}, x_0)$. □

Remark 3.12 The value ω^i, $i = 1, 2, \ldots, m$, coincides with the value of $F_x^i(s^{1*}, s^{2*}, \ldots, s^{m*})$, $i = 1, 2, \ldots, m$. If $\omega^i = 0$, then the ergodic cyclic game coincides with the dynamic c-game on the network $(G, \{X_i\}_{i=\overline{1,m}}, \{c^i\}_{i=\overline{1,m}}, x_0)$.

Note that for ergodic zero-sum games Theorem 3.11 gives necessary and sufficient conditions for the existence of saddle points, i.e., Theorem 3.11 extends results from [43] to cyclic games with m players.

3.1.6 Average Stochastic Positional Games on Networks

In a similar way as for average Markov decision problems we can apply the game-theoretical concept to average stochastic control problems on networks. This leads to a new game-theoretic control model on networks which we call *average stochastic positional games on networks*. We formulate this model as a stochastic version of the cyclic game. We assume that the graph of states' transitions $G = (X, E)$ along the subsets $X_1, X_2, \ldots, X_m$ contains a subset X^0 for which the transitions from the states $x \in X^0$ to another state $y \in X$ are made in a random way.

Thus, for the set of states X is given by a partition $X_1, X_2, \ldots, X_m, X^0$, where X_i represents the position set of the player $i \in \{1, 2, \ldots, m\}$ and X^0 is the subset of states of X for which the probability distributions $p_{x,y}$ of the system's transition from every $x \in X^0$ to another state $y \in X$ are given. This means that to the directed edges $e = (x, y)$ that originate in a state $x \in X^0$ the probabilities $p_{x,y}$ are associated, where

$$\sum_{y \in X(x)} p_{x,y} = 1, \ \forall x \in X^0.$$

In addition to each directed edge $e = (x, y) \in E$ m costs $c_e^1, c_e^2, \ldots c_e^m$ are associated, where c_e^i represents the cost of the system's transition from the state x to the state y for the player $i \in \{1, 2, \ldots, m\}$. In a similar way as for cyclic games we define here the stationary strategies of the players as m maps:

$$s^i : x \to y \in X(x) \ \text{ for } \ x \in X_i, \ \ i = 1, 2, \ldots, m.$$

Let $s^1, s^2, \ldots, s^m$ be a set of stationary strategies of the players. Then we can determine the graph $G_s{=}(X, \ E_s{\cup}E^0)$, induced by the situation $s = (s^1, s^2, \ldots, s^m)$, where

$$E_s = \{e = (x, y) \in E \,|\, x \in X_i, \ y = s^i(x), \ i = 1, 2, \ldots, m\}$$

and

$$E^0 = \{e = (x, y) \,|\, x \in X^0, \ y \in X\}.$$

This graph corresponds to a Markov process with the probability matrix $P^s = (p^s_{x,y})$, where

$$p^s_{x,y} = \begin{cases} p_{x,y}, & \text{if } x \in X^0 \text{ and } y = X; \\ 1, & \text{if } x \in X_i \text{ and } y = s^i(x); \\ 0, & \text{otherwise.} \end{cases}$$

In the considered game the transitions $(x, s^i(x))$ from the states $x \in X_i$ to the states $y = s^i(x) \in X$ are described by the probability $p_{x,s^i(x)} = 1$ if the players $1, 2, \ldots, m$ fix their stationary strategies $s^1, s^2, \ldots, s^m$, respectively. For this Markov process we can determine the average costs per transition $\omega_x^i(s^1, s^2, \ldots, s^m)$ with respect to each player $i \in \{1, 2, \ldots, m\}$ for an arbitrary fixed starting state $x \in X$. We regard these functions as payoff functions for the model of the control problem on the network, i.e.,

$$F^i_{x0}(s^1, s^2, \dots, s^m) = \omega^i_{x0}(s^1, s^2, \dots, s^m), \quad i = 1, 2, \dots, m.$$

We denote this game by $(G, \{X_i\}_{i=\overline{1,m}}, X^0, \{c^i\}_{i=\overline{1,m}}, p, x_0)$, where $G = (X, E)$ is the graph of states' transitions of the dynamical system and $p : E^0 \to [0, 1]$ is the probability transition function defined on the set $E^0 = \{e = (x, y) \in E \,|\, x \in X^0,\ y \in X\}$ that satisfies the condition $\sum_{y \in X(x)} p_{x,y} = 1,\ \forall x \in X^0$.

For this discrete game we can use the continuous game model with the continuous payoff functions (3.10), (3.11) and the compact sets $S^1, S^2, \dots, S^m$ determined by the set of the solutions of the systems (3.9). If in this continuous model we specify the expressions of the payoff function and the set of strategies in accordance with the structure of the network then we obtain the continuous model, where $S^1, S^2, \dots, S^m$ and $F^i_{x0}(s^1, s^2, \dots, s^m) = \psi^i(s^1, s^2, \dots, s^m)$ are defined as follows:

Each subset S^i, $i \in \{1, 2, \dots, m\}$ represents the set of solutions of the system

$$\begin{cases} \sum\limits_{y \in X(x)} s^i_{x,y} = 1, \quad \forall x \in X_i; \\ s^i_{x,y} \geq 0, \qquad \forall x \in X_i,\ y \in X(x), \end{cases} \tag{3.13}$$

and

$$\begin{aligned} \psi^i(s^1, s^2, \dots, s^m) = & \sum_{j=1}^{m} \sum_{x \in X_j} \sum_{y \in X(x)} c^i_{x,y} s^j_{x,y} q_x \\ & + \sum_{x \in X^0} \sum_{y \in X(x)} \mu^i_{x,y} q_x, \quad i = 1, 2, \dots, m, \end{aligned} \tag{3.14}$$

where q_x for $x \in X$ are determined from the following system of linear equations

$$\begin{cases} \sum\limits_{x \in X} \sum\limits_{y \in X(x)} p_{x,y} q_x = q_y, \quad \forall y \in X^0; \\ \sum\limits_{x \in X} s^i_{x,y} q_x = q_y, \quad \forall y \in X_i,\ i = 1, 2, \dots, m; \\ \sum\limits_{x \in X} q_x = 1. \end{cases} \tag{3.15}$$

According to Theorem 3.4 in the case of perfect Markov processes for the game with payoff functions (3.14), (3.15) on the compact sets determined by (3.13) there exists a Nash equilibrium $s^* = (s^{1^*}, s^{2^*}, \dots, s^m)$, where each s^{i^*} corresponds to a basic solution of the system (3.13). So, we can use the continuous model (3.13–3.15) for determining a Nash equilibrium for a discrete game on the network.

As a corollary from Theorem 3.8 we obtain the following result.

Theorem 3.13 *Let $(G, \{X_i\}_{i=\overline{1,m}}, X^0, \{c^i\}_{i=\overline{1,m}}, p, \overline{x})$ be a stochastic positional game with a given starting position $x \in X$ and average payoff functions*

$$F_{\overline{x}}^1(s^1, s^2, \dots, s^m),\ F_{\overline{x}}^2(s^1, s^2, \dots, s^m), \dots, F_{\overline{x}}^m(s^1, s^2, \dots, s^m)$$

of the players $1, 2, \dots, m$, respectively. Assume that for an arbitrary situation $s = (s^1, s^2, \dots, s^m)$ of the game the transition probability matrix $P^s = (p_{x,y}^s)$ induces a Markov unichain. Then there exist the functions

$$\varepsilon^i : X \to R,\ i = 1, 2, \dots, m$$

and the values $\omega^1, \omega^2, \dots, \omega^m$ that satisfy the following conditions:

(1) $\varepsilon_y^i - \varepsilon_x^i + c_{x,y}^i - \omega^i \ge 0, \quad \forall x \in X_i, \quad i = 1, 2, \dots, m;$

(2) $\min\limits_{y \in X(x)} \{\varepsilon_y^i - \varepsilon_x^i + c_{x,y}^i - \omega^i\} = 0, \quad \forall x \in X_i, \quad i = 1, 2, \dots, m;$

(3) $\varepsilon_y^i - \sum\limits_{y \in X} p_{x,y} \varepsilon_x^i + \mu_x^i - \omega^i = 0, \quad \forall x \in X^0, \quad i = 1, 2, \dots, m;$

(4) *On each position set X_i, $i \in \{1, 2, \dots, m\}$ there exists a map $s^{i^*} : X_i \to A$ such that*

$$s^{i^*}(x) = y^* \in \operatorname*{argmin}_{y \in A(x)} \left\{\varepsilon_y^i - \varepsilon_x^i + c_{x,y}^i - \omega^i\right\}, \quad \forall x \in X_i$$

and

$$\varepsilon_{y^*}^j - \varepsilon_x^j + c_{x,y^*}^j - \omega^j = 0, \quad \forall x \in X_i,\ j = 1, 2, \dots, m.$$

The set of maps $s^{1^*}, s^{2^*}, \dots, s^{m*}$ determines a Nash equilibrium situation $s^* = (s^{1^*}, s^{2^*}, \dots, s^{m*})$ for the stochastic positional game $(G, \{X_i\}_{i=\overline{1,m}}, X^0, \{c^i\}_{i=\overline{1,m}}, p, \overline{x})$ and

$$F_{\overline{x}}^i(s^{1^*}, s^{2^*}, \dots, s^{m*}) = \omega^i, \quad \forall \overline{x} \in X,\ i = 1, 2, \dots, m.$$

Moreover, the situation $s^* = (s^{1^*}, s^{2^*}, \dots, s^{m*})$ is a Nash equilibrium for an arbitrary starting position $\overline{x} \in X$.

Obviously, the average game control model on the network can be represented as an average stochastic positional game. On the other hand an arbitrary stochastic positional game can be reduced to a game-theoretic control model on networks using the construction from Sect. 2.3.2. Thus, the stochastic positional game is equivalent to the game-theoretic control problem on a certain network.

3.1.7 Saddle Point Conditions for Average Stochastic Antagonistic Positional Games and Determining the Optimal Strategies of the Players

We obtain the *stochastic antagonistic positional game with an average payoff function* from the game-theoretic model from Sect. 3.1 in the case $m = 2$ and $c = c^2 = -c^1$. So, this game is determined by the finite sets of strategies of the players $\mathbb{S}^1, \mathbb{S}^2$ and the payoff function $F_{x_0} : \mathbb{S}^1 \times \mathbb{S}^2 \to \mathbb{R}$, where $F_{x_0}(s^1, s^2)$ for a fixed strategies $s^1 \in \mathbb{S}^1, s^2 \in \mathbb{S}^2$ is equal to an average cost per transition in the Markov process induced by the strategies s^1 and s^2, if the system starts transitions in x_0, i.e.,

$$F_{x_0}(s^1, s^2) = \omega_{x_0}(s^1, s^2).$$

We denote this game by $(X, A, X_1, X_2, c, p, x_0)$, where $c = c^2 = -c^1$ and x_0 is the starting position of the game.

Based on Theorem 3.4 we may conclude that in the case of unichain Markov processes for stochastic antagonistic positional games a *saddle point* exists and the corresponding conditions for determining the optimal stationary strategies of the players can be derived from Theorem 3.8.

Below we show that in the stochastic antagonistic positional game the saddle point exists for an arbitrary Markov decision process [90]. So, we prove that for an arbitrary stochastic antagonistic positional game $(X, A, X_1, X_2, c, p, x_0)$ with an average payoff function $F_x(s^1, s^2)$ there exist the stationary strategies s^{1*}, s^{2*} for which

$$F_{x_0}(s^{1*}, s^{2*}) = \max_{s^1 \in \mathbb{S}^1} \min_{s^2 \in \mathbb{S}^2} F_{x_0}(s^1, s^2) = \min_{s^2 \in \mathbb{S}^2} \max_{s^1 \in \mathbb{S}^1} F_{x_0}(s^1, s^2). \tag{3.16}$$

We prove the existence of a saddle point for the payoff function $F_{x_0}(s^{1*}, s^{2*})$ by using the results from Sect. 2.3.5.

Theorem 3.14 *Let $(X, A, X_1, X_2, c, p, \overline{x})$ be an arbitrary stochastic positional game with an average payoff function $F_{\overline{x}}(s^1, s^2)$. Then the system of equations*

$$\begin{cases} \varepsilon_x + \omega_x = \max\limits_{a \in A(x)} \left\{ \mu_{x,a} + \sum\limits_{y \in X} p^a_{x,y} \varepsilon_y \right\}, & \forall x \in X_1; \\ \varepsilon_x + \omega_x = \min\limits_{a \in A(x)} \left\{ \mu_{x,a} + \sum\limits_{y \in X} p^a_{x,y} \varepsilon_y \right\}, & \forall x \in X_2 \end{cases} \tag{3.17}$$

has a solution under the set of solutions of the system of equations

$$\begin{cases} \omega_x = \max\limits_{a\in A(x)} \left\{ \sum\limits_{y\in X} p^a_{x,y}\omega_x \right\}, & \forall x \in X_1; \\ \omega_x = \min\limits_{a\in A(x)} \left\{ \sum\limits_{y\in X} p^a_{x,y}\omega_x \right\}, & \forall x \in X_2, \end{cases} \tag{3.18}$$

i.e., the system of equations (3.18) *has such a solution* ω^*_x, $x \in X$ *for which there exists a solution* ε^*_x, $x \in X$ *of the system of equations*

$$\begin{cases} \varepsilon_x + \omega^*_x = \max\limits_{a\in A(x)} \left\{ \mu_{x,a} + \sum\limits_{y\in X} p^a_{x,y}\varepsilon_y \right\}, & \forall x \in X_1; \\ \varepsilon_x + \omega^*_x = \min\limits_{a\in A(x)} \left\{ \mu_{x,a} + \sum\limits_{y\in X} p^a_{x,y}\varepsilon_y \right\}, & \forall x \in X_2. \end{cases} \tag{3.19}$$

The optimal stationary strategies of the players

$$\begin{aligned} s^{1*} &: x \to a^1 \in A(x) \quad \text{for } x \in X_1; \\ s^{2*} &: x \to a^2 \in A(x) \quad \text{for } x \in X_2 \end{aligned}$$

in the stochastic positional game can be found by fixing the arbitrary maps $s^{1*}(x) \in A(x)$ *for* $x \in X_1$ *and* $s^{1*}(x) \in A(x)$ *for* $x \in X_2$ *such that*

$$s^{1*}(x) \in \left(\operatorname*{argmax}_{a\in A(x)} \left\{ \sum_{y\in X} p^a_{x,y}\omega^*_x \right\} \right) \bigcap \left(\operatorname*{argmax}_{a\in A(x)} \left\{ \mu_{x,a} + \sum_{y\in X} p^a_{x,y}\varepsilon^*_y \right\} \right), \quad \forall x \in X_1 \tag{3.20}$$

and

$$s^{2*}(x) \in \left(\operatorname*{argmin}_{a\in A(x)} \left\{ \sum_{y\in X} p^a_{x,y}\omega^*_x \right\} \right) \bigcap \left(\operatorname*{argmin}_{a\in A(x)} \left\{ \mu_{x,a} + \sum_{y\in X} p^a_{x,y}\varepsilon^*_y \right\} \right). \quad \forall x \in X_2 \tag{3.21}$$

For the fixed optimal strategies s^{1*}, s^{2*} *the corresponding values of the payoff function* $F_{\overline{x}}(s^{1*}, s^{2*})$ *coincide with the values* $\omega^*_{\overline{x}}$ *for* $\overline{x} \in X$ *and* (3.16) *holds.*

Proof Let $\overline{x} \in X$ be an arbitrary state and consider the stationary strategies $\overline{s}^1 \in \mathbb{S}^1$, $\overline{s}^2 \in \mathbb{S}^2$ for which

$$F_{\overline{x}}(\overline{s}^1, \overline{s}^2) = \min_{s^2\in\mathbb{S}^2} \max_{s^1\in\mathbb{S}^1} F_{\overline{x}}(s^1, s^2).$$

We show that

$$F_{\overline{x}}(\overline{s}^1, \overline{s}^2) = \max_{s^1 \in \mathbb{S}^1} \min_{s^2 \in \mathbb{S}^2} F_{\overline{x}}(s^1, s^2),$$

i.e., we show that (3.16) holds and $\overline{s}^1 = s^{1*}$, $\overline{s}^2 = s^{2*}$.

According to Theorem 2.32 and Corollary 2.33 for the situation $\overline{s} = (\overline{s}^1, \overline{s}^2)$ the system of linear equations

$$\begin{cases} \varepsilon_x + \omega_x = \mu_{x,a} + \sum_{y \in X} p_{x,y}^a \varepsilon_y, & \forall x \in X_1, a = \overline{s}^1(x); \\ \varepsilon_x + \omega_x = \mu_{x,a} + \sum_{y \in X} p_{x,y}^a \varepsilon_y, & \forall x \in X_2, a = \overline{s}^2(x); \\ \omega_x = \sum_{y \in X} p_{x,y}^a \omega_x, & \forall x \in X_1, a = \overline{s}^1(x); \\ \omega_x = \sum_{y \in X} p_{x,y}^a \omega_x, & \forall x \in X_2, a = \overline{s}^2(x) \end{cases} \tag{3.22}$$

has the solution ε_x^*, ω_x^* $(x \in X)$ which for a fixed strategy $\overline{s}^2 \in \mathbb{S}^2$ satisfies the condition:

$$\begin{cases} \varepsilon_x^* + \omega_x^* \geq \mu_{x,a} + \sum_{y \in X} p_{x,y}^a \varepsilon_y^*, & \forall x \in X_1, \ a \in A(x); \\ \varepsilon_x^* + \omega_x^* = \mu_{x,a} + \sum_{y \in X} p_{x,y}^a \varepsilon_y^*, & \forall x \in X_2, \ a = \overline{s}^2(x); \\ \omega_x^* \geq \sum_{y \in X} p_{x,y}^a \omega_x^*, & \forall x \in X_1, \ a \in A(x); \\ \omega_x^* = \sum_{y \in X} p_{x,y}^a \omega_x^*, & \forall x \in X_2, \ a = \overline{s}^2(x) \end{cases}$$

and $F_x(\overline{s}^1, \overline{s}^2) = \omega_x^*, \ \forall x \in X$.

Taking into account that $F_x(\overline{s}^1, \overline{s}^2) = \min_{s^2 \in \mathbb{S}^2} F_x(\overline{s}^1, s^2)$ then for a fixed strategy $\overline{s}^1 \in \mathbb{S}^1$ the solution ε_x^*, ω_x^* $(x \in X)$ satisfies the condition

$$\begin{cases} \varepsilon_x^* + \omega_x^* = \mu_{x,a} + \sum_{y \in X} p_{x,y}^a \varepsilon_y^*, & \forall x \in X_1, \ a = \overline{s}^1(x); \\ \varepsilon_x^* + \omega_x^* \leq \mu_{x,a} + \sum_{y \in X} p_{x,y}^a \varepsilon_y^*, & \forall x \in X_2, \ a \in A(x); \\ \omega_x^* = \sum_{y \in X} p_{x,y}^a \omega_x^*, & \forall x \in X_1, \ a = \overline{s}^1(x); \\ \omega_x^* \leq \sum_{y \in X} p_{x,y}^a \omega_x^*, & \forall x \in X_2, \ a \in A(x) \end{cases}$$

So, the following system

$$\begin{cases} \varepsilon_x + \omega_x \geq \mu_{x,a} + \sum\limits_{y \in X} p_{x,y}^a \varepsilon_y, & \forall x \in X_1, \ a \in A(x); \\ \varepsilon_x + \omega_x \leq \mu_{x,a} + \sum\limits_{y \in X} p_{x,y}^a \varepsilon_y, & \forall x \in X_2, \ a \in A(x); \\ \omega_x \geq \sum\limits_{y \in X} p_{x,y}^a \omega_x, & \forall x \in X_1, \ a \in A(x)(x); \\ \omega_x \leq \sum\limits_{y \in X} p_{x,y}^a \omega_x, & \forall x \in X_2, \ a \in A(x) \end{cases}$$

has a solution, which satisfies condition (3.22).

This means that $s^{1*} = \overline{s}^1$, $s^{2*} = \overline{s}^2$ and

$$\max_{s^1 \in \mathbb{S}^1} \min_{s^2 \in \mathbb{S}^2} F_{\overline{x}}(s^1, s^2) = \min_{s^2 \in \mathbb{S}^2} \max_{s^1 \in \mathbb{S}^1} F_{\overline{x}}(s^1, s^2), \ \forall \overline{x} \in X,$$

i.e., the theorem holds. □

Thus, the optimal stationary strategies s^{1*}, s^{2*} of the players for an average antagonistic positional game can be found by using the solutions of the system of Eqs. (3.17), (3.19) and the conditions (3.20), (3.21). The solution of the system (3.17), (3.19) that satisfies the conditions (3.20), (3.21) can be found by using iterative algorithms like algorithms for decision problems from Sect. 2.4.

Algorithm 3.15 Determining the Optimal Stationary Strategies of the Players in Antagonistic Games with an Average Payoff Function

Preliminary step (*Step 0*): Fix the arbitrary stationary strategies

$$s_0^1 : x \to y \in X(x) \quad \text{for } x \in X_1,$$
$$s_0^2 : x \to y \in X(x) \quad \text{for } x \in X_2.$$

that determine the situation $s_0 = (s_0^1, s_0^2)$.

General step (*Step k*, $k \geq 1$): Determine the matrix $P^{s_{k-1}}$ that corresponds to the situation $s_k = (s_{k-1}^1, s_{k-1}^2)$ and find $\omega^{s_{k-1}}$ and $\varepsilon^{s_{k-1}}$ which satisfy the conditions

$$\begin{cases} (P^{s_{k-1}} - I)\omega^{s_{k-1}} = 0; \\ \mu^{s_{k-1}} + (P^{s_{k-1}} - I)\varepsilon^{s_{k-1}} - \omega^{s_{k-1}} = 0. \end{cases}$$

Then find a situation $s_k = (s_k^1, s_k^2)$ such that

$$s_k^1(x) \in \underset{a\in A(x)}{\operatorname{argmax}} \left\{ \sum_{y\in X} p_{x,y}^a \omega_x^{s_{k-1}^1} \right\}, \quad \forall x \in X_1;$$

$$s_k^2(x) \in \underset{a\in A(x)}{\operatorname{argmax}} \left\{ \sum_{y\in X} p_{x,y}^a \omega_x^{s_{k-1}^2} \right\}, \quad \forall x \in X_2$$

and set $s_k = s_{k-1}$ if

$$s_{k-1}^1(x) \in \underset{a\in A(x)}{\operatorname{argmax}} \left\{ \sum_{y\in X} p_{x,y}^a \omega_x^{s_{k-1}^1} \right\}, \quad \forall x \in X_1;$$

$$s_{k-1}^2(x) \in \underset{a\in A(x)}{\operatorname{argmax}} \left\{ \sum_{y\in X} p_{x,y}^a \omega_x^{s_{k-1}^2} \right\}, \quad \forall x \in X_2.$$

After that check if $s_k = s_{k-1}$? If $s_k = s_{k-1}$ then go to the next step $k+1$; otherwise choose a situation $s_k = (s_k^1, s_k^2)$ such that

$$s_k^1(x) \in \underset{a\in A(x)}{\operatorname{argmax}} \left\{ \mu_{x,a} + \sum_{y\in X} p_{x,y}^a \varepsilon_y^{s_{k-1}^1(x)} \right\} \quad \forall x \in X_1;$$

$$s_k^2(x) \in \underset{a\in A(x)}{\operatorname{argmax}} \left\{ \mu_{x,a} + \sum_{y\in X} p_{x,y}^a \varepsilon_y^{s_{k-1}^2(x)} \right\} \quad \forall x \in X_2$$

and set $s_k = s_{k-1}$ if

$$s_{k-1}^1(x) \in \underset{a\in A(x)}{\operatorname{argmax}} \left\{ \mu_{x,a} + \sum_{y\in X} p_{x,y}^a \varepsilon_y^{s_{k-1}^1(x)} \right\} \quad \forall x \in X_1;$$

$$s_{k-1}^2(x) \in \underset{a\in A(x)}{\operatorname{argmax}} \left\{ \mu_{x,a} + \sum_{y\in X} p_{x,y}^a \varepsilon_y^{s_{k-1}^2(x)} \right\} \quad \forall x \in X_2.$$

After that check if $s_k = s_{k-1}$? If $s_k = s_{k-1}$ then STOP and set $s^{1*} = s_{k-1}^1$, $s^{2*} = s_{k-1}^2$; otherwise go to the next step $k+1$.

Remark 3.16 If $p_{x,y} \in \{0.1\}$, $\forall x, y \in X$ then the algorithm is transformed for determining the optimal stationary strategies of the players in deterministic parity games.

The convergence of the algorithm described above can be determined in a similar way as the convergence of the iterative algorithm for determining the optimal solution of the Markov decision problem with an average cost optimization criteria (see [115, 140]).

3.1.8 Saddle Point Conditions for Average Stochastic Antagonistic Positional Games on Networks

We obtain the *average stochastic positional games on networks* from the game-theoretic control model on the network from Sect. 3.1.6 in the case $m = 2$ and $c = c^2 = -c^1$. This stochastic antagonistic game is determined by a tuple $(G, X_1, X_2, X^0, c^i, p, x)$, where $G = (X_1 \cup X_2 \cup X^0, E)$ is the graph of the states' transitions of the dynamical system with the cost function $c : E \to \mathbb{R}$, the set of positions of the first player X_1, the set of positions of the second player X_2 and the probability function $p : E^0 \to [0, 1]$ on $E^0 = \{e = (x, y) \in E \mid x \in X^0, y \in X\}$ that satisfy the condition $\sum_{y \in X(x)} p_{x,y} = 1, \ \forall x \in X^0$. The stationary strategies of the players are determined by the maps

$$s^1 : x \to y \in X(x) \text{ for } x \in X_1;$$
$$s^2 : x \to y \in X(x) \text{ for } x \in X_2$$

and for the average payoff function we have

$$F_x(s^1, s^2) = F_x^2(s^1, s^2) = -F_x^1(s^1, s^2).$$

Using the results from Sect. 3.1.6 we may conclude that the saddle point for this antagonistic game exists. Moreover, from Theorem 3.14 we obtain the following result.

Theorem 3.17 *Let an average stochastic antagonistic positional game* $(G, X_1, X_2, X^0, c^i, p, \overline{x})$ *be given. Then there exist a function* $\varepsilon : X \to \mathbb{R}$ *and the values* ω_x *for* $x \in X$ *that satisfy the following conditions:*

(1) $\varepsilon_y - \varepsilon_x + c_{x,y} - \omega_x \le 0, \ \ \forall x \in X_1;$

(2) $\max\limits_{y \in X(x)} \{\varepsilon_y - \varepsilon_x + c_{x,y} - \omega_x\} = 0, \qquad \forall x \in X_1;$

(3) $\varepsilon_y - \varepsilon_x + c_{x,y} - \omega_x \ge 0, \ \ \forall x \in X_2;$

(4) $\min\limits_{y \in X(x)} \{\varepsilon_y - \varepsilon_x + c_{x,y} - \omega_x\} = 0, \qquad \forall x \in X_2;$

(5) $\omega_x = \max\limits_{y \in X(x)} \omega_y, \ \ \forall x \in X_1;$

(6) $\omega_x = \min\limits_{y \in X(x)} \omega_y, \ \ \forall x \in X_2.$

The optimal stationary strategies s^{1^*} *and* s^{2^*} *of the players can be found by fixing the arbitrary maps*

$$s^{1^*} : x \to y \in X(x) \ \ \textit{for} \ \ x \in X_1;$$
$$s^{2^*} : x \to y \in X(x) \ \ \textit{for} \ \ x \in X_2$$

such that

$$(x, s^{1^*}(x)) \in E^1_\omega(x) \cap E^1_c(x), \quad \forall x \in X_1$$

and

$$(x, s^{2^*}(x)) \in E^2_\omega(x) \cap E^2_c(x), \quad \forall x \in X_2,$$

where

$$E^1_\omega(x) = \Big\{(x, y) \in E \mid y \in \underset{y \in X(x)}{\operatorname{argmax}}(\varepsilon_y - \varepsilon_x + c_{x,y} - \omega_x)\Big\}, \quad \textit{for} \quad x \in X_1;$$

$$E^1_c(x) = \Big\{(x, y) \in E \mid y \in \underset{y \in X(x)}{\operatorname{argmax}}\, \omega_y\Big\}, \quad \textit{for} \quad x \in X_1;$$

$$E^2_\omega(x) = \Big\{(x, y) \in E \mid y \in \underset{y \in X(x)}{\operatorname{argmin}}(\varepsilon_y - \varepsilon_x + c_{x,y} - \omega_x)\Big\}, \quad \textit{for} \quad x \in X_2;$$

$$E^2_c(x) = \Big\{(x, y) \in E \mid y \in \underset{y \in X(x)}{\operatorname{argmin}}\, \omega_y\Big\}, \quad \textit{for} \quad x \in X_2.$$

In addition $\omega_{\overline{x}} = F_{\overline{x}}(s^{1^*}, s^{2^*}), \ \forall \overline{x} \in X.$

We obtain the proof of this theorem if we specify the conditions of Theorem 3.17 for the game-theoretic control model on networks.

The considered stochastic antagonistic positional game on networks generalizes antagonistic deterministic cyclic games from [29, 43, 142]. The saddle point conditions for cyclic games proved in [43] can be obtained from Theorem 3.17 for the case $X^0 = \emptyset$.

3.1.9 Applications of Average Stochastic Positional Games for Studying Shapley Stochastic Games

In 1953 Lloyd Shapley introduced a class of stochastic games that generalizes Markov decision processes and repeated games (see [124]). Below we show that the stochastic positional games are tightly connected with Shapley stochastic games and can be used for studying some problems concerned with the existence of Nash equilibria in such games.

A stochastic game in the sense of Shapley is a dynamic game with probabilistic transitions played by several players in a sequence of stages, where the beginning of each stage corresponds to a state of the dynamical system. The game starts at a given state from the set of states of the system. At each stage the players select actions from their feasible sets of actions and each player receives a stage payoff that depends on the current state and the chosen actions. The game then moves to a new random state of which the distribution depends on the previous state and the actions chosen by the players. The procedure is repeated at the new state and the game continues for a finite or infinite number of stages. The total payoff of a player is either the limit inferior of the average of the stage payoffs or the discounted sum of the stage payoffs.

So, an average Shapley stochastic game with m players consists of the following elements:

1. A state space X (which we assume to be finite);
2. A finite set $A^i(x)$ of actions with respect to each player $i \in \{1, 2, \dots, m\}$ for an arbitrary state $x \in X$;
3. A stage payoff $f^i(x, a)$ with respect to each player $i \in \{1, 2, \dots, m\}$ for each state $x \in X$ and for an arbitrary action vector $a \in \prod_i A^i(x)$;
4. A transition probability function $p : X \times \prod_{x \in X} \prod_i A^i(x) \times X \rightarrow [0, 1]$ that gives the probability transitions $p^a_{x,y}$ from an arbitrary $x \in X$ to an arbitrary $y \in Y$ for a fixed action vector $a \in \prod_i A^i(x)$, where $\sum_{y \in X} p^a_{x,y} = 1, \ \forall x \in X, \ a \in \prod_i A^i(x)$;
5. A starting state $x_0 \in X$.

The stochastic game starts in state x_0. At stage t players observe state x_t and simultaneously choose actions $a^i_t \in A^i(x_t), \ i = 1, 2, \dots, m$. Then nature selects state x_{t+1} according to probability transitions $p^{a_t}_{x_t,y}$ for a fixed action vector $a_t = (a^1_t, a^2_t, \dots, a^m_t)$. A play of the stochastic game $x_0, a_0, x_1, a_1, \dots, x_t, a_t, \dots$ defines a stream of payoffs $f^i_0, f^i_1, f^i_2, \dots$, where

$$f^i_t = f^i(x_t, a_t), \quad t = 0, 1, 2, \dots.$$

The t-stage average stochastic game is the game where the payoff of player $i \in \{1, 2, \dots, m\}$ is

$$F^i_t = \frac{1}{t} \sum_{\tau=1}^{t-1} f^i_\tau.$$

The infinite average stochastic game is the game where the payoff of player $i \in \{1, 2, \dots, m\}$ is

$$\overline{F}^i = \lim_{t \to \infty} F^i_t.$$

In a similar way the Shapley stochastic game with expected discounted payoffs of the players can be defined. In such a game along to elements described above also a discount factor λ $(0 < \lambda < 1)$ is given and the total payoff of a player represents the expected discounted sum of the stage payoffs.

By comparison Shapley stochastic games with stochastic positional games we can observe the following: The probability transitions from a state to another state as well as the stage payoffs of the players in a Shapley stochastic game depend on the actions chosen by all players, while the probability transitions from a state to another state as well the stage payoffs (the immediate costs of the players) in a stochastic positional game depend only on the action of the player that controls the state in his position set. This means that a stochastic positional game can be regarded as a special case of the Shapley stochastic game. Nevertheless we can see that stochastic positional games can be used for studying some classes of Shapley stochastic games.

The main results concerned with determining Nash equilibria in Shapley stochastic games can be found in [35, 41, 65, 95, 104, 130]. The existence of Nash equilibria for such games are proven in the case of stochastic games with a finite set of stages and in the case of the games with infinite stages when the total payoff of each player is the discounted sum of stage payoffs. If the total payoff of a player represents the limit inferior of the average of the stage payoffs then the existence of Nash equilibrium in Shapley stochastic games is an open question.

Based on the results from previous sections we can show that in the case of the average non-antagonistic stochastic games a Nash equilibrium may not exist. In order to prove this we can use the average stochastic positional game $(X,\ A,\ \{X_i\}_{i=\overline{1,m}},\ \{c^i\}_{i=\overline{1,m}},\ p,\ x_0)$ from Sect. 3.1. It is easy to observe that this game can be regarded as a Shapley stochastic game with average payoff functions of the players, where for a fixed situation $s = (s^1, s^2, \ldots, s^m)$ the probability transition $p^s_{x,y}$ from a state $x = x(t) \in X_i$ to a state $y = x(t+1) \in X$ depends only on strategy s^i of player i and the corresponding stage payoff of player $i \in \{1, 2, \ldots, m\}$ is equal to $\mu^s_i = \sum_{y \in X} p^s_{x,y} c^i_{x,y}$. Taking into account that the cyclic game represents a particular case of the average stochastic positional game and for the cyclic game Nash equilibrium may not exist (see [43, 79]) we obtain that for the average non-antagonistic Shapley stochastic game Nash equilibrium may not exist.

Note that Theorem 3.4 can be extended for average Shapley stochastic games, i.e. we can ensure the existence of Nash equilibrium for such games in the case when an arbitrary situation induces a Markov unichain. Moreover, Theorems 3.6 and 3.8 and 3.14 also can be extended for Shapley stochastic games.

3.2 Stochastic Positional Games with Discounted Payoff Functions

In this section we apply the concept of noncooperative games to discounted Markov decision problems and formulate a new class of stochastic games which we call *stochastic positional games with discounted payoff functions* or *discounted stochastic positional games* [69, 91]. We show that for the considered class of games a Nash equilibrium always exists.

3.2.1 Problem Formulation

We formulate the stochastic positional game-theoretic model for the discounted Markov decision problem in a similar way as the game-theoretic model from the previous section. We apply the game-theoretical concept to a discounted Markov decision process (X, A, p, c) with finite set of states, finite set of actions A, the probability transition function $p : X \times X \times A \to \mathbb{R}$, the cost function $c : X \times X \to \mathbb{R}$ and a given discount factor γ, $0 < \gamma < 1$. So, for our game-theoretic model we assume that m transition cost functions $c^i : X \times X \to \mathbb{R}, i = 1, 2, \dots, m$, are given and the set of states X is divided into m disjoint subsets $X_1, X_2, \dots, X_m$, where X_i represents the positions set of the player $i \in \{1, 2, \dots, m\}$. Thus, the Markov process is controlled by m players, where each player $i \in \{1, 2, \dots, m\}$ fixes actions in his positions $x \in X_i$ using stationary strategies. We define the stationary strategies of the players in this game as m maps:

$$s^i : x \to a \in A(x) \quad \text{for} \quad x \in X_i; \quad i = 1, 2, \dots, m.$$

Let $s^1, s^2, \dots, s^m$ be a set of stationary strategies of the players that determine the situation $s = (s^1, s^2, \dots, s^m)$. Consider the matrix of probability transitions $P^s = (p^s_{x,y})$ which is induced by the situation s, i.e., each row of this matrix corresponds to a probability distribution $p^{s^i(x)}_{x,y}$ in the state x where $x \in X_i$. If the starting state x_0 is given, then for the Markov process with the matrix of probability transitions P^s we can determine the discounted expected total cost $\sigma^i_{x_0}(s^1, s^2, \dots, s^m)$ with respect to each player $i \in \{1, 2, \dots, m\}$ taking into account the corresponding matrix of transition costs $C^i = (c^i_{x,y})$. So, on the set of situations we can define the payoff functions of the players as follows:

$$\widehat{F}^i_{x_0}(s^1, s^2, \dots, s^m) = \sigma^i_{x_0}(s^1, s^2, \dots, s^m), \quad i = 1, 2, \dots, m.$$

In such a way we obtain a new discrete noncooperative game in normal form which is determined by the sets of strategies $\mathbb{S}^1, \mathbb{S}^2, \dots, \mathbb{S}^m$ of m players and the payoff functions defined above. In this game we are seeking for a Nash equilibrium, i.e., we are seeking for such strategies $s^{1*}, s^{2*}, \dots, s^{i-1*}, s^{i*}, s^{i+1*}, \dots, s^{m*}$ of the players for which it holds:

$$\begin{aligned} &\widehat{F}^i_{x_0}(s^{1*}, s^{2*}, \dots, s^{i-1*}, s^{i*}, s^{i+1*}, \dots, s^{m*}) \\ &\quad \le \widehat{F}^i_{x_0}(s^{1*}, s^{2*}, \dots, s^{i-1*}, s^{i}, s^{i+1*}, \dots, s^{m*}), \ \ \forall s^i \in \mathbb{S}^i, \ \ i = 1, 2, \dots, m. \end{aligned}$$

This game is determined uniquely by the set of states X, the positions sets X_1, $X_2, \dots, X_m$, the set of actions A, the cost functions $c^i : X \times X \to \mathbb{R}, i = 1, 2, \dots, m$, the probability function $p : X \times X \times A \to [0, 1]$ the discount factor γ and the starting position x_0. Therefore, we denote it by $(X, A, \{X_i\}_{i=\overline{1,m}}, \{c^i\}_{i=\overline{1,m}}, p, \gamma, x_0)$. As we have mentioned above we call this game stochastic positional game with discounted payoff functions.

3.2.2 *Determining Nash Equilibria for Stochastic Positional Games with Discounted Payoff Functions*

We show that for stochastic positional games with discounted payoff functions a Nash equilibrium always exists, if the discount factor γ satisfies the condition $0 < \gamma < 1$. To prove this result we shall use a continuous discounted game which represents the game-theoretic model for the following continuous optimization problem:
Maximize

$$\varphi_{x_0}(\sigma, s) = \sigma_{x_0} \tag{3.23}$$

subject to

$$\begin{cases} \sigma_x - \gamma \sum\limits_{y \in X} \sum\limits_{a \in A(x)} s_{x,a}\, p^a_{x,y}\, \sigma_y = \sum\limits_{a \in A(x)} s_{x,a}\, \mu_{x,a}, \quad \forall x \in X; \\ \sum\limits_{a \in A(x)} s_{x,a} = 1, \quad \forall x \in X; \\ s_{x,a} \geq 0, \quad \forall x \in X,\ a \in A(x), \end{cases} \tag{3.24}$$

where

$$\mu_{x,a} = \sum_{y \in X} p^a_{x,y}\, c_{x,y}.$$

This problem represents the continuous model for the discounted decision problem in the Markov decision process (X, A, p, c) with given discount factor γ and fixed starting state x_0. As we have shown the system of linear equations (3.24) has a unique solution with respect to σ_x and, therefore, the objective function (3.23) on the set of feasible solutions depends only on s. It is easy to observe that the optimal solution of the problem (3.23), (3.24) is not changed if the equations in (3.24) are replaced by inequalities ($\leq$). If we dualize after that (3.23), (3.24) with respect to σ_x for fixed s then we obtain the following problem:
Minimize

$$\overline{\varphi}_{x_0}(s, \beta) = \sum_{x \in X} \sum_{a \in A(x)} \mu_{x,a}\, s_{x,a} \beta_x \tag{3.25}$$

subject to

$$\begin{cases} \beta_y - \gamma \sum_{x \in X} \sum_{a \in A(x)} p_{x,y}^a \, s_{x,a} \beta_x = 0, \quad \forall y \in X \setminus \{x_0\}; \\ \beta_y - \gamma \sum_{x \in X} \sum_{a \in A(x)} p_{x,y}^a \, s_{x,a} \beta_x = 1 \ \text{ for } x = x_0; \\ \sum_{a \in A(x)} s_{x,a} = 1, \quad \forall x \in X; \\ \beta_y \ge 0 \ \forall y \in X; \quad s_{x,a} \ge 0, \quad \forall x \in X, \ a \in A(x). \end{cases} \tag{3.26}$$

Using elementary transformations in this problem and introducing the notations $\alpha_{x,a} = s_{x,a}\beta_x$, $\forall x \in X$, $a \in A(x)$ we obtain the following linear programming problem:
Minimize

$$\phi_{x_0}(s, \beta) = \sum_{x \in X} \sum_{a \in A(x)} \mu_{x,a} \, \alpha_{x,a} \tag{3.27}$$

subject to

$$\begin{cases} \beta_y - \gamma \sum_{x \in X} \sum_{a \in A(x)} p_{x,y}^a \, \alpha_x = 0, \quad \forall y \in X \setminus \{x_0\}; \\ \beta_y - \gamma \sum_{x \in X} \sum_{a \in A(x)} p_{x,y}^a \, \alpha_x = 1, \qquad x = x_0; \\ \sum_{a \in A(x)} \alpha_{x,a} = \beta_x, \quad \forall x \in X; \\ \beta_y \ge 0, \ \ \forall y \in X; \quad \alpha_{x,a} \ge 0, \quad \forall x \in X, \ a \in A(x). \end{cases} \tag{3.28}$$

If (α^*, β^*) is an optimal basic solution of the problem (3.27), (3.28) then the optimal stationary strategy s^* for the discounted Markov decision problem is determined as follows:

$$s_{x,a}^* = \begin{cases} 1, & \text{if } \alpha_{x,a}^* \neq 0; \\ 0, & \text{if } \alpha_{x,a}^* = 0. \end{cases} \tag{3.29}$$

and $\alpha_{x,a}^* = s_{x,a}^* \beta_x^*$, $\forall x \in X$, $a \in A(x)$.

It is easy to observe that $\beta_x > 0$, $\forall x \in X$ if for the considered Markov decision process there exists an action $a \in A(x_0)$ such that $p_{x_0,y} > 0$, $\forall y \in X$. Without loss of generality we may assume that such a condition for our problem holds; otherwise we can add a fictive action a' in the state x_0 for which $p_{x,y}^{a'} > 0$, $\forall y \in X$ ($\sum_{y \in X} p_{x,y}^{a'} = 1$) and $c_{x_0,y}^{a'} = K$, $\forall y \in X$, where K is a suitable big value.

For the continuous model of the discounted Markov decision problem we can prove a similar property as for the average Markov decision problem.

Lemma 3.18 *Let a Markov decision process (X, A, p, c) with a discount factor γ, $0 < \gamma < 1$ be given. Consider the function*

$$\varphi_{x_0}(s) = \sigma_{x_0},$$

where σ_x for $x \in X$ satisfies the condition

$$\sigma_x - \gamma \sum_{y\in X} \sum_{a\in A(x)} s_{x,a}\, p^a_{x,y}\, \sigma_y = \sum_{a\in A(x)} s_{x,a}\, \mu_{x,a}, \quad \forall x \in X. \tag{3.30}$$

Then the function $\varphi_{x_0}(s)$ depends only on $s_{x,a}$ for $x \in X$, $a \in A(x)$ and it is monotone on the set S of solutions of the following system:

$$\begin{cases} \sum\limits_{a\in A(x)} s_{x,a} = 1, & \forall x \in X; \\ s_{x,a} \geq 0, & \forall x \in X,\ a \in A(x). \end{cases}$$

The proof of this lemma is similar to the proof of Lemma 3.1.

Based on the results above we can formulate the continuous game model with m players for the discounted Markov decision problem as follows:

On the set $S = S^1 \times S^2 \times \cdots \times S^m$ we consider m payoff functions

$$\varphi^i_{x_0}(s^1, s^2, \ldots, s^m) = \sigma^i_{x_0}, \quad i = 1, 2, \ldots, m,$$

where σ^i_x for $x \in X$ satisfy the conditions

$$\sigma^i_x - \gamma \sum_{y\in X} \sum_{a\in A(x)} s^k_{x,a}\, p^a_{x,y} \sigma^i_y = \sum_{a\in A(x)} s^k_{x,a}\, \mu^i_{x,a}, \quad \forall x \in X_k;\ i, k = 1, 2, \ldots, m,$$

where

$$\mu^i_{x,a} = \sum_{y\in X} p^a_{x,y}\, c^i_{x,y}, \quad i, k = 1, 2, \ldots, m.$$

This game model possesses the same property as the previous continuous model:

- The set of Nash equilibria solutions of the continuous model is non empty if and only if the set of Nash equilibria solutions of the game in positional form is not empty;
- If $(s^1, s^2, \ldots, s^m)$ is an extreme point of S then $\widehat{F}^i_x(s^1, s^2, \ldots, s^m) = \varphi^i_x(s^1, s^2, \ldots, s^m)$, $\forall x \in X$, $i = 1, 2, \ldots, m$ and all Nash equilibria solutions for the continuous game model that correspond to extreme points in S represent Nash equilibria solutions for the game in positional form.

As a corollary from Lemma 3.18 we obtain the following result.

Lemma 3.19 *For an arbitrary discounted Markov decision process each function* $\varphi^i_{x_0}(s^1, s^2, \ldots, s^m)$, $i \in \{1, 2, \ldots, m\}$ *possesses the property that* $\varphi^i_{x_0}(\overline{s}^1, \overline{s}^2, \ldots, \overline{s}^{i-1}, s^i, \overline{s}^{i+1}, \ldots, \overline{s}^m)$ *is monotone with respect to* $s^i \in S^i$ *for an arbitrary fixed* $\overline{s}^k \in S^k$, $k = 1, 2, \ldots, i-1,\ i+1, \ldots, m$.

Using this lemma we can prove the following theorem.

Theorem 3.20 *Let* $(X, A, \{X_i\}_{i=\overline{1,m}}, \{c^i\}_{i=\overline{1,m}}, p, \gamma, x)$ *be a stochastic positional game with a given starting position* $x \in X$ *and discounted payoff functions*

$$\widehat{F}^1_x(s^1, s^2, \ldots, s^m),\ \widehat{F}^2_x(s^1, s^2, \ldots, s^m),\ \ldots,\ \widehat{F}^m_x(s^1, s^2, \ldots, s^m)$$

of the players $1, 2, \ldots, m$, *respectively. Then in this game there exists a Nash equilibrium* $s^* = (s^{1^*}, s^{2^*}, \ldots, s^{m*})$. *Moreover, for this game there exists a solution* $s^* = (s^{1^*}, s^{2^*}, \ldots, s^{m*})$ *which is a Nash equilibrium for an arbitrary starting position* $x \in X$.

The proof of this theorem is similar to the proof of Theorem 3.4.

Using Theorem 3.20 we can prove the existence of a saddle point for an arbitrary antagonistic stochastic positional game with a discounted payoff function. We obtain the antagonistic case of the game with a discounted payoff function from the general model of the discounted stochastic positional game if $m = 2$ and $c = c^2 = -c^1$. So, this game is determined by the finite sets of strategies of the players $\mathbb{S}^1, \mathbb{S}^2$ and the payoff function $\widehat{F}_{x_0} : \mathbb{S}^1 \times \mathbb{S}^2 \to \mathbb{R}$, where $\widehat{F}_{x_0}(s^1, s^2)$ for fixed strategies $s^1 \in \mathbb{S}^1$, $s^2 \in \mathbb{S}^2$ is equal to the expected total discounted cost in the Markov process induced by the strategies s^1 and s^2, if the system starts transitions in x_0, i.e.,

$$\widehat{F}_{x_0}(s^1, s^2) = \sigma_{x_0}(s^1, s^2).$$

We denote this game by $(X, A, X_1, X_2, c, p, \gamma, x_0)$, where $c = c^2 = -c^1$, x_0 is the starting position and γ, $(0 < \gamma < 1)$, is a given discount factor. From Theorem 3.20 we obtain the following result.

Corollary 3.21 *Let* $(X, A, X_1, X_2, c, p, \gamma, x)$ *be an antagonistic stochastic positional game with a given starting position* $x \in X$ *and a discounted payoff function* $\widehat{F}_x(s^1, s^2)$, *where* $s^1 \in \mathbb{S}^1$, $s^2 \in \mathbb{S}^2$. *Then for this game there exist the strategies* $s^{1^*} \in \mathbb{S}^1$, $s^{2^*} \in \mathbb{S}^2$ *for which*

$$\widehat{F}_x(s^{1^*}, s^{2^*}) = \max_{s^1 \in \mathbb{S}^1} \min_{s^2 \in \mathbb{S}^2} \widehat{F}_x(s^1, s^2) = \max_{s^2 \in \mathbb{S}^2} \min_{s^1 \in \mathbb{S}^1} \widehat{F}_x(s^1, s^2),\ \forall x \in X. \tag{3.31}$$

Theorem 3.20 can be extended for *discounted stochastic positional games on networks* $(G, \{X_i\}_{i=\overline{1,m}}, X^0, \{c^i\}_{i=\overline{1,m}}, p, \gamma, x)$, where G, X_i, p, c are defined in the same way as for the average game model on networks; γ is a discount factor that satisfies the condition $0 < \gamma < 1$.

The values of the payoff functions $\widehat{F}_x^i(s^1, s^2, \dots, s^m),\ i = 1, 2, \dots, m$ in this game are determined by the expected total discounted costs in the Markov process induced by the situation $s = (s^1, s^2, \dots, s^m)$ where each strategy s^i represents a map $s^i : x \to y \in X(x)$ for $x \in X_i$.

Respectively, Corollary 3.21 can be extended to antagonistic positional games on networks $(G, X_1, X_2, X^0, c, p, \gamma, x)$ with a discount factor γ and $c = c^2 = -c^1$.

The optimal stationary strategies of the players for the discounted game can be found using the continuous game-theoretic model in the same way as for the average stochastic positional game.

3.2.3 Discounted Stochastic Positional Games with Different Discount Factors for the Players

In the stochastic positional game model for discounted Markov decision problems formulated in Sect. 3.2.1 it is assumed that the discount factor γ is the same for all players. In general this game model can be formulated and studied for the case if for different players the discount factors may be different. So, if we assume that for each player $i \in \{1, 2, \dots, m\}$ his own discount factor γ_i is given then the corresponding payoff functions $\widehat{F}_{x_0}^i(s^1, s^2, \dots, s^m),\ i = 1, 2, \dots, m$ of the players for a fixed starting state x_0 are defined as follows:

Let $s^1, s^2, \dots, s^m$ be a set of strategies of the players. Then we calculate the expected total discounted costs $\sigma_{x_0}^i(s^1, s^2, \dots, s^m),\ i = 1, 2, \dots, m$ for the Markov process with the matrix of probability transitions P^s induced by the solution $s = (s^1, s^2, \dots, s^m)$ where the transition costs of the players are discounted with respect to the corresponding discount factors $\gamma_i,\ i = 1, 2, \dots, m$. After that we set

$$\widehat{F}_{x_0}^i(s^1, s^2, \dots, s^m) = \sigma_{x_0}^i(s^1, s^2, \dots, s^m) \quad i = 1, 2, \dots, m.$$

We denote this game by $(X, A, \{X_i\}_{i=\overline{1,m}}, \{c^i\}_{i=\overline{1,m}}, p, \{\gamma_i\}_{i=\overline{1,m}}, x_0)$.

For this game Lemma 3.18 also holds if in the corresponding continuous game model with different discount factors the payoff functions of the players on the compact set $S = S^1 \times S^2 \times \dots \times S^m$ are defined as follows:

$$\varphi_{x_0}^i(s^1, s^2, \dots, s^m) = \sigma_{x_0}^i, \quad i = 1, 2, \dots, m,$$

where σ_x^i for $x \in X$ satisfy the conditions

$$\sigma_x^i - \gamma_i \sum_{y \in X} \sum_{a \in A(x)} s_{x,a}^k\, p_{x,y}^a \sigma_y^i = \sum_{a \in A(x)} s_{x,a}^k\, \mu_{x,a}^i, \quad \forall x \in X_k;\ i, k = 1, 2, \dots, m.$$

This implies that Theorem 3.20 holds for the game with different discount factors of the players. So, for the considered game there exists a Nash equilibrium.

3.2.4 Determining Nash Equilibria for Discounted Stochastic Positional Games Using a Potential Transformation

Nash equilibria conditions for the discounted stochastic positional games can be formulated in the term of a potential transformation using a dual linear programming model (2.109), (2.110) and Theorem 2.44.

Theorem 3.22 *Let a discounted stochastic positional game $(X, A, \{X_i\}_{i=\overline{1,m}}, \{c^i\}_{i=\overline{1,m}}, p, \gamma, \overline{x})$ with a discount factor $0 < \gamma < 1$ be given. Then there exist the values σ_x^i, $i = 1, 2, \dots, m$, for $x \in X$ that satisfy the following conditions:*

(1) $\mu_{x,a}^i + \gamma \sum\limits_{y \in X} p_{x,y}^a \sigma_y^i - \sigma_x^i \geq 0, \quad \forall x \in X_i, \quad \forall a \in A(x), \quad i = 1, 2, \dots, m;$

(2) $\min\limits_{a \in A(x)} \left\{ \mu_{x,a}^i + \gamma \sum\limits_{y \in X} p_{x,y}^a \sigma_y^i - \sigma_x^i \right\} = 0, \quad \forall x \in X_i, \quad i = 1, 2, \dots, m;$

(3) *on each position set X_i, $i \in \{1, 2, \dots, m\}$ there exists a map $s^{i*} : X_i \to A$ such that*

$$s^{i*}(x) = a^* \in \underset{a \in A(x)}{\operatorname{argmin}} \left\{ \mu_{x,a}^i + \gamma \sum_{y \in X} p_{x,y}^a \sigma_y^i - \sigma_x^i \right\}, \quad \forall x \in X_i$$

and

$$\mu_{x,a^*}^j + \gamma \sum_{y \in X} p_{x,y}^{a^*} \sigma_y^j - \sigma_x^j = 0, \quad \forall x \in X_i, \quad j = 1, 2, \dots, m.$$

The set of maps $s^{1*}, s^{2*}, \dots, s^{m*}$ determines a Nash equilibrium situation $s^* = (s^{1*}, s^{2*}, \dots, s^{m*})$ for the discounted stochastic positional game $(X, A, \{X_i\}_{i=\overline{1,m}}, \{c^i\}_{i=\overline{1,m}}, p, \overline{x})$ and

$$\widehat{F}_{\overline{x}}^i(s^{1*}, s^{2*}, \dots, s^{m*}) = \sigma_{\overline{x}}^i, \ \forall \overline{x} \in X, \quad i = 1, 2, \dots, m.$$

Moreover, the solution $s^* = (s^{1*}, s^{2*}, \dots, s^{m*})$ is a Nash equilibrium for an arbitrary starting position $\overline{x} \in X$.

Proof According to Theorem 3.20 for a discounted stochastic positional game there exists a Nash equilibrium $s^* = (s^{1*}, s^{2*}, \dots, s^{m*})$, and

$$\sigma_x^i = \widehat{F}_x^i(s^{1*}, s^{2*}, \dots, s^{m*}), \quad \forall x \in X, \quad i = 1, 2, \dots, m.$$

Let us fix the strategies $s^{1*}, s^{2*}, \dots, s^{i-1*}, s^{i+1*}, \dots, s^{m*}$ of the players $1, 2, \dots, i-1, i+1, \dots, m$ and consider the problem of determining the expected total discounted cost with respect to the player i. Obviously, the optimal stationary strategy for this problem is s^{i^*}. The solution of this decision problem can be determined by using the dual linear programming model (2.109), (2.110). According to Theorem 2.44 for this problem there exist the values σ_x^i, $i = 1, 2, \dots, m$ for $x \in X$ that satisfy the conditions:

(1) $\mu_{x,a}^i + \gamma \sum_{y \in X} p_{x,y}^a \sigma_y^i - \sigma_x^i \ge 0, \;\; \forall x \in X_i, \; \forall a \in A(x), \;\; i = 1, 2, \dots, m;$

(2) $\min_{a \in A(x)} \left\{ \mu_{x,a}^i + \gamma \sum_{y \in X} p_{x,y}^a \sigma_y^i - \sigma_x^i \right\} = 0, \qquad \forall x \in X_i \;\; i = 1, 2, \dots, m.$

Moreover, for fixed strategies $s^{1^*}, s^{2^*}, \dots, s^{i-1^*}, s^{i^*}, s^{i+1^*}, \dots, s^{m*}$ of the corresponding players $1, 2, \dots, i-1, i+1, \dots, m$ we can select the strategy s^{i^*} of the player i where

$$s^{i^*}(x) \in \underset{a \in A(x)}{\operatorname{argmin}} \left\{ \mu_{x,a}^i + \gamma \sum_{y \in X} p_{x,y}^a \sigma_y^i - \sigma_x^i \right\}$$

and

$$\widehat{F}_{\overline{x}}^i(s^{1^*}, s^{2^*}, \dots, s^{m*}) = \sigma_{\overline{x}}^i, \;\; \forall \overline{x} \in X, \;\; i = 1, 2, \dots, m.$$

This means that the conditions (1)–3) of the theorem hold. □

Remark 3.23 For the game with different discount factors $\gamma_1, \gamma_2, \dots, \gamma_m$ Theorem 3.22 holds if in conditions (1)–(3) of the theorem we replace γ by γ_i.

Using the result proven above we can formulate saddle point conditions for *discounted stochastic antagonistic positional games*. We obtain such games from discounted stochastic positional games in the case $m = 2$ for $c = c^2 = -c^1$. As a corollary from Theorem 3.22 we obtain the following saddle point condition for a discounted stochastic antagonistic game.

Corollary 3.24 *Let $(X, A, X_1, X_2, c, p, \gamma, \overline{x})$ be an arbitrary stochastic antagonistic positional game with a discounted payoff function $\widehat{F}_{\overline{x}}(s^1, s^2)$. Then there exist the values σ_x for $x \in X$ that satisfy the conditions:*

(1) $\max_{a \in A(x)} \left\{ \mu_{x,a} + \gamma \sum_{y \in X} p_{x,y}^a \sigma_y - \sigma_x \right\} = 0, \qquad \forall x \in X_1,$ *where*

$\mu_{x,a} = \sum_{y \in X(x)} p_{x,y}^a c_{x,y};$

(2) $\min_{a \in A(x)} \left\{ \mu_{x,a} + \gamma \sum_{y \in X} p_{x,y}^a \sigma_y - \sigma_x \right\} = 0, \qquad \forall x \in X_2.$

The optimal stationary strategies s^{1^*}, s^{2^*} of the players in the game can be found by fixing the maps

$$s^{1^*}(x) = a^* \in \underset{a \in A(x)}{\operatorname{argmax}} \left\{ \mu_{x,a} + \gamma \sum_{y \in X} p^a_{x,y} \sigma_y - \sigma_x \right\}, \quad \forall x \in X_1;$$

$$s^{2^*}(x) = a^* \in \underset{a \in A(x)}{\operatorname{argmin}} \left\{ \mu_{x,a} + \gamma \sum_{y \in X} p^a_{x,y} \sigma_y - \sigma_x \right\}, \quad \forall x \in X_2,$$

where

$$\widehat{F}_{\overline{x}}(s^{1^*}, s^{2^*}) = \sigma_{\overline{x}}, \ \forall \overline{x} \in X.$$

Based on the results above we can propose the following iterative algorithm for determining the optimal stationary strategies of the players for discounted antagonistic positional games.

Algorithm 3.25 Determining the Optimal Stationary Strategies of the Players in an Antagonistic Game with a Discounted Payoff Function

Preliminary step (Step 0): Fix the arbitrary stationary strategies

$$s_0^1 : x_i \to a \in A(x_i) \quad \text{for} \;\; x_i \in X_1;$$
$$s_0^1 : x_i \to a \in A(x_i) \quad \text{for} \;\; x_i \in X_1.$$

and determine the situation $s_0 = (s_0^1, s_0^2)$.

General step (Step k, $k > 0$): Calculate

$$\mu_{x_i, s_{k-1}} = \sum_{y \in X(x_i)} p_{x_i,y}^{s_{k-1}(x_i)} c_{x_i,y}^{s_{k-1}(x_i)}$$

for every $x_i \in X$. Then solve the system of linear equations

$$\sigma_{x_i} = \mu_{x_i, s_{k-1}(x_i)} + \gamma \sum_{x_j \in X} p_{x_i,x_j}^{s_{k-1}(x_i)} \sigma_{x_j}, \quad i = 1, 2, \ldots, n$$

and find the solution $\sigma_{x_1}^{k-1}, \sigma_{x_2}^{k-1}, \ldots, \sigma_{x_n}^{k-1}$. After that determine the new strategies

$$s_k^1 : x_i \to a \in A(x_i) \quad \text{for} \;\; x_i \in X_1,$$
$$s_k^2 : x_i \to a \in A(x_i) \quad \text{for} \;\; x_i \in X_2,$$

and the corresponding solution $s_k = (s_k^1, s_k^2)$, where

$$s_k^1(x_i) = \underset{a \in A(x_i)}{\operatorname{argmax}} \left[\mu_{x_i,a} + \gamma \sum_{x_j \in X} p^a_{x_i,x_j} \sigma_{x_i}^{k-1} \right] \quad \text{for} \;\; x_i \in X_1;$$

$$s_k^2(x_i) = \operatorname*{argmin}_{a\in A(x_i)} \left[\mu_{x_i,a} + \gamma \sum_{x_j\in X} p^a_{x_i,x_j} \sigma^{k-1}_{x_i} \right] \quad \text{for} \quad x_i \in X_2.$$

Check if the following condition holds

$$\begin{aligned} s_k^1(x_i) &= s_{k-1}^1(x_i), \ \ \forall x_i \in X_1; \\ s_k^1(x_i) &= s_{k-1}^1(x_i), \ \ \forall x_i \in X_2? \end{aligned} \tag{3.32}$$

If the condition (3.32) holds then fix

$$\begin{aligned} s^{1*} &= s_k^1; \ \ \sigma^*_{x_i} = \sigma^k_{x_i}, \ \ \forall x_i \in X_1; \\ s^{2*} &= s_k^1; \ \ \sigma^*_{x_i} = \sigma^k_{x_i}, \ \ \forall x_i \in X_2; \end{aligned}$$

otherwise go to the next step $k+1$. The strategies s^{1*} and s^{2*} represent the optimal stationary strategies of the players in the game.

The convergence of this algorithm can be derived in a similar way as the convergence of the iterative algorithm for determining the optimal solution of the Markov decision problem with a discounted optimization criterion (see [115, 140]).

3.2.5 Nash Equilibria Conditions for Discounted Games on Networks

From Theorem 3.22 we can derive Nash equilibria conditions in the term of potential transformation for a discounted game-theoretic model on networks.

Theorem 3.26 *Let a discounted stochastic positional game on the network* $(G, \{X_i\}_{i=\overline{1,m}}, X^0, \{c^i\}_{i=\overline{1,m}}, p, \gamma, \overline{x})$ *with a discount factor* γ, $0 < \gamma < 1$, *where* $G = (X, E)$ *is the graph of the state's transition of the dynamical system and* $X = \bigcup_{i=1}^m X_i \cup X^0$ *be given. Then there exist the values* σ^i_x, $i = 1, 2, \dots m$, *for* $x \in X$ *that satisfy the following conditions:*

(1) $c^i_{x,y} + \gamma\sigma^i_y - \sigma^i_x \geq 0, \ \ \forall x \in X_i, \ \forall y \in X(x), \ \ i = 1, 2, \dots, m;$

(2) $\min\limits_{y\in X(x)} \{c^i_{x,y} + \gamma\sigma^i_y - \sigma^i_x\} = 0, \quad \forall x \in X_i, \quad i = 1, 2, \dots, m;$

(3) $\mu^i_x + \gamma \sum\limits_{y\in X(x)} p_{x,y}\sigma^i_y - \sigma^i_x = 0, \ \ \forall x \in X^0, \ \ i = 1, 2, \dots, m;$

(4) *on each position set* X_i, $i \in \{1, 2, \dots, m\}$ *there exists a map* $s^{i*} : X_i \to X$ *such that*

$$s^{i*}(x) = y^* \in \operatorname*{argmin}_{y\in X(x)}\{\mu^i_x + \gamma\sigma^i_y - \sigma^i_x\}, \ \forall x \in X_i$$

and

$$c^j_{x,y^*} + \gamma\sigma^j_{y^*} - \sigma^j_x = 0, \quad \forall x \in X_i, \ j = 1, 2, \ldots, m.$$

The set of maps $s^{1^*}, s^{2^*}, \ldots, s^{m*}$ *determines a Nash equilibrium situation* $s^* = (s^{1^*}, s^{2^*}, \ldots, s^{m*})$ *for the discounted stochastic positional game on the network* $(G, X, \{X_i\}_{i=\overline{1,m}}, \{c^i\}_{i=\overline{1,m}}, \gamma, p, \overline{x})$ *and*

$$\widehat{F}^i_{\overline{x}}(s^{1^*}, s^{2^*}, \ldots, s^{m*}) = \sigma^i_{\overline{x}}, \quad \forall \overline{x} \in X, \ \ i = 1, 2, \ldots, m.$$

Moreover, the situation $s^* = (s^{1^*}, s^{2^*}, \ldots, s^{m*})$ *is a Nash equilibrium for an arbitrary starting position* $x \in X$.

In the case $m = 2$, $c = c^2 = -c^1$ from Theorem 3.26 we obtain the following result.

Corollary 3.27 *Let a discounted antagonistic stochastic positional game on the network* $(G, X_1, X_2, X^0, c, p, \gamma, \overline{x})$ *with a discount factor* $0 < \gamma < 1$, *where* $G = (X, E)$ *is the graph of the states' transition of the dynamical system and* $X = X_1 \cup X_2 \cup X^0$ *be given. Then there exist the values* σ_x, *for* $x \in X$ *that satisfy the following conditions*:

(1) $\max\limits_{y \in X(x)} \{c_{x,y} + \gamma\sigma_y - \sigma_x\} = 0, \quad \forall x \in X_1;$

(2) $\min\limits_{y \in X(x)} \{c_{x,y} + \gamma\sigma_y - \sigma_x\} = 0, \quad \forall x \in X_2;$

(3) $\mu_x + \gamma \sum\limits_{y \in X(x)} p_{x,y}\sigma_y - \sigma_x = 0, \quad \forall x \in X^0.$

The optimal stationary strategies s^{1^*}, s^{2^*} *in the antagonistic game that satisfy the condition*

$$\widehat{F}_{\overline{x}}(s^{1^*}, s^{2^*}) = \max_{s^1 \in \mathbb{S}^1} \min_{s^2 \in \mathbb{S}^2} \widehat{F}_{\overline{x}}(s^1, s^2) = \min_{s^2 \in \mathbb{S}^2} \max_{s^1 \in \mathbb{S}^1} \widehat{F}_{\overline{x}}(s^1, s^2), \quad \forall \overline{x} \in X$$

can be found by fixing

$$s^{1^*}(x) = y^* \in \underset{y \in X(x)}{\operatorname{argmax}}\{c_{x,y} + \gamma\sigma_y - \sigma_x\}, \quad \forall x \in X_1$$

and

$$s^{2^*}(x) = y^* \in \underset{y \in X(x)}{\operatorname{argmin}}\{c_{x,y} + \gamma\sigma_y - \sigma_x\}, \quad \forall x \in X_2.$$

3.3 Nash Equilibria Conditions for Stochastic Positional Games with Stopping States and Determining Optimal Strategies of the Players in Dynamic c-Games

In this section we apply the game-theoretical concept to a Markov decision problem with a stopping state and formulate the game-theoretic model for this problem. We specify Nash equilibria conditions for the considered game using the discounted positional game and develop algorithms for determining the optimal stationary strategies of the players in a dynamic c-game on networks.

3.3.1 Problem Formulation and the Main Results

We consider the *stochastic positional game with a stopping state* for a perfect Markov processes with stopping state. We obtain this game from a discounted stochastic positional with $0<\gamma\le 1$ assuming that the perfect Markov decision process contains an absorbing stopping state $z \in X$. We denote the stochastic positional game with a stopping state by $(X, A, \{X_i\}_{i=\overline{1,m}}, \{c^i\}_{i=\overline{1,m}}, p, \gamma, x_0, z)$. Here x_0 is the starting state of the decision process and z is the state in which the process stops as soon as this state is reached. In the considered game for the absorbing stopping state z we will assume that $c^i_{z,z} = 0,\ i = 1, 2, \dots, m$.

If $0 < \gamma < 1$ then Nash equilibria conditions for the positional game with an absorbing stopping state can be derived from Theorem 3.22 considering $c^i_{z,z} = 0,\ i = 1, 2, \dots, m$.

Below we assume that $\gamma = 1$ and formulate Nash equilibria conditions for the game with an absorbing stopping state if the cost functions $c^i,\ i = 1, 2, \dots, m$, satisfy the condition

$$c^i_{x,y} > 0,\ \forall x \in X \setminus \{z\},\ y \in X,\ i = 1, 2, \dots, m;\ c^i_{z,z} = 0,\ \ i = 1, 2, \dots, m,$$

and for the players there exists at least one solution $s = (s^1, s^2, \dots, s^m)$ which generates a unichain Markov process with an absorbing state z and $c^i_{z,z} = 0$, $i = 1, 2, \dots, m$. We define the payoff functions

$$\widehat{F}^1_{x_0}(s^1, s^2, \dots, s^m),\ \widehat{F}^2_{x_0}(s^1, s^2, \dots, s^m),\ \dots,\ \widehat{F}^m_{x_0}(s^1, s^2, \dots, s^m)$$

of the players for the game with the conditions mentioned above in the following way: Let $s^1, s^2, \dots, s^m$ be a fixed set of strategies of the players $1, 2, \dots, m$ and assume that $\gamma \in (0, 1]$. Then for a fixed starting state x the situation $s = (s^1, s^2, \dots, s^m)$ induces a Markov process for which we can determine the expected discounted total cost

$$\sigma^{\gamma}{}^i_{x_0}(s^1, s^2, \dots, s^m),\ \ \forall \gamma \in (0, 1),\ \ i = 1, 2, \dots, m.$$

For each $i \in \{1, 2, \ldots, m\}$ we define

$$\widehat{F}^i_{x_0}(s^1, s^2, \ldots, s^m) = \lim_{\gamma \to 1} \sigma^{\gamma}{}^{i}_{x_0}(s^1, s^2, \ldots, s^m),$$

where $\widehat{F}^i_{x_0}(s^1, s^2, \ldots, s^m)$ represents the value of the payoff function for player i if the corresponding strategies $s^1, s^2, \ldots, s^m$ of the players $1, 2, \ldots, m$ are fixed. Obviously, $\widehat{F}^i_x(s^1, s^2, \ldots, s^m) > 0,\ \forall x \in X \setminus \{z\},\ i = 1, 2, \ldots, m$, and $\widehat{F}^i_x(s^1, s^2, \ldots, s^m) < \infty$, for $x \in X \setminus \{z\},\ i = 1, 2, \ldots, m$ if the situation $s = (s^1, s^2, \ldots, s^m)$ with fixed starting state x generates a Markov chain with absorbing state z; otherwise $\widehat{F}^i_x(s^1, s^2, \ldots, s^m) = \infty$ for $x \in X \setminus \{z\},\ i = 1, 2, \ldots, m$. We obtain a characterization of Nash equilibria situations in the considered game on the bases of the following theorem.

Theorem 3.28 *Let a stochastic positional game $(X, A, \{X_i\}_{i=\overline{1,m}}, \{c^i\}_{i=\overline{1,m}}, p, \gamma, \overline{x}, z)$ with a stopping state z where $\gamma = 1$ and the cost functions c^i, $i = 1, 2, \ldots, m$, satisfy the condition*

$$c^i_{x,y} > 0,\ \forall x \in X \setminus \{z\},\ y \in X,\ i = 1, 2, \ldots, m;\ c^i_{z,z} = 0,\ i = 1, 2, \ldots, m$$

be given. Additionally, assume that for the considered game there exists at least one solution $s = (s^1, s^2, \ldots, s^m)$ that generates a unichain Markov process with an absorbing state z. Then for the payoff functions

$$\widehat{F}^1_{\overline{x}}(s^1, s^2, \ldots, s^m),\ \widehat{F}^2_{\overline{x}}(s^1, s^2, \ldots, s^m),\ \ldots,\ \widehat{F}^m_{\overline{x}}(s^1, s^2, \ldots, s^m)$$

on the set of strategies $\mathbb{S}^1, \mathbb{S}^2, \ldots, \mathbb{S}^m$ of the players in this game there exists a Nash equilibrium $s^{1*}, s^{2*}, \ldots, s^{m*}$.

Moreover, there exist the values σ^i_x, $i = 1, 2, \ldots, m$, for $x \in X$ that satisfy the following conditions:

(1) $\sigma^i_z = 0,\ i = 1, 2, \ldots, m;$

(2) $\mu^i_{x,a} + \sum_{y \in X} p^a_{x,y}\sigma^i_y - \sigma^i_x \geq 0,\ \forall x \in X_i \setminus \{z\},\ \forall a \in A(x),\ i = 1, 2, \ldots, m;$

(3) $\min_{a \in A(x)} \left\{\mu^i_{x,a} + \sum_{y \in X} p^a_{x,y}\sigma^i_y - \sigma^i_x\right\} = 0, \quad \forall x \in X_i \setminus \{z\},\ i = 1, 2, \ldots, m;$

(4) *on each position set X_i, $i \in \{1, 2, \ldots, m\}$ there exists a map $s^{i*} : X_i \to A$ such that*

$$s^{i*}(x) = a^* \in \operatorname*{argmin}_{a \in A(x)} \left\{\mu^i_{x,a} + \sum_{y \in X} p^a_{x,y}\sigma^i_y - \sigma^i_x\right\},\ \forall x \in X_i$$

and

$$\mu^j_{x,a^*} + \sum_{y \in X} p^{a^*}_{x,y} \sigma^j_y - \sigma^j_x = 0, \;\; \forall x \in X_i, \;\; j = 1, 2, \ldots, m.$$

The set of maps $s^{1*}, s^{2*}, \ldots, s^{m*}$ *determines a Nash equilibrium solution* $s^* = (s^{1*}, s^{2*}, \ldots, s^{m*})$ *for the stochastic positional game* $(X, A, \{X_i\}_{i=\overline{1,m}}, \{c^i\}_{i=\overline{1,m}}, p, \overline{x}, z)$ *with a final state* z*, where*

$$\widehat{F}^i_{\overline{x}}(s^{1*}, s^{2*}, \ldots, s^{m*}) = \sigma^i_{\overline{x}}, \;\; \forall \overline{x} \in X, \quad i = 1, 2, \ldots, m.$$

The situation $s^* = (s^{1*}, s^{2*}, \ldots, s^{m*})$ *is a Nash equilibrium of the game for an arbitrary starting position* $\overline{x} \in X$.

Proof According to Theorem 3.20 for the discounted stochastic positional game $(X, A, \{X_i\}_{i=\overline{1,m}}, \{c^i\}_{i=\overline{1,m}}, p, \gamma, \overline{x}, z)$ there exists a Nash equilibrium situation $s^*(\gamma) = (s^{1*}(\gamma), s^{2*}(\gamma), \ldots, s^{m*}(\gamma))$ for an arbitrary $\gamma \in (0, 1)$. If an arbitrary situation $s = (s^1, s^2, \ldots, s^m)$ does not generate a unichain Markov process with the absorbing state z then taking into account that

$$c^i_{x,y} > 0, \forall x \in X \setminus \{z\}, \;\; y \in X, \;\; i = 1, 2, \ldots, m; \;\; c^i_{z,z} = 0, \;\; i = 1, 2, \ldots, m$$

we have

$$\widehat{F}^i_{\overline{x}}(s^1, s^2, \ldots, s^m) = \lim_{\gamma \to 1} \sigma^{\gamma}{}^i_{\overline{x}}(s^1, s^2, \ldots, s^m) = \infty, \; i = 1, 2, \; \ldots, \; m.$$

However, for an arbitrary situation which generates a unichain Markov process with the absorbing state z it holds

$$\widehat{F}^i_{\overline{x}}(s^1, s^2, \ldots, s^m) = \sigma^{\gamma}{}^i_{\overline{x}}(s^1, s^2, \ldots, s^m) < \infty, \; i = 1, 2, \; \ldots, \; m$$

for an arbitrary $\gamma \in (0, 1]$ because $c_{z,z} = 0$.

This means that there exists $\gamma_0 \in (0, 1)$ such that for an arbitrary $\gamma \geq \gamma_0$ $(\gamma \leq 1)$ the corresponding Nash equilibrium solution $s^{1*}(\gamma), s^{2*}(\gamma), \ldots, s^{m*}(\gamma)$ of the game with discount factor γ induces a unichain Markov process with the absorbing state z. So, we obtain a Nash equilibrium solution $s^* = (s^{1*}, s^{2*}, \ldots, s^{m*})$ for the stochastic positional game with stopping state z in the case $\gamma = 1$ as follows

$$\widehat{F}^i_{\overline{x}}(s^{1*}, s^{2*}, \ldots, s^{m*}) = \lim_{\gamma \to 1} \widehat{F}^i_{\overline{x}}(s^{1*}(\gamma), s^{2*}(\gamma), \ldots, s^{m*}(\gamma)), \;\; i = 1, 2, \ldots, m.$$

The conditions (1)–(4) of the theorem can be obtained from the conditions of Theorem 3.22 for $\gamma = 1$ in the case if an arbitrary Nash equilibrium solution s^* of the game generates a Markov unichain with the absorbing state z.

For the absorbing final state we obtain $\sigma_z^i = 0,\ i = 1, 2, \ldots, m$ because $c_{z,z}^i = 0,\ i = 1, 2, \ldots, m$. Moreover, the values $\sigma_x,\ x \in X$ for optimal stationary strategies $s^{1*},\ s^{2*}, \ldots, s^{m*}$ of the players in the game with the absorbing stopping state z coincide with the values $\varepsilon_x^*,\ x \in X$ for the game with average payoff function, where $F_x^i(s^{1*}, s^{2*}, \ldots, s^{m*}) = 0,\ i = 1, 2, \ldots, m$ □

It is easy to see that the result described above can be extended for some classes of stochastic positional games with a stopping state if γ is an arbitrary real value. Nash equilibria conditions for such games can be derived also on the basis of Theorem 3.22.

3.3.2 Stochastic Positional Games with a Stopping State on Networks and Determining Nash Equilibria

We formulate the stochastic positional game with a stopping state on networks using the network model of discounted stochastic positional games from Sect. 3.2.5 in the case that the graph $G = (X, E)$ contains a vertex z which can be reached from every $x \in X$. In terms of stationary strategies of the players this means that for the discounted stochastic positional game $(G, \{X_i\}_{i=\overline{1,m}}, X^0, \{c^i\}_{i=\overline{1,m}}, p, \gamma, \overline{x}, z)$ with a stopping state z there exists at least one solution $s = (s^1, s^2, \ldots, s^m)$ which generates a Markov unichain with an absorbing state z. We formulate Nash equilibria conditions for this game in the case that $\gamma = 1$ and the cost functions $c^i : E \to \mathbb{R},\ i = 1, 2, \ldots, m$, satisfy the conditions: $c_e^i > 0,\ \forall e \in E,\ i = 1, 2, \ldots, m$, and $c_{e_z}^i = 0$ for $e_z = (z, z),\ i = 1, 2, \ldots, m$.

Theorem 3.29 *Let* $(G, \{X_i\}_{i=\overline{1,m}}, X^0, \{c^i\}_{i=\overline{1,m}}, p, \gamma, \overline{x}, z)$ *be a stochastic positional game on networks with a stopping state z where $\gamma = 1$ and the cost function $c^i,\ i = 1, 2, \ldots, m$, satisfy the condition*

$$c_{x,y}^i > 0,\ \forall x \in X \setminus \{z\},\ y \in X(x),\ i = 1, 2, \ldots, m;\ c_{z,z}^i = 0,\ i = 1, 2, \ldots, m.$$

Additionally, assume that for the considered game there exists at least one solution $s = (s^1, s^2, \ldots, s^m)$ that generates a unichain Markov process with an absorbing state z.

Then for the payoff functions

$$\widehat{F}_{\overline{x}}^1(s^1, s^2, \ldots, s^m),\ \widehat{F}_{\overline{x}}^2(s^1, s^2, \ldots, s^m),\ \ldots,\ \widehat{F}_{\overline{x}}^m(s^1, s^2, \ldots, s^m)$$

on the set of strategies $\mathbb{S}^1, \mathbb{S}^2, \ldots, \mathbb{S}^m$ of the players in this game there exists a Nash equilibrium $s^{1*}, s^{2*}, \ldots, s^{m*}$. Moreover, there exist the values $\varepsilon_x^i,\ i = 1, 2, \ldots, m$, for $x \in X$ that satisfy the following conditions:

(1) $\varepsilon_z^i = 0, \quad i = 1, 2, \ldots, m;$

(2) $c_{x,y}^i + \varepsilon_y^i - \varepsilon_x^i \geq 0,\ \ \forall x \in X_i \setminus \{z\},\ \ \forall y \in X(x), \quad i = 1, 2, \ldots, m;$

(3) $\min\limits_{y \in X(x)} \{c^i_{x,y} + \varepsilon^i_y - \varepsilon^i_x\} = 0, \quad \forall x \in X_i \setminus \{z\}, \quad i = 1, 2, \dots, m;$

(4) *for each* $x \in X^0$ *it holds*

$$\mu^i_x + \sum_{y \in X} p_{x,y} \varepsilon^i_y - \varepsilon^i_x = 0, \ i = 1, 2, \dots, m;$$

(5) *on each position set* X_i, $i \in \{1, 2, \dots, m\}$ *there exists a map* $s^{i^*} : X_i \to X$ *such that*

$$s^{i^*}(x) = y^* \in \operatorname*{argmin}_{y \in X(x)} \{c^i_{x,y} + \varepsilon^i_y - \varepsilon^i_x\}, \ \ \forall x \in X_i$$

and

$$c^j_{x,y^*} + \varepsilon^j_{y^*} - \varepsilon^j_x = 0, \forall x \in X_i, \ i, j = 1, 2, \dots, m.$$

The set of maps $s^{1^*}, s^{2^*}, \dots, s^{m*}$ *determines a Nash equilibrium solution* $s^* = (s^{1^*}, s^{2^*}, \dots, s^{m*})$ *for the stochastic positional game* $(X, A, \{X_i\}_{i=\overline{1,m}}, \{c^i\}_{i=\overline{1,m}}, p, \overline{x}, z)$ *with a stopping state z, where*

$$\widehat{F}^i_{\overline{x}}(s^{1^*}, s^{2^*}, \dots, s^{m*}) = \varepsilon^i_{\overline{x}}, \ \ \forall \overline{x} \in X, \quad i = 1, 2, \dots, m.$$

The solution $s^* = (s^{1^*}, s^{2^*}, \dots, s^{m*})$ *is a Nash equilibrium for an arbitrary starting position* $\overline{x} \in X$.

We obtain the proof of this theorem from Theorem 3.28 if we regard the game as a particular case of the discounted stochastic positional game from the previous section, where the strategies of the players are defined as maps $s^i : x \to y \in X(x)$ for $x \in X_i$, $i = 1, 2, \dots, m$. Here $\varepsilon^i_x = \sigma^i_x$, $i = 1, 2, \dots, m$.

3.3.3 Determining Optimal Stationary Strategies of the Players in Dynamic c-Games

In [9, 12, 77, 83] the following dynamic game on networks has been studied. Let $G = (X, E)$ be a directed graph of the states' transitions of the dynamical system $\mathbb{L}$ with a given partition $X_1, X_2, \dots, X_m$ of the vertex set X, where the corresponding subsets X_i, $i = 1, 2, \dots, m$ are regarded as position sets of m players $1, 2, \dots, m$.

Graph G possesses the property that each vertex contains at least one leaving directed edge and on the edge set E there are defined m cost functions

$$c^i : E \to \mathbb{R}, \ \ i = 1, 2, \dots, m.$$

These functions represent for each directed edge $e = (x, y) \in E$ the corresponding costs $c^1_{x,y}, c^2_{x,y}, \dots, c^m_{x,y}$, for the players if the dynamical system makes a transition from the state x to the state y through the directed edge $e = (x, y)$. Additionally, in G two vertices x_0 and x_f are distinguished, where x_0 represents the starting position of the game and x_f is a position in which the game stops.

Thus, the game starts in the position x_0 at the moment of time $t = 0$. If x_0 belongs to the set of positions X_{k_0} of the player $k_0 \in \{1, 2, \dots, m\}$ then the move is done by player k_0. Move means that the player k_0 fixes an outgoing directed edge $e_1 = (x_0, x_1)$ and the dynamical system makes a transition from the state x_0 to the state $x_1 \in X(x_0)$. After the first move the dynamical system is in the state x_1 at the moment of time $t = 1$. If x_1 belongs to the set of positions X_{k_1} of the player k_1 then the move is made by player $k_1 \in \{1, 2, \dots, m\}$, i.e., player k_1 selects a new position x_2 for the state's transition of the system and so on. The game stops at the moment of time t if the state x_f is reached, i.e., if $x_t = x_f$. This game may be finite or infinite. If the final state x_f is reached at a finite moment of time t then the game is finite; otherwise the game is infinite. Each player in the game has the aim to minimize his own integral cost $\sum_{\tau=0}^{t-1} c^i_{e_t}$. In this game we are seeking for a Nash equilibrium.

In [77, 83] the considered game is called *dynamic c-game* and it is denoted by $(G, \{X_i\}_{i=\overline{1,m}}, \{c^i\}_{i=\overline{1,m}}, x_0, x_f)$. Below we formulate conditions for determining Nash equilibria in a dynamic c-game. We show that if in G the vertex x_f is attainable from the vertex x_0 and the cost functions are positive then a Nash equilibrium exists. Moreover, we show that the optimal strategies of the players in the game can be found in the set of stationary strategies $\mathbb{S}_1, \mathbb{S}_2, \dots, \mathbb{S}_m$. We define the stationary strategies of the players $1, 2, \dots, m$ as m maps:

$$s^i : x \to y \in X(x) \quad \text{for} \quad x \in X_i, \quad i = 1, 2, \dots, m.$$

In the terms of stationary strategies the dynamic c-game in normal form with payoff functions

$$H^1_{x_0x_f}(s^1, s^2, \dots, s^m), \; H^2_{x_0x_f}(s^1, s^2, \dots, s^m), \; \dots, \; H^m_{x_0x_f}(s^1, s^2, \dots, s^m)$$

of the players $1, 2, \dots, m$ can be defined in the following way.

Let $s^1, s^2, \dots, s^m$ be a fixed set of strategies of the players $1, 2, \dots, m$. Then in G the subset of directed edges $E^i_s = \{(x, s^i(x)) \in E \mid x \in X_i\}$ corresponds to the set of transitions of the dynamical system in the states $x \in X_i$ controlled by player $i \in \{1, 2, \dots, m\}$.

The subset $E_s = \bigcup_{i=1}^m E^i_s$ in G generates a subgraph $G_s = (X, E_s)$ in which either a unique directed path $P_s(x_0, x_f)$ from x_0 to x_f exists or such directed path in G does not exist.

If $s^1, s^2, \dots, s^m$ generate in G a subgraph G_s, which contains a unique directed path $P_s(x_0, x_f)$ from x_0 to x_f, then we put

$$H^i_{x_0x_f}(s^1, s^2, \ldots, s^m) = \sum_{e \in E(P_s(x_0,x_f))} c^i_e, \quad i = 1, 2, \ldots, m. \tag{3.33}$$

where $E(P_s(x_0, x_f))$ represents the set of directed edges of the directed path $P_s(x_0, x_f)$.

If in G_s there is no directed path from x_0 to x_f then a unique directed cycle C_s with a set of edges $E(C_s)$ can be obtained wif we pass through directed edges from x_0. Therefore, there exists a unique directed cycle C_s, which we can get from x_0 and a unique directed path $P'_s(x_0, x')$, which connects x_0 and C_s (the vertex x' is a unique common vertex of $P'_s(x_0, x')$ and C_s). In this case $H^i_{x_0x_f}(s^1, s^2, \ldots, s^m)$, $i = 1, 2, \ldots, m$ are defined as follows:

$$H^i_{x_0x_f}(s^1, s^2, \ldots, s^m) = \begin{cases} +\infty, & \text{if } \sum_{e \in E(C_s)} c^i_e > 0; \\ \sum_{e \in E(P'_s(x_0,x'))} c^i_e, & \text{if } \sum_{e \in E(C_s)} c^i_e = 0; \\ -\infty, & \text{if } \sum_{e \in E(C_s)} c^i_e < 0. \end{cases} \tag{3.34}$$

It is easy to observe that the dynamic c-game represents a particular case of the network model of a stochastic positional game with a stopping state $z = x_f$. We obtain the dynamic c-game from a stochastic positional game in the case $X^0 = \emptyset$. Therefore, from Theorem 3.28 we obtain the following result.

Theorem 3.30 *Let $(G, \{X_i\}_{i=\overline{1,m}}, \{c^i\}_{i=\overline{1,m}}, x_0, x_f)$ be a dynamic network for which the vertex x_f in G is attainable from every $x \in X$. Assume that the vectors $c^i = (c^i_{e_1}, c^i_{e_2}, \ldots, c^i_{e_{|E|}})$, $i \in \{1, 2, \ldots, m\}$ have positive and constant components. Then in a dynamic c-game on the network $(G, \{X_i\}_{i=\overline{1,m}}, \{c^i\}_{i=\overline{1,m}}, x_0, x_f)$ for the players $1, 2, \ldots, m$ there exists an optimal solution $s^{1*}, s^{2*}, \ldots, s^{m*}$ in the sense of Nash, which satisfies the following properties:*

- *the graph $G_{s^*} = (X, E_{s^*})$ generated by $s^{1*}, s^{2*}, \ldots, s^{m*}$ has the structure of a directed tree with a sink vertex x_f;*
- *$s^{1*}, s^{2*}, \ldots, s^{m*}$ represents the solution of a dynamic c-game on the network $(G, \{X_i\}_{i=\overline{1,m}}, \{c^i\}_{i=\overline{1,m}}, x, x_f)$ with an arbitrary starting position $x \in X$ and a given final position x_f.*

This theorem has been formulated and proved in [9]. Theorem 3.29 represents a generalization of this theorem for stochastic positional games. In [9] the following result has also been proved.

Theorem 3.31 *Let $(G, \{X_i\}_{i=\overline{1,m}}, \{c^i\}_{i=\overline{1,m}}, x_0, x_f)$ be a network for which the vertex x_f in G is attainable from every $x \in X$. Assume that the vectors $c^i = (c^i_{e_1}, c^i_{e_2}, \ldots, c^i_{e_{|E|}})$, $i \in \{1, 2, \ldots, m\}$ have positive and constant components. Then on the vertex set X of the network game there exist m real functions*

$$\varepsilon^1 : X \to \mathbb{R}^1, \ \varepsilon^2 : X \to \mathbb{R}^1, \ \ldots, \ \varepsilon^m : X \to \mathbb{R}^1,$$

which satisfy the conditions:

(1) $\varepsilon_y^i - \varepsilon_x^i + c_{x,y}^i \geq 0, \quad \forall (x, y) \in E_i, \ i = 1, 2, \ldots, m,$ *where* $E_i = \{e = (x, y) \in E \mid x \in X_i, \ y \in X\}$*;*

(2) $\min\limits_{y \in X(x)} \{\varepsilon_y^i - \varepsilon_x^i + c_{x,y}^i\} = 0, \quad \forall x \in X_i, \quad i = 1, 2, \ \ldots, \ m;$

(3) *the subgraph* $G^0 = (X, E^0)$ *generated by the edge set* $E^0 = E_1^0 \cup E_2^0 \cup \cdots \cup E_m^0$, $E_i^0 = \{e = (x, y) \in E_i \mid \varepsilon_y^i - \varepsilon_x^i + c_{x,y}^i = 0\}$, $i = 1, 2, \ldots, m$, *has the property that the vertex* x_f *is attainable from any vertex* $x \in X$ *and* G^0 *contains a subgraph* $\overline{G}^0 = (X, \overline{E}^0)$, $\overline{E}^0 \subset E$, *which possesses the same property and besides that*

$$\varepsilon_y^i - \varepsilon_x^i + c_{x,y}^i = 0, \ \forall (x, y) \in \overline{E}^0, \quad i = 1, 2, \ldots, m.$$

If $\varepsilon^1, \varepsilon^2, \ldots, \varepsilon^m$ *are arbitrary real functions, which satisfy the conditions* (1)–(3), *then the optimal solution characterized by Nash strategies in the dynamic c-game on the network* $(G, \{X_i\}_{i=\overline{1,m}}, \{c^i\}_{i=\overline{1,m}}, x_0, x_f)$ *can be found as follows: Choose in* $\overline{G}^0$ *an arbitrary directed tree* $GT = (X, E^*)$ *with sink vertex* x_f *and fix in* GT *the following maps:*

$$\begin{aligned} s^{1*} &: x \to y \in X_{GT}(x) \quad \textit{for} \quad x \in X_1; \\ s^{2*} &: x \to y \in X_{GT}(x) \quad \textit{for} \quad x \in X_2; \\ &\vdots \\ s^{m*} &: x \to y \in X_{GT}(x) \quad \textit{for} \quad x \in X_m, \end{aligned}$$

where $X_{GT}(x) = \{y \in X \mid (x, y) \in E^*\}$.

Proof According to Theorem 3.30 for the considered dynamic c-game $(G, \{X_i\}_{i=\overline{1,m}}, \{c^i\}_{i=\overline{1,m}}, x_0, x_f)$ there exists an optimal solution characterized by Nash strategies $s^{1^*}, s^{2^*}, \ldots, s^{m*}$ of the players $1, 2, \ldots, m$ and these strategies generate in G a directed tree $GT_{s^*} = (X, E_{s^*})$ with sink vertex x_f. In this tree we find the functions

$$\varepsilon^1 : X \to \mathbb{R}^1, \ \varepsilon^2 : X \to \mathbb{R}^1, \ \ldots, \ \varepsilon^m : X \to \mathbb{R}^1,$$

where $\varepsilon_x^i = H_{xx_f}^i(s^{1^*}, s^{2^*}, \ldots, s^{m*}), \ \forall x \in X, i = 1, 2, \ldots, m$. It is easy to verify that $\varepsilon^1, \varepsilon^2, \ldots, \varepsilon^m$ satisfy the conditions (1) and (2).

Additionally, we can see that in G^0 there exists the graph $\overline{G}^0 = (X, \overline{E}^0)$, which satisfies condition (3), because $GT \subseteq \overline{G}^0$. Moreover, if in $\overline{G}^0$ a directed tree $GT_{s'} = (X, E_{s'})$, which is different from GT_{s^*}, with sink vertex is chosen, then

$GT_{s'}$ generates another optimal solution characterized by a Nash solution s'^1, s'^2, $\dots$, s'^m.

Now let us show that if

$$\varepsilon^1 : X \to \mathbb{R}^1, \ \ \varepsilon^2 : X \to \mathbb{R}^1, \ \dots, \ \varepsilon^m : X \to \mathbb{R}^1,$$

are arbitrary functions, which verify the conditions (1)–(3), then an arbitrary directed tree $GT = (X, E_{s^*})$ of $\overline{G}^0$ generates the maps:

$$\begin{array}{ll} s^{1^*} : x \to y \in X_{GT}(x) & \text{for} \quad x \in X_1; \\ s^{2^*} : x \to y \in X_{GT}(x) & \text{for} \quad x \in X_2; \\ & \vdots \\ s^{m*} : x \to y \in X_{GT}(x) & \text{for} \quad x \in X_m, \end{array}$$

which correspond to an optimal solution characterized by a Nash solution.

We use the induction on the number m of the players in the dynamic c-game. In the case that $m = 1$ the statement is true, because $X_1 = X$ and the conditions (1)–(3) for positive c_e^1 provide the existence of a tree $GT = (X, E_{s^*})$ of optimal paths, which corresponds to the solution s^{1^*} for the problem of finding the shortest paths from $x \in X$ to x_f in G.

Assume that the statement holds for $m \le k, k \ge 1$, and let us prove it for $m = k+1$. We consider that the first player fixes his strategy $s^1 = s^{1^*}$ and consider the problem of finding an optimum by Nash strategies in the network game with respect to other players. The obtained game in the positional form can be interpreted as a c-game with $m - 1$ players, since the positions of the first player can be considered as the positions of any other player. Furthermore, we consider them as the positions of the second player.

Thus, if we fix $s^1 = s^{1^*}$, then we obtain a new game with $m - 1$ players, where we may consider the position of the first player as the position of an arbitrary other player; we will consider these positions as the positions of the second player. This means that we obtain a new c-game $(G^1, X_2^1, X_3, \dots, X_m, c_1^2, c_1^3, \dots, c_1^m, x_0, x_f)$, where X_2^1, G^1 and the functions c_1^i, $i = 2, \dots, m$, are defined as follows:

$$X_2^1 = X_1 \cup X_2, \ \ G = (X, (E \setminus E_1) \cup E^1_{s^{1*}})$$

and the cost function c_1^2 on $(E \setminus E_1) \cup E^1_{s^{1*}}$ is induced by the cost function c^2.

In the normal form this game is determined by the payoff functions

$$H^2_{x_0 x_f}(s^{1^*}, s^2, \dots, s^m), \ H^3_{x_0 x_f}(s^{1^*}, s^2, \dots, s^m), \ \dots, \ H^m_{x_0 x_f}(s^{1^*}, s^2, \dots, s^m),$$

where $s^2 \in \mathbb{S}_2$, $s^3 \in \mathbb{S}_3$, $\dots$, $s^m \in \mathbb{S}_m$; $\mathbb{S}_2, \mathbb{S}_3, \dots, \mathbb{S}_m$ are the respective sets of feasible stationary strategies of the players $2, 3, \dots, m$.

In this new dynamic c-game $(G^1, X_2^1, X_3, \ldots, X_m, c_1^2, c_1^3, \ldots, c_1^m, x_0, x_f)$ we consider $m-1$ functions

$$\varepsilon^2 : X \to \mathbb{R}^1, \quad \varepsilon^3 : X \to \mathbb{R}^1, \quad \ldots, \quad \varepsilon^m : X \to \mathbb{R}^1,$$

which satisfy the conditions:

(1) $\varepsilon_y^i - \varepsilon_x^i + c_{1(x,y)}^i \geq 0, \quad \forall (x,y) \in E_i^1, \quad i = 2, \ldots, m,$ where
$E_2^1 = \{e = (x,y) \in E^1 \mid x \in X_2^1, \ y \in X\},$
$E_i^1 = \{e = (x,y) \in E^1 \mid x \in X_i, \ y \in X\}, \quad i = 3, \ldots, m;$

(2) $\min\limits_{y \in X_{G^1}(x)} \{\varepsilon_y^2 - \varepsilon_x^2 + c_{1(x,y)}^2\} = 0, \quad \forall x \in X_2^1,$
$\min\limits_{y \in X_{G^1}(x)} \{\varepsilon_y^2 - \varepsilon_x^2 + c_{1(x,y)}^i\} = 0, \quad \forall x \in X_i, \quad i = 3, \ldots, m;$

(3) the subgraph $G^{1^0} = (X, E^{1^0})$ generated by the edge set $E^{1^0} = E_2^{1^0} \cup E_3^0 \cup \cup \cdots \cup E_m^0$, $E_2^{1^0} = \{e = (x,y) \in E_2^1 \mid \varepsilon_y^2 - \varepsilon_x^2 + c_{1(x,y)}^2 = 0\}$, $E_i^0 = \{e = (x,y) \in E_i \mid \varepsilon_y^i - \varepsilon_x^i + c_{1(x,y)}^i = 0\}$, $i = 2, \ldots, m$, has the property that the vertex x_f is attainable from any vertex $x \in X$ and G^{1^0} contains a subgraph $\overline{G}^{1^0} = (X, \overline{E}^{1^0})$, which possesses the same property and besides that

$$\varepsilon_y^i - \varepsilon_x^i + c_{1(x,y)}^i = 0, \quad \forall (x,y) \in \overline{E}^{1^0}, \quad i = 2, \ldots, m.$$

According to the induction assumption, in the dynamic c-game $(G^1, X_2^1, X_3, \ldots, X_m, c_1^2, c_1^3, \ldots, c_1^m, x_0, x_f)$ the stationary strategies $\overline{s}^{2*}, s^{3*}, \ldots, s^{m*}$ induced by the directed tree $GT = (X, E_{s^*})$,

$$\begin{aligned}
\overline{s}^{2*} &: x \to y \in X_{GT}(x) \quad \text{for} \quad x \in X_2^1; \\
s^{3*} &: x \to y \in X_{GT}(x) \quad \text{for} \quad x \in X_3; \\
&\vdots \\
s^{m*} &: x \to y \in X_{GT}(x) \quad \text{for} \quad x \in X_m,
\end{aligned}$$

where $\overline{s}^{2*}(x) = s^{1*}(x)$ for $x \in X_1$ and $\overline{s}^{2*}(x) = s^{2*}(x)$ for $x \in X_2$, determine a Nash equilibrium.

Thus,

$$\begin{aligned}
&H_{xx_f}^i(s^{1*}, s^{2*}, s^{3*}, \ldots, s^{i-1*}, s^{i*}, s^{i+1*}, \ldots, s^{m*}) \\
&\leq H_{xx_f}^i(s^{1*}, s^{2*}, s^{3*}, \ldots, s^{i-1*}, s^i, s^{i+1*}, \ldots, s^{m*}), \quad \forall s^i \in \mathbb{S}_i, \ 2 \leq i \leq m.
\end{aligned}$$

Also, it is easy to verify that

$$H^1_{xx_f}(s^{1*}, s^{2*}, \dots, s^{m*}) \le H^1_{xx_f}(s^1, s^{2*}, \dots, s^{m*}), \quad \forall s^1 \in S_1,$$

because for fixed $s^{2*}, s^{3*}, \dots, s^{m*}$ in G the problem of finding

$$\min_{s^1 \in \mathbb{S}_1} H^1_{xx_f}(s^1, s^{2*}, \dots, s^{m*}) \text{ for } x \in X$$

becomes the problem of finding the shortest paths from x to x_f in the graph $G' = (X, E')$, generated by a set E_1 and the edges $(x, s_i^*(x))$, $x \in X_i$, $i = 2, 3, \dots, m$, with costs c_e^1 on the edges $e \in E'$. On this graph the following condition is satisfied:

$$\varepsilon^1_y - \varepsilon^1_x + c^1_{x,y} \ge 0; \quad \forall (x, y) \in E',$$

which implies

$$H^1_{xx_f}(s^{1*}, s^{2*}, \dots, s^{m*}) \le H^1_{xx_f}(s^1, s^{2*}, \dots, s^{m*}), \quad \forall s^1 \in \mathbb{S}_1,$$

because $H^1_{xx_f}(s^{1*}, s^{2*}, \dots, s^{m*}) = \varepsilon^1_x, \forall x \in X$.

Hence $s^{1*}, s^{2*}, \dots, s^{m*}$ is a Nash equilibrium for the dynamic c-game. □

Remark 3.32 Let

$$\varepsilon^1 : X \to \mathbb{R}^1, \ \ \varepsilon^2 : X \to \mathbb{R}^1, \ \ \dots, \ \ \varepsilon^m : X \to \mathbb{R}^1,$$

be arbitrary real functions on X in G and $\overline{c}^1, \overline{c}^2, \dots, \overline{c}^m$ are m new cost functions on the edges $e \in E$ obtained from $c^1, c^2, \dots, c^m$ as follows:

$$\overline{c}^i_{x,y} = \varepsilon^i_y - \varepsilon^i_x + c^i_{x,y}, \quad \forall (x, y) \in E, \ i = \overline{1, m}. \tag{3.35}$$

Then the dynamic c-games determined on the networks $(G, X_1, X_2, \dots, X_m, c^1, c^2, \dots, c^m, x_0, x_f)$ and $(G, X_1, X_2, \dots, X_m, \overline{c}^1, \overline{c}^2, \dots, \overline{c}^m, x_0, x_f)$, respectively, are equivalent because the payoff functions $H^i_{xx_f}(s^1, s^2, \dots, s^m)$ and $\overline{H}^i_{xx_f}(s^1, s^2, \dots, s^m)$, $i = 1, 2, \dots, m$ in such games differ only by a constant, i.e.,

$$H^i_{xx_f}(s^1, s^2, \dots, s^m) = \overline{H}^i_{xx_f}(s^1, s^2, \dots, s^m) + \varepsilon^i(x_f) - \varepsilon^i(x).$$

In [79, 83] the transformation (3.35) is named the *potential transformation* of the edges' costs of the players in the game.

Remark 3.33 The conditions of Theorem 3.31 ensure the existence of the optimal stationary strategies $s^{1*}, s^{2*}, \dots, s^{m*}$ of the players $1, 2, \dots, m$ for every starting position $x \in X$ in a dynamic c-game on the network $(G, X_1, X_2, \dots, X_m, c^1, c^2, \dots, c^m, x, x_f)$ with positive and constant cost functions $c^1, c^2, \dots, c^m$. If $c^1, c^2, \dots, c^m$ are arbitrary constant functions then the conditions of Theorem 3.31 represent necessary and sufficient conditions for the existence of optimal stationary strategies $s^{1*}, s^{2*}, \dots, s^{m*}$ in the dynamic c-game on the network $(G, X_1, X_2, \dots, X_m, c^1, c^2, \dots, c^m, x, x_f)$ for every starting position $x \in X$.

On the basis of the obtained results we can propose the following algorithm for determining Nash equilibria in the considered dynamic game with constant costs on the edges of the network.

Algorithm 3.34 Determining Nash Equilibria for the Dynamic c-Game on an Acyclic Network
Let us consider a dynamic c-game for which the graph $G = (X, E)$ has the structure of an acyclic directed graph with sink vertex x_f. Then a Nash equilibrium for the dynamic c-game on the network can be found as follows:

Preliminary step (Step 0): Fix $X^0 = \{x_f\}$ and put $\varepsilon^i(x_f) = 0$, $i = 1, 2, \dots, m$;

General step (Step k, k ≥ 1): If $X \setminus X^{k-1} = \emptyset$ then STOP; otherwise find a vertex $x^k \in X \setminus X^{k-1}$ for which $X_G(x^k) \subseteq X^{k-1}$, where $X_G(x^k) = \{y \in X \mid (x^k, y) \in E\}$. If $x^k \in X_{i_k}$, $i_k \in \{1, 2, \dots, m\}$, then find an edge (x^k, y^k) for which

$$\varepsilon^{i_k}_{y^k} + c^{i_k}_{x^k, y^k} = \min_{y \in X_G(x^k)} \{\varepsilon^{i_k}_y + c^{i_k}_{x^k, y}\}.$$

After that put

$$\varepsilon^i_{x^k} = \varepsilon^i_{y^k} + c^i_{x^k, y^k}, \quad i = 1, 2, \dots, m$$

and

$$X^k = X^{k-1} \cup \{x^k\}.$$

Then go to the next step.

If the functions ε^i, $i = 1, 2, \dots, m$, are known, then the optimal strategies of the players $s^{1*}, s^{2*}, \dots, s^{m*}$ can be found as follows:
Find a tree $GT_{s^*} = (X, E_{s^*})$ in the graph $\overline{G}^0 = (X, \overline{E}^0)$ and fix the strategies

$$s^i(x) : x \to y \in X_i, \quad (x, y) \in E_{s^*}, \quad i = 1, 2, \dots, m.$$

Example Let a dynamic c-game be given on an acyclic network with two players represented in Fig. 3.1, i.e., the network consists of a directed graph $G = (X, E)$ with partition $X = X_1 \cup X_2$, $X_1 = \{0, 1, 4, 5\}$, $X_2 = \{2, 3\}$, a starting position $x_0 = 0$, a final position $x_f = 5$ and costs of the players 1 and 2 as given in parenthesis in Fig. 3.1. In Fig. 3.1 the positions of the first player are represented by circles and the

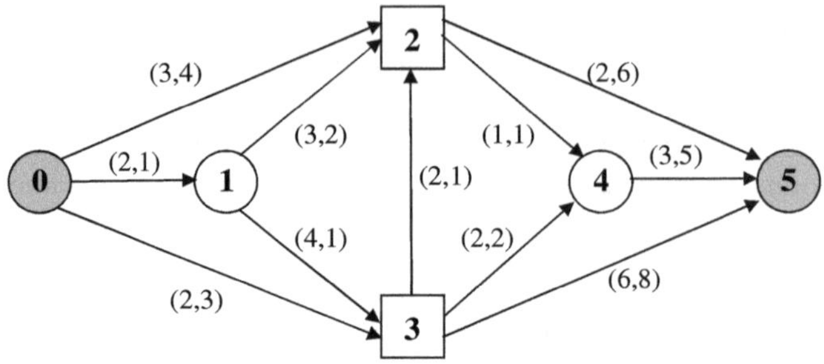

Fig. 3.1 The network for the acyclic c-game

positions of the second player are represented by squares. We consider the problem of determining optimal stationary strategies of the players in this dynamic c-game with an arbitrary starting position $x \in X$ and a fixed stopping position $x_f = 5$.

If we apply Algorithm 3.34 we obtain

Step 0
$X^0 = \{5\},\ \varepsilon_5^1 = 0,\ \varepsilon_5^2 = 0.$

Step 1
$X \setminus X^0 \neq \emptyset$, therefore, find a vertex $x^1 \in X \setminus X^0$ such that $X_G(x^1) \subseteq X^0$, i.e., $x^1 = 4$. Vertex 4 belongs to the set of positions of the first player and we calculate

$$\varepsilon^1(4) = \varepsilon_5^1 + 3 = 3; \quad \varepsilon_4^2 = \varepsilon_5^2 + 5 = 5;$$
$$X^1 = X^0 \cup \{4\} = \{5, 4\}.$$

Step 2
$X \setminus X^1 \neq \emptyset$ and find vertex $x^2 \in X \setminus X^1$ such that $X_G(x^2) \subseteq X^1$, i.e., $x^2 = 2$. Vertex 2 belongs to the set of positions of the second player and we calculate

$$\min\{\varepsilon_5^2 + c_{2,5}^2, \varepsilon_4^2 + c_{2,4}^2\} = \min\{6, 6\} = 6.$$

So, we obtain this minimum for the edges $(2, 4)$ and $(2, 5)$. Here we can fix an arbitrary edge from $\{(2, 4), (2, 5)\}$. For example, we fix edge $(2, 5)$. Then at step 2 we obtain

$$\varepsilon_2^2 = 6; \quad \varepsilon_2^1 = \varepsilon_5^1 + c_{2,5}^1 = 2;$$
$$X^2 = X^1 \cup \{2\} = \{2, 4, 5\}.$$

Step 3
$X \setminus X^2 \neq \emptyset$; $x^3 = 3$. Vertex 3 belongs to the set of positions of the second player, therefore, we find

$$\min\{\varepsilon_2^2 + c_{3,2}^2,\ \ \varepsilon_4^2 + c_{3,4}^2,\ \ \varepsilon^2(5) + c_{(3,5)}^2\} = 7.$$

So, we obtain this minimum for $e = (3, 2)$. We calculate

$$\varepsilon_3^2 = \varepsilon_2^2 + c_{3,2}^2 = 7;\quad \varepsilon_3^1 = \varepsilon_2^1 + c_{3,2}^1 = 4;$$
$$X^3 = X^2 \cup \{3\} = \{2, 3, 4, 5\}.$$

Step 4
$X \setminus X^3 \neq \emptyset$; $x^4 = 1$. Vertex 1 belongs to the set of positions of the first player, therefore, we find

$$\min\{\varepsilon_2^1 + c_{1,2}^1,\ \ \varepsilon_2^1 + c_{1,3}^1\} = 5.$$

So, we obtain this minimum for $e = (1, 2)$. We calculate

$$\varepsilon_1^1 = \varepsilon_2^1 + c_{1,2}^1 = 5;\quad \varepsilon_1^2 = \varepsilon_2^2 + c_{1,2}^2 = 8;$$
$$X^4 = X^3 \cup \{1\} = \{1, 2, 3, 4, 5\}.$$

Step 5
$X \setminus X^4 \neq \emptyset$; $x^5 = 0$. Vertex 1 belongs to the set of positions of the first player, therefore, we find

$$\min\{\varepsilon_2^1 + c_{0,2}^1,\ \ \varepsilon_1^1 + c_{0,1}^1,\ \ \varepsilon_3^1 + c_{0,3}^1\} = 5.$$

We determine

$$\varepsilon_0^1 = \varepsilon_2^1 + c_{0,2}^1 = 5;\quad \varepsilon_0^2 = \varepsilon_2^2 + c_{0,2}^2 = 10;$$
$$X^5 = X^4 \cup \{0\} = \{0, 1, 2, 3, 4, 5\}.$$

Step 6
$X \setminus X^5 = \emptyset$ STOP.

Thus, according to Theorem 3.31 we can determine a tree of the optimal paths $GT_{s^*} = (X, E_{s^*})$ that corresponds to the optimal stationary strategies of the players:

$$s^{1^*} : 0 \to 2;\quad 1 \to 2;\quad 4 \to 5;$$
$$s^{2^*} : 2 \to 5;\quad 3 \to 2.$$

The graph $GT_{s^*} = (X, E_{s^*})$, generated by the corresponding edges that determine ε_x^i, is represented in Fig. 3.2.

Note that at step 2 the minimal value ε_2^2 is not determined uniquely because $\varepsilon_5^2 + c_{2,5}^2 = \varepsilon_4^2 + c_{2,4}^2 = 6$. Therefore, if at this step we select the directed edge $(2, 4)$, then in the following the calculation procedure leads to another set of

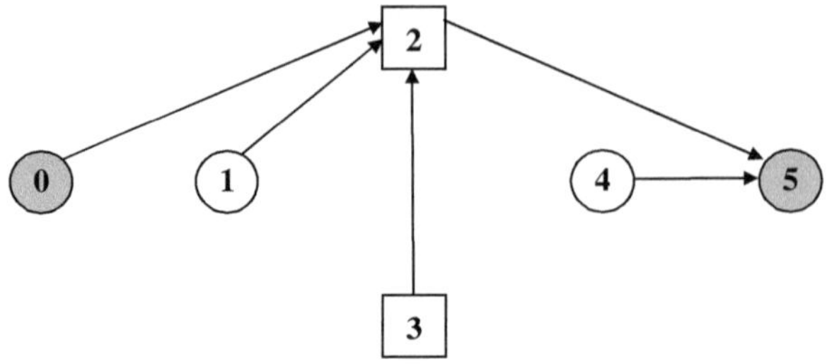

Fig. 3.2 The graph induced by the optimal strategies of the players

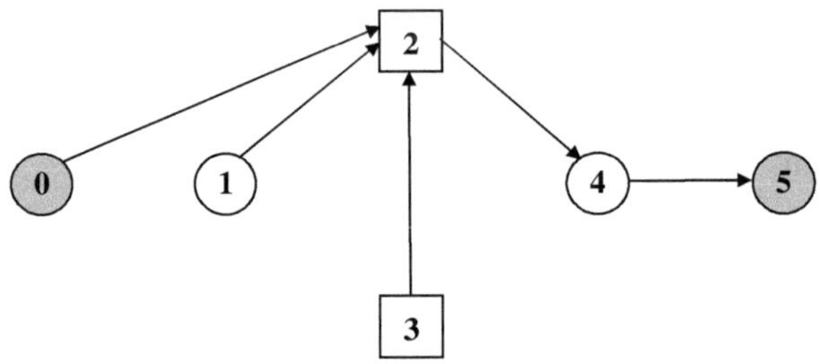

Fig. 3.3 The directed tree $GT_{s^*} = (X, E_{s^*})$ for the second solution

optimal stationary strategies of the players. Below the result of the algorithm with the mentioned alternative at Step 2 is presented.

Step 0
$\varepsilon_5^1 = 0, \quad \varepsilon_5^2 = 0.$

Step 1
$\varepsilon_4^1 = 3, \quad \varepsilon_4^2 = 5.$

Step 2
$\varepsilon_2^1 = 4, \quad \varepsilon_2^2 = 6.$

Step 3
$\varepsilon_3^1 = 6, \quad \varepsilon_3^2 = 7.$

Step 4
$\varepsilon_1^1 = 7, \quad \varepsilon_1^2 = 8.$

Step 5
$\varepsilon_0^1 = 7, \quad \varepsilon_0^2 = 10.$

In this case the corresponding directed tree $GT_{s^*} = (X, E_{s^*})$ is represented in Fig. 3.3. This directed tree corresponds to the optimal stationary strategies:

$$s^{1*}: 0 \to 2; \quad 1 \to 2; \quad 4 \to 5;$$
$$s^{2*}: 2 \to 4; \quad 3 \to 2.$$

Algorithm 3.35 Determining Nash Equilibria in a Dynamic c-Game on an Arbitrary Network Based on the Reduction in Comparison to the Case with an Acyclic Network

Let us consider a dynamic c-game with m players and let the directed graph G have an arbitrary structure, i.e., G may contain directed cycles. Moreover, we consider that for x_f there are no leaving edges $(x_f, x) \in E$. We show that in this case the problem can be reduced to the problem of finding optimal strategies in an auxiliary game with a network without directed cycles.

We construct an auxiliary directed graph $\overline{G} = (Z, \overline{E})$ without directed cycles, where Z and $\overline{E}$ are defined as follows:

$$Z = Z^0 \cup Z^1 \cup Z^2 \cup \cdots \cup Z^{|X|-1},$$

where

$$Z^j = \{z_0^j, z_1^j, z_2^j, \ldots, z_{|X|-1}^j\}, \quad j = 0, 1, 2, \ldots, |X| - 1,$$

so, $Z^0, Z^1, \ldots, Z^{|X|-1}$ represent the copies of the set X;

$$\overline{E} = E^0 \cup E^1 \cup E^2 \cup \cdots \cup E^{|X|-2} \cup E^f,$$

where

$$E^j = \{(z_k^j, z_l^{j+1}) \mid (x_k, x_l) \in E\}, \quad j = 0, 1, 2, \ldots, |X| - 2;$$
$$E^f = \{(z_k^j, z_f^{|X|-1}) \mid (x_k, x_f) \in E, \quad j = 0, 1, 2, \ldots, |X| - 3\}.$$

It is easy to observe that the vertex $z_f^{|X|-1}$ is attainable in this graph from any $z_k^0 \in Z^0$. If we delete in $\overline{G}$ all vertices z_k^i, for which there is no directed path from z_k^i to z_f^i, then we obtain an acyclic directed graph $\overline{G}' = (Z', \overline{E}')$ with sink vertex $z_f^{|X|-1}$. In the following we divide the vertex set Z' into m subsets $Z_1', Z_2', \ldots, Z_m'$ corresponding to the position sets of the players $1, 2, \ldots, m$, respectively:

$$Z_1' = \{z_k^j \in Z' \mid x_k \in X_1, \ j = 0, 1, 2, \ldots, |X| - 1\};$$
$$Z_2' = \{z_k^j \in Z' \mid x_k \in X_2, \ j = 0, 1, 2, \ldots, |X| - 1\};$$
$$\vdots$$
$$Z_m' = \{z_k^j \in Z' \mid x_k \in X_m, \ j = 0, 1, 2, \ldots, |X| - 1\}.$$

We define on the edge set $\overline{E}'$ the cost functions as follows:

$$\overline{c}^i_{z_k^j, z_l^{j+1}} = c^i_{x_k, x_l}, \ \forall (z_k^j, z_l^{j+1}) \in E^j, \ j = 0, 1, 2, \ldots, |X| - 2, \ i = 1, 2, \ldots, m;$$

$$\overline{c}^i_{z_k^j, z_f^{|X|-1}} = c^i_{x_k, x_f}, \ \forall (z_k^j, z_f^{|X|-1}) \in E^f, \ j = 0, 1, 2, \ldots, |X| - 3.$$

After that we consider a dynamic c-game on the network $(\overline{G},' Z_1', Z_2', \ldots, Z_m', \overline{c}^1, \overline{c}^2, \ldots, \overline{c}^m, z_0^0, z_f^{|X|-1})$, where $\overline{G}'$ is an acyclic directed graph with sink vertex $z_f^{|X|-1}$. If we use Algorithm 3.35, then we determine the values $\varepsilon^i_{z_k^j}$, $\forall z_k^j \in Z'$, $i = 1, 2, \ldots, m$. It is easy to observe that if we put $\varepsilon^i_{x_f} = 0$, $i = 1, 2, \ldots, m$, and $\varepsilon^i_{x_k} = \varepsilon^i_{z_k^{|X|-1}}, \forall x_k \in X \setminus \{x_f\}, i = 1, 2, \ldots, m$, then we obtain functions $\varepsilon^i : X \to \mathbb{R}$, which satisfy the conditions (1)–(3) from Theorem 3.31. Thus, we find the tree $GT = (X, E_s)$, which corresponds to the optimal strategies $s_1^*, s_2^*, \ldots, s_m^*$ of the players in our dynamic c-game.

Algorithm 3.35 is inconvenient because of the great number of vertices in the auxiliary network.

Furthermore, we present a simpler algorithm for finding optimal strategies of the players.

Algorithm 3.36 Determining Nash Equilibria for the Dynamic c-Game with an Arbitrary Network

Preliminary step: Assign to every vertex $x \in X$ a set of labels $\varepsilon_x^1, \varepsilon_x^2, \ldots, \varepsilon_x^m$ as follows:

$$\begin{aligned} \varepsilon^i_{x_f} &= 0, \quad \forall i = 1, 2, \ldots, m, \\ \varepsilon^i_x &= \infty, \quad \forall x \in X \setminus \{x_f\}, \ i = 1, 2, \ldots, m. \end{aligned}$$

General step (step k (k ≥ 1)): For every vertex $x \in X \setminus \{x_f\}$ change the labels $\varepsilon^i(x)$, $i = 1, 2, \ldots, m$, in the following way: If $x \in X_k$ then find vertex $\overline{x}$ for which

$$\varepsilon^k_{\overline{x}} + c^k_{x,\overline{x}} = \min_{y \in X(x)} \{\varepsilon^k_y + c^k_{x,y}\}.$$

If $\varepsilon^k_x > \varepsilon^k_{\overline{x}} + c^k_{x,\overline{x}}$, then replace $\varepsilon^i(x)$ by $\varepsilon^i_{\overline{x}} + c^i_{x,\overline{x}}$, $i = 1, 2, \ldots, p$. If $\varepsilon^k_x \leq \varepsilon^k_{\overline{x}} + c^k_{x,\overline{x}}$, then do not change the labels.

Repeat the general step $n-1$ times. Then the labels ε^i_x, $i = 1, 2, \ldots, m$ for $x \in X$ become constant.

Remark 3.37 In the algorithm the labels $\varepsilon^i(x)$, $i = 1, 2, \ldots, m$, may become constant after less than $n-1$ steps. So, the algorithm stops if the labels become constant.

Let us state that these labels satisfy the conditions of Theorem 3.31. Hence, using the labels ε^i_x, $i = 1, \ldots, m$, $x \in X$, and Theorem 3.31, we construct an optimal solution

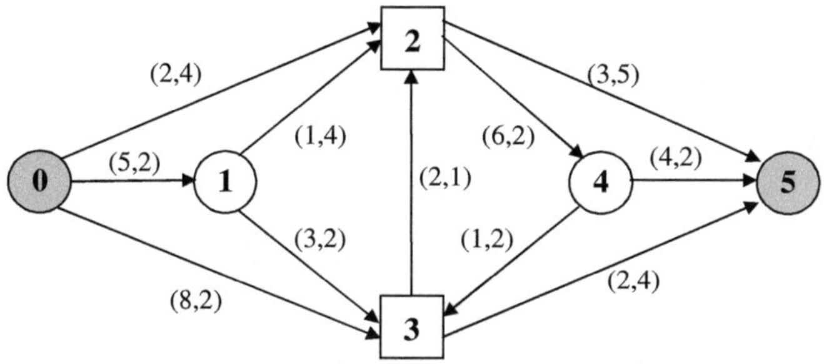

Fig. 3.4 The network for c-game that contains a directed cycle

characterized by Nash strategies of the players $1, 2, \ldots, m$. Algorithm 3.36 has the computational complexity $O(m|X|^2|E|)$.

Example Let a dynamic c-game of two players on the network represented by Fig. 3.4 be given. This network consists of a directed graph $G = (X, E)$ with sink vertex $x_f = 5$, given partition $X = X_1 \cup X_2$, $X_1 = \{0, 1, 4, 5\}$, $X_2 = \{2, 3\}$, and the costs for the players 1 and 2 written respectively in parenthesis in Fig. 3.4. We are seeking for optimal stationary strategies of the players in the dynamic c-game with an arbitrary starting position $x \in X$ and fixed stopping state $x_f = 5$.

Step 0 (Preliminary step)

Fix $\varepsilon_5^1 = 0$, $\varepsilon_5^2 = 0$;

$\varepsilon_0^1 = \varepsilon_1^1 = \varepsilon_2^1 = \varepsilon_3^1 = \varepsilon_4^1 = \infty$;

$\varepsilon_0^2 = \varepsilon_1^2 = \varepsilon_2^2 = \varepsilon_3^2 = \varepsilon_4^2 = \infty$.

We repeat the general step 5 times. At each step we examine each vertex $x \in X$ and update its labels $\varepsilon_x^1, \varepsilon_x^2$ by using the condition of the algorithm; we will examine the vertices according to their numerical order.

Step 1

Vertex $0 \in X_1$, therefore, calculate $\varepsilon_0^1 = \infty$; this implies $\varepsilon_0^2 = \infty$;

vertex $1 \in X_1$, therefore, calculate $\varepsilon_0^1 = \infty$; this implies $\varepsilon_1^2 = \infty$;

vertex $2 \in X_2$, therefore, calculate

$\varepsilon_5^2 + c_{2,5}^2 = \min\{\varepsilon_5^2 + c_{2,5}^2, \varepsilon_4^2 + c_{2,4}^2\} = \min\{5, \infty\} = 5$;

so, $\varepsilon_2^2 = 5$; this implies $\varepsilon_2^1 = \varepsilon_5^1 + c_{2,5}^1 = 3$;

vertex $3 \in X_2$, therefore, calculate

$\varepsilon_5^2 + c_{3,5}^2 = \min\{\varepsilon_5^2 + c_{3,5}^2, \varepsilon_2^2 + c_{3,2}^2\} = \min\{0 + 4, 5 + 1\} = 4$;

so, $\varepsilon_3^2 = 4$; this implies $\varepsilon_3^1 = \varepsilon_5^1 + c_{3,5}^1 = 0 + 2 = 2$;

vertex $4 \in X_1$, therefore, calculate

$\varepsilon_3^1 + c_{4,3}^1 = \min\{\varepsilon_5^1 + c_{4,5}^1, \varepsilon_3^1 + c_{4,3}^1\} = \min\{0+4, 2+1\} = 3;$
so, $\varepsilon_4^1 = 3$; this implies $\varepsilon_4^2 = \varepsilon_3^2 + c_{4,3}^2 = 4 + 2 = 6;$
vertex $5 \in X_1$; $\varepsilon_5^1 = 0$, $\varepsilon_5^2 = 0.$

Step 2

Vertex $0 \in X_1$, therefore, calculate $\varepsilon_0^1 = \infty$; this implies $\varepsilon_0^2 = \infty$;
$\varepsilon_2^1 + c_{0,2}^1 = \min\{\varepsilon_2^1 + c_{0,2}^1, \varepsilon_1^1 + c_{0,1}^1, \varepsilon_3^1 + c_{0,3}^1\} = \min\{3+2, \infty+5, 2+8\} = 5;$
so, $\varepsilon_0^1 = 5$; this implies $\varepsilon_2^2 = \varepsilon_2^2 + c_{0,2}^2 = 4 + 5 = 9;$
vertex $1 \in X_1$, therefore, calculate
$\varepsilon_2^1 + c_{1,2}^1 = \min\{\varepsilon_2^1 + c_{1,2}^1, \varepsilon_3^1 + c_{1,3}^1\} = \min\{3+1, 2+3\} = 4;$
so, $\varepsilon_1^1 = 4$; this implies $\varepsilon_1^2 = \varepsilon_2^2 + c_{1,2}^2 = 5 + 4 = 9;$
vertex $2 \in X_2$, therefore, calculate
$\varepsilon_5^2 + c_{2,5}^2 = \min\{\varepsilon_5^2 + c_{2,5}^2, \varepsilon_4^2 + c_{2,4}^2\} = \min\{5, 8\} = 5;$
so, $\varepsilon_2^2 = 5$; this implies $\varepsilon_2^1 = \varepsilon_5^1 + c_{2,5}^2 = 3;$
vertex $3 \in X_2$, therefore, calculate
$\varepsilon_5^2 + c_{3,5}^2 = \min\{\varepsilon_5^2 + c_{3,5}^2, \varepsilon_2^2 + c_{3,2}^2\} = \min\{4, 6\} = 4;$
so, $\varepsilon_3^2 = 4$; this implies $\varepsilon_3^1 = \varepsilon_5^1 + c_{3,5}^1 = 2;$
vertex $4 \in X_1$, therefore, calculate
$\varepsilon_3^1 + c_{1,3}^1 = \min\{\varepsilon_5^1 + c_{4,5}^1, \varepsilon_3^1 + c_{4,3}^1\} = \min\{4, 3\} = 3;$
so, $\varepsilon_4^1 = 3$ and $\varepsilon_4^2 = 6;$
vertex $5 \in X_1$; $\varepsilon_5^1 = 0$, $\varepsilon_5^2 = 0.$

Step 3

Vertex $0 \in X_1$, therefore, calculate
$\varepsilon_1^1 + c_{0,1}^1 = \min\{\varepsilon_2^1 + c_{0,2}^1, \varepsilon_1^1 + c_{0,1}^1, \varepsilon_3^1 + c_{0,3}^1\} = \min\{5, 9, 10\} = 5;$
so, $\varepsilon_0^1 = 5$ and $\varepsilon_0^2 = \varepsilon_2^2 + c_{0,2}^2 = 9;$
vertex $1 \in X_1$, therefore, calculate
$\varepsilon_2^1 + c_{1,2}^1 = \min\{\varepsilon_2^1 + c_{1,2}^1, \varepsilon_3^1 + c_{1,3}^1\} = \min\{3+1, 2+3\} = 4;$
so, $\varepsilon_1^1 = 4$ and $\varepsilon_1^2 = \varepsilon_2^2 + c_{1,2}^2 = 5 + 4 = 9;$
vertex $2 \in X_2$, therefore, calculate
$\varepsilon_5^2 + c_{2,5}^2 = \min\{\varepsilon_5^2 + c_{2,5}^2, \varepsilon_4^2 + c_{2,4}^2\} = \min\{5, 8\} = 5;$
so, $\varepsilon_2^2 = 5$ and $\varepsilon_2^1 = 3;$
vertex $3 \in X_2$, therefore, calculate
$\varepsilon_5^2 + c_{3,5}^2 = \min\{\varepsilon_5^2 + c_{3,5}^2, \varepsilon_2^2 + c_{3,2}^2\} = \min\{4, 6\} = 4;$
so, $\varepsilon_3^2 = 4$ and $\varepsilon^1(3) = 2;$

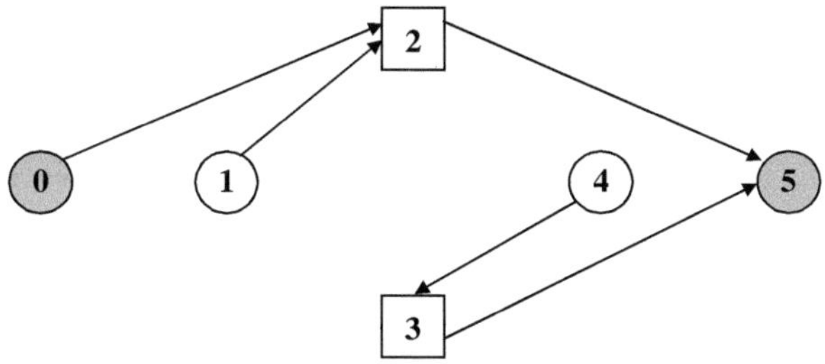

Fig. 3.5 The tree of the optimal paths

vertex $4 \in X_1$, therefore, calculate

$\varepsilon_3^1 + c_{4,3}^1 = \min\{\varepsilon_5^1 + c_{4,5}^1, \varepsilon_3^1 + c_{4,3}^1\} = \min\{4, 3\} = 3;$

so, $\varepsilon_4^1 = 4$ and $\varepsilon_4^2 = 2$.

After step 3 we observe that the labels coincide with the labels after step 2. So, the labels become constant and we finish the algorithm. Finally we have obtained

$$\varepsilon_0^1 = 5, \ \varepsilon_1^1 = 4, \ \varepsilon_2^1 = 3, \ \varepsilon_3^1 = 2, \ \varepsilon_4^1 = 3, \ \varepsilon_5^1 = 0;$$
$$\varepsilon_0^2 = 9, \ \varepsilon_1^2 = 9, \ \varepsilon_2^2 = 5, \ \varepsilon_3^2 = 4, \ \varepsilon_4^2 = 6, \ \varepsilon_5^2 = 0.$$

Thus, if we make the potential transformation of the edges' costs then we can select the tree $GT_{s^*} = (X, E_{s^*})$ with zero cost of edges that satisfy the conditions of Theorem 3.31. This tree is represented in Fig. 3.5.

So, the optimal stationary strategies of the players are the following:

$$s^{1*} : 0 \to 2; \quad 1 \to 2; \quad 4 \to 3;$$
$$s^{2*} : 2 \to 5; \quad 3 \to 5.$$

In [76, 77, 79] the non-stationary version of a dynamic c-game has been considered. A similar result for the non-stationary game is obtained and algorithms for determining optimal strategies of the players are derived. Moreover, in [79] this game model has been formulated and studied with the condition that the stopping position x_f should be reached at the moment of time $t(x_f)$, such that $t_1 \leq t(x_f) \leq t_2$, where t_1 and t_2 are given. However, efficient polynomial time algorithms for determining Nash equilibria in such games can be derived only for the non-stationary case.

3.3.4 On Determining Nash Equilibria for Stationary Dynamic c-Games and Multicriteria Problems with Restrictions on the Number of Moves

The stationary dynamic c-game with restrictions on the number of moves of the players has been formulated in [79]. In this game each player has the aim to minimize his integral-time cost and to reach the final state x_f at the moment of time

$t(x_f)$, such that $t_1 \le t(x_f) \le t_2$, where t_1 and t_2 are given. The payoff functions $H^i_{x_0x_f}(s^1, s^2, \dots, s^m)$, $i = 1, 2, \dots, m$ in this game are defined as follows.

Let $s^1, s^2, \dots, s^m$ be the set of stationary strategies of the players and $G_s = (X, E_s)$ be the graph generated by these strategies. Then we set

$$H^i_{x_0x_f}(s^1, s^2, \dots, s^m) = \sum_{e \in E(P_s(x_0, x_f))} c^i_e, \quad i = 1, 2, \dots, m$$

if $t_1 \le |E(P_s(x_0, x_f))| \le t_2$; otherwise we put $H^i_{x_0x_f}(s^1, s^2, \dots, s^m) = +\infty$. In this case the problem of determining a Nash equilibrium in a dynamic c-game is NP-hard [38, 79]. This problem is complicate even in the case $m = 1$ because for $t_1 = t_2 = |X| - 1$ and unit costs on the edges it is transformed into the Hamiltonian path problem in a directed graph [38]. We have the same situation for the stationary multicriteria problems of determining Pareto optima with a fixed number of transitions. For the non-stationary versions of the considered problems there exist polynomial time algorithms for determining Nash equilibria and Pareto optima. Such algorithms are elaborated and formulated in [79].

3.4 Determining Pareto Solutions for Multicriteria Markov Decision Problems

We formulate multicriteria models for Markov decision problems with average and discounted optimization criteria applying the concept of cooperative games to the framework of a Markov decision process (X, A, p, c) with a finite set of states X, a finite set of actions A, a transition probability function $p : X \times X \times A \to [0, 1]$ that satisfies the condition

$$\sum_{y \in X} p^a_{x,y} = 1, \quad \forall x \in X, \quad \forall a \in A$$

and a transition cost function $c : X \times X \to \mathbb{R}$ which gives the costs $c_{x,y}$ if the dynamical system makes a transition from $x \in X$ to $y \in X$.

We assume that in the decision process m actors (players) participate and for each player $i \in \{1, 2, \dots, m\}$ a transition cost function $c^i : X \times X \to \mathbb{R}$, is given that determines the costs $c^i_{x,y}$ for the player i if the system makes transitions from x to y. Additionally, we assume that the starting state x_0 is known and that the players use a common stationary strategy

$$s : x \to a \in A(x) \quad \text{for } x \in X$$

of fixing the actions in the states.

A fixed strategy s generates a Markov process for which we can determine the average cost $\omega^i_{x_0}(s)$ and the expected discounted cost $\sigma^i_{x_0}(s)$ for a given discount factor γ with respect to each player $i \in \{1, 2, \dots, m\}$.

If we denote $F^i_{x_0}(s) = \omega^i_{x_0}(s)$, $\widehat{F}^i_{x_0}(s) = \omega^i_{x_0}(s)$, $i = 1, 2, \dots, m$, then on the set of a stationary strategies S we obtain the vector functions

$$F_{x0}(s) = (F^1_{x0}(s), F^2_{x0}(s), \dots, F^m_{x0}(s)),$$
$$\widehat{F}_{x0}(s) = (\widehat{F}^1_{x0}(s), \widehat{F}^2_{x0}(s), \dots, \widehat{F}^m_{x0}(s)).$$

In such a way we obtain multicriteria decision problems for which Pareto solutions must be found. Each of the considered multicriteria decision problems can be reduced to a single-objective problem using on S the corresponding convolution criteria

$$\min \rightarrow F'_{x_0}(s) = \sum_{i=1}^{m} \theta_i F^i_{x_0}(s),$$
$$\min \rightarrow \widehat{F}'_{x_0}(s) = \sum_{i=1}^{m} \theta_i \widehat{F}^i_{x_0}(s),$$

where $\theta_i > 0$, $i = 1, 2, \dots, m$; $\sum_{i=1}^{m} \theta_i = 1$. Note that such a reduction of the multicriteria decision problems to single-objective problems allows us to find Pareto solutions, however, the approach mentioned above may not find all Pareto solutions of the corresponding multi-objective problems (see [30]). Some classes of the multicriteria Markov decision problems and approaches for determining Pareto solutions are described in [27, 40, 133, 137–139].

3.5 Deterministic Antagonistic Positional Games on Networks and Algorithms for Finding Optimal Strategies of the Players

The mathematical tool we develop in this section is related to deterministic antagonistic positional games on networks. We propose polynomial-time algorithms for finding max-min paths in networks and for determining optimal strategies of the players in antagonistic positional games. These algorithms are applied for studying and solving cyclic games. The computational complexity of the proposed algorithms is analyzed.

3.5.1 Zero-Sum Games on Networks and Polynomial Time Algorithms for Max-Min Path Problems

In the previous section we have studied dynamic c-games with positive cost functions on the edges. Therefore, we cannot use those results for zero-sum games. In the following we study zero-sum games of two players with arbitrary cost functions on the edges and propose polynomial-time algorithms for their solving. The main results related to this problem have been obtained in [29, 43, 65–67, 83, 142]. At first we study a max-min path problem on networks, which generalizes classical combinatorial problems of the shortest and the longest paths in weighted directed graphs.

This max-min paths problem arose as an auxiliary one if optimal stationary strategies of the players in cyclic games are searched. In addition, we shall use the considered dynamic c-game for studying and solving zero-sum control problems with alternate player's control [83]. The main results are concerned with the existence of polynomial-time algorithms for determining max-min paths in networks as well as with an elaboration of such algorithms.

Let $G = (X, E)$ be a directed graph with a vertex set X and an edge set E. Assume that G contains a vertex $x_f \in X$ such that it is attainable from each vertex $x \in X$, i.e., x_f is a sink vertex in G. On the edge set E a function $c : E \to \mathbb{R}$, which assigns a cost c_e to each edge $e \in E$ is given. In addition the vertex set is divided into two disjoint subsets X_1 and X_2 ($X = X_1 \cup X_2$, $X_1 \cap X_2 = \emptyset$), which we regard as position sets of two players.

On G we consider a game of two players. The game starts at the position $x_0 \in X$. If $x_0 \in X_1$, then the move is done by the first player, otherwise it is done by the second one. The move indicates the passage from the position x_0 to the neighbour position x_1 through the edge $e_1 = (x_0, x_1) \in E$. After that if $x_1 \in X_1$, then the move is done by the first player, otherwise it is done by the second one and so on. As soon as the final position is reached the game is over. The game can be finite or infinite. If the final position x_f is reached in finite time, then the game is finite. In the case if the final position x_f is not reached, the game is infinite. The first player in this game has the aim to maximize $\sum_i c_{e_i}$ while the second one has the aim to minimize $\sum_i c_{e_i}$.

The considered game in normal form in the terms of stationary strategies can be defined as follows: We identify the stationary strategies s^1 and s^2 of the players with the maps

$$s^1 : x \to y \in X(x) \quad \text{for} \quad x \in X_1;$$
$$s^2 : x \to y \in X(x) \quad \text{for} \quad x \in X_2,$$

where $X(x)$ represents the set of extremities of edges $e = (x, y) \in E$, i.e. $X(x) = \{y \in X \mid e = (x, y) \in E\}$. Since G is a finite graph then the sets of strategies of the players

$$\mathbb{S}^1 = \{s^1 : x \to y \in X(x) \quad \text{for} \quad x \in X_1\};$$
$$\mathbb{S}^2 = \{s^2 : x \to y \in X(x) \quad \text{for} \quad x \in X_2\}$$

are finite sets. The payoff function $H_{x_0}(s^1, s^2)$ on $\mathbb{S}^1 \times \mathbb{S}^2$ is defined by the following way:

Let in G be a subgraph $G_s = (X, E_s)$ generated by edges of the form $(x, s^1(x))$ for $x \in X_1$ and $(x, s^2(x))$ for $x \in X_2$. Then either a unique directed path $P_s(x_0, x_f)$ from x_0 to x_f exists in G_s or such a path does not exist in G_s. In the second case in G_s there exists a unique directed cycle C_s, which can be reached from x_0.

For given s^1 and s^2 we set

$$H_{x_0}(s^1, s^2) = \sum_{e \in E(P_s(x_0, x_f))} c_e,$$

if in G_s there exists a directed path $P_s(x_0, x_f)$ from x_0 to x_f, where $E(P_s(x_0, x_f))$ is a set of edges of the directed path $P_s(x_0, x_f)$. If in G there is no directed path from x_0 to x_f, then we define $H_{x_0}(s^1, s^2)$ as follows: Let $P_s'(x_0, y')$ be a directed path, which connects the vertex x_0 with the cycle C_s and $P_s'(x_0, y')$ has no other common vertices with C_s except y'. Then we put

$$H_{x_0}(s^1, s^2) = \begin{cases} +\infty, & \text{if} \sum_{e \in E(C_s)} c_e > 0; \\ \sum_{e \in E(P_s'(x_0, y'))} c_e, & \text{if} \sum_{e \in E(C_s)} c_e = 0; \\ -\infty, & \text{if} \sum_{e \in E(C_s)} c_e < 0. \end{cases}$$

This game is related to zero-sum positional games of two players and it is determined by the graph G with the sink vertex x_f, the partition $X = X_1 \cup X_2$, the cost function $c : E \to \mathbb{R}$ and the starting position x_0. We denote the network, which determines this game, by $(G, X_1, X_2, c, x_0, x_f)$. In the case when the dynamic c-game is considered for an arbitrary starting position $x \in X$ we shall use the notation (G, X_1, X_2, c, x_f).

In [83, 85] it is shown that if G does not contain directed cycles, then for every $x \in X$ the following equality holds

$$v(x) = \max_{s^1 \in \mathbb{S}^1} \min_{s^2 \in \mathbb{S}^2} H_x(s^1, s^2) = \min_{s^2 \in \mathbb{S}^2} \max_{s^1 \in \mathbb{S}^1} H_x(s^1, s^2), \tag{3.36}$$

which means the existence of optimal strategies of the players in the considered game. Moreover, in [83, 85] it is shown that in G there exists a tree $GT^* = (X, E^*)$ with the sink vertex x_f, which gives the optimal strategies of the players in the game for an arbitrary starting position $x_0 \in X$. The strategies of the players are obtained by fixing

$$s^{1^*}(x) = y, \quad \text{if} \quad (x, y) \in E^* \text{ and } x \in X_1 \setminus \{x_f\};$$
$$s^{2^*}(x) = y, \quad \text{if} \quad (x, y) \in E^* \text{ and } x \in X_2 \setminus \{x_f\}.$$

In the general case for an arbitrary graph G equality (3.36) may fail to hold. Therefore we formulate necessary and sufficient conditions for the existence of optimal strategies of the players in this game and propose a polynomial-time algorithm for determining the tree of max-min paths from every $x \in X$ to x_f. Furthermore we show that our max-min paths problem on the network can be regarded as a zero value ergodic cyclic game. So, the proposed algorithm can be used for solving such games.

In [43, 65, 83] the formulated game on network $(G, X_1, X_2, c, x_0, x_f)$ is named the dynamic c-game. Some preliminary results related to this problem have been obtained in [83, 85]. More general models of positional games on networks with p players have been studied in [9, 74–76, 81–83].

The considered max-min paths problem can be used for the zero-sum control problem with an alternate players' control (see Sect. 3.9).

3.5.2 An Algorithm for Solving the Problem on Acyclic Networks

The formulated problem for acyclic networks has been studied in [76, 83, 85].

Let $G = (X, E)$ be a finite directed graph without directed cycles and a given sink vertex x_f. The partition $X = X_1 \cup X_2$ $(X_1 \cap X_2 = \emptyset)$ of the vertex set of G is given and the cost function $c : E \to \mathbb{R}$ on the edges is defined. We consider the dynamic c-game on G with a given starting position $x \in X$.

It is easy to observe that for fixed strategies of players $s^1 \in \mathbb{S}^1$ and $s^2 \in \mathbb{S}^2$ the subgraph $G_s = (X, E_s)$ has a structure of a directed tree with sink vertex $x_f \in X$. This means that the value $H_x(s^1, s^2)$ is determined uniquely by the sum of edge costs of the unique directed path $P_s(x, x_f)$ from x to x_f. In [83, 85] it is proved that for an acyclic c-game on network (G, X_1, X_2, c, x, x_f) there exist the strategies of players s^{1^*}, s^{2^*} such that

$$\begin{aligned} v(x) = H_x(s^{1^*}, s^{2^*}) &= \max_{s^1 \in \mathbb{S}^1} \min_{s^2 \in S^2} H_x(s^1, s^2) \\ &= \min_{s^2 \in \mathbb{S}^2} \max_{s^1 \in \mathbb{S}^1} H_x(s^1, s^2) \end{aligned} \tag{3.37}$$

and s^{1^*}, s^{2^*} do not depend on a starting position $x \in X$, i.e. (3.37) holds for every $x \in X$.

The equality (3.37) is evident in the case if $\text{ext}(c, x) = 0$, $\forall x \in X \setminus \{x_f\}$, where

$$\operatorname{ext}(c,x) = \begin{cases} \max\limits_{y \in X(x)} c_{(x,y)}, & x \in X_1; \\ \min\limits_{y \in X(x)} c_{(x,y)}, & x \in X_2. \end{cases}$$

In this case $v(x) = 0, \forall x \in X$ and the optimal strategies of the players can be obtained by fixing the maps $s^{1^*} : X_1 \setminus \{x_f\} \to X$ and $s^{2^*} : X_2 \setminus \{x_f\} \to X$ such that $s^{1^*} \in \operatorname{VEXT}(c,x)$ for $x \in X_1 \setminus \{x_f\}$ and $s^{2^*} \in \operatorname{VEXT}(c,x)$ for $x \in X_2 \setminus \{x_f\}$, where

$$\operatorname{VEXT}(c,x) = \{y \in X(x) \mid c_{(x,y)} = \operatorname{ext}(c,x)\}.$$

If the network $(G, X_1, X_2, c, x_0, x_f)$ has the property that $\operatorname{ext}(c,x) = 0$, $\forall x \in X \setminus \{x_f\}$, then it is named the network in canonic form. So, for the acyclic c-game on the network in canonic form equality (3.37) holds and $v(x) = 0, \forall x \in X$.

In the general case equality (3.37) can be proved by using properties of the potential transformation $c'_{(x,y)} = c_{(x,y)} + \varepsilon_y - \varepsilon_x$ on the edges $e = (x,y)$ of the network, where $\varepsilon : X \to \mathbb{R}$ is an arbitrary real function on X (the potential transformation for antagonistic positional games has been introduced in [43, 76]). The fact is that such a transformation of the costs on the edges of the acyclic network in a c-game does not change the optimal strategies of the players, although values $v(x)$ of positions $x \in X$ are changed by $v(x) + \varepsilon_{x_f} - \varepsilon_x$. It means that for an arbitrary function $\varepsilon : X \to \mathbb{R}$ the optimal strategies of the players in acyclic c-games on the networks $(G, X_1, X_2, c, x_0, x_f)$ and $(G, X_1, X_2, c', x_0, x_f)$ are the same.

We assume that the vertices $x \in X$ of the acyclic graph G are numbered with $1, 2, \ldots, |X|$, such that if $x > y$ then in G there is no directed path from y to x. For acyclic graphs such a numbering of the vertices is possible and can be made starting from a sink vertex. Therefore, we can use the following recursive formula

$$\varepsilon(x) = \begin{cases} \max\limits_{y \in X(x)} \{c_{(x,y)} + \varepsilon_y\} & \text{for } x \in X_1 \setminus \{x_f\}; \\ \min\limits_{y \in X(x)} \{c_{(x,y)} + \varepsilon_y\} & \text{for } x \in X_2 \setminus \{x_f\} \end{cases} \tag{3.38}$$

to tabulate the values $\varepsilon_x, \forall x \in X$ starting with $\varepsilon(x_f) = 0$. It is evident that the transformation $c'_{(x,y)} = c_{(x,y)} + \varepsilon_y - \varepsilon_x$ satisfies the condition $\operatorname{ext}(c',x) = 0$, $\forall x \in X$. This means that the following theorem holds.

Theorem 3.38 *For an arbitrary acyclic network* $(G, X_1, X_2, c, x_0, x_f)$ *with a sink vertex* x_f *there exists a function* $\varepsilon : X \to \mathbb{R}$ *which determines the potential transformation* $c'_{(x,y)} = c_{(x,y)} + \varepsilon_y - \varepsilon_x$ *on the edges* $e = (x,y)$ *such that the network* $(G, X_1, X_2, c, x_0, x_f)$ *has the canonic form. The values* $\varepsilon_x, x \in X$, *which determine function* $\varepsilon : X \to \mathbb{R}$, *can be found by using recursive formula* (3.38).

On the basis of this theorem the following algorithm for determining optimal strategies of the players in the c-game is proposed in [76].

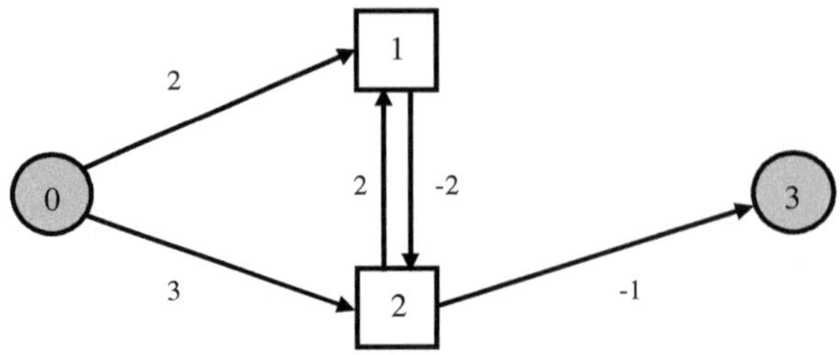

Fig. 3.6 The network for an antagonistic game that has no saddle points

Algorithm 3.39 Determining Optimal Strategies for the Players on an Acyclic Network

1. Determine the values $\varepsilon(x), x \in X,$ according to the recursive formula (3.38) and the corresponding potential transformation $c'_{(x,y)} = c_{(x,y)} + \varepsilon_y - \varepsilon_x$ on the edges $(x, y) \in E$.
2. Fix arbitrary maps $s^{1*}(x) \in \text{VEXT}(c', x)$ for $x \in X_1 \setminus \{x_f\}$ and $s^{2*}(x) \in \text{VEXT}(c', x)$ for $x \in X_2 \setminus \{x_f\}$.

Remark 3.40 The values $\varepsilon_x, x \in X,$ represent the values of the acyclic c-game on $(G, X_1, X_2, c, x_0, x_f)$ with starting position x, i.e. $\varepsilon(x) = v(x), \forall x \in X$. Algorithm 3.39 needs $O(|X|^2)$ elementary operations because the tabulation of the values $\varepsilon(x), x \in X,$ using formula (3.38) for acyclic networks requires this number of operations.

3.5.3 The Main Results for the Problem on an Arbitrary Network

First of all we give an example which shows that equality (3.36) may fail to hold. In Fig. 3.6 the network with the starting position $x_0 = 0$ and the final position $x_f = 3$ is given, where positions of the first player are represented by circles and positions of the second player are represented by squares; values of cost functions on the edges are given alongside them.

It is easy to observe that

$$\max_{s^1 \in \mathbb{S}^1} \min_{s^2 \in \mathbb{S}^2} H_0(s^1, s^2) = 2, \quad \min_{s^2 \in \mathbb{S}^2} \max_{s^1 \in \mathbb{S}^1} H_0(s^1, s^2) = 3.$$

The following theorem gives conditions for the existence of a saddle point with finite $v(x)$ for each $x \in X$ in the c-game.

Theorem 3.41 *Let* $(G, X_1, X_2, c, x_0, x_f)$ *be an arbitrary network with the sink vertex* $x_f \in X$. *In addition assume that* $\sum_{e \in E(C_s)} c_e \neq 0$ *for every directed cycle* C_s *from* G. *Then for the* c*-game on* $(G, X_1, X_2, c, x_0, x_f)$ *condition* (3.36) *with finite*

$v(x)$ holds for every $x \in X$ if and only if there exists a function $\varepsilon : X \to \mathbb{R}$, which determines the potential transformation $c'_{(x,y)} = c_{(x,y)} + \varepsilon_y - \varepsilon_x$ on edges $(x, y) \in E$ such that $\text{ext}(c', x) = 0$, *$\forall x \in X$. Moreover, if in G there exists the potential transformation $c'_{(x,y)} = c_{(x,y)} + \varepsilon_y - \varepsilon_x$ on edges $(x, y) \in E$ such that* $\text{ext}(c', x) = 0$, $\forall x \in X \setminus \{x_f\}$, *then $v(x) = \varepsilon_x - \varepsilon_{x_f}$, $\forall x \in X$.*

Proof $\Longrightarrow$ Let us consider that $\sum_{e \in E(C_s)} c_e \neq 0$ for every directed cycle C_s in G and condition (3.36) holds for every $x \in X$. Moreover, we consider that $v(x)$ is a finite value for every $x \in X$.

Taking into account that the potential transformation does not change the cost of the cycles, we obtain that this transformation does not change the optimal strategies of the players although values $v(x)$ of the positions $x \in X$ are changed by $v(x) - \varepsilon_x + \varepsilon_{x_f}$. It is easy to observe that if we put $\varepsilon_x = v(x)$ for $x \in X$, then the function $\varepsilon : X \to \mathbb{R}$ determines the potential transformation $c'_{(x,y)} = c_{(x,y)} + \varepsilon_y - \varepsilon_x$ on the edges $(x, y) \in E$ such that $\text{ext}(c', x) = 0$, $\forall x \in X$.

$\Longleftarrow$ Let us consider that there exists the potential transformation $c'_{(x,y)} = c_{(x,y)} + \varepsilon_y - \varepsilon_x$ on the edges $(x, y) \in E$ such that $\text{ext}(c', x) = 0$, $\forall x \in X$. The value $v(x)$ of the game after the potential transformation is zero for every $x \in X$ and the optimal strategies of the players can be found by fixing s^{1*} and s^{2*} such that $s^{1*}(x) \in \text{VEXT}(c', x)$ for $x \in X_1 \setminus \{x_f\}$ and $s^{2*}(x) \in \text{VEXT}(c', x)$ for $x \in X_2 \setminus \{x_f\}$. Since the potential transformation does not change the optimal strategies of the players we put $v(x) = \varepsilon_x - \varepsilon_{x_f}$ and obtain (3.36). □

Corollary 3.42 *If for every directed cycle C_s in G the condition $\sum_{e \in E(G_s)} c_e \neq 0$ and equality* (3.36) *hold then there exists the potential transformation $\varepsilon : X \to \mathbb{R}$ such that* $\text{ext}(c', x) = 0$, $\varepsilon(x_f) = 0$ *and* $v(x) = \varepsilon_x$, $\forall x \in X$.

Corollary 3.43 *If for every directed cycle C_s in G the condition $\sum_{e \in E(G_s)} c_e \neq 0$ holds then the existence of the potential transformation $c'_{(x,y)} = c_{(x,y)} + \varepsilon_y - \varepsilon_x$ on edges $(x, y) \in E$ such that*

$$\text{ext}(c', x) = 0, \forall x \in X \tag{3.39}$$

represents necessary and sufficient conditions for the validity of the equality (3.36) *for every $x \in X$. In the case if in G there exists a cycle C_s with $\sum_{e \in E(C_s)} c_e = 0$ condition* (3.39) *becomes only a necessary one for the validity of the equality* (3.36) *for every $x \in X$.*

Corollary 3.44 *If in the c-game there exist the strategies s^{1*} and s^{2*}, for which* (3.36) *holds for every $x \in X$, and these strategies generate in G a tree $T_{s^*} = (X, E_{s^*})$ with the sink vertex x_f, then there exists the potential transformation $c'_{(x,y)} = c_{(x,y)} + \varepsilon_y - \varepsilon_x$ on the edges $(x, y) \in E$ such that the graph $G^0 = (X, E^0)$, generated by the set of the edges $E^0 = \{(x, y) \in E | c'_{(x,y)} = 0\}$, contains the tree T_{s^*} as a subgraph.*

Taking into account the results mentioned above we propose an algorithm for determining the optimal strategies of the players in a c-game based on the construction of the tree of max-min paths. This algorithm works if such a tree in G exists.

3.5.4 A Polynomial Time Algorithm for Determining the Optimal Strategies of the Players in an Antagonistic Dynamic c-Game

We consider the dynamic c-game determined by the network (G, X_1, X_2, c, x_f) where the graph G has a sink vertex x_f. At first we assume that for an arbitrary vertex there exists the value $v(x)$ which satisfies condition (3.39) and $v(x) \neq \pm\infty$. So, we assume that in G there exists a tree of max-min paths from $x \in X$ to x_f. We show that for determining the optimal strategies of the players in the considered game there exists a polynomial time algorithm. In this section we propose such an algorithm based on the reduction to an auxiliary dynamic c-game with an acyclic network $(\overline{G}, W_1, W_2, \overline{c}, w_f^0)$, where graph $\overline{G} = (W, \overline{E})$ with $W = W_1 \cup W_1$ is obtained from $G = (X, E)$ in the following way:

The set of vertices W consists of $n - 1$ copies of the vertex set X and the sink vertex w_f^0, i.e.

$$W = \{w_f^0\} \cup W^1 \cup W^2 \cup \dots \cup W^{n-1},$$

where $W^i = \{w_0^i, w_1^i, \dots, w_{n-1}^i\}$, $i = 1, 2, \dots, n-1$. Here $W^i \cap W^j = \emptyset$ for $i \neq j$ and vertices $w_k^i \in W^i$, $i = 1, 2, \dots, n-1$, correspond to vertex x_k from $X = \{x_0, x_1, x_2, \dots, x_{n-1}\}$.

The set of edges $\overline{E}$ is defined by the following way:

$$\begin{aligned}
\overline{E} &= E^0 \cup E^1 \cup E^2 \cup \dots \cup E^{n-1};\\
E^i &= \{(w_k^{i+1}, w_l^i) \mid (x_k, x_l) \in E\},\ i = 1, 2, \dots, n-2;\\
E^0 &= \{(w_k^i, w_f^0) \mid (x_k, x_f) \in E,\ i = 1, 2, \dots, n-1\}.
\end{aligned}$$

In $\overline{G}$ the edge subset $E^i \subseteq \overline{E}$ connects vertices of the set W^{i+1} with vertices of the set W^i by the edges (w_k^{i+1}, w_l^i) if in G there exists a directed edge (x_k, x_l). In addition in $\overline{G}$ each vertex w_k^i, $i = 1, 2, \dots, n-1$, is connected with a sink vertex w_f^0 by the edge (w_k^i, w_f^0) if in G there exists a directed edge (x_k, x_f).

The subsets W_1, W_2 and the cost function $\overline{c}\colon \overline{E} \to \mathbb{R}$ are defined as follows:

$$W_1 = \{w_k^i \in W \mid x_k \in X_1\},\ W_2 = \{w_k^i \in W \mid x_k \in X_2\};$$

$$\begin{aligned}
\overline{c}_{(w_k^{i+1}, w_l^i)} &= c_{(x_k, x_l)},\ (x_k, x_l) \in E \ \text{ and } \ (w_k^{i+1}, w_l^i) \in E^i;\ i = 1, 2, \dots, n-2;\\
\overline{c}_{(w_k^i, w_f^0)} &= c_{(x_k, x_f)},\ (x_k, x_f) \in E \ \text{ and } \ (w_k^i, w_f^0) \in E^0;\ i = 1, 2, \dots, n-1.
\end{aligned}$$

From $\overline{G}$ we delete all vertices w_k^i for which there are no directed paths from w_k^i to w_f^0. For the obtained directed graph we will preserve the same notation and we will keep in mind that $\overline{G}$ does not contain such vertices.

Let us consider the dynamic c-game determined by the acyclic network $(\overline{G}, W_1, W_2, \overline{c}, w_f^0)$ with sink vertex w_f^0. So, we consider the problem of determining the values $v'(w_k^i)$ of the game for every $w_k^i \in W$.

We show that if $v'(w_k^1), v'(w_k^2), \ldots, v'(w_k^{n-1})$ are the corresponding values of the vertices $w_k^1, w_k^2, \ldots, w_k^{n-1}$ in the auxiliary game, then there exists $i \in \{1, n-1\}$ such that $v(x_k) = v'(w_k^i)$. We seek the vertex w_k^i among $w_k^{n-1}, w_k^{n-2}, \ldots, w_k^2, w_k^1$ that starts with the highest level set W^{n-1}.

We consider in $\overline{G}$ the max-min path

$$P_{\overline{G}}(w_k^{n-1}, w_f^0) = \{w_k^{n-1}, w_{k_1}^{n-2}, w_{k_2}^{n-3}, \ldots, w_{k_r}^{n-r-1}, w_f^0\}$$

from w_k^{n-1} to w_f^0 generated by directed edges $e = (w_{k_i}^{n-i-1}, w_{k_{i+1}}^{n-i})$ for which

$$\varepsilon'_{(w_{k_{i+1}}^{n-i})} - \varepsilon'_{(w_{k_i}^{n-i-1})} + \overline{c}_{(w_{k_i}^{n-i-1}, w_{k_{i+1}}^{n-i})} = 0,$$

where

$$\varepsilon'_{(w_k^j)} = v'(w_k^j), \quad \forall w_k^j \in \{w_k^{n-1}, w_{k_1}^{n-2}, w_{k_2}^{n-3}, \ldots, w_{k_r}^{n-r-1}, w_f^0\}.$$

The directed path $P_{\overline{G}}(w_k^{n-1}, w_f^0)$ corresponds in G to a directed path

$$P_G(x_k, x_f) = \{x_k, x_{k_1}, x_{k_2}, \ldots, x_{k_r}, x_f\}$$

from x_k to x_f. In G we consider the subgraph $G_k^{n-1} = (X_k^{n-1}, E_k^{n-1})$ induced by the set of vertices $X_k^{n-1} = \{x_k, x_{k_1}, x_{k_2}, \ldots, x_{k_r}, x_f\}$. For vertices x_{k_i} and x_k we put $v(x_{k_i}) = v'(w_{k_i}^{n-i-1})$, $v(x_k) = v'(w_k^{n-1})$ and determine $\varepsilon_{x_{k_i}} = v(x_{k_i})$, $\varepsilon_{x_k} = v(x_k)$. Then we verify if in G_k^{n-1} the following condition holds:

$$\text{ext}(c', z) = 0, \ \forall z \in X_k^{n-1}, \tag{3.40}$$

where $c'_{(z,x)} = \varepsilon_x - \varepsilon_z + c_{(z,x)}$ for $e = (z, x) \in E_k^{n-1}$.

If condition (3.40) holds and G_k^{n-1} does not contain directed cycles then we may conclude that for the dynamic c-game on G with starting position x_k it holds $v(x_k) = v'(w_k^i)$. Note that for every vertex x_{k_i} of the directed path $P_0(x_{k_i}, x_f)$ we obtain $v(x_{k_i}) = v'(w_{k_i}^{n-i-1})$. If the condition mentioned above does not take place, then $v(x_k) \neq v'(w_k^{n-1})$ and we delete w_k^{n-1} from $\overline{G}$. After that we consider the vertex w_k^{n-2}, construct the graph $G_k^{n-2} = (X_k^{n-2}, E_k^{n-2})$ and in the same way verify if $v(x_k) = v'(w_k^{n-1})$. Finally we obtain that at least for an vertex w_k^i the directed path $P_{\overline{G}}(w_k^i, w_f)$ does not contain a directed cycle and condition (3.40) holds, i.e. $v(x_k) = v(w_k^i)$. In such a way we obtain $v(x_k)$ for every $x_k \in X$.

If $v(x)$ is known for every $x \in X$ then we fix $\varepsilon_x = v(x)$ and define the potential transformation $c'_{(z,x)} = c_{(z,x)} + \varepsilon_x - \varepsilon_z$ on edges $(z, x) \in E$. After that find the graph $G^0 = (V, E^0)$, generated by the set of edges $E^0 = \{(z, x) \in E \mid c'_{(z,x)} = 0\}$.

In G^0 we fix an arbitrary tree $T^* = (V, E^*)$, which determines the optimal strategies of the players as follows:

$$s^{1*}(z) = x, \quad \text{if } (z, x) \in E^* \text{ and } z \in X_1 \setminus \{x_f\};$$
$$s^{2*}(z) = x, \quad \text{if } (z, x) \in E^* \text{ and } z \in V_B \setminus \{x_f\}.$$ □

The correctness of the algorithm is based on the following theorem.

Theorem 3.45 *Let $v(x_k)$ be the value of the vertex x_k in the dynamic c-game on G and*

$$P_G(x_k, x_f) = \{x_k, x_{k_1}, x_{k_2}, \ldots, x_{k_r}, x_f\}$$

be the max-min path from x_k to x_f in G. Then $v'(w_k^{r+1}) = v(x_k)$.

Proof The construction described above allows us to conclude that between the set of max-min directed paths from x_k to x_f with no more than $r + 1$ edges in G and the set of max-min directed paths from w_k^{r+1} to w_f^0 with no more than $r + 1$ edges in $\overline{G}$ there exists a bijective mapping which preserves the sum of costs of the edges. Therefore $v'(w_k^{r+1}) = v(x_k)$. □

Remark 3.46 If $P_G(x_k, x_f) = \{x_k, x_{k_1}, x_{k_2}, \ldots, x_{k_r}, x_f\}$ is the max-min path from x_k to x_f in G then in $\overline{G}$ several vertices $w_k^{r+i} \in W$ for which $v'(w_k^{r+i}) = v(x_k)$, where $i \geq 1$ may exist. If $v'(w_k^{r+i}) = v(x_k)$, then in $\overline{G}$ the max-min path $P_{\overline{G}}(w_k^{r+1}, w_f^0) = \{w_k^{r+i}, w_{k_1}^{r+i-1}, w_{k_2}^{r+i-2}, \ldots, w_{k_r}^{i}, w_f^0\}$ corresponds to the max-min path $P_G(x_k, x_f)$ in G.

It is easy to observe that the running time of the algorithm is $O(|X|^4)$. Indeed, the values of the positions of the game on the auxiliary acyclic network can be calculated in time $O(N^2)$, where N is the number of vertices of the auxiliary network. Taking into account that $N \approx \mathrm{X}^2$ for our auxiliary network we obtain that the running time of the algorithm is $O(|X|^4)$.

Note that the proposed algorithm can also be applied for the c-game if the tree of max-min paths in G may not exist but there exists a max-min path from a given vertex $x_0 \in X$ to x_f. Then the algorithm finds in G the max-min path with a given starting position x_0 and a final position x_f.

An important problem for the dynamic c-game is how to determine vertices $x \in X$ for which $v(x) = +\infty$ and vertices $x \in X$ for which $v(x) = -\infty$. Taking into account that the final position x_f in such games cannot be reached we may delete vertices x of the graph G for which there exist max-min paths from x to x_f. In order to specify the algorithm for this case we need to study the infinite dynamic c-game where the graph G has no sink vertex x_f. This means that the outcome of the game

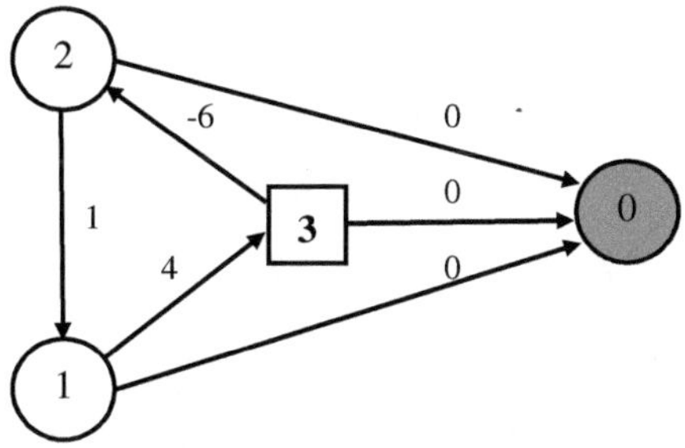

Fig. 3.7 The network for the dynamic c-game

is a cycle which may have positive, negative or zero sum costs of the edges. For determining the outcome of the game in this case we can use the same approach based on the reduction to the acyclic c-game.

The algorithm for finding the optimal strategies of the players in an infinite dynamic c-games is similar to the algorithm for finding the optimal strategies of the players in cyclic games. We describe such an algorithm in Sect. 3.7.5 and we can state that for an arbitrary position $x \in X$ the value of the cyclic game is positive if and only if the value $v(x)$ of the infinite dynamic c-game is positive. In addition we can state that efficient polynomial time algorithms for solving cyclic games can be elaborated if a polynomial time algorithm for solving the infinite dynamic c-game exists.

In the following we give an example which illustrates the details of the algorithm proposed above.

Example Consider the dynamic c-game determined by the network (G, X_1, X_2, c, x_f) given in Fig. 3.7. The position set X_1 of the first player is represented by circles and the position set X_2 of the second player is represented by squares; $x_f = 0$. The costs of the edges are given alongside them.

The auxiliary acyclic network for our dynamic c-game is represented in Fig. 3.8. Each vertex in Fig. 3.8 is represented by double numbers where the first one represents the number of the copy in G and the second one corresponds to the number of the vertex in G. Alongside the edges there are given their costs and alongside the vertices there are given values of the dynamic c-game on an auxiliary network.

Let us fix vertex $w_k^{n-1} = 33$ as the starting position of the dynamic c-game on the auxiliary network. Then we obtain $v'(33) = -5$. In order to verify if $v(3) = -5$ we find the max-min path $P_{\overline{G}}(33, 00) = \{33, 22, 11, 00\}$ from 33 to 00 and the values $v'(33) = -5$, $v'(22) = 1$, $v'(11) = 0$, $v'(00) = 0$. The path $P_{\overline{G}}(33, 00)$ in G corresponds to the path $P_G(3, 0) = \{3, 2, 1, 0\}$. For vertices 3, 2, 1, 0 in G we fix $\varepsilon_3 = v'(33) = -5$, $\varepsilon_2 = v'(22) = 1$, $\varepsilon_1 = v'(11) = 0$, $\varepsilon_0 = v'(00) = 0$. After that find the graph $G_3^3 = (X_3^3, E_3^3)$ generated by the set of vertices $X^3 = \{3, 2, 1, 0\}$. In this case graph G_3^3 coincides with graph G. Then we make a potential transformation $c'_{(x,y)} = \varepsilon_y - \varepsilon_x + c_{(x,y)} = 0$ with given $\varepsilon_3 = -5$, $\varepsilon_2 = 1$, $\varepsilon_1 = 0$, $\varepsilon_0 = 0$,

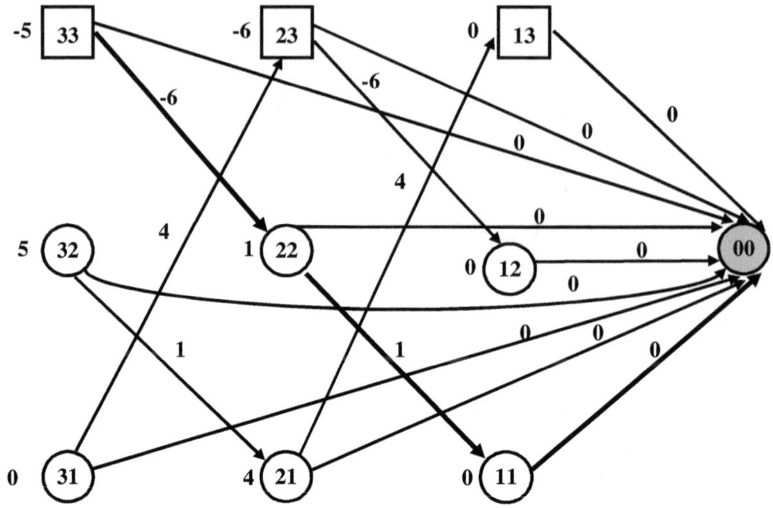

Fig. 3.8 The auxiliary network for the dynamic c-game

$$
\begin{aligned}
c'_{(1,0)} &= \varepsilon_0 - \varepsilon_1 + c_{(1,0)} = 0 - 0 + 0 = 0, \\
c'_{(2,0)} &= \varepsilon_0 - \varepsilon_2 + c_{(2,0)} = 0 - 1 + 0 = -1, \\
c'_{(3,0)} &= \varepsilon_0 - \varepsilon_3 + c_{(3,0)} = 0 - (-5) + 0 = 5, \\
c'_{(1,3)} &= \varepsilon_3 - \varepsilon_1 + c_{(1,3)} = -5 - 0 + 4 = -1, \\
c'_{(2,1)} &= \varepsilon_1 - \varepsilon_2 + c_{(2,1)} = 0 - 1 + 1 = 0, \\
c'_{(3,2)} &= \varepsilon_2 - \varepsilon_3 + c_{(3,2)} = 1 - (-5) - 6 = 0.
\end{aligned}
$$

So, after the potential transformation $c'_{(x,y)} = \varepsilon_y - \varepsilon_x + c_{(x,y)}$, $\forall (x, y) \in E$, we obtain the network given in Fig. 3.9 with new costs on the edges.

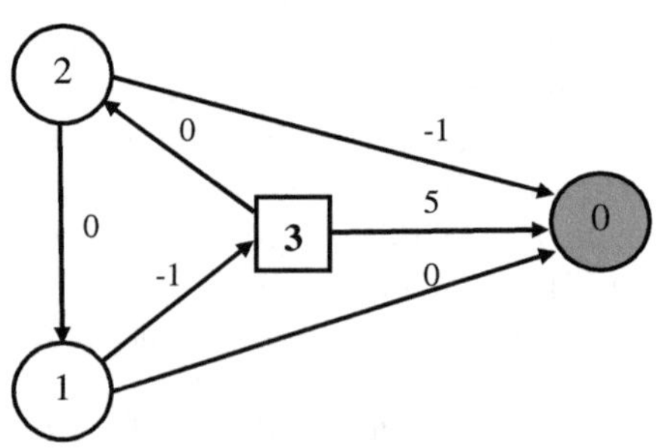

Fig. 3.9 The network after the potential transformation

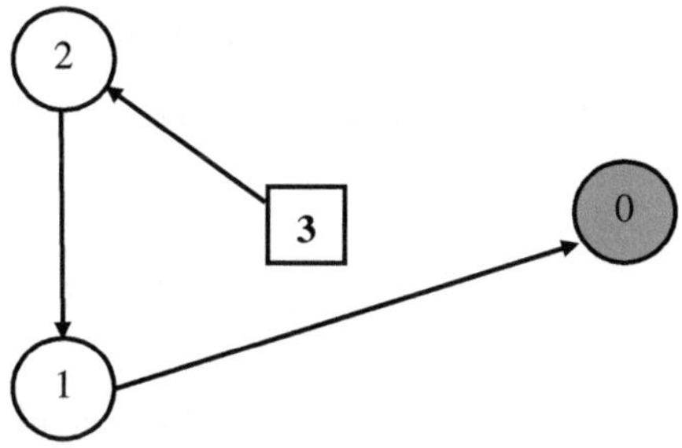

Fig. 3.10 The tree of max-min paths

If we select the tree with zero cost edges we obtain the tree of max-min paths, represented in Fig. 3.10.

If we start with vertex $w_k^{n-1} = 32$ then we obtain the subgraph $G_2^3 = (X_2^3, E_2^3)$ which coincides with the graph $G = (X, E)$ for which we determine $\varepsilon_2 = v'(32) = 5$, $\varepsilon_1 = v'(21) = 4$, $\varepsilon_3 = v'(13) = 0$, $\varepsilon_0 = v'(00) = 0$. It is easy to see that in this case the condition

$$\text{extr}(c', x) = 0, \ \forall x \in X,$$

is not satisfied.

3.5.5 Pseudo-polynomial Time Algorithms for Solving Antagonistic Dynamic c-Games

In this section we describe an algorithm for solving dynamic c-games which is based on a special recursive procedure for the calculation of the values $v(x)$. From the practical point of view the proposed algorithm may be more useful than the algorithm from the previous section although its computational complexity is $O(|X|^3 \sum_{e\in E} |c_e|)$ ($c : E \to \mathbb{R}$ is an integer function).

We assume that in G there exists the tree of max-min paths.

Preliminary step (*Step 0*): Set $X^* = \{x_f\}$, $\varepsilon_{x_f} = 0$.

General step (*Step k*): Find the set of vertices

$$X' = \{z \in X \setminus X^* \mid (z, x) \in E, x \in X^*\}.$$

For each $z \in X'$ we calculate

$$\varepsilon_z = \begin{cases} \max\limits_{x\in O_{X^*}(z)} \{\varepsilon_x + c_{(z,x)}\}, & z \in X_1 \cap X'; \\ \min\limits_{x\in O_{X^*}(z)} \{\varepsilon_x + c_{(z,x)}\}, & z \in X_2 \cap X', \end{cases} \tag{3.41}$$

where $O_{X^*}(z) = \{x \in X^* \mid (z, x) \in E\}$, and then do the following points (a) and (b):

(a) Fix $\beta(z) = \varepsilon_x$ for $z \in X' \cup X^*$ and then for every $x \in X' \cup X^*$ calculate

$$\beta(z) = \begin{cases} \max\limits_{x \in O_{X^* \cup X'}(z)} \{\varepsilon_x + c_{(z,x)}\}, & z \in X_1 \cap (X' \cup X^*); \\ \min\limits_{x \in O_{X^* \cup X'}(z)} \{\varepsilon_x + c_{(z,x)}\}, & z \in X_2 \cap (X' \cup X^*) \end{cases} \tag{3.42}$$

(b) Check if $\beta(z) = \varepsilon_z$ for every $z \in X' \cup X^*$. If this condition is not satisfied then fix $\varepsilon_z = \beta(z)$ for every $z \in X' \cup X^*$ and go to point (a).

If $\beta(z) = \varepsilon_z$ for every $z \in X' \cup X^*$ then in $X' \cup X^*$ we find the subset

$$Y^k = \left\{z \in X^* \cup X' \mid \text{extr}_{x \in O_{X^* \cup X'}(z)} \{\varepsilon_x - \varepsilon_z + c_{(z,x)}\} = 0\right\},$$

where

$$\text{ext}_{x \in O_{X^* \cup X'}(z)} \{\varepsilon_x - \varepsilon_z + c_{(z,x)}\} = \begin{cases} \max\limits_{x \in O_{X^* \cup X'}(z)} \{\varepsilon_x - \varepsilon_z + c_{(z,x)}\}, & z \in (X' \cup X^*) \cap X_1; \\ \min\limits_{x \in O_{X^* \cup X'}(z)} \{\varepsilon_x - \varepsilon_z + c_{(z,x)}\}, & z \in (X' \cup X^*) \cap X_2. \end{cases}$$

After that we change X^* by Y^k and check if $X^* = X$? If $X^* \neq X$, then go to the next step. If $X^* = X$, then define the potential transformation $c'_{(z,x)} = c_{(z,x)} + \varepsilon_x - \varepsilon_z$ on edges $(z, x) \in E$ and find the graph $G^0 = (X, E^0)$, generated by the set of edges $E^0 = \{(z, x) \in E \mid c'(z, x) = 0\}$. In G^0 fix an arbitrary tree $T^* = (X, E^*)$, which determines the optimal strategies of the players as follows:

$$s^{1*}(z) = x, \text{ if } (z, x) \in E^* \text{ and } z \in X_1 \setminus \{x_f\};$$
$$s^{2*}(z) = x, \text{ if } (z, x) \in E^* \text{ and } z \in X_2 \setminus \{x_f\}.$$

Let us show that this algorithm finds the tree of max-min paths $T^* = (X, E^*)$ if such a tree exists in G.

Theorem 3.47 *If there exists in G the tree of max-min paths $T^* = (X, E^*)$ with sink vertex x_f then the algorithm finds it using $O\left(|X|^3 \sum_{e \in E} |c_e|\right)$ elementary operations.*

Proof Consider the set Y^{k-1} obtained after $k - 1$ steps of the algorithm and assume that at step k after points (a) and (b) the condition

$$\beta(z) = \varepsilon_z \text{ for every } z \in X' \cup X^*$$

holds. This condition is equivalent to the condition

$$\text{ext}_{x \in O_{X^* \cup X'}(z)} \{\varepsilon_x - \varepsilon_z + c_{(z,x)}\} = 0, \ \forall z \in X' \cup X^*$$

which involves $Y^{k-1} \subset Y^k$.

Therefore, in the following we obtain that if for every step k of the algorithm the corresponding calculation procedure (3.42) is convergent then $Y^0 \subset Y^1 \subset Y^2 \subset \cdots \subset Y^r = X$, where $r < n$. This means that after $r < n$ steps the algorithm finds the values $\varepsilon(x)$ for $x \in X$ and the potential transformation $c'_{(y,x)} = \varepsilon_x - \varepsilon_y + c_{(y,x)}$ for edges $e = (y, x) \in E$ such that $\text{ext}(c', y) = 0, \forall x \in X$, i.e. the algorithm constructs the tree $T^* = (X, E^*)$. So, for a complete proof of the theorem we have to show the convergence of the calculation procedure based on formula (3.42) for an arbitrary step k of the algorithm.

Assume that at step k of the algorithm the following condition

$$\text{ext}_{x \in O_{X^* \cup X'}(z)} \{\varepsilon_x - \varepsilon_z + c_{(z,x)}\} \neq 0 \ \text{ for every } \ z \in X'$$

holds. Consider the set of edges $E' = \{e = (z, x') \in E \mid \beta(z) = \varepsilon_{x'} + c_{(z,x')}, z \in X', \ x' \in x \in O_{X^* \cup X'}(z)\}$ where x' corresponds to vertex z such that

$$\varepsilon(x') + c(z, x') = \begin{cases} \max\limits_{x \in O_{X^* \cup X'}(z)} \{\varepsilon_x + c_{(z,x)}\}, \ z \in X_1 \cap (X' \cup X^*); \\ \min\limits_{x \in O_{X^* \cup X'}(z)} \{\varepsilon_x + c_{(z,x)}\}, \ z \in X_2 \cap (X' \cup X^*). \end{cases}$$

The calculation on the basis of (a) and (b) can be treated as follows. Players improve the values ε_z of vertices $z \in X'$ using passages from z to corresponding vertices $x' \in O_{X^* \cup X'}(z)$. At each iteration of this calculation procedure the players can improve their income by $\beta(z) - \varepsilon(z)$ units for every position $z \in X$.

Denote by $\widetilde{X}$ the subset of vertices $z' \in X'$ for which in $G' = (X', E')$ there exist directed paths from $z' \in \widetilde{X}$ to vertices from X^{k-1}. Then the improvements of the players mentioned above are possible for an arbitrary vertex $z \in \widetilde{X}$. This means that if procedure (a), (b) at step k is applied then after using one iteration of this procedure we obtain $\beta(z) = \varepsilon_z, \forall z \in \widetilde{X}$. In the following we can see that in order to achieve $\varepsilon_z = \beta(z)$ for the rest of the vertices $z \in X' \setminus \widetilde{X}$ it is necessary to apply more than one iteration.

Let us consider in G' the subset $\widetilde{X}' = X' \setminus \widetilde{X}$. Then in G' there are no directed edges $e = (z, x')$ such that $z \in \widetilde{X}'$ and $x' \in X^{k-1}$. Without loss of generality we may consider that in G the subset $\widetilde{X}'$ generates a directed cycle C'. Denote by $n(C')$ the number of the vertices of this cycle and assume that the sum of its edges is equal to θ (θ may be positive or negative). We can see that if we apply formula (3.42) then after each $n(G)$ iterations of the calculation procedure the values ε_z of the vertices $z \in C'$ will decrease at least by $|\theta|$ units if $\theta < 0$; if $\theta > 0$ then these values will increase by θ. Therefore the first player will preserve passages from vertices $z \in C'$ to vertices x' of the cycle C' if $\beta(z) - \varepsilon(z) > 0$; otherwise the first player will change the passage from one vertex $z^0 \in C'$ to a vertex $x'' \in O_{X^* \cup X'}(z)$ which may belong to X^{k-1}.

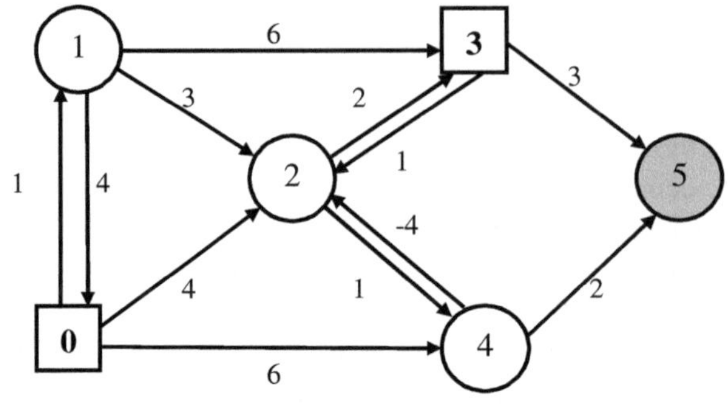

Fig. 3.11 The network with sink vertex $x_f = 5$

In an analogous way the second player will preserve passages from vertices $z \in C'$ to vertices x' of cycle C' if $\beta(z) - \varepsilon(z) < 0$; otherwise the second player will change the passage from one vertex $z^0 \in X$ to a vertex x'' which may belong to X^{k-1}. So, if in G there exists the tree of max-min paths then after a finite number of iterations of the procedure (a), (b) we obtain $\beta(z) = \varepsilon(z)$ for $z \in X'$. Taking into account that the values $\beta(z)$ will decrease (or increase) after each $n(G)$ iterations by integer units $|\theta|$ we may conclude that the number of iterations of the procedure is comparable with $|X|^2 \cdot \max_{z \in X'} |\beta(z) - \varepsilon_z|$. In the worst case these quantities are limited by $|X|^2 \sum_{e \in E} |c_e|$. This involves that the computational complexity of the algorithm is $O\Big(|X|^3 \sum_{e \in E} |c(e)|\Big)$. □

Remark 3.48 The algorithm for an acyclic network can be applied without points (a) and (b) because the condition $\beta(z) = \varepsilon_z, \forall z \in X'$ holds at every step k. In general, the version of the algorithm without the points (a) and (b) can be used if $Y^{k-1} \neq Y^k$ at every step k. In this case the running time of the algorithm is $O(|X|^3)$.

The algorithm described above can be modified for the dynamic c-game in general form if the network contains vertices x for which $v(x) = \pm\infty$. In order to detect such vertices in point (a) it is necessary to introduce a new condition which allows us to select vertices $z \in X'$ with great values $\beta(z)$ (positive and negative). But in this case the algorithm becomes more difficult than the algorithm for finite games.

Below we present two examples which illustrate the details of the algorithm. The first example illustrates the work of the algorithm if it is not necessary to use points (a) and (b). The second example illustrates the details of the recursive calculation procedure in the points (a) and (b).

Example 1 Consider the problem of determining the optimal stationary strategies on the network which may contain cycles. The corresponding network with sink vertex $x_f = 5$ is given in Fig. 3.11. In this network the positions of the first player are represented by circles and the positions of the second one are represented by squares, i.e. $X_1 = \{1, 2, 4, 5\}$, $X_2 = \{0, 3\}$.

The values of the cost functions of the edges are given in parenthesis alongside them. We can see that for the given network there exists a tree of max-min paths which can be found by using the algorithm.

Step 0
$X^* = \{5\}; \quad \varepsilon_5 = 0.$

Step 1
Find the set of vertices $X' = \{3, 4\}$ for which there exist directed edges $(3, 5)$ and $(4, 5)$ from vertices 3 and 4 to vertex 5. Then we calculate according to (3.41) values $\varepsilon_3 = 3$, $\varepsilon_4 = 2$. It is easy to check that for vertices 3 and 4 the following condition holds:

$$\text{ext}_{y \in X_{X^* \cup X'}(x)} \{\varepsilon_y - \varepsilon_x + c_{(x,y)}\} = 0.$$

So, $Y^1 = \{3, 4, 5\}$. Therefore if we change X^* by Y^1, after step 1 we obtain $X^* = \{3, 4, 5\}$.

Step 2
Find the set of vertices $X' = \{0, 1, 2\}$ for which there exist directed edges from vertices $x \in X'$ to vertices $y \in X^*$. Then according to (3.41) we calculate

$$\varepsilon_2 = \max_{y \in X^*(2)} \{\varepsilon_3 + c_{(2,3)},\ \varepsilon_4 + c_{(2,4)}\} = \max\{5, 3\} = 5;$$

$$\varepsilon_1 = \varepsilon_3 + c_{(1,3)} = 9;$$

$$\varepsilon_0 = \varepsilon_4 + 6 = 8.$$

So, $\varepsilon_0 = 8,\ \varepsilon_1 = 9, \varepsilon_2 = 5,\ \varepsilon_3 = 3,\ \varepsilon_4 = 2,\ \varepsilon_5 = 0.$

It is easy to check that $Y^2 = \{0, 2, 3, 4, 5\}$. Indeed,

$$\text{ext}_{y \in X_{X^* \cup X'}(3)} \{\varepsilon_y - \varepsilon_3 + c_{(3,y)}\}$$

$$= \min\{\varepsilon_5 - \varepsilon_3 + c_{(3,5)},\ \varepsilon_2 - \varepsilon_3 + c_{(3,2)}\} = \min\{0 - 3 + 3,\ 5 - 3 + 1\} = 0;$$

$$\text{ext}_{y \in X_{X^* \cup X'}(2)} \{\varepsilon_y - \varepsilon_3 + c_{(2,y)}\}$$

$$= \max\{\varepsilon_3 - \varepsilon_2 + c_{(2,3)},\ \varepsilon_4 - \varepsilon_2 + c_{(2,4)}\} = \max\{3 - 5 + 2,\ 2 - 5 + 1\} = 0;$$

$$\text{ext}_{y \in X_{X^* \cup X'}(1)} \{\varepsilon_y - \varepsilon_1 + c_{(1,y)}\}$$

$$= \max\{\varepsilon_3 - \varepsilon_1 + c_{(1,3)},\ \varepsilon_2 - \varepsilon_1 + c_{(1,2)},\ \varepsilon_0 - \varepsilon_1 + c_{(0,1)}\}$$

$$= \max\{3 - 9 + 6,\ 3 - 9 + 5,\ 8 - 9 + 4\} = 3;$$

$$\text{ext}_{y \in X_{X^* \cup X'}(0)} \{\varepsilon_y - \varepsilon_0 + c_{(0,y)}\}$$

$$= \min\{\varepsilon_4 - \varepsilon_0 + c_{(0,4)},\ \varepsilon_2 - \varepsilon_0 + c_{(0,2)},\ \varepsilon_1 - \varepsilon_0 + c_{(0,1)}\}$$

$$= \min\{2 - 8 + 6,\ 5 - 8 + 4,\ 9 - 8 + 1\} = 0;$$

$$\text{ext}_{y \in X_{X^* \cup X'}(4)}\{\varepsilon_y - \varepsilon_4 + c_{(4,y)}\}$$

$$= \max\{\varepsilon_5 - \varepsilon_4 + c_{(4,5)},\ \varepsilon_2 - \varepsilon_4 + c_{(4,2)}\} = \max\{0 - 2 + 2,\ 5 - 2 - 4\} = 0.$$

So, the set of vertices for which $\text{ext}_{y \in X_{X^* \cup X'}(x)}\{\varepsilon_y - \varepsilon_x + c_{(x,y)}\} = 0$ consists of vertices 0, 2, 3, 4, 5.

Step 3

Find the set of vertices $X' = \{1\}$ and calculate

$$\varepsilon(1) = \max_{y \in X_{X^*}(1)}\{\varepsilon_y + c_{(1,y)}\} = \max\{\varepsilon_3 + c_{(1,3)},\ \varepsilon_2 + c_{(1,2)},\ \varepsilon_0 + c_{(1,0)}\}$$

$$= \max\{3 + 6,\ 5 + 3,\ 8 + 4\} = 12.$$

Now we can see that the obtained values $\varepsilon_0 = 8,\ \varepsilon_1 = 12,\ \varepsilon_2 = 5,\ \varepsilon_3 = 3,\ \varepsilon_4 = 2,\ \varepsilon_5 = 0$ satisfy the conditions

$$\varepsilon_y - \varepsilon_x + c_{(x,y)} \le 0 \text{ for every } (x, y) \in E,\ x \in X_1;$$
$$\varepsilon_y - \varepsilon_x + c_{(x,y)} \ge 0 \text{ for every } (x, y) \in E,\ x \in X_2.$$

The directed tree $GT = (X, E^*)$ generated by edges $(x, y) \in E$ for which $\varepsilon_y - \varepsilon_x + c_{(x,y)} = 0$ is represented in Fig. 3.12.

The optimal strategies of the players are:

$$s^1 : 1 \to 0;\ 2 \to 3;\ 4 \to 5;$$

$$s^2 : 0 \to 4;\ 3 \to 5.$$

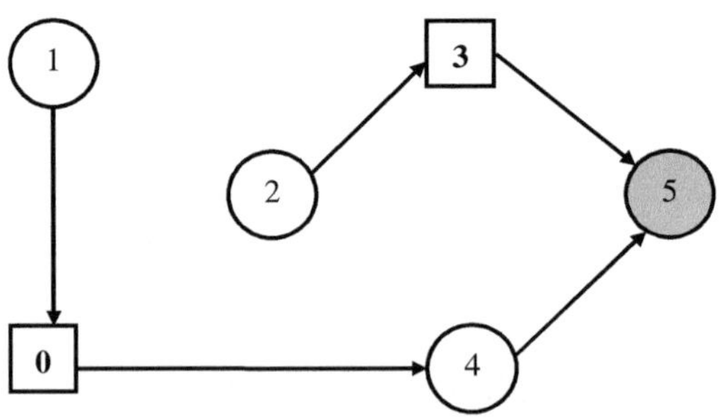

Fig. 3.12 The tree induced by the optimal strategies of the players

Table 3.1 The results of the iteration procedure for $\beta(0)$, $\beta(1)$, $\beta(2)$, $\beta(3)$

i	$\beta(0)$	$\beta(1)$	$\beta(2)$	$\beta(3)$
0	0	0	0	0
1	0	4	1	−6
2	0	0	5	−5
3	0	0	1	−1
4	0	3	1	−5
5	0	0	4	−5
6	0	0	1	−2
7	0	2	1	−5
8	0	0	3	−5
9	0	0	1	−3
10	0	1	1	−5
11	0	0	2	−5
12	0	0	1	−4
13	0	0	1	−5
14	0	0	1	−5

Example 2 Consider the problem of determining the tree of max-min paths $T^* = (X, E^*)$ for the network given in Fig. 3.8 with the same costs of the edges as in the previous section.

If we apply the algorithm described above then we use only one step ($k = 1$). But this step consists of two items (a) and (b) that make calculations on the basis of formula (3.42). In Table 3.1 the values $\beta(0)$, $\beta(1)$, $\beta(2)$, $\beta(3)$ at each iteration of the calculation procedure on the bases of formula (3.42) are given.

We can see that the convergence of the calculation procedure is obtained at iteration 14. Therefore we conclude that $\varepsilon_0 = 0$, $\varepsilon_1 = 0$, $\varepsilon_2 = 1$, $\varepsilon_3 = -5$. If we make a potential transformation we obtain the network in Fig. 3.9. In Fig. 3.10 it is presented the tree of max-min paths $T^* = (X, E^*)$.

3.6 A Polynomial Time Algorithm for Solving Acyclic l-Games on Networks

An acyclic l-game on networks has been introduced in [66, 76] as an auxiliary problem for studying and solving special cyclic games, which we will consider in the next section.

3.6.1 Problem Formulation

Let $(G, X_1, X_2, c, x_0, x_f)$ be a network, where $G = (X, E)$ represents a directed acyclic graph with the sink vertex $x_f \in X$. On E it is defined a function $c: E \to \mathbb{R}$ and on X it is given a partition $X = X_1 \cup X_2$ $(X_1 \cap X_2 = \emptyset)$ where X_1 and X_2 correspond to positions sets of two players 1 and 2, respectively.

We consider the following acyclic game from [76]. Again we define the strategies of the players as maps

$$s^1: x \to y \in X(x) \quad \text{for} \quad x \in X_1 \setminus \{x_f\};$$
$$s^2: x \to y \in X(x) \quad \text{for} \quad x \in X_2 \setminus \{x_f\}.$$

We define the payoff function $\overline{H}_{x_0}: S^1 \times S^2 \to \mathbb{R}$ in this game as follows:

Let $s^1 \in S^1$ and $s^2 \in S^2$ be fixed strategies of the players. Then the graph $G_s = (X, E_s)$, generated by the edges $(x, s^1(x)), x \in X \setminus \{x_f\}$, and $(x, s^2(x))$, $x \in X \setminus \{x_f\}$, has a structure of a directed tree with the sink vertex x_f. Therefore, it contains a unique directed path $P_s(x_0, x_f)$ with $n(P_s(x_0, x_f))$ edges. We put

$$\overline{H}_{x_0}(s^1, s^2) = \frac{1}{n(P_s(x_0, x_f))} \sum_{e \in E(P_s(x_0, x_f))} c_e.$$

The payoff function $\overline{H}_{x_0}(s^1, s^2)$ on $S^1 \times S^2$ defines a game in normal form, which is determined by the network $(G, X_1, X_2, c, x_0, x_f)$.

We consider the problem of finding the strategies s^{1^*} and s^{2^*}, for which

$$\overline{v}(x_0) = \overline{H}_{x_0}(s^{1^*}, s^{2^*}) = \max_{s^1 \in \mathbb{S}^1} \min_{s^2 \in \mathbb{S}^2} \overline{H}_{x_0}(s^1, s^2).$$

3.6.2 The Main Properties of Optimal Strategies in Acyclic l-Games

First of all let us show that for the considered max-min problem there exists a saddle point.

Denote

$$\overline{\overline{v}}(x_0) = \overline{H}_{x_0}(s^{1^0}, s^{2^0}) = \min_{s^2 \in \mathbb{S}^2} \max_{s^1 \in \mathbb{S}^1} \overline{H}_{x_0}(s^1, s^2)$$

and let us show that $\overline{v}(x_0) = \overline{\overline{v}}(x_0)$.

Theorem 3.49 *For an arbitrary acyclic l-game the following equality holds:*

$$\overline{v}(x_0) = \overline{H}_{x_0}(s^{1^*}, s^{2^*}) = \max_{s^1 \in \mathbb{S}^1} \min_{s^2 \in \mathbb{S}^2} \overline{H}_{x_0}(s^1, s^2) = \min_{s^2 \in \mathbb{S}^2} \max_{s^1 \in \mathbb{S}^1} \overline{H}_{x_0}(s^1, s^2).$$

Proof First of all let us note the following property of an acyclic l-game, determined by $(G, X_1, X_2, c, x_0, x_f)$: If the cost function c is changed by $c' = c + h$ (h is an arbitrary real number), then we obtain an equivalent acyclic l-game determined by $(G, X_1, X_2, c', x_0, x_f)$ for which $\overline{v}'(x_0) = \overline{v}(x_0) + h$ and $\overline{\overline{v}}''(x_0) = \overline{\overline{v}}(x_0) + h$. If we denote by $\overline{H}'_{x_0}(s^1, s^2)$ the payoff function of the l-game after the transformation mentioned above then we have

$$\overline{H}_{x_0}(s^1, s^2) = \overline{H}'_{x_0}(s^1, s^2) + h, \forall s^1 \in \mathbb{S}^1, \forall s^2 \in \mathbb{S}^2$$

It is easy to observe that if $h = -\overline{v}(x_0)$ then for the acyclic l-game with network $(G, X_1, X_2, c', x_0, x_f)$ we obtain $\overline{v}'(x_0) = 0$. This means that an acyclic l-game becomes an acyclic c-game with max-min value of the game $\overline{H}'_{x_0}(s^{1^0}, s^{2^0})$. Therefore if after the transformation of the game we regard it as an acyclic c-game then the following property holds:

$$0 = \overline{v}'(x_0) = \max_{s^1 \in \mathbb{S}^1} \min_{s^2 \in \mathbb{S}^2} \overline{H}'_{x_0}(s^1, s^2) = \min_{s^2 \in \mathbb{S}^2} \max_{s^1 \in \mathbb{S}^1} \overline{H}'_{x_0}(s^1, s^2) = 0.$$

Taking into account that

$$\overline{H}'_{x_0}(s^1, s^2) = \overline{H}_{x_0}(s^1, s^2) - \overline{v}(x_0)$$

we obtain that

$$\min_{s^2 \in \mathbb{S}^2} \max_{s^1 \in \mathbb{S}^1} \left(\overline{H}_{x_0}(s^1, s^2) - \overline{v}(x_0)\right) = \max_{s^1 \in \mathbb{S}^1} \min_{s^2 \in \mathbb{S}^2} \left(\overline{H}_{x_0}(s^1, s^2) - \overline{v}(x_0)\right)$$
$$= \overline{\overline{v}}(x_0) - \overline{v}(x_0),$$

i.e. $\overline{\overline{v}}(x_0) - \overline{v}(x_0) = 0$. So, $\overline{\overline{v}}(x_0) = \overline{v}(x_0)$. □

Theorem 3.50 *Let an acyclic l-game determined by the network* $(G, X_1, X_2, c, x_0, x_f)$ *with the starting position* x_0 *be given. Then there exists the value* $\overline{v}(x_0)$ *and the function* $\varepsilon: X \to \mathbb{R}$, *which determines the potential transformation* $c'_{(x,y)} = c_{(x,y)} + \varepsilon_x - \varepsilon_y$ *of costs on the edges* $e = (x, y) \in E$ *such that the following conditions hold*

(a) $\overline{v}(x_0) = \text{ext}(c', x), \ \forall x \in X \setminus \{x_f\}$;

(b) $\varepsilon_{x_0} = \varepsilon_{x_f}$.

The optimal strategies of the players in an acyclic l-game can be found as follows: Fix the arbitrary maps $s^{1^*}: X_1 \setminus \{x_f\} \to X$ *and* $s^{2^*}: X_2 \setminus \{x_f\} \to X$ *such that* $s^{1^*}(x) \in \text{VEXT}(c', x)$ *for* $x \in X_1 \setminus \{x_f\}$ *and* $s^{2^*}(x) \in \text{VEXT}(c', x)$ *for* $x \in X_2 \setminus \{x_f\}$.

Proof The proof of the theorem follows from Theorem 3.38 if we regard the acyclic l-game as an acyclic c-game on the network $(G, X_1, X_2, c', x_0, x_f)$ with the cost function $c' = c - \overline{v}(x_0)$. □

Corollary 3.51 *The difference* $\varepsilon_x - \varepsilon_{x_0}$, $x \in X$, *represents the costs of max-min path from* x *to* x_f *in the acyclic c-game on the network* $(G, X_1, X_2, c', x_0, x_f)$ *with* $c'_{(x,y)} = c_{(x,y)} - \overline{v}(x_0)$, $\forall (x, y) \in E$.

3.6.3 A Polynomial Time Algorithm for Finding the Value and the Optimal Strategies in the Acyclic l-Game

The algorithm, which we describe below, is based on the results from Sect. 3.6.2. In this algorithm we shall use the following properties:

1. The value $\overline{v}(x_0)$ of an acyclic l-game on the network $(G, X_1, X_2, c, x_0, x_f)$ is nonnegative if and only if the value $v(x_0)$ of an acyclic l-game on the network $(G, X_1, X_2, c, x_0, x_f)$ is nonnegative; moreover $\overline{v}(x_0) = 0$ if and only if $v(x_0) = 0$.
2. If $M^1 = \min_{e \in E} c_e$ and $M^2 = \max_{e \in E} c_e$, then $M^1 \le \overline{v}(x_0) \le M^2$.
3. If in the network $(G, X_1, X_2, c, x_0, x_f)$ the cost function $c : E \to \mathbb{R}$ is changed by the function $c^h : E \to \mathbb{R}$, where
$$c_e^h = c_e - h, \forall e \in E \tag{3.43}$$
(h is an arbitrary constant), then the acyclic l-games on $(G, X_1, X_2, c, x_0, x_f)$ and $(G, X_1, X_2, c^h, x_0, x_f)$, respectively, have the same optimal strategies s^{1*}, s^{2*}. In addition, the values $\overline{v}(x_0)$ and $\overline{v}_h(x_0)$ of these games differ by a constant h: $\overline{v}_h(x_0) = \overline{v}(x_0) - h$. So, the acyclic l-games on $(G, X_1, X_2, c, x_0, x_f)$ and $(G, X_1, X_2, c^h, x_0, x_f)$ are equivalent.

According to the properties mentioned above, if $\overline{v}(x_0)$ is known, then the acyclic l-game can be reduced to the acyclic c-game by using transformation (3.38) with $h = \overline{v}(x_0)$. After that we can find the optimal strategies in the game with network $(G, X_1, X_2, c^h, x_0, x_f)$ by using Algorithm 3.39. The most important phase in the proposed algorithm represents the problem of finding the value h, for which $\overline{v}_h(x_0) = 0$. Taking into account properties 1 and 2, we will seek for this value by using the dichotomy method on segment $[M^1, M^2]$, such that at each step of this method we will solve a dynamic c-game with network $(G, X_1, X_2, c^k, x_0, x_f)$, where $c^k = c - h_k$. The main idea of the general step of the algorithm is the following. We make transformation (3.43) with $h = h_k$, where h_k is a midpoint of the segment $[M_k^1, M_k^2]$ at step k. After that we apply Algorithm 3.39 for the dynamic c-game on network $(G, X_1, X_2, c^{h_k}, x_0, x_f)$ and find $v_{h_k}(x_0)$. If $v_{h_k}(x_0) > 0$ then we fix segment $[M^1_{k+1}, M^2_{k+1}]$, where $M^1_{k+1} = M^1_k$ and $M^2_{k+1} = \left(M^1_k + M^2_k\right)/2$; otherwise we put $M^1_{k+1} = \left(M^1_k + M^2_k\right)/2$ and $M^2_{k+1} = M^2_k$. If $v_{h_k}(x_0) = 0$ then STOP. The detailed description of the algorithm is the following.

Algorithm 3.52 Determining the Value and Optimal Strategies of the Players for the Acyclic l-Game
Let us assume that the cost function $c : E \to \mathbb{R}$ is integer and $\max_{e\in E} |c_e| \neq 0$.

Preliminary step (step 0): Find the value $v(x_0)$ and optimal strategies s^{1*} and s^{2*} of the dynamic c-game on $(G, X_1, X_2, c, x_0, x_f)$ by using Algorithm 3.39. If $v(x_0) = 0$ then fix s^{1*} and s^{2*} as the solution of the l-game, put $\overline{v}(x_0) = 0$ and STOP; otherwise fix $M^1_1 = \min_{e\in E} c_e$, $M^2_1 = \max_{e\in E} c_e$, $L = \max_{e\in E} |c_e| + 1$.

General step (step k, $k \geq 1$): Find $h_k = \left(M^1_k + M^2_k\right)/2$ and make the transformation of the edges' costs

$$c^k_e = c_e - h_k \quad \text{for} \quad e \in E.$$

Solve the dynamic c-game on the network $(G, X_1, X_2, c^k, x_0, x_f)$ and find the value $v_k(x_0)$ and the optimal strategies s^{1*}, s^{2*}.

If $v_k(x_0) = 0$ then fix the optimal strategies s^{1*} and s^{2*} and put $\overline{v}(x_0) = h_k$. If $|v_k(x_0)| \leq 1/\left(4|X|^2L\right)$ then fix s^{1*} and s^{2*}; find $\overline{v}(x_0) = \overline{H}_{x_0}(s^{1*}, s^{2*})/n(P_{s^*}(x_0, x_f))$ and STOP. If $v_k(x_0) > 1/\left(4|X|^2L\right)$ then fix $M^1_{k+1} = M^1_k$, $M^2_{k+1} = h_k$ and go to step $k+1$. If $v_k(x_0) < -1/\left(4|X|^2L\right)$ then fix $M^1_{k+1} = h_k$, $M^2_{k+1} = M^2_k$ and go to step $k+1$.

Theorem 3.53 *Let $(G, X_1, X_2, c, x_0, x_f)$ be a network with an integer cost function $c : E \to \mathbb{R}$, and $L = \max_{e\in E} |c_e|$. Then Algorithm 3.52 finds correctly the value $\overline{v}(x_0)$ and optimal strategies s^{1*}, s^{2*} in the acyclic l-game. The running time of the algorithm is $O(|X|^2 \log L + 2|X|^2 \log |X|)$.*

Proof Let $(G, X_1, X_2, c^k, x_0, x_f)$ be a network after the final step k of Algorithm 3.52. If we solve the corresponding acyclic c-game, then

$$|v_k(x_0)| \leq \frac{1}{4|X|^2L}$$

and the values ε^k_x, $x \in X$, determined according to Algorithm 3.39 represents the approximation solution of the system

$$\begin{cases} \varepsilon_y - \varepsilon_x + c^k_{(x,y)} \leq 0 & \text{for} \quad x \in X_1, \ (x, y) \in E; \\ \varepsilon_y - \varepsilon_x + c^k_{(x,y)} \geq 0 & \text{for} \quad x \in X_2, \ (x, y) \in E; \\ \varepsilon_{x_0} = \varepsilon_{x_f}. \end{cases}$$

This means that ε^k_x, $x \in X$, and h_k represents the approximative solution of the system

$$\begin{cases} \varepsilon_y - \varepsilon_x + c_{(x,y)} \le h_k & \text{for } x \in X_1, \ (x,y) \in E; \\ \varepsilon_y - \varepsilon_x + c_{(x,y)} \ge h_k & \text{for } x \in X_2, \ (x,y) \in E; \\ \varepsilon_{x_0} = \varepsilon_{x_f}. \end{cases}$$

According to [57, 58], the exact solution $h = \overline{v}(x)$, ε_x, $x \in X$, of this system can be obtained from h_k, ε_x^k, $x \in X$, by using the special round-off procedure in time $O(\log(L+1))$. Therefore the strategies s^{1*}, s^{2*} after the final step k of the algorithm correspond to the optimal solution of the acyclic l-game.

Taking into account that the tabulation of the values $\varepsilon(x)$, $x \in X$, in G needs $O(|X|^2)$ operations and the number of iterations of the algorithm is $O(\log L + 2\log|X|)$, we obtain that the running time of the algorithm is $O(|X|^2 \log L + 2|X|^2 \log|X|)$. □

3.7 Algorithms for Finding the Optimal Strategies of the Players in a Cyclic Game

Cyclic games have been studied in [29, 43, 66, 67, 99, 142]. Later in [68, 79, 83] this game has been used for studying the game variant of the infinite horizon discrete control problem with average cost criterion by a trajectory. Here we show that the problem of finding optimal strategies of the players in a cyclic game is tightly connected with the problem of finding optimal strategies of the players in the dynamic c-game and the acyclic l-game. On the basis of these results we propose algorithms for determining the value and the optimal strategies in a cyclic game.

3.7.1 Problem Formulation and the Main Properties

Let $G = (X, E)$ be a finite directed graph in which every vertex $x \in X$ has at least one leaving edge $e = (x, y) \in E$. On the edge set E it is given a function $c : E \to \mathbb{R}$ which assigns a cost c_e to each edge $e \in E$. In addition the vertex set X is divided into two disjoint subsets X_1 and $X_2 (X = X_1 \cup X_2, X_1 \cap X_2 = \emptyset)$ which we will regard as position sets of the two players. On G we consider the following two-person game from [29, 43, 79, 131, 142]: The game starts at position $x_0 \in X$. If $x_0 \in X_1$ then the move is done by the first player, otherwise it is done by the second one. The move means that the passage from the position x_0 to the neighbor position x_1 through the edge $e_1 = (x_0, x_1) \in E$. After that if $x_1 \in X_1$ then the move is done by the first player, otherwise it is done by the second one and so on indefinitely. The first player has the aim to maximize $\lim_{t\to\infty} \inf \frac{1}{t} \sum_{i=1}^{t} c_{e_i}$ while the second player has the aim to minimize $\lim_{t\to\infty} \sup \frac{1}{t} \sum_{i=1}^{t} c_{e_i}$. In [29] it is proved that for this game there exists a value $\overline{v}(x_0)$ such that the first player has a strategy of moves that insures $\lim_{t\to\infty} \inf \frac{1}{t} \sum_{i=1}^{t} c_{e_i} \ge \overline{v}(x_0)$ and the second player has a strategy of moves that insures $\lim_{t\to\infty} \sup \frac{1}{t} \sum_{i=1}^{t} c_{e_i} \le \overline{v}(x_0)$.

Furthermore in [29] it is shown that the players can achieve the value $\overline{v}(x_0)$ applying the strategies of moves which do not depend on t. This means that the considered game can be formulated in terms of stationary strategies. Such a statement of the game in [43] is named a cyclic game. In [131, 142] this game is called *parity game* .

The strategies of the players in the cyclic game are defined as maps

$$s^1: x \to y \in X(x) \quad \text{for} \quad x \in X_1,$$
$$s^2: x \to y \in X(x) \quad \text{for} \quad x \in X_2,$$

where $X(x) = \{y \in X \mid e = (x, y) \in E\}$. Since G is a finite graph then the sets of strategies of the players

$$\mathbb{S}^1 = \{s^1: x \to y \in X(x) \quad \text{for} \quad x \in X_1\};$$
$$\mathbb{S}^2 = \{s^2: x \to y \in X(x) \quad \text{for} \quad x \in X_2\}$$

are finite sets.

The payoff function $\overline{H}_{x_0}: \mathbb{S}^1 \times \mathbb{S}^2 \to \mathbb{R}$ in the cyclic game is defined as follows:

Let $s^1 \in \mathbb{S}^1$ and $s^2 \in \mathbb{S}^2$ be fixed strategies of the players. Denote by $G_s = (X, E_s)$ the subgraph of G generated by edges of the form $(x, s^1(x))$ for $x \in X_1$ and $(x, s^2(x))$ for $x \in X_2$. Then G_s contains a unique directed cycle C_s which can be reached from x_0 through the edges $e \in E_s$. We consider that the value $\overline{H}_{x_0}(s^1, s^2)$ is equal to mean edges cost of cycle C_s, i.e.,

$$\overline{H}_{x_0}(s^1, s^2) = \frac{1}{n(C_s)} \sum_{e \in E(C_s)} c_e,$$

where $E(C_s)$ represents the set of edges of cycle C_s and $n(C_s)$ is a number of the edges of C_s. So, the cyclic game is determined uniquely by the network (G, X_1, X_2, c, x_0), where x_0 is the given starting position of the game. If we consider the problem of finding the optimal strategies of the players for an arbitrary starting position $x \in X$, then we will use the notation (G, X_1, X_2, c). In [29, 43] it is proved that there exist the strategies $s^{1^*} \in S^1$ and $s^{2^*} \in S^2$ such that

$$\overline{v}(x) = \overline{H}_x(s^{1^*}, s^{2^*}) = \max_{s^1 \in \mathbb{S}^1} \min_{s^2 \in \mathbb{S}^2} \overline{H}_x(s^1, s^2)$$
$$= \min_{s^2 \in \mathbb{S}^2} \max_{s^1 \in \mathbb{S}^1} \overline{H}_x(s^1, s^2), \quad \forall x \in X.$$

So, the optimal strategies s^{1^*}, s^{2^*} of the players in cyclic games do not depend on a starting position x_0 although for different positions $x, y \in X$ the values $\overline{v}(x)$ and $\overline{v}(y)$ may be different. It means that the positions set X can be divided into several classes $X = X^1 \cup X^2 \cup \cdots \cup X^k$ according to values of positions $\overline{v}^1, \overline{v}^2, \ldots, \overline{v}^k$, i.e., $x, y \in X^i$ if and only if $\overline{v}^i = \overline{v}(x) = \overline{v}(y)$. In the case $k = 1$ the network

(G, X_1, X_2, c) is named the ergodic network [43]. In [79, 83] it is shown that every cyclic game with an arbitrary network (G, X_1, X_2, c, x_0) with given starting position x_0 can be reduced to a cyclic game on an auxiliary ergodic network (G', X'_A, X'_B, c').

It is well-known [54, 131, 142] that the decision problem associated to the cyclic game is in $NP \cap$ co-NP. Some exponential and pseudo-polynomial algorithms for finding the value and the optimal strategies of the players in cyclic game are proposed in [142]. The computational complexity of the problem of determining the optimal stationary strategies for stochastic games is studied in [22]. Our aim is to propose polynomial time algorithms for determining optimal strategies of the players in cyclic games. We argue such algorithms on the basis of results which have been announced in [83, 85].

3.7.2 Some Preliminary Results

First of all we need to remind some preliminary results from [43, 68, 76, 79, 83, 85]. Let (G, X_1, X_2, c) be a network with the properties described in Sect. 3.7.1. In an analogous way as for dynamic c-games here we denote

$$\text{ext}(c, x) = \begin{cases} \max\limits_{y \in X(x)} c_{(x,y)} & \text{for} \quad x \in X_1, \\ \min\limits_{y \in X(x)} c_{(x,y)} & \text{for} \quad x \in X_2, \end{cases}$$

$$\text{VEXT}(c, x) = \{y \in X(x) \mid c_{(x,y)} = \text{ext}(c, x)\}.$$

We shall use the potential transformation $c'_{(x,y)} = c_{(x,y)} + \varepsilon(y) - \varepsilon(x)$ for costs on the edges $e = (x, y) \in E$, where $\varepsilon \colon X \to \mathbb{R}$ is an arbitrary function on the vertex set X. In [43] it is shown that the potential transformation does not change the value and the optimal strategies of the players in cyclic games.

Theorem 3.54 *Let (G, X_1, X_2, c) be an arbitrary network with the properties described in* Sect. 3.7.1. *Then there exists the value $\overline{v}(x)$, $x \in X$ and the function $\varepsilon \colon X \to \mathbb{R}$ which determines the potential transformation $c'_{(x,y)} = c_{(x,y)} + \varepsilon(y) - \varepsilon(x)$ for costs on the edges $e = (x, y) \in E$, such that the following properties hold*

(a) $\overline{v}(x) = \text{ext}(c', x)$ *for* $x \in X$,
(b) $\overline{v}(x) = \overline{v}(y)$ *for* $x \in X_1 \cup X_2$ *and* $y \in \text{VEXT}(c', x)$,
(c) $\overline{v}(x) \geq \overline{v}(y)$ *for* $x \in X_1$ *and* $y \in X_G(x)$,
(d) $\overline{v}(x) \leq \overline{v}(y)$ *for* $x \in X_2$ *and* $y \in X_G(x)$,
(e) $\max\limits_{e \in E} |c'_e| \leq 2|X| \max\limits_{e \in E} |c_e|$.

The values $\overline{v}(x), x \in X$ on the network (G, X_1, X_2, c) are determined uniquely and the optimal strategies of the players can be found by the following way: Fix the arbitrary strategies $s^{1*} \colon X_1 \to X$ and $s^{2*} \colon X_2 \to X$ such that $s^{1*}(x) \in \text{VEXT}(c', x)$ for $x \in X_1$ and $s^{2*}(x) \in \text{VEXT}(c', x)$ for $x \in X_2$.

This theorem follows from Theorem 3.17. In [43] a constructive proof of Theorem 3.54 is given. In [79] it is shown that the conditions of this theorem can be obtained from the continuous optimal mean cost cycle problem in a weighted directed graph.

Further we shall use Theorem 3.54 in the case of the ergodic network (G, X_1, X_2, c), i.e., we shall use the following corollary:

Corollary 3.55 *Let (G, X_1, X_2, c) be an ergodic network. Then there exists the value $\overline{v}$ and the function $\varepsilon: X \to \mathbb{R}$ which determines the potential transformation $c'_{(x,y)} = c_{(x,y)} + \varepsilon_y - \varepsilon_x$ for costs of the edges $e = (x, y) \in E$ such that $\overline{v} = \mathrm{ext}(c', x)$ for $x \in X$. The optimal strategies of the players can be found as follows: Fix arbitrary strategies $s^{1^*}: X_1 \to X$ and $s^{2^*}: X_2 \to X$ such that $s^{1^*}(x) \in \mathrm{VEXT}(c', x)$ for $x \in X_1$ and $s^{2^*}(x) \in \mathrm{VEXT}(c', x)$ for $x \in X_2$.*

3.7.3 The Reduction of Cyclic Games to Ergodic Ones

Let us consider an arbitrary network (G, X_1, X_2, c, x_0) with a given starting position $x_0 \in X$ which determines a cyclic game. In [79, 83, 85] it is shown that this game can be reduced to a cyclic game on an auxiliary ergodic network $(G', W_1, W_2, \overline{c})$, $G' = (W, E')$ with the same value $\overline{v}(x_0)$ of the game as the initial one, where $x_0 \in W = X \cup U \cup Z$.

The graph $G' = (W, E')$ is obtained from G if each edge $e = (x, y)$ is changed by a triple of edges $e^1 = (x, u)$, $e^2 = (u, z)$, $e^3 = (z, y)$ with the costs $\overline{c}_{e^1} = \overline{c}_{e^2} = \overline{c}_{e^3} = c_e$. Here $u \in U$, $z \in Z$ and $x, y \in X$; $W = X \cup U \cup Z$. In addition, in G' each vertex u is connected with x_0 by edge (u, x_0) with the cost $\overline{c}_{(u,x_0)} = M$ (M is a great value) and each vertex z is connected with x_0 by edge (z, x_0) with the cost $\overline{c}_{(z,x_0)} = -M$. In $(G', W_1, W_2, \overline{c})$ the sets W_1 and W_2 are defined as follows: $W_1 = X_1 \cup Z$; $W_2 = X_2 \cup U$.

It is easy to observe that this reduction can be done in linear time.

3.7.4 A Polynomial Time Algorithm for Solving Ergodic Zero-Value Cyclic Games

Let us consider an ergodic zero value cyclic game determined by the network (G, X_1, X_2, c, x_0), where $G = (X, E)$. Then according to Theorem 3.54 there exists the function $\varepsilon : X \to \mathbb{R}$ which determines the potential transformation $c'_{(x,y)} = c_{(x,y)} + \varepsilon_y - \varepsilon_x$ on the edges $(x, y) \in E$ such that

$$\mathrm{ext}(c, x) = 0, \quad \forall x \in X. \tag{3.44}$$

This means that if x_f is a vertex of the cycle C_{s^*} determined by optimal strategies s^{1^*} and s^{2^*} then the problem of finding the function $\varepsilon : X \to \mathbb{R}$ which determines the canonic potential transformation is equivalent to the problem of finding the values ε_x, $x \in X$ in a max-min path problem on G with the sink vertex x_f where $\varepsilon_{x_f} = 0$.

So, in order to solve the zero value cyclic game we fix at each time-step a vertex $x \in X$ as a sink vertex ($x_f = x$) and solve a max-min path problem on G with the sink vertex x_f. If for the given $x_f = x$ the function $\varepsilon : X \to \mathbb{R}$ obtained on the basis of the algorithm from Sects. 3.5.4 and 3.5.5 determines the potential transformation which satisfies (3.44) then we fix s^{1^*} and s^{2^*} such that $s^{1^*}(x) \in \text{VEXT}(c', x)$ for $x \in X_1$ and $s^{2^*}(x) \in \text{VEXT}(c', x)$ for $x \in X_2$. If for the given x the function $\varepsilon : X \to \mathbb{R}$ does not satisfy (3.44) then we select another vertex $x \in X$ as a sink vertex and so on. This means that the optimal strategies of the players in zero value ergodic cyclic games can be found in time $O(|X|^4)$.

Example Consider the ergodic zero-sum cyclic game determined by the network given in Fig. 3.13 with starting position $x_0 = 0$. Positions of the first player are represented by circles and positions of the second one are represented by squares, i.e., $X_1 = \{1, 2, 4, 5\}$, $X_2 = \{0, 3\}$.

The network in Fig. 3.13 is obtained from the network in Fig. 3.13 by adding the edge (5, 2) with the cost $c_{(5,2)} = -5$. It is easy to check that the value of a cyclic game on this network for an arbitrary fixed starting position is equal to zero.

The max-min mean cycle which determines a way with zero cost is $2 \to 3 \to 5 \to 2$. Therefore, if we fix a vertex of this cycle as a sink vertex (as example $x = 5$) then we can find the potential function $\varepsilon : X \to \mathbb{R}$ which determines the potential transformation $c'_{(x,y)} = \varepsilon_y - \varepsilon_x + c_{(x,y)}$ such that $ext(c', x) = 0, \forall x \in X$. This function $\varepsilon : X \to \mathbb{R}$ can be found by using the algorithm from example from Sects. 3.5.4 and 3.5.5, i.e. we find costs of min-max paths from every $x \in X$ to vertex 5. So, $\varepsilon_0 = 8$, $\varepsilon_1 = 12$, $\varepsilon_2 = 5$, $\varepsilon_3 = 3$, $\varepsilon_4 = 2$, $\varepsilon_5 = 0$. After the potential transformation we obtain the network with the following costs of edges:

$$
\begin{aligned}
c'_{(3,5)} &= \varepsilon_5 - \varepsilon_3 + c_{(3,5)} = 0 - 3 + 3 = 0, \\
c'_{(4,5)} &= \varepsilon_5 - \varepsilon_4 + c_{(4,5)} = 0 - 2 + 2 = 0, \\
c'_{(5,2)} &= \varepsilon_2 - \varepsilon_5 + c_{(5,2)} = 5 - 0 - 5 = 0, \\
c'_{(2,3)} &= \varepsilon_3 - \varepsilon_2 + c_{(2,3)} = 3 - 5 + 2 = 0,
\end{aligned}
$$

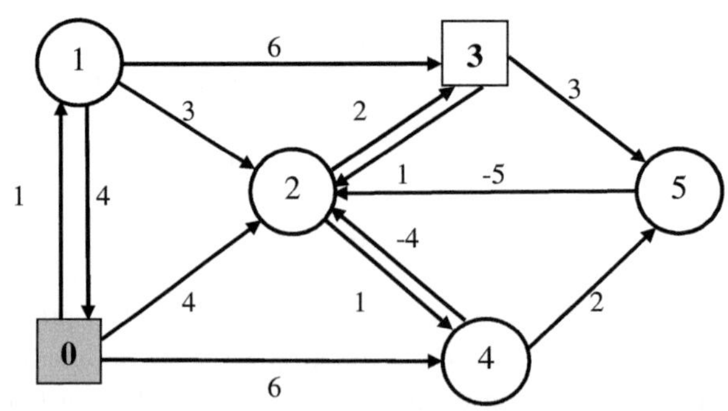

Fig. 3.13 The network for the ergodic zero-sum cyclic game

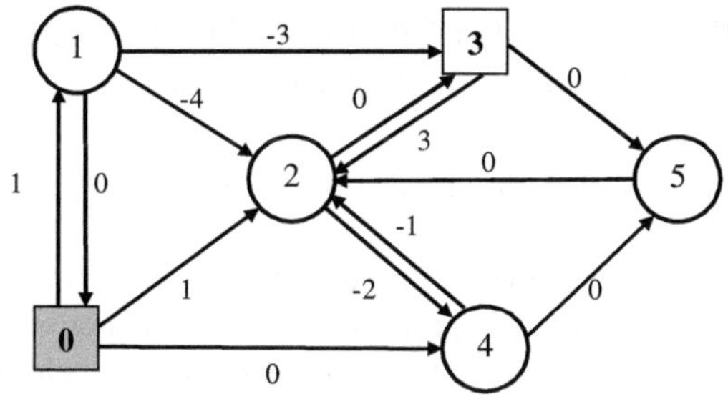

Fig. 3.14 The network after the potential transformation

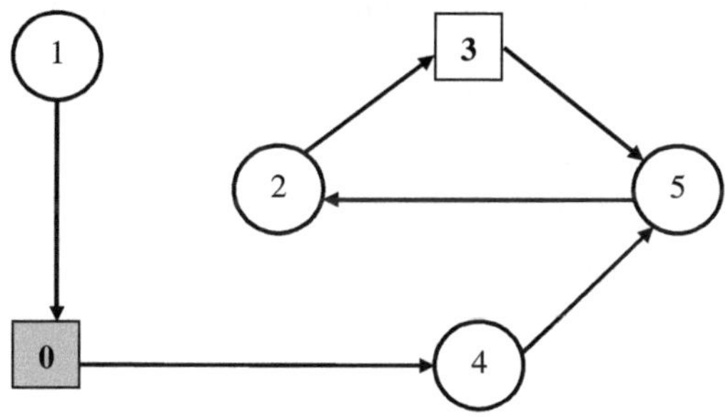

Fig. 3.15 The network induced by the optimal strategies of the players

$$
\begin{aligned}
c'_{(3,2)} &= \varepsilon_2 - \varepsilon_3 + c_{(3,2)} = 5 - 3 + 1 = 3,\\
c'_{(4,2)} &= \varepsilon_2 - \varepsilon_4 + c_{(4,2)} = 5 - 2 - 4 = -1,\\
c'_{(2,4)} &= \varepsilon_4 - \varepsilon_2 + c_{(2,4)} = 2 - 5 + 1 = -2,\\
c'_{(0,4)} &= \varepsilon_4 - \varepsilon_0 + c_{(0,4)} = 2 - 8 + 6 = 0,\\
c'_{(0,2)} &= \varepsilon_2 - \varepsilon_0 + c_{(0,2)} = 5 - 8 + 4 = 1,\\
c'_{(1,3)} &= \varepsilon_3 - \varepsilon_1 + c_{(1,3)} = 3 - 12 + 6 = -3,\\
c'_{(1,2)} &= \varepsilon_2 - \varepsilon_1 + c_{(1,2)} = 5 - 12 + 3 = -4,\\
c'_{(1,0)} &= \varepsilon_0 - \varepsilon_1 + c_{(1,0)} = 8 - 12 + 4 = 0,\\
c'_{(0,1)} &= \varepsilon_1 - \varepsilon_0 + c_{(0,1)} = 12 - 8 + 1 = 5.
\end{aligned}
$$

The network after the potential transformation is given in Fig. 3.14. We can see that $\text{ext}(c', x) = 0, \forall x \in X$. Therefore, the edges with zero cost determine the optimal strategies of the players:

$$
s^{1*} : 1 \to 0;\;\; 2 \to 3;\;\; 4 \to c;\;\; 5 \to 2;
$$
$$
s^{2*} : 0 \to 4;\;\; 3 \to 5.
$$

The graph $G_{s^*} = (X, E_{s^*})$ generated by these strategies is represented in Fig. 3.15.

3.7.5 A Polynomial Time Algorithm for Solving Cyclic Games Based on a Reduction to Acyclic l-Games

On the basis of the obtained results we can propose a polynomial time algorithm for solving cyclic games.

We consider an acyclic game on the ergodic network (G, X_1, X_2, c, x_0) with the given starting position x_0. The graph $G = (X, E)$ is considered to be strongly connected and $X = \{x_0, x_1, x_2, \dots, x_{n-1}\}$. Assume that x_0 belongs to the cycle C_{s^*} determined by the optimal strategies of the players s^{1^*} and s^{2^*}. If in G there are several of such cycles we consider one of them with the minimum number of edges.

We construct an auxiliary acyclic graph $GT_r = (\overline{W}_r, \overline{E}_r)$, where

$$\begin{aligned}
\overline{W}_r &= \{w_0^0\} \cup W^1 \cup W^2 \cup \dots \cup W^r, \quad W^i \cap W^j = \emptyset, \;\; i \neq j;\\
W^i &= \{w_0^i, w_1^i, \dots, w_{n-1}^i\}, \quad i = 1, 2, \dots, r;\\
\overline{E}_r &= E^0 \cup E^1 \cup E^2 \cup \dots \cup E^{r-1};\\
E^i &= \{(w_k^{i+1}, w_l^i) \mid (x_k, x_l) \in E\}, \quad i = 1, 2, \dots, r-1;\\
E^0 &= \{(w_k^i, w_0^0) \mid (x_k, x_0) \in E, \quad i = 1, 2, \dots, r\}.
\end{aligned}$$

The vertex set $\overline{W}_r$ of GT_r is obtained from X if it is doubled r times and then a sink vertex w_0^0 is added. The edge subset $E^i \subseteq \overline{E}$ in GT_r connects the vertices of the set W^{i+1} and the vertices of the set W^i in the following way:

If in G there exists an edge $(x_k, x_l) \in E$ then in GT_r we add the edge (w_k^{i+1}, w_l^i). The edge subset $E^0 \subseteq \overline{E}$ in GT_r connects the vertices $w_k^i \in W^1 \cup W^2 \cup \dots \cup W^r$ with the sink vertex w_0^0, i.e., if there exists an edge $(x_k, x_0) \in E$ then in GT_r we add the edges $(w_k^i, w_0^0) \in E^0$, $i = 1, 2, \dots, r$.

After that we define the acyclic network $(GT_r', W_1, W_2, c', w_0^0)$, $GT_r' = (W_r, E_r)$ where GT_r' is obtained from GT_r by deleting the vertices $w_k^i \in \overline{W}_r$ from which the vertex w_0^0 can not be attainable. The sets W_1, W_2 and the cost function $c'\colon E_r \to \mathbb{R}$ are defined as follows:

$$\begin{aligned}
W_1 &= \{w_k^i \in W_r \mid x_k \in X_1\}, \quad W_2 = \{w_k^i \in W_r \mid x_k \in X_2\};\\
c'_{(w_k^{i+1}, w_l^i)} &= c_{(x_k, x_l)} \text{ if } (x_k, x_l) \in E \text{ and } (w_k^{i+1}, w_l^i) \in E^i; \;\; i = 1, 2, \dots, r-1;\\
c'_{(w_k^i, w_0^0)} &= c_{(x_k, x_0)} \text{ if } (x_k, x_0) \in E \text{ and } (w_k^i, w_0^0) \in E^0; \;\; i = 1, 2, \dots, r.
\end{aligned}$$

Now we consider the acyclic c-game on the acyclic network $(GT_r', W_1, W_2, c', w_0^r, w_0^0)$ with the sink vertex w_0^0 and the starting position w_0^r.

Lemma 3.56 *Let $\overline{v} = \overline{v}(x_0)$ is a value of the ergodic cyclic game on G and the number of edges of the max-min cycle C_{s^*} in G is equal to r. In addition, let $\overline{v}_r(w_0^r)$ be the value of the l-game on (GT_r', W_1, W_2, c') with the starting position w_0^r. Then $\overline{v}(x_0) = \overline{v}_r(w_0^r)$.*

Proof It is evident that there exists a bijective mapping between the set of cycles with no more than r edges (which contains the vertex x_0) in G and the set of directed paths with no more than r edges from w_0^r to w_0^0 in GT_r'. Therefore $\overline{v}(x_0) = \overline{v}_r(w_0^r)$. □

On the basis of this lemma we can propose the following algorithm for finding the optimal strategies of the players in cyclic games.

Algorithm 3.57 Determining the Optimal Stationary Strategies of the Players in Cyclic Games with a Known Vertex of a Max-Min Cycle of the Network
We construct the acyclic networks (GT_r', W_1, W_2, c'), $r = 2, 3, \ldots, n$, and for each of them solve a l-game. In such a way we find the values $\overline{v}_2(w_0^2), \overline{v}_3(w_0^3), \ldots, \overline{v}_n(w_0^n)$ for these l-games.

Then we consecutively fix $\overline{v} = \overline{v}_2(w_0^2)$, $\overline{v}_3(w_0^3)$, $\ldots$, $\overline{v}_n(w_0^n)$ and at each time solve the c-game on the network (G, X_1, X_2, c''), where $c'' = c - \overline{v}$. Fixing at each time the values $\varepsilon'(x_k) = v(x_k)$ for $x_k \in X$ we check if the following condition

$$\mathrm{ext}(c^r, x_k) = 0, \quad \forall x_k \in X$$

is satisfied, where $c^r_{(x_k,x_l)} = c''_{(x_k,x_l)} + \varepsilon(x_l) - \varepsilon(x_k)$. We find r for which this condition holds and fix the respective maps s^{1^*} and s^{2^*} such that $s^{1^*}(x_k) \in \mathrm{VEXT}(c'', x_k)$ for $x_k \in X_1$ and $s^{2^*}(x_k) \in \mathrm{VEXT}(c'', x_k)$ for $x_k \in X_2$. So, s^{1^*} and s^{2^*} represent the optimal strategies of the players in cyclic games on G.

Remark 3.58 Algorithm 3.57 finds the value $\overline{v}(x_0)$ and optimal strategies of the players in time $O(|X|^5 \log L + 4|X|^3 \log |X|)$, because Algorithm 3.52 needs $O(|X|^4 \log L + 4|X|^2 \log |X|)$ elementary operations for solving an acyclic l-game on the network (GT_r', W_1, W_2, c'), where $L = \max_{e \in E} |c_e| + 1$.

In the general case, if the belonging of x_0 to the max-min cycle is unknown then we use the following algorithm.

Algorithm 3.59 Determining the Optimal Strategies of the Players in Ergodic Cyclic Games (General case)

Preliminary step (step 0): Fix $Y_1 = X$.

General step (step k): Select a vertex $y \in Y_1$, fix $x_0 = y$ and apply Algorithm 3.57. If there exists $r \in \{2, 3, \ldots, n\}$ such that $\mathrm{ext}(c^r, x) = 0$, $\forall x \in X$, then fix $s^{1^*} \in \mathrm{VEXT}(c^k, x)$ for $x \in X_1$ and $s^{2^*} \in \mathrm{VEXT}(c^k, x)$ for $x \in X_2$ and STOP; otherwise put $Y_{k+1} = Y_k \setminus \{y\}$ and go to next step $k + 1$.

Remark 3.60 Algorithm 3.59 finds the value $\overline{v}$ and optimal strategies of the players in time $O(|X|^6 \log L + 4|X|^4 \log |X|)$, because in the worst case Algorithm 3.57 is repeated $|X|$ times.

The algorithm for solving cyclic games allows us to determine the sign of value $v(x_0)$ in an infinite dynamic c-game on G with starting position x_0. In order to determine $\mathrm{sign}(v(x_0))$ we solve on G a cyclic game with starting position x_0 and determine $\overline{v}(x_0)$. Then $\mathrm{sign}(v(x_0)) = \mathrm{sign}(\overline{v}(x_0))$.

3.7.6 An Approach for Solving Cyclic Games Based on the Dichotomy Method and Solving a Dynamic c-Game

In this section we describe an approach for solving cyclic games considering that there exist efficient algorithms for solving a dynamic c-game (including infinite dynamic c-games).

Consider the ergodic cyclic game determined by the ergodic network (G, X_1, X_2, c, x_0) where the value of the game may be different from zero. The graph G is assumed to be strongly connected.

At first we show how to determine the value of the game and optimal strategies of the players in the case if the vertex x_0 belongs to a max-min cycle in G induced by optimal strategies of the players.

To our ergodic cyclic game we associate a dynamic c-game determined by an auxiliary network $(\overline{G}, X_1, X_2 \cup \{x_0'\}, \overline{c}, x_0, x_0')$, where the graph $\overline{G} = (X \cup \cup \{x_0'\}, \overline{E})$ is obtained from G by adding a copy x_0' of vertex x_0 together with copies $e' = (x, x_0')$ of the edges $e = (x, x_0) \in E$ with costs $\overline{c}_{e'} = c_e$. So, for x_0' there are no leaving edges (x_0', x).

It is evident that if the value $\overline{v} = \overline{v}(x_0)$ of the ergodic cyclic game on (G, X_1, X_2, c, x_0) is known then the problem of finding the optimal strategies of the players is equivalent to the problem of finding optimal strategies of the players in a dynamic c-game on the network $(\overline{G}, X_1, X_2 \cup \{x_0'\}, \overline{c}', x_0, x_0')$ with the cost functions

$$\overline{c}_e' = \overline{c}_e - \overline{v}(x_0) \quad \text{for} \quad e \in \overline{E}.$$

Moreover, if s^{1*} and s^{2*} are optimal strategies of the players in the dynamic c-game on $(\overline{G}, X_1, X_2 \cup \{x_0'\}, \overline{c}', x_0, x_0')$, then the optimal strategies $\overline{s}_A^*$ and $\overline{s}_B^*$ of the players in the ergodic cyclic game can be found as follows:

$$\begin{aligned} \overline{s}_1^*(x) &= s^1(x) \quad \text{for} \quad x \in X_1 \quad \text{if} \quad s^1(x) \neq x_0'; \\ \overline{s}_2^*(x) &= s^2(x) \quad \text{for} \quad x \in X_2 \quad \text{if} \quad s^2(x) \neq x_0' \end{aligned}$$

and

$$\begin{aligned} \overline{s}_1^*(x) &= x_0 \quad \text{if} \quad s^1(x) = x_0'; \\ \overline{s}_2^*(x) &= x_0' \quad \text{if} \quad s^2(x) = x_0'. \end{aligned}$$

It is easy to observe that for the considered problems the following properties hold.

1. The value $\overline{v}(x_0)$ of the ergodic cyclic game on the network (G, X_1, X_2, c, x_0) is nonnegative if and only if the value $v(x_0)$ of the dynamic c-game on the network $(\overline{G}, X_1, X_2 \cup \{x_0'\}, \overline{c}, x_0, x_0')$ is nonnegative; moreover $\overline{v}(x_0) = 0$ if and only if $v(x_0) = 0$.

2. If $M^1 = \min_{e \in E} c_e$ and $M^2 = \max_{e \in E} c_e$, then $M^1 \leq \overline{v}(x_0) \leq M^2$.
3. If in the network (G, X_1, X_2, c, x_0) the cost function $c : E \to \mathbb{R}$ is changed by $c' = c + h$, then the optimal strategies of the players in the ergodic cyclic game on the network (G, X_1, X_2, c', x_0) do not change although the value $\overline{v}(x_0)$ is changed by $\overline{v}'(x_0) = \overline{v}(x_0) + h$.

On the basis of these properties we seek for the unknown value $\overline{v}(x_0) = v(x)$, which we denote by h, using the dichotomy method on the segment $[M^1, M^2]$ such that at each step of this method we will solve a dynamic c-game with network $(\overline{G}, X_1, X_2 \cup \{x_0'\}, \overline{c}^h, x_0, x_0')$, where $\overline{c}^h = \overline{c} - h$. So, the main idea of the general step of the algorithm is the following: We make the transformation

$$\overline{c}^k = \overline{c} - h_k \quad \text{for} \quad e \in E,$$

where h_k is a midpoint of the segment $[M_k^1, M_k^2]$ at step k.

After that we apply the algorithm from Sect. 3.5.5 for the dynamic c-game on the network $(\overline{G}, X_1, X_2 \cup \{x_0'\}, \overline{c}^k, x_0, x_0')$ and find $v_{h_k}(x_0)$. If $v_{h_k}(x_0) > 0$ then we fix the segment $[M_{k+1}^1, M_{k+1}^2]$, where $M_{k+1}^1 = M_k^1$ and $M_{k+1}^2 = \left(M_k^1 + M_k^2\right)/2$; otherwise we put $M_{k+1}^1 = \left(M_k^1 + M_k^2\right)/2$ and $M_{k+1}^2 = M_k^2$. If $v_{h_k}(x_0) = 0$ then STOP.

So, using the dichotomy method in an analogous way as for the acyclic l-game we determine the value of the acyclic game. If this value of the dynamic c-game is known then we determine the strategies of the players by using algorithms from Sect. 3.5.4 or Sect. 3.5.5.

In the case if x_0 may not belong to a max-min cycle determined by optimal strategies of the players in the cyclic game we solve $|X|$ problems by fixing at each time the starting position $x_0 = x$ for $x \in X$. Then at least for a position $x_0 = x \in X$ we obtain the value of the cyclic game and the optimal strategies of the players.

3.8 On Determining Pareto Optima for Cyclic Games with *m* Players

To determine a Pareto solution for the cyclic game with m players we can use the continuous model from Sect. 2.2.4 and extend it for the multi-objective case of the problem in the following way:
Minimize the vector function

$$\overline{H}(\alpha) = (\overline{H}^1(\alpha), \overline{H}^2(\alpha), \ldots, \overline{H}^m(\alpha))$$

subject to

$$\begin{cases} \sum\limits_{e \in E^-(x)} \alpha_e - \sum\limits_{e \in E(x)} \alpha_e = 0, \ \forall x \in X; \\ \sum\limits_{e \in E} \alpha_e = 1; \\ \qquad \alpha_e \geq 0, \ e \in E, \end{cases}$$

where

$$\overline{H}^i(\alpha) = \sum_{e \in E} c_e^i \alpha_e, \quad i = 1, 2, \ldots, m;$$
$$E^-(x) = \{e = (y, x) \mid (y, x) \in E\}; \quad E(x) = \{e = (x, y) \mid (x, y) \in E\}.$$

Pareto optima for this multicriterion problem can be found by using the approach from [14–16, 30, 111]. Solutions of this continuous problem will correspond to solutions of the discrete multicriterion problem on a given strongly connected graph $G = (X, E)$ with cost functions $c^i : E \to \mathbb{R}$, $i = 1, 2, \ldots, m$.

Note that a Pareto solution for the cyclic game with m players on G does not depend on the partition $X = X_1 \cup X_2 \cup \cdots \cup X_m$.

3.9 Multi-objective Control Based on the Concept of Noncooperative Games: Nash Equilibria

Consider a dynamic system $\mathbb{L}$ with a finite set of states X, where at every time-step t the state of $\mathbb{L}$ is $x(t) \in X$. For the system $\mathbb{L}$ two states $x_0, \ x_f \in X$ are given where x_0 represents the starting state of $\mathbb{L}$, i.e., $x_0 = x(0)$, and x_f represents the state in which the dynamical system must be brought, i.e. x_f is the final state of $\mathbb{L}$. We assume that the dynamic system should reach the final state x_f at the moment of time $t(x_f)$ such that $t_1 \leq t(x) \leq t_2$ where t_1 and t_2 are given. The dynamics of the system $\mathbb{L}$ is controlled by m players and it is described as follows:

$$x(t+1) = g_t(x(t), u^1(t), u^2(t), \ldots, u^m(t)), \quad t = 0, 1, 2, \ldots \tag{3.45}$$

where

$$x(0) = x_0$$

is a starting point of the system $\mathbb{L}$ and $u^i(t) \in \mathbb{R}^{n_i}$ represents the vector of the control parameters of the player $i, i \in \{1, 2, \ldots, m\}$. The state $x(t+1)$ of the system $\mathbb{L}$ at the time-step $t+1$ is obtained uniquely if the state $x(t)$ at the time-step t is known and the players $1, 2, \ldots, m$ fix their vectors of the control parameters $u^1(t), u^2(t), \ldots, u^m(t)$

independently, respectively. For each player i, $i \in \{1, 2, \dots, m\}$ the admissible sets $U_t^i(x(t))$ for the vectors of the control parameters $u^i(t)$ are given, i.e.,

$$u^i(t) \in U_t^i(x(t)), \quad t = 0, 1, 2, \dots; \; i = 1, 2, \dots, m. \tag{3.46}$$

We assume that $U_t^i(x(t)),\ t = 0, 1, 2, \dots;\ \ i = 1, 2, \dots, m$ are non-empty finite sets and

$$U_t^i(x(t)) \cap U_t^j(x(t)) = \emptyset, \quad i \neq j, \;\; t = 0, 1, 2, \dots.$$

Let us consider that the players $1, 2, \dots, m$ fix their vectors of the control parameters

$$u^1(t), u^2(t), \dots, u^m(t); \quad t = 0, 1, 2, \dots,$$

respectively, and the starting state $x(0) = x_0$ as well as the final state x_f are known. Then for the fixed vectors of the control parameters $u^1(t), u^2(t), \dots, u^m(t)$ either a unique trajectory

$$x_0 = x(0), x(1), x(2), \dots, x(t(x_f)) = x_f$$

from x_0 to x_f exists and $t(x_f)$ represents the time-moment if the state x_f is reached, or such a trajectory from x_0 to x_f does not exist.

We denote by

$$F^i_{x_0 x_f}(u^1(t), u^2(t), \dots, u^m(t)) = \sum_{t=0}^{t(x_f)-1} c_t^i(x(t), g_t(x(t), u^1(t), u^2(t), \dots, u^m(t)))$$

the integral-time cost of the system's transition from x_0 to x_f for player i, $i \in \{1, 2, \dots, m\}$ if the vectors $u^1(t), u^2(t), \dots, u^m(t)$ satisfy condition (3.46) and generate a trajectory

$$x_0 = x(0), x(1), x(2), \dots, x(t(x_f)) = x_f$$

from x_0 to x_f such that

$$t_1 \leq t(x_f) \leq t_2;$$

otherwise we put

$$F^i_{x_0 x_f}(u^1(t), u^2(t), \dots, u^m(t)) = \infty.$$

Note that $c_t^i(x(t), g_t(x(t), u^1(t), u^2(t), \ldots, u^m(t))) = c_t^i(x(t), x(t+1))$ represents the cost of the system's passage from state $x(t)$ to state $x(t+1)$ at the stage $[t, t+1]$ for player i.

Problem 3.61 Find vectors of control parameters

$$u^{1^*}(t),\ u^{2^*}(t),\ \ldots,\ u^{i-1^*}(t),\ u^{i^*}(t),\ u^{i+1^*}(t),\ \ldots,\ u^{m^*}(t),$$

which satisfy the condition

$$\begin{aligned} F^i_{x_0x_f}(u^{1^*}(t),\ u^{2^*}(t),\ \ldots,\ u^{i-1^*}(t),\ u^{i^*}(t),\ u^{i+1^*}(t),\ \ldots,\ u^{m^*}(t)) \le \\ \le F^i_{x_0x_f}(u^{1^*}(t),\ u^{2^*}(t),\ \ldots,\ u^{i-1^*}(t),\ u^{i}(t),\ u^{i+1^*}(t),\ \ldots,\ u^{m^*}(t)) \\ \forall\, u^i(t) \in \mathbb{R}^{m_i},\ \ i = 0, 1, 2, \ldots, m. \end{aligned}$$

So, we consider the problem of finding the solution in the sense of Nash [2,79, 100, 102].

The problems formulated above, can be regarded as mathematical models for dynamical systems controlled by several players who do not inform each other which vectors of control parameters they use in the control process.

An important particular case of Problem 3.61 is represented by the zero-sum control problem of two players with the given costs

$$c_t(x(t),\ x(t+1)) = c_t^2(x(t),\ x(t+1)) = -c_t^1(x(t),\ x(t+1))$$

of the system's passage from state $x(t)$ to state $x(t+1)$, which determine the payoff function

$$F_{x_0x_f}(u^1(t),\ u^2(t)) = F^2_{x_0x_f}(u^1(t),\ u^2(t)) = -F^1_{x_0x_f}(u^1(t),\ u^2(t)).$$

In this case we seek for a saddle point $(u^{1*}(t), u^{2*}(t))$ of the function $F_{x_0x_f}(u^1(t), u^2(t))$ [103], i.e. we consider the following max-min control problem:

Problem 3.62 Find vectors of control parameters $u^{1^*}(t)$, $u^{2^*}(t)$ such that

$$\begin{aligned} F_{x_0x_f}(u^{1^*}(t),\ u^{2^*}(t)) &= \max_{u^1(t)} \min_{u^2(t)} F_{x_0x_f}(u^1(t),\ u^2(t)) \\ &= \min_{u^2(t)} \max_{u^1(t)} F_{x_0x_f}(u^1(t),\ u^2(t)). \end{aligned}$$

So, for this max-min control problem we are seeking for a saddle point [103].

The results obtained in previous sections allows us to formulate conditions for the existence of Nash equilibria in such dynamic games. Moreover, we describe a class of game control problems for which the dynamic programming technique can be used for determining Nash equilibria.

3.9.1 Stationary and Non-stationary Multi-objective Control Models

The multi-objective control model formulated above is related to the non-stationary case. In this model the functions g_t^i, $t = 0, 1, 2, \ldots$, may be different for different moments of time and the players in the control process can change their vectors of control parameters for an arbitrary state $x = x(t)$ at different moments in time t. Additionally, for a given state $x = x(t)$ the admissible sets $U_t^i(x(t)), i = 1, 2, \ldots, m$, can be different for different moments in time t. Moreover, the costs of the system's transition $c_t(x(t), x(t+1))$ from state $x = x(t)$ to state $y = x(t+1)$ are varying in time for given x and y.

Stationary versions of the considered control problems correspond to the case if the functions $g_t^i, t = 0, 1, 2, \ldots$, do not change in time, i.e., $g_t^i \equiv g^i, t = 0, 1, 2, \ldots$, and the players preserve the same vectors of control parameters in time for given states $x \in X$. Additionally, we consider that the admissible sets $U_t^i(x(t)), t = 0, 1, 2, \ldots$, for vectors of control parameters do not change in time, i.e., $U_t^i(x(t)) = U^i(x)$, $t = 0, 1, 2, \ldots, \quad i = 1, 2, \ldots, m$.

In general, for non-stationary control problems the players can use non-stationary strategies although the functions $g_t^i, t = 0, 1, 2, \ldots$, and the admissible sets of the control parameters $U_t^i(x(t)), t = 0, 1, 2, \ldots$, may not change in time, i.e., $g_t^i \equiv g^i$, $t = 0, 1, 2, \ldots$ and $U_t^i(x(t)) = U^i(x), t = 0, 1, 2, \ldots, i = 1, 2, \ldots, m$.

3.9.2 Multi-objective Control Problems with Infinite Time Horizon

For the control problems with infinite time horizon the final state is not given and the control process is made indefinitely on discrete moments in time $t = 0, 1, 2, \ldots$. Mainly two classes of multi-objective problems in this topic are considered.

In the first class of problems each player $i \in \{1, 2, \ldots, m\}$ has the aim to minimize his own objective function

$$F_{x0}^i(u^1(t), u^2(t), \ldots, u^m(t)) = \lim_{T \to \infty} \frac{1}{T} \sum_{t=0}^{T} c_t^i(x(t), g_t(x(t), u^1(t), u^2(t), \ldots, u^m(t)))$$

that expresses the average cost per transitions by a trajectory determined by all players together. For the second class each player $i \in \{1, 2, \ldots, m\}$ has to minimize the discounted objective function

$$\widehat{F}^i_{x_0}(u^1(t), u^2(t), \ldots, u^m(t)) = \sum_{t=0}^{\mathcal{T}} \gamma^t c^i_t(x(t), g_t(x(t), u^1(t), u^2(t), \ldots, u^m(t))),$$

with a given discount factor γ, where $0 < \gamma < 1$.

As we have noted for the single-objective control problems with infinite time horizon and constant transition costs there exists the optimal stationary control. Based on the results obtained for positional games we may derive conditions and algorithms for determining Nash equilibria for the stationary game control problem with average and discounted objective functions.

If $m = 2$ and

$$c^2_t(x(t), g_t(x(t), u^1(t), u^2(t))) = -c^1_t(x(t), g_t(x(t), u^1(t), u^2(t))),$$

then we obtain a zero-sum control problem with infinite time horizon. For such a game we are seeking for a saddle point.

3.10 Hierarchical Control and Stackelberg's Optimization Principle

Now we shall use the concept of hierarchical control and assume that in (3.45) for an arbitrary state $x(t)$ at every moment in time, the players fix their vectors of control parameters successively one after another according to their numerical order. Moreover, we assume that each player fixing his vectors of control parameters informs posterior players which vector of control parameters has been chosen at the given moment in time for a given state. So, we consider the following hierarchical control process:

Let $\mathbb{L}$ be a dynamical system with a finite set of states X and a fixed starting point $x(0) = x_0 \in X$. The dynamics of system $\mathbb{L}$ is defined by the system of difference equations (3.45) and it is controlled by p players using the corresponding vectors of the control parameters $u^1(t), u^2(t), \ldots, u^m(t)$.

For each vector of control parameters $u^i(t)$ the feasible set (3.46) is defined for an arbitrary state $x(t)$ at every discrete moment in time t. Additionally, we assume that for an arbitrary state $x(t) \in X$ at every moment in time t the players fix their vectors of control parameters successively one after another according to a given order. For simplicity, we will consider that the players fix their vectors of control parameters in the order corresponding to their numbers. Each player, after fixing his vectors of control parameters, informs posterior players which vector of control parameters has been chosen at the given moment in time for a given state. Finally, if the vectors of control parameters $u^1(t), u^2(t), \ldots, u^m(t)$ and the starting state $x(0) = x_0$ are known then the cost $F^i_{x_0x_f}(u^1(t), u^2(t), \ldots, u^m(t))$ of the system's passage from the starting state x_0 to the final state x_f for player $i \in \{1, 2, \ldots, m\}$ is defined in the same way as in Sect. 3.9.

In this hierarchical control process we are looking for Stackelberg strategies [78, 79, 126], i.e. we consider the following hierarchical control problem:

Problem 3.63 Find vectors of control parameters $u^{1^*}(t), u^{2^*}(t), \ldots, u^{m^*}(t)$, for which

$$u^{1^*}(t) = \operatorname*{argmin}_{\substack{u^1(t)\in U^1 \\ (u^i(t)\in R_i(u^1,\ldots,u^{i-1}))_{2\le i\le m}}} F^1_{x_0x_f}(u^1(t),\ u^2(t),\ \ldots,\ u^m(t));$$

$$u^{2^*}(t) = \operatorname*{argmin}_{\substack{u^2(t)\in R_2(u^{1^*}) \\ \left(u^i(t)\in R_i(u^{1^*},u^2,\ldots,u^{i-1})\right)_{3\le i\le m}}} F^2_{x_0x_f}(u^{1*}(t),\ u^2(t),\ \ldots,\ u^m(t));$$

$$u^{3^*}(t) = \operatorname*{argmin}_{\substack{u^3(t)\in R_3(u^{1^*},u^{2^*}) \\ \left(u^i(t)\in R_i(u^{1^*},u^{2^*},u^3,\ldots,u^{i-1})\right)_{4\le i\le m}}} F^3_{x_0x_f}(u^{1^*}(t),\ u^{2^*}(t),\ \ldots,\ u^m(t));$$

$$\vdots$$

$$u^{m^*}(t) = \operatorname*{argmin}_{u^m(t)\in R_m(u^{1^*},u^{2^*},\ldots,u^{m-1^*})} F^m_{x_0x_f}(u^{1^*}(t), u^{2^*}(t), \ldots, u^{m-1^*}(t), u^m(t));$$

where $R_k(u^1, u^2, \ldots, u^{k-1})$ represents the best responses of player k if the players $1, 2, \ldots, k-1$ have already fixed their vectors $u^1(t),\ u^2(t),\ \ldots,\ u^{k-1}(t)$, i.e.,

$$R_2(u^1) = \operatorname*{argmin}_{\substack{u^2(t)\in U^2 \\ (u^i(t)\in R_i(u^1,\ldots,u^{i-1}))_{3\le i\le m}}} F^2_{x_0x_f}(u^1(t), u^2(t), \ldots, u^m(t));$$

$$R_3(u^1, u^2) = \operatorname*{argmin}_{\substack{u^3(t)\in U^3 \\ (u^i(t)\in R_i(u^1,\ldots,u^{i-1}))_{4\le i\le m}}} F^3_{x_0x_f}(u^1(t), u^2(t), \ldots, u^m(t));$$

$$\vdots$$

$$R_m(u^1, u^2, \ldots, u^{m-1}) = \operatorname*{argmin}_{u^m(t)\in U^m} F^m_{x_0x_f}(u^1(t),\ u^2(t),\ \ldots,\ u^m(t));$$

$$U^i = \prod_{x(t)}\prod_{t} U^i_t(x(t)), \quad t = 0, 1, 2, \ldots; \quad i = 0, 1, 2, \ldots, m.$$

It is easy to observe that if the solution $u^{1^*}(t),\ u^{2^*}(t),\ \ldots,\ u^{m^*}(t)$ of Problem 3.61 does not depend on the order of fixing the vectors of control parameters of the players $1, 2, \ldots, m$ then $u^{1^*}(t),\ u^{2^*}(t),\ \ldots,\ u^m(t)$ is a solution in the sense of Nash.

If $c^2_t(x(t),\ x(t+1)) = -c^1_t(x(t),\ x(t+1)) = c_t(x(t),\ x(t+1))$, then we obtain the max-min control problem of two players with the payoff functions

$$F_{x_0 x_f}(u^1(t), u^2(t)) = F^2_{x_0 x_f}(u^1(t), u^2(t)) = -F^1_{x_0 x_f}(u^1(t), u^2(t)).$$

In this case we are seeking for the vector of the control parameters $u^{1^*}(t)$, $u^{2^*}(t)$ such that

$$F_{x_0 x_f}(u^{1^*}(t), u^{2^*}(t)) = \max_{u^1(t)} \min_{u^2(t)} F_{x_0 x_f}(u^1(t), u^2(t)).$$

For the considered class of problems we will also develop an algorithm based on dynamic programming.

3.10.1 Multi-objective Control Based on the Concept of Cooperative Games: Pareto Optima

We consider a dynamical system $\mathbb{L}$, which is controlled by m players $1, 2, \ldots, m$ and formulate the control model which is based on the concept of cooperative games. Assume that the players coordinate their actions in the control processes by using common vectors of control parameters $u(t) = (u^1(t), u^2(t), \ldots, u^m(t))$ (see [11, 17, 60, 100, 101, 107]). So, the dynamics of the system is described by the following system of difference equations

$$x(t+1) = g_t(x(t), u(t)), \quad t = 0, 1, 2, \ldots$$

where

$$x(0) = x_0 \quad \text{and} \quad u(t) \in U_t(x(t)), \quad t = 0, 1, 2, \ldots.$$

Additionally, we assume that system $\mathbb{L}$ should reach the final state at the time moment $t(x_f)$ such that $t_1 \le t(x_f) \le T_2$.

Let

$$u(0), u(1), u(2), \ldots, u(t-1), \ldots$$

be a players' control, which generates a trajectory

$$x(0), x(1), x(2), \ldots, x(t), \ldots.$$

Then either this trajectory passes through state x_f at the finite moment $T(x_f)$ or it does not pass through x_f. We denote by

$$F^i_{x_0 x_f}(u(t)) = \sum_{t=0}^{t(x_f)-1} c^i_t(x(t), g_t(x(t), u(t))), \quad i = 1, 2, \ldots, m$$

the integral-time cost of the system's passage from x_0 to x_f if

$$t_1 \le t(x_f) \le t_2;$$

otherwise we put

$$F^i_{x_0 x_f}(u(t)) = \infty.$$

Here $c^i_t(x(t), g_t(x(t), u(t))) = c^i_t(x(t), x(t+1))$ represents the cost of the system's passage from state $x(t)$ to state $x(t+1)$ at the stage $[t, t+1]$ for player i, $i \in \{1, 2, \ldots, m\}$.

Problem 3.64 Find vectors of control parameters $u^*(t)$ such that there is no other control vector $u(t) \neq u^*(t)$, for which

$$\begin{aligned}&(F^1_{x_0 x_f}(u(t)),\ F^2_{x_0 x_f}(u(t)),\ \ldots,\ F^m_{x_0 x_f}(u(t)))\\&\quad \le (F^1_{x_0 x_f}(u^*(t)),\ F^2_{x_0 x_f}(u^*(t)),\ \ldots,\ F^m_{x_0 x_f}(u^*(t)))\end{aligned}$$

and for any $i_0 \in \{1, 2, \ldots, m\}$

$$F^{i_0}_{x_0 x_f}(u(t)) < F^{i_0}_{x_0 x_f}(u^*(t)).$$

So, we consider the problem of finding a Pareto solution [100, 101, 107].

Unlike Nash equilibria, Pareto optima for multi-objective discrete control always exist if there is an admissible solution $u(t)$, $t = 0, 1, 2, \ldots, t(x_f)$, which generates a trajectory $x_0 = x(0),\ x(1),\ x(2),\ \ldots,\ x(t(x_f)) = x_f$ from x_0 to x_f.

3.11 Alternate Players' Control Condition and Nash Equilibria for Dynamic Games in Positional Form

In order to formulate the theorem of the existence of Nash equilibria for the considered multi-objective control problem from Sect. 3.9 we will apply the following condition:

We assume that an arbitrary state $x(t) \in X$ of the dynamic system $\mathbb{L}$ at the time-moment t represents a position $(x, t) \in X \times \{0, 1, 2, \ldots\}$ of one of the players $i \in \{1, 2, \ldots, m\}$. This means that in the control process the next state $x(t+1) \in X$ is determined (chosen) by player i if the dynamic system $\mathbb{L}$ at the time-moment t has the state $x(t)$, which corresponds to the position (x, t) of player i. This situation corresponds to the case if the expression

$$g_t(x(t),\ u^1(t),\ u^2(t),\ \ldots,\ u^{i-1}(t),\ u^i(t),\ u^{i+1}(t),\ \ldots,\ u^m(t))$$

in (3.45) for a given position (x, t) of player i only depends on the control vector $u^i(t)$, i.e.,

$$g_t(x(t),\ u^1(t),\ u^2(t),\ \ldots,\ u^{i-1}(t),\ u^i(t),\ u^{i+1}(t),\ \ldots,\ u^m(t)) = g_t^i(x(t),\ u^i(t)).$$

So, the notations (x, t) and $x(t)$ have the same meaning.

Definition 3.65 We say that the alternate players' control condition is satisfied for the multi-objective control problems if for any fixed $(x, t) \in X \times \{0, 1, 2, \ldots\}$ the Eq. (3.45) only depend on one of the vectors of control parameters. The multi-objective control problems with such an additional condition are called game control models in positional form.

The following lemma presents a necessary and sufficient condition that holds for the alternate players' control.

Lemma 3.66 *The alternate players' control condition for the multi-objective control problem holds if and only if at every time-step* $t = 0, 1, 2, \ldots$ *for the set of states* X *there exists a partition*

$$X = X_1(t) \cup X_2(t) \cup \cdots \cup X_m(t); \quad (X_i(t) \cap X_j(t) = \emptyset,\ i \neq j) \tag{3.47}$$

such that the equations (3.45) *can be represented as follows:*

$$\begin{aligned} x(t+1) = g_t^i(x(t),\ u^i(t)) \ \ &\textit{if} \ \ x(t) \in X_i(t); \\ t = 0, 1, 2, \ldots; \ \ &i = 1, 2, \ldots, m, \end{aligned} \tag{3.48}$$

i.e.,

$$\begin{aligned} &g_t(x(t),\ u^1(t),\ u^2(t),\ \ldots,\ u^i(t),\ u^{i+1}(t),\ \ldots,\ u^m(t)) \\ &\quad = g_t^i(x(t),\ u^i(t)) \ \textit{if} \ x(t) \in X_i(t); \quad t = 0, 1, 2, \ldots; \quad i = 1, 2, \ldots, m. \end{aligned}$$

Here, $X_i(t)$ *corresponds to the set of the positions of player* i *at the time-step* t *(note that some of* $X_i(t)$ *in* (3.47) *can be empty sets).*

Proof $\Rightarrow$ Let us assume that the alternate players' control condition holds for a multi-objective control problem. Then for a fixed time-step t the equations (3.45) depend on only one of the vectors of control parameters $u^i(t)$, $i \in \{1, 2, \ldots, m\}$. Therefore, if we denote by $X_i(t)$ the set of the states of the dynamical system which corresponds to the positions of player i at time-step t, equation (3.45) can be regarded as a solution which satisfies (3.48).

$\Leftarrow$ Let us assume that the partition (3.47) is given for any $t = 0, 1, 2, \ldots$, and the expression in (3.45) is represented in the form (3.48). This means that at every time-step t this equation depends on only one of the vectors of the control parameters. □

On the basis of these results we can prove the important fact that the set of the positions can be characterized in the following way:

Corollary 3.67 *If the alternate players' control condition for the multi-objective control problem holds then the set of the positions $Z_i \subseteq X \times \{0, 1, 2, \dots\}$ of player i can be represented as follows:*

$$Z_i = \bigcup_t (X_i(t), t), \quad i = 1, 2, \dots, m.$$

Let us assume that the alternate players' control for the problem from Sect. 3.9 holds. Then the set of possible system's transitions of the dynamical system $\mathbb{L}$ can be described by a directed graph $\overline{G} = (Z, \overline{E})$ with the set of vertices $Z = \bigcup_{i=1}^{m} Z_i$, where Z_i, $i = 1, 2, \dots, m$, represents the set of the positions of player i. An arbitrary vertex $z \in Z$ in $\overline{G}$ corresponds to a position (x, t) of one of the players $i \in \{1, 2, \dots, m\}$ and a directed edge $e = (z', z'')$ reflects the possibility of the system's transition from state $z' = (x, t)$ to state $z'' = (y, t + 1)$ determined by $x(t)$ and the control vector $u^i(t) \in U_t^i(x(t))$ such that

$$y = x(t + 1) = g_t^i(x(t), u^i(t)) \quad \text{if} \quad x(t) \in Z_i.$$

We associate to the edges $((x, t), (y, t + 1))$ of the graph G the costs $c^i((x, t), (y, t + 1)) = c_t^i(x(t), g_t^i(x(t), u^i(t)))$, $i = 1, 2, \dots, m$.

Graph $\overline{G}$ is represented in Fig. 3.16. This graph contains t_2+1 copies of the set of states $X(t) = (X, t)$, where $X(t) = X_1(t) \cup X_2(t) \cup \dots \cup X_m(t)$, $t = 0, 1, 2, \dots, t_2$. In $\overline{G}$ there are also the edges $((x, t), (x_f, t_2))$ if $t_1 - 1 \le t \le t_2 - 1$ and for a given position $(x, t) = x(t) \in X_i(t)$ of the player i there exists a control $u^i(t) \in U_t^i(x(t))$ such that

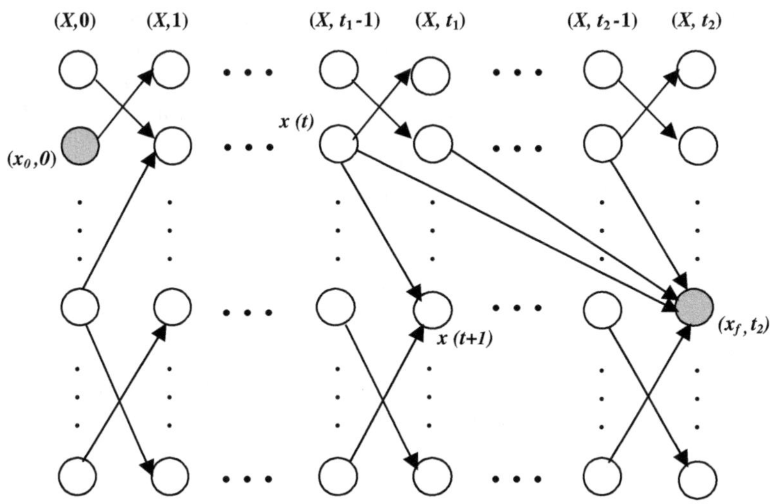

Fig. 3.16 The network for the multi-objective control problem

$$x_f = x(t+1) = g_t^i(x(t),\ u^i(t)).$$

To these edges $((x,t),\ (x_f,t_2))$ we associate the costs $c^i((x,t),\ (x_f,t_2)) = c_t^i(x(t),\ g_t^i(x(t),\ u^i(t)))$, $i = 1, 2, \ldots, m$.

It is easy to observe that $\overline{G}$ is an acyclic directed graph in which an arbitrary directed path from $(x_0, 0)$ to (x_f, t_2) contains $t(x_f)$ edges such that $t_1 \le t(x_f) \le t_2$. So, a directed path from $(x_0, 0)$ to (x_f, t_2) corresponds to a feasible trajectory of the dynamical system from x_0 to x_f.

This means that our multi-objective problem with alternate players' control condition can be regarded as a dynamic noncooperative game on a network.

Based on such a representation of the dynamics of system $\mathbb{L}$ in [74, 76, 77, 79] the following result is grounded.

Theorem 3.68 *Let us assume that for the multi-objective control problem there exists a trajectory*

$$x_0 = x(0),\ x(1),\ x(2),\ \ldots,\ x(t(x_f)) = x_f$$

from a starting state x_0 to a final state x_f generated by vectors of control parameters

$$u^1(t),\ u^2(t),\ \ldots,\ u^m(t),\quad t = 0, 1, 2, \ldots, t(x_f) - 1,$$

where $u^i(t) \in U_t^i(x(t))$, $i = 1, 2, \ldots, m$, $t = 0, 1, 2, \ldots, t(x_f) - 1$ and $t_1 \le t(x_f) \le t_2$. Moreover, we assume that the alternate players' control condition is satisfied. Then for this problem there exists an optimal solution $u^{1^*}(t),\ u^{2^*}(t),\ \ldots,\ u^{m^*}(t)$ in the sense of Nash.

The correctness of this theorem follows from Theorem 3.30 because the problem of determining Nash equilibria for the game-theoretic control problem with alternate player' control condition is equivalent to the problem of determining Nash equilibria in the dynamic c-game on an auxiliary constructed time-expanded network.

As an important result from Theorem 3.68 we obtain the following corollary:

Corollary 3.69 *Assume that for any $u^1(t) \in U_t^1(x(t))$, $t = 0, 1, 2, \ldots$ in the max-min control problem there exists a control $u^2(t) \in U_t^2(x(t))$, $t = 0, 1, 2, \ldots, t(x_f) - 1$ such that $u^1(t)$ and $u^2(t)$ generate a trajectory*

$$x_0 = x(0),\ x(1),\ x(2),\ \ldots,\ x(t(x_f)) = x_f$$

from a starting state x_0 to a final state x_f, where $t_1 \le t(x_f) \le t_2$. Moreover, we assume that the alternate players' control condition is satisfied. Then for the payoff function $F_{x_0x_f}(u^1(t),\ u^2(t))$ in the max-min control problem there exists a saddle point $(u^{1^*}(t),\ u^{2^*}(t))$, i.e.,

$$F_{x_0 x_f}(u^{1^*}(t),\ u^{2^*}(t)) = \max_{u^1(t)} \min_{u^2(t)} F_{x_0 x_f}(u^1(t),\ u^2(t))$$
$$= \min_{u^2(t)} \max_{u^1(t)} F_{x_0 x_f}(u^1(t),\ u^2(t)).$$

All results related to the existence theorems and algorithms for solving the problems on networks can be referred to the problems from Sects. 3.3–3.5.

3.12 Determining a Stackelberg Solution for Hierarchical Control Problems

We consider the hierarchical control problem from Sect. 3.10. In order to develop a dynamic programming technique for determining a Stackelberg solution we study the static case of the hierarchical problem and analyze the computational complexity of this problem. Additionally, we formulate the hierarchical control problem on the network and propose a dynamic programming algorithm for its solving. Based on these results we extend the dynamic programming technique for the hierarchical control problem from Sect. 3.10.

3.12.1 A Stackelberg Solution for Static Games

Let a static game of m players $\Gamma = (\{\mathbb{S}i\}_{i=\overline{1,m}}, \{F_i\}_{i=\overline{1,m}})$ be given, where $\mathbb{S}_i$, $i = 1, 2, \ldots, m$, represent nonempty finite sets of the strategies of the players and

$$F_i : \mathbb{S}_1 \times \mathbb{S}_2 \times \mathbb{S}_m \to \mathbb{R}^1, \quad i = 1, 2, \ldots, m,$$

are the corresponding payoff functions in Γ. We consider the problem of determining a Stackelberg solution in this game, i.e., we are seeking for strategies $s^{1*}, s^{2*}, \ldots, s^{m*}$ such that it holds:

$$s^{1*} = \operatorname*{argmin}_{\substack{s^1 \in \mathbb{S}^1, \\ (s^i \in R_2(s^1, \ldots, s^{i-1}))_{2 \le i \le m}}} F_1(s^1, s^2, \ldots, s^m);$$

$$s^{2*} = \operatorname*{argmin}_{\substack{s^2 \in R_2(s^{1*}), \\ \left(s^i \in R_i(s^{1*}, s^2, \ldots, s^{i-1})\right)_{3 \le i \le m}}} F_2(s^{1*}, s^2, \ldots, s^m);$$

$$s_3^* = \operatorname*{argmin}_{\substack{s^3 \in R_3(s^{1*}, s^{2*}), \\ \left(s^i \in R_i(s^{1*}, s^{2*}, s^3, \ldots, s_{i-1})\right)_{4 \le i \le m}}} F_3(s^{1*}, s^{2*}, \ldots, s^m);$$

$$\vdots$$

$$s^{m*} = \operatorname*{argmin}_{s^m \in R_m(s^{1*}, s^{2*}, \ldots, s^{m-1*})} F_m(s^{1*}, s^{2*}, \ldots, s^{m-1*}, s^m),$$

where $R_k(s^1, s^2, \ldots, s^{k-1})$ is the set of the best responses of player k if the players $1, 2, \ldots, k-1$ have already fixed their strategies $s^1, s^2, \ldots, s^{k-1}$, i.e.,

$$R_2(s^1) = \operatorname*{argmin}_{\substack{s^2 \in \mathbb{S}_2, \\ (s^i \in R_i(s^1, \ldots, s^{i-1}))_{3 \le i \le m}}} F_2(s^1, s^2, \ldots, s^m);$$

$$R_3(s^1, s^2) = \operatorname*{argmin}_{\substack{s^3 \in \mathbb{S}_3, \\ (s^i \in R_i(s^1, s^2, \ldots s^{i-1}))_{4 \le i \le m}}} F_3(s^1, s^2, \ldots, s^m);$$

$$\vdots$$

$$R_m(s^1, s^2, \ldots, s^{m-1}) = \operatorname*{argmin}_{s^m \in \mathbb{S}_m} F_m(s^1, s^2, \ldots, s^m).$$

In this game the players fix their strategies successively according to their numerical order. Therefore, if the order of fixing the strategies of the players is changed then the best responses of the players will correspond to a Stackelberg solution with respect to a new order of the players.

Lemma 3.70 *Let $s^{1*}, s^{2*}, \ldots, s^{m*}$ be a Stackelberg solution of the game Γ. If this solution remains the same for an arbitrary order of fixing strategies of the players, then $s^{1*}, s^{2*}, \ldots, s^{m*}$ is a Nash equilibrium.*

Proof Assume that $s^{1*}, s^{2*}, \ldots, s^{i-1*}, s^{i*}, s^{i+1*}, \ldots, s^{m*}$ is a Stackelberg solution for an arbitrary order of fixing strategies of the players. Then we may consider that an arbitrary player i fixes his strategy in the last order and therefore

$$F_i(s^{1*}, s^{2*}, \ldots, s^{i-1*}, s^{i*}, s^{i+1*}, \ldots, s^{m*}) \le$$
$$\le F_i(s^{1*}, s^{2*}, \ldots, s^{i-1*}, s^i, s^{i+1*}, \ldots, s^{m*}), \quad \forall s^i \in \mathbb{S}_i, \ \ i = 1, 2, \ldots, m.$$

So, $s^{1*}, s^{2*}, \ldots, s^{i-1*}, s^{i*}, s^{i+1*}, \ldots, s^{m*}$ is a Nash equilibrium. □

The computational complexity of determining pure Nash equilibria in discrete games has been studied in [31, 42, 105]. Based on the results from [31] we can conclude that finding a Stackelberg solution in the considered games is NP-hard if the number of players m acts as input data parameter of the problem. In the case that m is fixed (i.e. m does not act as input data parameter of the problem), then a Stackelberg solution can be found in polynomial time. In the case of a small number of players, especially in the case of two or three players, exhaustive search allows us to determine Stackelberg strategies for large dimensioned finite games. Indeed, if we calculate $s^{1*}, s^{2*}, \ldots, s^{m*}$ according to the condition from the definition of a Stackelberg solution we use $|\mathbb{S}_1| \times |\mathbb{S}_2| \times \cdots \times |\mathbb{S}_m|$ steps. So, in the case of two players we can determine a Stackelberg solution using $O(|\mathbb{S}_1||\mathbb{S}_2|)$ elementary operations (here we do not take into account the number of operations for calculating the values $F_i(s^1, s^2)$ for given (s^1, s^2). We can use this fact for solving hierarchical control problems with two or three players.

3.12.2 Hierarchical Control on Networks and Determining Stackelberg Stationary Strategies

Let $G = (X, E)$ be the graph of states' transitions for a time-discrete system $\mathbb{L}$. So, X corresponds to the set of states of $\mathbb{L}$ and an arbitrary directed edge $e = (x, y) \in E$ means the possibility of system $\mathbb{L}$ to pass from state $x = x(t)$ to state $y = x(t+1)$ at every moment in time $t = 0, 1, 2, \ldots$ Assume that the system $\mathbb{L}$ is controlled by m players and on the edge set the following m functions are defined:

$$c^i : E \to \mathbb{R}^1, \quad i = 1, 2, \ldots, m,$$

which assign m costs $c_e^1, c_e^2, \ldots, c_e^m$ to each edge $e \in E$. For player i the quantity c_e^i expresses the cost of system $\mathbb{L}$ to pass through edge $e = (x, y)$ from state $x = x(t)$ to state $y = x(t+1)$ at every moment in time $t = 0, 1, 2, \ldots$ On G the players use only stationary strategies and intend to minimize their integral-time costs by a trajectory $x_0 = x(0), x(1), x(2), \ldots, x(t(x_f)) = x_f$ from a starting state x_0 to a final state x_f, where $t_1 \le t(x_f) \le t_2$. We define the stationary strategies of the players as m multi-value functions

$$\begin{aligned}
s^1 &: x \to X_1^{j_1}(x) \in A^1(x) \quad \text{for} \quad x \in X \setminus \{x_f\};\\
s^2 &: x \to X_2^{j_2}(x) \in A^2(x) \quad \text{for} \quad x \in X \setminus \{x_f\};\\
&\qquad \vdots\\
s^m &: x \to X_m^{j_m}(x) \in A^m(x) \quad \text{for} \quad x \in X \setminus \{x_f\};
\end{aligned}$$

which satisfy the condition

$$|s^1(x) \cap s^2(x) \cap \cdots \cap s^m(x)| = 1, \quad \forall x \in X, \tag{3.49}$$

where $A^i(x)$, $i = 1, 2, \ldots, m$, are given sets of subsets from $X(x) = \{y \in X \mid e = (x, y) \in E\}$, i.e., $A^i(x) = \{X_i^1(x), X_i^2(x), \ldots, X_i^{K_i(x)}(x)\}$, $X_i^j(x) \subseteq X_i(x)$, $j = 1, 2, \ldots, K_i(x)$.

The strategies $s^1(x), s^2(x), \ldots, s^m(x)$ for a given $x \in X \setminus \{x_f\}$ correspond to vectors of control parameters $u^1(t), u^2(t), \ldots, u^m(t)$ at a given state $x = x(t)$ in the control Problem 3.62 and reflect the fact that the set of control parameters at the given state $x = x(t)$ uniquely determines the next state $y = x(t+1) \in X$ at every moment in time $t = 0, 1, 2, \ldots$. Therefore, here we use condition (3.49) and consider that $y = \bigcap_{i=1}^{m} s^i(x)$.

If $|\cap s^j(x)| \neq 1$ then the set of strategies $s^1, s^2, \ldots, s^m$ is not feasible. In the following we consider that the players use only feasible strategies.

An arbitrary set $X_i^j(x) \in A^i(x)$ in our control problem represents a possible set of next states $y = x(t+1) \in X$ in which player i prefers to transfer system $\mathbb{L}$ if at the moment in time t the state of the dynamical system is $x = x(t)$. This set for control Problem 3.62 can be treated as a set of possible next states $y = x(t+1) \in X$ if player i fixes a feasible vector of control parameters $u^i(t) \in U_t^i(x(t))$. Therefore, if we treat $X_i^j(x)$ as preferences of the next possible sets of the states from $A^i(x)$ for player $i \in \{1, 2, \ldots, m\}$ then the unique next state y represents the intersection of preferences $s^1(x), s^2(x), \ldots, s^m(x)$ of the players $1, 2, \ldots, m$, i.e. $y = \bigcap_{i=1}^{m} s^i(x)$, where $s^i : x \to X_i^{j_i} \in A^i(x)$ for $x \in X \setminus \{x_f\}$, $i = 1, 2, \ldots, m$.

Let $s^1, s^2, \ldots, s^m$ be a fixed set of feasible strategies of the players $1, 2, \ldots, m$. Denote by $G_s = (X, E_s)$ the subgraph generated by edges $e = (x, y) \in E$ such that $y = \bigcap_{i=1}^{m} s^i(x)$ for $x \in X \setminus \{x_f\}$. Then in G_s either a unique directed path $P_s(x_0, x_f)$ in G_s exists or such a path does not exist. Therefore, for $s^1, s^2, \ldots, s^m$ and fixed starting and final states $x_0, x_f \in X$ we can define the quantities

$$\widehat{H}_{x_0 x_f}^1(s^1, s^2, \ldots, s^m),\ \widehat{H}_{x_0 x_f}^2(s^1, s^2, \ldots, s^m),\ \ldots,\ \widehat{H}_{x_0 x_f}^m(s^1, s^2, \ldots, s^m)$$

in the following way. We put

$$\widehat{H}_{x_0 x_f}^i(s^1, s^2, \ldots, s^m) = \sum_{e \in E(P_s(x_0, x_f))} c_e^i, \quad i = 1, 2, \ldots, m$$

if

$$t_1 \leq |E(P_s(x_0, x_f))| \leq t_2;$$

otherwise we put

$$\widehat{H}_{x_0 x_f}^i(s^1, s^2, \ldots, s^m) = +\infty.$$

Note that in this control process the players fix their strategies successively one after another according to their numerical order at each moment in time $t = 0, 1, 2, \ldots$ for every state $x \in X \setminus \{x_f\}$.

Additionally, we assume that each player fixing his strategies informs posterior players which strategy has been chosen.

In the considered hierarchical control problem we are seeking for Stackelberg stationary strategies, i.e. we are seeking for strategies $s^{1*}, s^{2*}, \ldots, s^{m*}$, for which

$$s^{1*} = \operatorname*{argmin}_{\substack{s^1 \in \mathbb{S}_1, \\ (s^i \in R_i(s^1,\ldots,s^{i-1}))_{2\le i\le m}}} \widehat{H}^1_{x_0 x_f}(s^1, s^2, \ldots, s^m);$$

$$s^{2*} = \operatorname*{argmin}_{\substack{s^2 \in R_2(s^{1*}), \\ \left(s^i \in R_3(s^{1*},s^2,\ldots,s^{i-1})\right)_{3\le i\le m}}} \widehat{H}^2_{x_0 x_f}(s^{1*}, s^2, \ldots, s^m);$$

$$s^{3*} = \operatorname*{argmin}_{\substack{s^3 \in R_3(s^{1*},s^{2*}), \\ \left(s^i \in R_i(s^{1*},s^{2*},s^3,\ldots,s^{i-1})\right)_{4\le i\le m}}} \widehat{H}^3_{x_0 x_f}(s^{1*}, s^{2*}, \ldots, s^m);$$

$$\vdots$$

$$s^{m*} = \operatorname*{argmin}_{s^m \in R_m(s^{1*},s^{2*},\ldots,s^{m-1*})} \widehat{H}^m_{x_0 x_f}(s^{1*}, s^{2*}, \ldots, s^{m-1*}, s^m),$$

where $R_k(s^1, s^2, \ldots, s^{k-1})$ is the set of best responses of player k if the players $1, 2, \ldots, k-1$ have already fixed their strategies $s_1, s_2, \ldots, s_{k-1}$, i.e.,

$$R_2(s^1) = \operatorname*{argmin}_{\substack{s^2 \in \mathbb{S}_2, \\ (s^i \in R_i(s^1,\ldots,s^{i-1}))_{3\le i\le m}}} \widehat{H}^2_{x_0 x_f}(s^1, s^2, \ldots, s^m);$$

$$R_3(s^1, s^2) = \operatorname*{argmin}_{\substack{s^3 \in \mathbb{S}_3, \\ (s^i \in R_i(s^1,\ldots,s^{i-1}))_{4\le i\le m}}} \widehat{H}^3_{x_0 x_f}(s^1, s^2, \ldots, s^m);$$

$$\vdots$$

$$R_m(s^1, s^2, \ldots, s^{m-1}) = \operatorname*{argmin}_{s^m \in \mathbb{S}_m} \widehat{H}^m_{x_0 x_f}(s^1, s^2, \ldots, s^m).$$

where $\mathbb{S}_1, \mathbb{S}_2, \ldots, \mathbb{S}_m$ represent the corresponding admissible sets of stationary strategies of the players $1, 2, \ldots, m$.

Remark 3.71 In general the stationary strategies $s^1, s^2, \ldots, s^m$ of the players in the hierarchical control problem on G can be defined as arbitrary multi-value functions

$$s^i : x \to X_i^{j_i}(x) \in A^i(x) \ \text{ for } x \in X \setminus \{x_f\}, \quad i = 1, 2, \ldots, m.$$

If the conditions (3.49) for $x \in X \setminus \{x_f\}$ do not take place, i.e. if at least for a state $x \in X \setminus \{x_f\}$ the following condition holds:

$$|s^1(x) \cap s^2(x) \cap \cdots \cap s^m(x)| \neq 1,$$

then we put $\widehat{H}^i_{x_0 x_f}(s^1, s^2, \ldots, s^m) = +\infty$.

So, the hierarchical control problem is determined by the dynamic network $(G, A^1, A^2, \ldots, A^m, c^1, c^2, \ldots, c^m, x_0, x_f, T_1, T_2)$, where $A^i = \bigcup_{x\in X} A^i(x)$

and $c^i = (c^i_{e_1}, c^i_{e_2}, \ldots, c^i_{e_{|E|}})$, $i = 1, 2, \ldots, m$. In the case $T_2 = \infty$, $T_1 = 0$ we denote the corresponding network by $(G, A^1, A^2, \ldots, A^m, c^1, c^2, \ldots, c^m, x_0, x_f)$. The following theorem allows us to describe a class of multi-objective hierarchical control problems for which an arbitrary Stackelberg solution is also a Nash equilibrium.

Theorem 3.72 *Let the hierarchical control problem on the network* $(G, A^1, A^2, \ldots, A^m, c^1, c^2, \ldots, c^m, x_0, x_f)$ *be given, where* G *has the property that for an arbitrary vertex* $x \in X$ *there exist a directed path from* x *to* x_f*. Additionally, assume that the sets* $A^1(x), A^2(x), \ldots, A^m(x)$ *satisfy the following condition: For an arbitrary vertex* $x \in X \setminus \{x_f\}$ *there exists* $i_x \in \{1, 2, \ldots, m\}$ *such that* $A^{i_x}(x) = \big\{\{y\} \mid y \in X_G(x)\big\}$ *and* $A^{i_x}(x) = \{X_G(x)\}$ *if* $i \in \{1, 2, \ldots, m\} \setminus \{i_x\}$*. Then for the hierarchical control problem on* G *there exists a Nash equilibrium.*

Proof First of all it is easy to observe that if the conditions of the theorem hold then in the multi-objective control problem on G there exist stationary strategies $s^1, s^2, \ldots, s^m$ which generate a trajectory $x_0, x_1, x_2, \ldots, x_f$ from a starting state x_0 to a final state x_f for an arbitrary given starting vertex $x_0 = x \in X$. This means that a Stackelberg solution for the hierarchical control problem on G exists. Additionally, we can see that the dynamic c-game from Sect. 3.3.3 (the multi-objective control problem in positional form) represents a particular case of the problem formulated above if the sets $A^1(x),\ A^2(x),\ \ldots,\ A^m(x)$ satisfy the condition that for an arbitrary $x \in X \setminus \{x_f\}$ there exists $i_x \in \{1, 2, \ldots, m\}$ such that $A^{i_x}(x) = \big\{\{y\} \mid y \in X_G(x)\big\}$ and $A^i(x) = \{X_G(x)\}$ if $i \in \{1, 2, \ldots, m\} \setminus \{i_x\}$. In this case a Stackelberg solution of the hierarchical control problem does not depend on the order of fixing strategies by the players, and therefore, on the bases of Lemma 3.70 an arbitrary Stackelberg solution of the multi-objective control problem on G is a Nash equilibrium. □

3.12.3 An Algorithm for Determining Stackelberg Stationary Strategies on Acyclic Networks

We consider the hierarchical control problem on an acyclic network $(G, A^1, A^2, \ldots, A^m, c^1, c^2, \ldots, c^m, x_0, x_f)$, i.e., $G = (X, E)$ is an acyclic directed graph and $t_1 = 0$, $t_2 = \infty$. We also assume that in G vertex x_f is attainable from every vertex $x \in X$.

Algorithm 3.73 Determining Stackelberg Strategies on Acyclic Networks

Preliminary step (Step 0): Fix $X^0 = \{x_f\}$, $E^0 = \emptyset$ and put $\varepsilon^i(x_f) = 0$, $i = 1, 2, \ldots, m$.

General step (Step k, $k \geq 1$): If $X \setminus X^{k-1} = \emptyset$ then STOP; otherwise find a vertex $x^k \in X \setminus X^{k-1}$ for which $X_G(x^k) \subseteq X^{k-1}$, where $X_G(x^k) = \{y \in X \mid (x^k, y) \in E\}$. With respect to vertex x^k we consider the static problem of finding Stackelberg strategies $s^{1*}(x^k), s^{2*}(x^k), \ldots, s^{m*}(x^k)$ in the game

$$\Gamma(x^k) = (S_1(x^k),\ S_2(x^k),\ \dots,\ S_m(x^k),\ F_1,\ F_2,\ \dots,\ F_m),$$

where the sets of strategies of the players $S_1(x^k),\ S_2(x^k),\ \dots,\ S_m(x^k)$ and the payoff functions $F_1, F_2, \dots, F_m$ are defined as follows:

$$S_i(x^k) = \{s^i(x^k) \mid s^i : x^k \to X_i^{j_j}(x^k) \in A^i(x^k)\}, \quad i = 1, 2, \dots m;$$

$$F_i(s^1, s^2, \dots, s^m) = \begin{cases} \varepsilon^i(y) + c^i_{(x^k,y)}, & \text{if} \quad y = \bigcap_{i=1}^m s^i(x^k); \\ +\infty, & \text{if} \quad \left|\bigcap_{i=1}^m s^i(x^k)\right| \neq 1. \end{cases} \tag{3.50}$$

After that find a Stackelberg solution $s^{1^*}(x^k),\ s^{2^*}(x^k),\ \dots,\ s^{m*}(x^k)$ for the static game $\Gamma(x^k)$ and determine the vertex $y^* = \bigcap_{i=1}^m s^{i^*}(x^k)$.

Then calculate

$$\varepsilon^i(x^k) = \varepsilon^i(y^*) + c^i_{(x^k,y^*)}, \quad i = 1, 2, \dots, m$$

and fix $\widehat{H}_{x^k x_f}(s^{1^*}, s^{2^*}, \dots, s^{m*}) = \varepsilon^i(x^k), i = 1, 2, \dots, m.$

Put $X^k = X^{k-1} \cup \{x^k\}$, $E^k = E^{k-1} \cup \{(x^k, y)\}$, $GT^k = (X^k, E^k)$ and go to the next step.

This algorithm determines Stackelberg stationary strategies $s^{1^*}, s^{2^*}, \dots, s^{m*}$ for an arbitrary starting position $x_0 = x$ and the fixed final position x_f. The corresponding optimal values of integral costs of the system's passage from a starting state $x_0 = x$ to a final state x_f are $\widehat{H}_{x_0 x_f}(s^{1^*}, s^{2^*}, \dots, s^{m*})$. The algorithm determines the tree $GT^{|X|-1} = (X, E^{|X|-1})$ of optimal strategies with a sink vertex x_f which gives Stackelberg strategies for an arbitrary starting position $x_0 = x \in X$. It is easy to observe that if for a given starting position x_0 of the considered dynamic game a Nash equilibrium exists then the algorithm determines this equilibrium. Note that the proposed algorithm can also be adapted for the problem if for different moments in time and for different states the order of fixing stationary strategies of the players can be different. We should take into account this order of fixing strategies of the players calculating values $h_i^*(x^k, y)$ for a given state $x^k(t)$.

Example Consider a hierarchical control problem on a network with two players where the corresponding graph $G = (X, E)$ is presented in Fig. 3.17.

This network has the structure of a directed acyclic graph with a given starting vertex $x_0 = x(0)$ and final vertex $x_f = 4$. To each vertex it is given a set of subsets $A^i(x) = \{X_i^j\}$, where

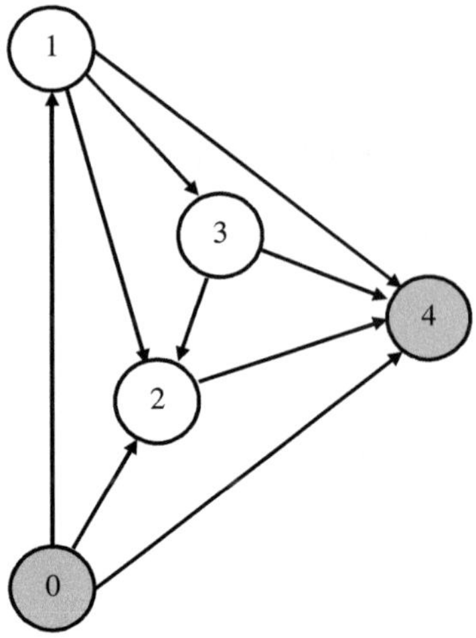

Fig. 3.17 The network for the hierarchial control problem

$$A^1(0) = \{\{1,2\},\{1,4\}\};\quad A^2(0) = \{\{2,4\},\{1,4\}\};$$
$$A^1(1) = \{\{2,3\},\{3,4\}\};\quad A^2(1) = \{\{2,4\},\{2,3\}\};$$
$$A^1(2) = \{\{4\}\};\quad A^2(2) = \{\{4\}\};$$
$$A^1(3) = \{\{2\},\{2,4\}\};\quad U^2(2) = \{\{4\},\{2,4\}\}.$$

On the edges $e \in E$ there are defined cost functions $c^1 : E \to \mathbb{R}^1$ and $c^2 : E \to \mathbb{R}^1$, where

$$c^1_{(0,1)} = 2;\quad c^1_{(0,2)} = 1;\quad c^1_{(0,4)} = 5;$$
$$c^2_{(0,1)} = 3;\quad c^2_{(0,2)} = 2;\quad c^2_{(0,4)} = 6;$$
$$c^1_{(1,2)} = 3;\quad c^1_{(1,3)} = 4;\quad c^1_{(1,4)} = 3;$$
$$c^2_{(1,2)} = 3;\quad c^2_{(1,3)} = 1;\quad c^2_{(1,4)} = 5;$$
$$c^1_{(3,2)} = 1;\quad c^1_{(3,4)} = 2;$$
$$c^2_{(3,2)} = 1;\quad c^2_{(3,4)} = 4;$$
$$c^1_{(2,4)} = 1;\quad c^2_{(2,4)} = 2.$$

If we use Algorithm 3.73 then we obtain:

Step 0
Fix $X^0 = \{4\},\ \varepsilon^1(4) = 0,\ \varepsilon^2(4) = 0,\ E^0 = \emptyset.$

Step 1
$X \setminus X^0 \neq \emptyset$ and $X_G(2) \subseteq X^0$, therefore, fix $x^1 = 2$ and solve the static game

$$\Gamma(2) = (S_1(2),\ S_2(2),\ F_1,\ F_2)$$

where

$$S_1(2) = \{s^1 : 2 \to \{4\}\},\quad S_2(2) = \{s^2 : 2 \to \{4\}\}$$

and $F_1(s^1, s^2) = 1$; $F_2(s^1, s^2) = 2$. For this game we have a trivial solution $s^{1*}(2) = \{4\}$; $s^{2*}(2) = \{4\}$. We calculate $\varepsilon^1(2) = 0 + c^1_{(2,4)} = 1$; $\varepsilon^2(2) = 0 + c^2_{(2,4)} = 2$ and put $\widehat{H}^1_{2,4}(s^{1*}, s^{2*}) = 1,\ \widehat{H}^2_{2,4}(s^{1*}, s^{2*}) = 2$.

Fix $X^1 = X^0 \cup \{2\} = \{2, 4\}$; $E^1 = E^0 \cup \{(2, 4)\} = \{(2, 4)\}$; $GT^1 = (\{2, 4\},\ \{(2, 4)\})$.

Step 2
$X \setminus X^1 \neq \emptyset$ and $X_G(3) \subseteq X^1$, therefore, fix $x^2 = 3$ and solve the static game

$$\Gamma(3) = (S_1(3),\ S_2(3),\ F_1,\ F_2)$$

where

$$S_1(3) = \{s^1_1 : 3 \to \{2\};\quad s^1_2 : 3 \to \{2, 4\}\},$$
$$S_2(3) = \{s^2_1 : 3 \to \{2, 4\};\quad s^2_2 : 3 \to \{4\}\}$$

and $F_i(s^1_j, s^2_j)$ are defined according to (3.50), i.e.,

$$\begin{aligned}
&F_1(s^1_1, s^2_1) = 2;\quad F_2(s^1_1, s^2_1) = 3\quad (s^1_1(3) \cap s^2_1(3) = 2);\\
&F_1(s^1_1, s^2_2) = F_2(s^1_1, s^2_2) = \infty \qquad (s^1_1(3) \cap s^2_2(3) = \emptyset);\\
&F_1(s^2_1, s^1_2) = F_2(s^2_1, s^1_2) = \infty\ (s^2_1(3) \cap s^1_2(3) = \{2, 4\},\\
&\qquad\qquad\qquad\qquad\qquad\qquad \text{i.e. } |\{2, 4\}| \neq 1);\\
&F_1(s^2_1, s^2_2) = 2;\quad F_2(s^2_1, s^2_2) = 4\quad (s^2_1(3) \cap s^2_2(3) = 4).
\end{aligned}$$

If we solve this game we find a Stackelberg solution $s^{1*}(3) = \{2\}$; $s^{2*}(3) = \{2, 4\}$. We calculate $\varepsilon^1(3) = \varepsilon^1(2) + c^1_{(3,2)} = 2$; $\varepsilon^2(3) = \varepsilon^2(2) + c^2_{(3,2)} = 3$ and put $\widehat{H}^1_{2,4}(s^{1*}, s^{2*}) = 1,\ \widehat{H}^2_{2,4}(s^{1*}, s^{2*}) = 2$.

Fix $X^2 = X^1 \cup \{3\} = \{2, 3, 4\}$; $E^2 = E^1 \cup \{(3, 2)\} = \{(2, 4), (3, 2)\}$; $GT^2 = (\{2, 3, 4\},\ \{(3, 2),\ (2, 4)\})$.

Step 3
$X \setminus X^2 \neq \emptyset$ and $X_G(1) \subseteq X^2$, therefore, fix $x^3 = 1$ and solve the static game

$$\Gamma(1) = (S_1(1),\ S_2(1),\ F_1,\ F_2)$$

where

$$S_1(1) = \{s_1^1 : 1 \to \{2,3\};\quad s_2^1 : 1 \to \{3,4\}\},$$
$$S_2(3) = \{s_1^2 : 1 \to \{2,4\};\quad s_2^2 : 1 \to \{3,4\}\}$$

and $F_i(s_j^1, s_j^2)$ are defined according to (3.50), i.e.

$$\begin{aligned}
&F_1(s_1^1, s_1^2) = 4; \quad F_2(s_1^1, s_1^2) = 5 \quad (s_1^1(1) \cap s_1^2(1) = 2);\\
&F_1(s_1^1, s_2^2) = F_2(s_1^1, s_2^2) = \infty \qquad (s_1^1(1) \cap s_2^2(1) = \emptyset);\\
&F_1(s_2^1, s_1^2) = 3; \quad F_2(s_2^1, s_1^2) = 5 \quad (s_2^1(1) \cap s_1^2(1) = 4);\\
&F_1(s_2^1, s_2^2) = 5; \quad F_2(s_2^1, s_2^2) = 3 \quad (s_2^1(1) \cap s_2^2(1) = 3).
\end{aligned}$$

If we solve this game we find a Stackelberg solution $s^{1*}(1) = s_1^1(1)$; $s^{2*}(1) = s_1^2(1)$, i.e. $s^{1*}(1) = \{2,3\}$; $s^{2*}(1) = \{2,4\}$. We calculate $\varepsilon^1(1) = \varepsilon^1(2) + c^1_{(3,2)} = 4$; $\varepsilon^2(1) = \varepsilon^2(1) + c^2_{(3,2)} = 5$ and put $\widehat{H}^1_{2,4}(s^{1*}, s^{2*}) = 4$, $\widehat{H}^2_{2,4}(s^{1*}, s^{2*}) = 5$.

Fix $X^3 = X^2 \cup \{1\} = \{1,2,3,4\}$; $E^3 = E^2 \cup \{(1,2)\} = \{(1,2),(2,4),(3,2)\}$; $GT^3 = (\{1,2,3,4\}, \{(1,2), (3,2), (2,4)\})$.

Step 4
$X \setminus X^3 \neq \emptyset$ and $X_G(0) \subseteq X^3$, therefore, fix $x^4 = 0$ and solve the static game

$$\Gamma(0) = (S_1(0), S_2(0), F_1, F_2)$$

where

$$S_1(0) = \{s_1^1 : 0 \to \{1,2\};\ s_2^1 : 0 \to \{1,4\}\},$$
$$S_2(0) = \{s_1^2 : 0 \to \{2,4\};\ s_2^2 : 0 \to \{1,4\}\}$$

and $F_i(s_j^1, s_j^2)$ are defined according to (3.50), i.e.

$$\begin{aligned}
&F_1(s_1^1, s_1^2) = 2; \quad F_2(s_1^1, s_1^2) = 4 \quad (s_1^1(0) \cap s_1^2(0) = 2);\\
&F_1(s_1^1, s_2^2) = F_2(s_1^1, s_2^2) = \infty \qquad (s_1^1(0) \cap s_2^2(0) = \emptyset);\\
&F_1(s_2^1, s_1^2) = 5; \quad F_2(s_2^1, s_1^2) = 6 \quad (s_2^1(0) \cap s_1^2(0) = 4);\\
&F_1(s_2^1, s_2^2) = F_2(s_2^1, s_2^2) = \infty \quad (s_2^1(0) \cap s_2^2(0) = \{1,4\},\\
&\qquad\qquad\qquad\qquad\qquad\qquad\qquad\qquad |\{1,4\}| \neq 1).
\end{aligned}$$

If we solve this game we find a Stackelberg solution $s^{1*} = s_1^1$; $s^{2*} = s_2^2$, i.e. $s^{1*} = \{1,2\}$; $s^{2*} = \{2,4\}$. We calculate $\varepsilon^1(0) = \varepsilon^1(2) + c^1_{(0,2)} = 2$; $\varepsilon^2(0) = \varepsilon^2(0) + c^2_{(0,2)} = 4$ and put $\widehat{H}^1_{2,4}(s^{1*}, s^{2*}) = 2$, $\widehat{H}^2_{2,4}(s^{1*}, s^{2*}) = 4$.

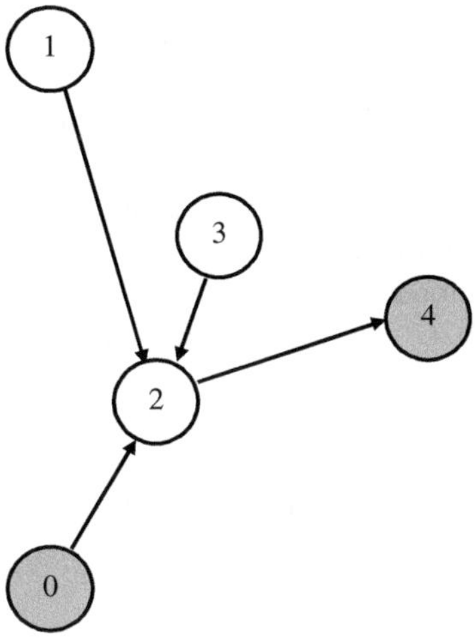

Fig. 3.18 The tree of the solution for the hierarchial control problem

Fix $X^4 = X^3 \cup \{0\} = \{0, 1, 2, 3, 4\}$; $E^4 = E^3 \cup \{(0, 2)\} = \{(0, 2), (1, 2), (2, 4), (3, 2)\}$; $GT^4 = (\{0, 1, 2, 3, 4\}, \{(0, 2), (1, 2), (3, 2), (2, 4)\})$.

Step 5

We obtain $X \setminus X^4 = \emptyset$, therefore STOP.

So, the optimal stationary strategies of the players are the following:

$$s^{1^*} : 0 \to \{1, 2\}; \quad 1 \to \{2, 3\}; \quad 2 \to \{4\}; \quad 3 \to \{2\};$$
$$s^{2^*} : 0 \to \{2, 4\}; \quad 1 \to \{2, 4\}; \quad 2 \to \{4\}; \quad 3 \to \{2, 4\}.$$

The set of stationary strategies in G generates the tree given in Fig. 3.18.

In this tree the strategies

$$s^{1^*}(0), s^{2^*}(0) \text{ generate the transition } (0, 1);$$
$$s^{1^*}(1), s^{2^*}(1) \text{ generate the transition } (1, 2);$$
$$s^{1^*}(2), s^{2^*}(2) \text{ generate the transition } (2, 4);$$
$$s^{1^*}(3), s^{2^*}(3) \text{ generate the transition } (3, 4).$$

So, this tree gives the optimal stationary strategies of the players for an arbitrary starting position $x_0 = x$.

In this example, for $x_0 = 0$ we obtain a Stackelberg solution which is also a Nash equilibrium. If we fix $x_0 = 1$ this Stackelberg solution is not a Nash equilibrium.

3.12.4 Algorithms for Solving Hierarchical Control Problems

Based on the results from Sect. 3.12.3 we can propose an algorithm for solving the multi-objective hierarchical control problem from Sect. 3.10. First of all we show that the hierarchical control problem from Sect. 3.10 can be reduced to a stationary hierarchical control problem on an auxiliary network $(\overline{G}, c^1, c^2, \dots, c^m, y_0, y_T)$ for which Stackelberg stationary strategies should be found.

We construct the graph $\overline{G} = (Y, E)$ of the network in the following way:

$$Y = Y^0 \cup Y^1 \cup Y^2 \cup \dots \cup Y^{t_1} \cup Y^{t_1+1} \cup \dots \cup Y^{t_2} \quad (t^k \cap Y^l = \emptyset,\ k \neq l);$$

where $Y^t = (X, t)$ corresponds to the set of states $x(t) \in X$ of system $\mathbb{L}$ at time moment t $(t = 0, 1, 2, \dots, t_2)$:

$$E = E^0 \cup E^1 \cup E^2 \cup \dots \cup E^{t_1} \cup E^{t_1+1} \cup \dots \cup E^{t_2-1} \cup E^f;$$

where E^t, $t = 0, 1, 2, \dots, t_2 - 1$, represents the set of directed edges in $\overline{G}$ which connects vertices from Y^t with vertices from Y^{t+1}.

We include an arbitrary directed edge $((x, t), (y, t + 1))$ in $E^t, t = 0, 1, 2, \dots, t_2 - 1$, if in the control process at time moment t for a given state $x = x(t)$ there exist vectors of control parameters $u^1(t)$, $u^2(t), \dots, u^m(t)$ from corresponding feasible sets $U_t^1(x(t))$, $U_t^2(x(t))$, $\dots$, $U_t^m(x(t))$ such that

$$y(t + 1) = g_t(x(t),\ u^1(t),\ u^2(t),\ \dots,\ u^m(t)).$$

In an analogous way, we define the set E^f. We include an arbitrary edge $((x, t), (x_f, t_2))$ in E^f, $t = t_1 - 1,\ t_1,\ t_1 + 1,\ \dots,\ t_2 - 1$, if at time moments $t \in [t_1 - 1,\ t_2 - 1]$ for a state $x(t)$ there exist vectors of control parameters $u^1(t), u^2(t), \dots, u^m(t)$ from corresponding feasible sets $U_t^1(x(t)), U_t^2(x(t)), \dots, U_t^m(x(t))$ such that

$$x_f(t + 1) = g_t(x(t),\ u^1(t),\ u^2(t),\ \dots,\ u^m(t)).$$

Additionally, to each vertex (x, t) we associate a set of subsets $A^i(x, t) = \{X_i^j(x, t + 1),\ j = 1, 2, \dots, K_i(x)\}$, where an arbitrary set $X_i^j(x, t + 1)$ represents the set of possible next states $x(t + 1)$ if player i fixes a feasible vector of control parameters $u(t) \in U_t^i(x(t))$, i.e. $|A^i(x, t)| = |U_t(x(t))|$.

We define in $\overline{G}$ the cost functions $c^1, c^2, \dots, c^m$ as follows:

To each edge $e_t = ((x, t),\ (y, t + 1)) \in E^t$ we associate the costs

$$c^i_{e_t} = c^i_t(x(t),\ y(t + 1)),\ \ i = 1, 2,\ \dots,\ m,\ \ t = 0, 1, 2, \dots, t_2 - 1$$

and to edge $e_t = ((x, t),\ (x_f, T_2)) \in E^f$ we associate the costs

$$c^i_{e_t} = c^i_t(x(t),\ x_f(t + 1)),\ \ i = 1, 2, \dots, m,\ \ t = t_1 - 1,\ t_1,\ t_1 + 1,\ \dots,\ t_2 - 1.$$

After that we use the algorithm from Sect. 3.12.3 and determine Stackelberg stationary strategies on $\overline{G}$ with a fixed starting state $y_0 = (x_0, t)$ and a final state $y_t = (x_f, t_2)$. Taking into account that there exists a bijection between the set of Stackelberg stationary strategies of the players on $\overline{G}$ and the Stackelberg solution of the hierarchical control problem from Sect. 3.10 we can find a Stackelberg solution of the problem.

3.13 Extensions and Generalizations of the Dynamic Decision Problem Based on Concept of Multi-objective Games

The considered control problems and the corresponding game models from the previous sections can be generalized applying the concept of multi-objective games [79]. In this section we consider multi-objective games, which extend noncooperative ones [59, 102, 103] and Pareto multi-criteria problems [79, 107, 111]. The payoff functions of the players in such games are presented as vector functions, where the players intend to optimize them in the sense of Pareto on their sets of strategies. At the same time in our game-theoretic model it is assumed that the players are interested to preserve a Nash optimality principle if they interact between them on the set of situations. Such an aspect of the game leads to a new equilibria notion which we call Pareto-Nash equilibria [13, 92, 105]. Such a concept can be used for multi-objective control problems and algorithms for their solving can be derived.

3.13.1 Problem Formulation

The multi-objective game with p players is denoted by $\overline{\Gamma} = (X_1, X_2, \dots, X_m, \overline{F}_1, \overline{F}_2, \dots, \overline{F}_m)$, where X_i is the set of strategies of the player i, $i = 1, 2, \dots, m$, and $\overline{F}_i = (F_i^1, F_i^2, \dots, F_i^{r_i})$ is the vector payoff function of player i, defined on the set of situations $X = X_1 \times X_2 \times \cdots \times X_m$:

$$\overline{F}_i : X_1 \times X_2 \times \cdots \times X_m \to \mathbb{R}^{r_i}, \quad i = 1, 2, \dots, m.$$

Each component F_i^k of $\overline{F}_i$ corresponds to a partial criterion of player i and represents a real function defined on the set of situations $X = X_1 \times X_2 \times \cdots \times X_m$:

$$F_i^k : X_1 \times X_2 \times \cdots \times X_m \to \mathbb{R}^1, \quad k = \overline{1, r_i}, \; i = 1, 2, \dots, m.$$

We call a solution of the multi-objective game $\overline{\Gamma} = (X_1, X_2, \dots, X_m, \overline{F}_1, \overline{F}_2, \dots, \overline{F}_m)$ a Pareto-Nash equilibrium and define it in the following way:

Definition 3.74 The situation $x^* = (x_1^*, x_2^*, \ldots, x_m^*) \in X$ is called a Pareto-Nash equilibrium for a multi-objective game $\overline{\Gamma} = (X_1, X_2, \ldots, X_p, \overline{F}_1, \overline{F}_2, \ldots, \overline{F}_m)$ if for every $i \in \{1, 2, \ldots, p\}$ strategy x_i^* represents a Pareto solution for the following multi-criteria problem:

$$\max_{x_i \in X_i} \rightarrow \overline{f}^i_{x^*}(x_i) = (f^{i1}_{x^*}(x_i), f^{i2}_{x^*}(x_i), \ldots, f^{ir_i}_{x^*}(x_i)),$$

where

$$f^{ik}_{x^*}(x_i) = F_i^k(x_1^*, x_2^*, \ldots, x_{i-1}^*, x_i, x_{i+1}^*, \ldots, x_m^*),$$
$$k = 1, 2, \ldots, m, \quad i = 1, 2, \ldots, m.$$

This definition generalizes the well-known Nash equilibria notation for classical noncooperative games (single-objective games) and Pareto optima for multi-criteria problems. If $r_i = 1$, $i = 1, 2, \ldots, m$, then $\overline{\Gamma}$ becomes a classical noncooperative game, where x^* represents a Nash equilibria solution; in case $p = 1$ the game $\overline{\Gamma}$ becomes a Pareto multi-criteria problem, where x^* is a Pareto solution.

An important special class of multi-objective games represents zero-sum games of two players. This class is obtained from the general case of a multi-objective game $\overline{\Gamma} = (X_1, X_2, \ldots, X_m, \overline{F}_1, \overline{F}_2, \ldots, \overline{F}_m)$ if $m = 2$, $r_1 = r_2 = r$ and $\overline{F}_2(x_1, x_2) = -\overline{F}_1(x_1, x_2)$.

The zero-sum multi-objective game is denoted by $\overline{\Gamma} = (X_1, X_2, \overline{F})$, where $\overline{F}(x_1, x_2) = \overline{F}_2(x_1, x_2) = -\overline{F}_1(x_1, x_2)$. A Pareto-Nash equilibrium for this game corresponds to a saddle point $x^* = (x_1^*, x_2^*) \in X_1 \times X_2$ for the following max-min multi-objective problem:

$$\max_{x_1 \in X_1} \min_{x_2 \in X_2} \rightarrow \overline{F}(x_1, x_2) = (F^1(x_1, x_2), F^2(x_1, x_2), \ldots, F^r(x_1, x_2)). \tag{3.51}$$

Strictly, we define a saddle point $x^* = (x_1^*, x_2^*) \in X_1 \times X_2$ for the zero-sum multi-objective problem (3.51) in the following way:

Definition 3.75 The situation $(x_1^*, x_2^*) \in X_1 \times X_2$ is called a saddle point for the max-min multi-objective problem (3.51) (i.e. for the zero-sum multi-objective game $\overline{\Gamma} = (X_1, X_2, \overline{F})$) if x_1^* is a Pareto solution for the multi-criteria problem:

$$\max_{x_1 \in X_1} \rightarrow \overline{F}(x_1, x_2^*) = (F^1(x_1, x_2^*), F^2(x_1, x_2^*), \ldots, F^r(x_1, x_2^*)),$$

and x_2^* is a Pareto solution for the multi-criteria problem:

$$\min_{x_2 \in X_2} \rightarrow \overline{F}(x_1^*, x_2) = (F^1(x_1^*, x_2), F^2(x_1^*, x_2), \ldots, F^r(x_1^*, x_2)).$$

If $r = 1$ this notion corresponds to a classical saddle point notation for min-max problems, i.e. we obtain the saddle point notation for classical zero-sum games of two players.

In this section we show that the theorems of Nash [102] and Neumann [100, 103] related to classical noncooperative games can be extended for our multi-objective case of games. Moreover, we show that all results related to discrete multi-objective games, especially matrix games can be developed in an analogous way as for classical ones. Algorithms for determining optimal strategies of the players in the considered games will be developed.

3.13.2 Main Results

At first we formulate the main theorem which represents an extension of the Nash theorem for our multi-objective version of the game.

Theorem 3.76 *Let* $\overline{\Gamma} = (X_1, X_2, \ldots, X_m, \overline{F}_1, \overline{F}_2, \ldots, \overline{F}_m)$ *be a multi-objective game, where* $X_1, X_2, \ldots, X_m$ *are convex compact sets and* $\overline{F}_1, \overline{F}_2, \ldots, \overline{F}_m$ *represent continuous vector payoff functions. Moreover, let us assume that for every* $i \in \{1, 2, \ldots, m\}$ *each component* $F_i^k(x_1, x_2, \ldots, x_{i-1}, x_i, x_{i+1}, \ldots, x_m)$, $k \in \{1, 2, \ldots, r_i\}$, *of the vector function* $\overline{F}_i(x_1, x_2, \ldots, x_{i-1}, x_i, x_{i+1}, \ldots, x_m)$ *represents a concave function with respect to* x_i *on* X_i *for fixed* $x_1, x_2, \ldots, x_{i-1}, x_{i+1}, \ldots, x_m$. *Then for the multi-objective game* $\overline{\Gamma} = (X_1, X_2, \ldots, X_m, \overline{F}_1, \overline{F}_2, \ldots, \overline{F}_m)$ *there exists a Pareto-Nash equilibria situation* $x^* = (x_1^*, x_2^*, \ldots, x_m^*) \in X_1 \times X_2 \times \cdots \times X_m$.

Proof Let $\alpha_{11}, \alpha_{12}, \ldots, \alpha_{1r_1}, \alpha_{21}, \alpha_{22}, \ldots, \alpha_{2r_2}, \ldots, \alpha_{m1}, \alpha_{m2}, \ldots, \alpha_{mr_m}$ be an arbitrary set of real numbers which satisfy the following condition

$$\begin{cases} \sum_{k=1}^{r_i} \alpha_{ik} = 1, \ i = 1, 2, \ldots, m; \\ \alpha_{ik} > 0, \ k = 1, 2, \ldots, r_i, \quad i = 1, 2, \ldots, m. \end{cases} \tag{3.52}$$

We consider an auxiliary noncooperative game (single-objective game) $\Gamma = (X_1, X_2, \ldots, X_m, f_1, f_2, \ldots, f_m)$, where

$$f_i(x_1, x_2, \ldots, x_m) = \sum_{k=1}^{r_i} \alpha_{ik} F_i^k(x_1, x_2, \ldots, x_m), \quad i = 1, 2, \ldots, m.$$

It is evident that $f_i(x_1, x_2, \ldots, x_{i-1}, x_i, x_{i+1}, \ldots, x_m)$ for every $i \in \{1, 2, \ldots, m\}$ represents a continuous and concave function with respect to x_i on X_i for fixed $x_1, x_2, \ldots, x_{i-1}, x_{i+1}, \ldots, x_m$ because $\alpha_{11}, \alpha_{12}, \ldots, \alpha_{1r_1}, \alpha_{21}, \alpha_{22}, \ldots, \alpha_{2r_2}, \ldots, \alpha_{m1}, \alpha_{m2}, \ldots, \alpha_{mr_m}$ satisfy condition (3.52) and $F_i^k(x_1, x_2, \ldots, x_{i-1}, x_i, x_{i+1},$

$\ldots, x_m)$ is a continuous and concave function with respect to x_i on X_i for fixed $x_1, x_2, \ldots, x_{i-1}, x_{i+1}, \ldots, x_m, k = 1, 2, \ldots, r_i, i = 1, 2, \ldots, m$.

According to the Nash theorem [102] for the noncooperative game $\Gamma = (X_1, X_2, \ldots, X_p, f_1, f_2, \ldots, f_m)$ there exists a Nash equilibrium $x^* = (x_1^*, x_2^*, \ldots, x_m^*)$, i.e.,

$$\begin{aligned} & f_i(x_1^*, x_2^*, \ldots, x_{i-1}^*, x_i, x_{i+1}^*, \ldots, x_m^*) \\ & \quad \leq f_i(x_1^*, x_2^*, \ldots, x_{i-1}^*, x_i^*, x_{i+1}^*, \ldots, x_m^*), \quad \forall x_i \in X_i, \ i = 1, 2, \ldots, m. \end{aligned}$$

Let us show that $x^* = (x_1^*, x_2^*, \ldots, x_m^*)$ is a Pareto-Nash equilibria solution for the multi-objective game $\overline{\Gamma} = (X_1, X_2, \ldots, X_m, \overline{F}_1, \overline{F}_2, \ldots, \overline{F}_m)$.

Indeed, for every $x_i \in X_i$ we have

$$\begin{aligned} & \sum_{k=1}^{r_i} \alpha_{ik} F_i^k(x_1^*, x_2^*, \ldots, x_{i-1}^*, x_i, x_{i+1}^*, \ldots, x_m^*) \\ & \quad = f_i(x_1^*, x_2^*, \ldots, x_{i-1}^*, x_i, x_{i+1}^*, \ldots, x_m^*) \\ & \quad \leq f_i(x_1^*, x_2^*, \ldots, x_{i-1}^*, x_i^*, x_{i+1}^*, \ldots, x_m^*) \\ & \quad = \sum_{k=1}^{r_i} \alpha_{ik} F_i^k(x_1^*, x_2^*, \ldots, x_{i-1}^*, x_i^*, x_{i+1}^*, \ldots, x_m^*), \\ & \qquad \forall x_i \in X_i, \ i = 1, 2, \ldots, m. \end{aligned}$$

So,

$$\begin{aligned} & \sum_{k=1}^{r_i} \alpha_{ik} F_i^k(x_1^*, x_2^*, \ldots, x_{i-1}^*, x_i, x_{i+1}^*, \ldots, x_m^*) \\ & \quad \leq \sum_{k=1}^{r_i} \alpha_{ik} F_i^k(x_1^*, x_2^*, \ldots, x_{i-1}^*, x_i^*, x_{i+1}^*, \ldots, x_m^*), \\ & \qquad \forall x_i \in X_i; \quad i = 1, 2, \ldots, m \end{aligned} \tag{3.53}$$

for given $\alpha_{11}, \alpha_{12}, \ldots, \alpha_{1r_1}, \alpha_{21}, \alpha_{22}, \ldots, \alpha_{2r_2}, \ldots, \alpha_{m1}, \alpha_{n2}, \ldots, \alpha_{mr_m}$ which satisfy (3.52).

Taking in account that the functions $f_{x^*}^{ik} = F_i^k(x_1^*, x_2^*, \ldots, x_{i-1}^*, x_i, x_{i+1}^*, \ldots, x_m^*)$, $k = 1, 2, \ldots, r_i$, are concave functions with respect to x_i on a convex set X_i and $\alpha_{i1}, \alpha_{i2}, \ldots, \alpha_{ik}$ satisfy the condition $\sum_{k=1}^{r_i} \alpha_{ik} = 1$, $\alpha_{ik} > 0$, $k = 1, 2, \ldots, r_i$, then according to the theorem from [30] (see also [14–16]) condition (3.53) implies that x_i^* is a Pareto solution for the following multi-criteria problem:

$$\max_{x_i \in X_i} \rightarrow \overline{f}_{x^*}^i(x_i) = (f_{x^*}^{i1}(x_i), f_{x^*}^{i2}(x_i), \ldots, f_{x^*}^{ir_i}(x_i)), \quad i \in \{1, 2, \ldots, m\}.$$

This means that $x^* = (x_1^*, x_2^*, \dots, x_p^*)$ is a Pareto-Nash equilibria solution for the multi-objective game $\overline{\Gamma} = (X_1, X_2, \dots, X_m, \overline{F}_1, \overline{F}_2, \dots, \overline{F}_m)$. □

So, if the conditions of Theorem 3.76 are satisfied then a Pareto-Nash equilibria solution for the multi-objective game can be found by using the following algorithm:

Algorithm 3.77 Determining Pareto-Nash Equilibria of a Multi-objective Game

1. Fix an arbitrary set of real numbers $\alpha_{11}, \alpha_{12}, \dots, \alpha_{1r_1}, \alpha_{21}, \alpha_{22}, \dots, \alpha_{2r_2}, \dots, \alpha_{m1}, \alpha_{m2}, \dots, \alpha_{mr_m}$, which satisfy condition (3.52);
2. Form the single-objective game $\Gamma = (X_1, X_2, \dots, X_m, f_1, f_2, \dots, f_m)$, where

$$f_i(x_1, x_2, \dots, x_m) = \sum_{k=1}^{r_i} \alpha_{ik} F_i^k(x_1, x_2, \dots, x_m), \quad i = 1, 2, \dots, m;$$

3. Find Nash equilibria $x^* = (x_1^*, x_2^*, \dots, x_m^*)$ for the noncooperative game $\Gamma = (X_1, X_2, \dots, X_m, f_1, f_2, \dots, f_m)$ and fix x^* as a Pareto-Nash equilibria solution for the multi-objective game $\overline{\Gamma} = (X_1, X_2, \dots, X_m, \overline{F}_1, \overline{F}_2, \dots, \overline{F}_m)$.

Remark 3.78 Algorithm 3.77 finds only one of the solutions for the multi-objective game $\overline{\Gamma} = (X_1, X_2, \dots, X_p, \overline{F}_1, \overline{F}_2, \dots, \overline{F}_m)$. In order to find all solutions in the sense of Pareto-Nash, it is necessary to apply Algorithm 3.77 for every $\alpha_{11}, \alpha_{12}, \dots, \alpha_{1r_1}, \alpha_{21}, \alpha_{22}, \dots, \alpha_{2r_2}, \dots, \alpha_{m1}, \alpha_{m2}, \dots, \alpha_{mr_m}$ which satisfy (3.52) and to form the union of all obtained solutions.

Note that the proof of Theorem 3.76 is based on a reduction of the multi-objective game $\overline{\Gamma}=(X_1, X_2, \dots, X_m, \overline{F_1}, \overline{F_2}, \dots, \overline{F_m})$ to an auxiliary one $\Gamma = (X_1, X_2, \dots, X_m, f_1, f_2, \dots, f_m)$ for which the Nash theorem [102] can be applied. In order to reduce the multi-objective game $\overline{\Gamma}$ to an auxiliary one Γ a linear convolution criteria for vector payoff functions in the proof of Theorem 3.76 has been used. Perhaps a similar reduction of the multi-objective game to a classical one can be used also applying other convolution procedures for vector payoff functions of the players, for example, the standard procedure for the multi-criteria problem from [30, 111].

For the zero-sum multi-objective game of two players the following theorem holds:

Theorem 3.79 *Let $\overline{\Gamma} = (X_1, X_2, \overline{F})$ be a zero-sum multi-objective game of two players, where X_1, X_2 are convex compact sets and $\overline{F}(x_1, x_2)$ is a continuous vector function on $X_1 \times X_2$. Moreover, let us assume that each component $F^k(x_1, x_2)$, $k \in \{1, 2, \dots, r\}$, of $\overline{F}(x_1, x_2)$ for fixed $x_1 \in X_1$ represents a convex function with respect to x_2 on X_2 and for every fixed $x_2 \in X_2$ it is a concave function with respect to x_1 on X_1. Then for the zero-sum multi-objective game $\overline{\Gamma} = (X_1, X_2, \overline{F})$ there exists a saddle point $x^* = (x_1^*, x_2^*) \in X_1 \times X_2$, i.e. x_1^* is a Pareto solution for the multi-criteria problem*:

$$\max_{x_1 \in X_1} \rightarrow \overline{F}(x_1, x_2^*) = (F^1(x_1, x_2^*), F^2(x_1, x_2^*), \ldots, F^r(x_1, x_2^*))$$

and x_2^* is a Pareto solution for the multi-criteria problem:

$$\min_{x_2 \in X_2} \rightarrow \overline{F}(x_1^*, x_2) = (F^1(x_1^*, x_2), F^2(x_1^*, x_2), \ldots, F^r(x_1^*, x_2)).$$

Proof The proof of Theorem 3.79 can be obtained as a corollary from Theorem 3.76 if we regard our zero-sum game as a game of two players of the form $\overline{\Gamma} = (X_1, X_2, \overline{F}_1(x_1, x_2), \overline{F}_2(x_1, x_2)$, where $\overline{F}_2(x_1, x_2) = -\overline{F}_1(x_1, x_2) = \overline{F}(x_1, x_2)$.

The proof of Theorem 3.79 can be obtained also by reducing our zero-sum multi-objective game $\overline{\Gamma} = (X_1, X_2, \overline{F})$ to the classical single-objective case $\Gamma = (X_1, X_2, f)$ and applying Neumann's theorem from [103], where

$$f(x_1, x_2) = \sum_{k=1}^{r} \alpha_k F^k(x_1, x_2)$$

and $\alpha_1, \alpha_2, \ldots, \alpha_r$ are arbitrary real numbers, such that

$$\sum_{k=1}^{r} \alpha_k = 1; \quad \alpha_k > 0, \ k = 1, 2, \ldots, r.$$

It is easy to show that if $x^* = (x_1^*, x_2^*)$ is a saddle point for the zero-sum game $\Gamma = (X_1, X_2, f)$ then $x^* = (x_1^*, x_2^*)$ represents a saddle point for the zero-sum multi-objective game $\overline{\Gamma} = (X_1, X_2, \overline{F})$. □

So, if the conditions of Theorem 3.79 are satisfied then a solution of a zero-sum multi-objective game $\overline{\Gamma} = (X_1, X_2, \overline{F})$ can be found by using the following algorithm:

Algorithm 3.80 Determining the Saddle Point of Payoff Functions in a Zero-Sum Multi-objective Game

1. Fix an arbitrary set of real numbers $\alpha_1, \alpha_2, \ldots, \alpha_r$, such that

$$\sum_{k=1}^{r} \alpha_k = 1; \quad \alpha_k > 0, \ k = 1, 2, \ldots, r;$$

2. Form the zero-sum game $\Gamma = (X_1, X_2, f)$, where

$$f(x_1, x_2) = \sum_{k=1}^{r} \alpha_k F^k(x_1, x_2).$$

3. Find a saddle point $x^* = (x_1^*, x_2^*)$ for the single-objective zero-sum game $\Gamma = (X_1, X_2, f)$. Then fix $x^* = (x_1^*, x_2^*)$ as a saddle point for the zero-sum multi-objective game $\overline{\Gamma} = (X_1, X_2, \overline{F})$.

Remark 3.81 Algorithm 3.80 finds only one solution for the given zero-sum multi-objective game $\overline{\Gamma} = (X_1, X_2, \overline{F})$. In order to find all saddle points it is necessary to apply Algorithm 3.80 for every $\alpha_1, \alpha_2, \ldots, \alpha_r$ satisfying the conditions $\sum_{k=1}^{r} \alpha_k = 1;\ \alpha_k > 0,\ k = 1, 2, \ldots, r$, and then to form the union of the obtained solutions.

Note, that for reducing the zero-sum multi-objective games to classical ones another convolution criteria for vector payoff functions can be used, i.e. the standard procedure from [30, 111].

3.13.3 Discrete and Matrix Multi-objective Games

Discrete multi-objective games are determined by the discrete structure of the sets of strategies $X_1, X_2, \ldots, X_m$. If $X_1, X_2, \ldots, X_m$ are finite sets then we may consider $X_i = J_i,\ J_i = \{1, 2, \ldots, l_i\},\ i = 1, 2, \ldots, m$. In this case the multi-objective game is determined by vectors

$$\overline{F}_i = (F_i^1, F_i^2, \ldots, F_i^{r_i}), \quad i = 1, 2, \ldots, m,$$

where each component F_i^k, $k = 1, 2, \ldots, r_i$, represents a m-dimensional matrix of size $l_1 \times l_2 \times \cdots \times l_m$.

If $m = 2$ then we have a bi-matrix multi-objective game and if $F_2 = -F_1$ then we obtain a matrix multi-objective one. In an analogous way as for single-objective matrix games here we can interpret the strategies $j_i \in J_i$, $i = 1, 2, \ldots, m$, of the players as pure strategies.

It is evident that for such matrix multi-objective games Pareto-Nash equilibria may not exist, because Nash equilibria may not exist for bi-matrix and matrix games in pure strategies. But with each finite discrete multi-objective game we can associate a continuous multi-objective game $\overline{\overline{\Gamma}} = (Y_1, Y_2, \ldots, Y_m, \overline{f}_1, \overline{f}_2, \ldots, \overline{f}_m)$ by introducing mixed strategies $y_i = (y_{i1}, y_{i2}, \ldots, y_{ir_i}) \in Y_i$ of player i and vector payoff functions $\overline{f}_1, \overline{f}_2, \ldots, \overline{f}_m$, which we define in the following way:

$$Y_i = \left\{ y_i = (y_{i1}, y_{i2}, \ldots, y_{il_i}) \in \mathbb{R}^{l_i} \;\middle|\; \sum_{j=1}^{l_i} y_{ij} = 1,\ y_{ij} \geq 0,\ j = 1, 2, \ldots, l_i \right\};$$

$$\overline{f}_i = (f_i^1, f_i^2, \ldots, f_i^{r_i}),$$

where

$$f_i^k(y_{11}, y_{12}, \ldots, y_{1r_1}, y_{21}, y_{22}, \ldots, y_{2r_2}, \ldots, y_{m1}, y_{m2}, \ldots, y_{mr_m})$$
$$= \sum_{j_1=1}^{l_1} \sum_{j_2=1}^{l_2} \cdots \sum_{j_m=1}^{l_m} F_i^k(j_1, j_2, \ldots, j_m)\, y_{ij_1} y_{ij_2} \cdots y_{ij_m};$$
$$k = 1, 2, \ldots, r_i,\ i = 1, 2, \ldots, m.$$

It is easy to observe that for an auxiliary multi-objective game $\overline{\overline{\Gamma}} = (Y_1, Y_2, \ldots, Y_m, \overline{f}_1, \overline{f}_2, \ldots, \overline{f}_m)$ the conditions of Theorem 3.76 are satisfied and, therefore, Pareto-Nash equilibria $y^* = (y^*_{11}, y^*_{12}, \ldots, y^*_{1r_1}, y^*_{21}, y^*_{22}, \ldots, y^*_{2r_2}, \ldots, y^*_{m1}, y^*_{m2}, \ldots, y^*_{ml_m})$ exist.

In the case of matrix games the auxiliary zero-sum multi-objective game of two players is defined as follows:

$$\overline{\overline{\Gamma}} = (Y_1, Y_2, \overline{f});$$

$$Y_1 = \left\{ y_1 = (y_{11}, y_{12}, \ldots, y_{1l_1}) \in \mathbb{R}^{l_1} \,\middle|\, \sum_{j=1}^{l_1} y_{1j} = 1;\ y_{1j} \geq 0,\ j = 1, 2, \ldots, l_1 \right\};$$

$$Y_2 = \left\{ y_2 = (y_{21}, y_{22}, \ldots, y_{2l_2}) \in \mathbb{R}^{l_2} \,\middle|\, \sum_{j=1}^{l_2} y_{2j} = 1;\ y_{2j} \geq 0,\ j = 1, 2, \ldots, l_2 \right\};$$

$$\overline{f} = (f^1, f^2, \ldots, f^r),$$

where

$$f^k(y_{11}, y_{12}, \ldots, y_{1r}, y_{21}, y_{22}, \ldots, y_{2r}) = \sum_{j_1=1}^{l_1} \sum_{j_2=1}^{l_2} F^k(j_1, j_2) y_{1j_1} y_{2j_2};$$

$$k = 1, 2, \ldots, r.$$

The game $\overline{\overline{\Gamma}} = (Y_1, Y_2, \overline{f})$ satisfies the conditions of Theorem 3.79 and, therefore, a saddle point $y^* = (y^*_1, y^*_2) \in Y_1 \times Y_2$ exists.

So, the results related to discrete and matrix games can be extended for the multi-objective case of the game and can be interpreted in an analogous way as for single-objective games. In order to solve these associated multi-objective games, Algorithms 3.77 and 3.80 can be applied.

3.13.4 Some Comments on Multi-objective Games

The considered multi-objective games extend the classical ones and represent a combination of cooperative and noncooperative games. Indeed, the player i in a multi-objective game $\overline{\Gamma} = (X_1, X_2, \ldots, X_m, \overline{F}_1, \overline{F}_2, \ldots, \overline{F}_m)$ can be regarded as a union of r_i sub-players with payoff functions $F_i^1, F_i^2, \ldots, F_i^{r_i}$, respectively. So, the game $\overline{\Gamma}$ represents a game with m coalitions $1, 2, \ldots, m$ which interact between them on the set of situations $X_1 \times X_2 \times \cdots \times X_m$.

The introduced Pareto-Nash equilibria notation uses the concept of cooperative games because according to this notation sub-players of the same coalitions should optimize in the sense of Pareto their vector functions $\overline{F}_i$ on the set of strategies X_i.

On the other hand the Pareto-Nash equilibria notation takes also into account the concept of noncooperative games, because the coalitions interact between them on the set of situations $X_1 \times X_2 \times \cdots \times X_m$ and are interested to maintain Nash equilibria between coalitions.

The obtained results allow us to describe a class of multi-objective games for which a Pareto-Nash equilibria exists. Moreover, a suitable algorithm for finding Pareto-Nash equilibria is proposed.

3.13.5 Determining a Pareto-Stackelberg Solution for Multi-objective Games

For the multi-objective game $\overline{\Gamma} = (X_1, X_2, \ldots, X_p, \overline{F}_1, \overline{F}_2, \ldots, \overline{F}_p)$ a hierarchical optimization principle can be applied in an analogous way as for the classical static game from Sect. 3.12. This allows us to define a Pareto-Stackelberg solution x_1^*, $x_2^*, \ldots, x_m^*$ for the multi-objective game $\overline{\Gamma}$ if we assume that the players fix their strategies successively one after another according to their numerical order. Each player optimizes his vector payoff function in the sense of Pareto and, fixing his optimal strategy x_i^*, informs posterior players which strategy has been applied.

So, a Pareto-Stackelberg solution for the considered multi-objective game $\overline{\Gamma} = (X_1, X_2, \ldots, X_p, \overline{F}_1, \overline{F}_2, \ldots, \overline{F}_m)$ can be defined in the same way as the Stackelberg solution for the classical static game if at the corresponding level player i optimizes his vector payoff function in the sense of Pareto taking into account that the previous players have already fixed their strategies and the posterior players will act optimally.

It is easy to show that if a set of strategies $x_1^*, x_2^*, \ldots, x_m^*$ is a Pareto-Stackelberg solution of the multi-objective game for an arbitrary order of fixing strategies of the players, then $x_1^*, x_2^*, \ldots, x_m^*$ is a Pareto-Nash solution.

Chapter 4
Dynamic Programming Algorithms for Finite Horizon Control Problems and Markov Decision Processes

In this chapter we study stochastic discrete control problems and Markov decision processes with finite time horizon. We assume that the set of states of dynamical system is finite and the starting and the final states are fixed. For finite horizon control problems we use the same concept as for infinite horizon decision models, i.e we assume that in the control process the dynamical system may admit dynamical states where the vector of control parameters is changing in a random way according to given distribution functions of the probabilities on given feasible dynamical sets. So, we consider the control problems for which the dynamics may contain controllable states as well the uncontrollable ones. We show that in general form the stochastic control problems can be formulated on networks and the dynamic programming algorithms for determining the optimal solution of the problems can be developed. We extend the corresponding dynamic programming technique for finite horizon Markov decision problems.

4.1 Problem Formulation

We consider a time-discrete system $\mathbb{L}$ with a finite set of states $X \subset \mathbb{R}^n$. At every time-step $t = 0, 1, 2, \ldots,$ the state of the system $\mathbb{L}$ is $x(t) \in X$. Two states x_0 and x_f are given in X, where $x_0 = x(0)$ represents the starting state of system $\mathbb{L}$ and x_f is the state in which the system $\mathbb{L}$ must be brought, i.e., x_f is the final state of $\mathbb{L}$. We assume that the system $\mathbb{L}$ should reach the final state x_f at the time-moment $t(x_f)$ such that $t_1 \leq t(x_f) \leq t_2$, where t_1 and t_2 are given. The dynamics of the system $\mathbb{L}$ is described as follows

$$x(t+1) = g_t(x(t), u(t)), \quad t = 0, 1, 2, \ldots, \tag{4.1}$$

where

$$x(0) = x_0 \tag{4.2}$$

D. Lozovanu and S. Pickl, *Optimization of Stochastic Discrete Systems and Control on Complex Networks*, Advances in Computational Management Science 12,
DOI 10.1007/978-3-319-11833-8_4

and $u(t) = (u_1(t), u_2(t), \dots, u_m(t)) \in \mathbb{R}^m$ represents the vector of control parameters.

For any time-step t and an arbitrary state $x(t) \in X$ a feasible finite set $U_t(x(t)) = \{u^1_{x(t)}, u^2_{x(t)}, \dots, u^{k(x(t))}_{x(t)}\}$, for the vector of control parameters $u(t)$ is given, i.e.,

$$u(t) \in U_t(x(t)), \quad t = 0, 1, 2, \dots. \tag{4.3}$$

We assume that in (4.1) the vector functions $g_t(x(t), u(t))$ are uniquely determined by $x(t)$ and $u(t)$, i.e., the state $x(t+1)$ is determined uniquely by $x(t)$ and $u(t)$ at every time-step $t = 0, 1, 2, \dots$. In addition we assume that at each moment of time t the cost $c_t(x(t), x(t+1)) = c_t(x(t), g_t(x(t), u(t)))$ of system's transition from the state $x(t)$ to the state $x(t+1)$ is known.

Let

$$x_0 = x(0), x(1), x(2), \dots, x(t), \dots$$

be a trajectory generated by given vectors of control parameters

$$u(0), u(1), \dots, u(t-1), \dots.$$

Then either this trajectory passes through the state x_f at the time-moment $t(x_f)$ or it does not pass through x_f.

We denote

$$F_{x_0 x_f}(u(t)) = \sum_{t=0}^{t(x_f)-1} c_t(x(t), g_t(x(t), u(t))) \tag{4.4}$$

the integral-time cost of system's transitions from x_0 to x_f if $t_1 \le t(x_f) \le t_2$; otherwise we put $F_{x_0 x_f}(u(t)) = \infty$.

In [6, 11, 59, 79] the following problem have been formulated and studied: Determine the vectors of control parameters $u(0), u(1), \dots, u(t), \dots$ which satisfy conditions (4.1)–(4.3) and minimize functional (4.4).

This problem represents a finite horizon control model with controllable states of a dynamical system where for an arbitrary state $x(t)$ at every moment of time the choosing of the vector of control parameter $u(t) \in U_t(x(t))$ is assumed to be at our disposition.

In the following we consider the stochastic versions of the control model formulated above. We assume that the dynamical system $\mathbb{L}$ may admit uncontrollable states, i.e., for the system $\mathbb{L}$ there exist dynamical states in which we are not able to control the dynamics of the system and the vector of control parameters $u(t) \in U_t(x(t))$ for such states is changing in a random way according to a given distribution function

$$p : U_t(x(t)) \to [0, 1], \quad \sum_{i=1}^{k(x(t))} p(u^i_{x(t)}) = 1 \tag{4.5}$$

on the corresponding dynamical feasible sets $U_t(x(t))$, where $k(x(t)) = |U_t(x(t))|$.

If we regard an arbitrary dynamic state $x(t)$ of the system $\mathbb{L}$ at a given moment of time t as position (x, t) then the set of positions

$$Z = \{(x, t) \mid x \in X,\ t = 0, 1, 2, \ldots, t_2\}$$

of the dynamical system can be divided into two disjoint subsets

$$Z = Z^C \cup Z^N \quad (Z^C \cap Z^N = \emptyset),$$

where Z^C represents the set of controllable positions of $\mathbb{L}$ and Z^N represents the set of positions $(x, t) = x(t)$ for which the distribution function (4.5) of the vectors of control parameters $u(t) \in U_t(x(t))$ are given. This means that we obtain the following behavior of the dynamical system $\mathbb{L}$ in the control process. If the starting points belong to controllable positions then the decision maker fixes a vector of control parameters and we obtain the state $x(1)$. If the starting state belongs to the set of uncontrollable positions then the system passes to the next state in the random way. After that if at the time-moment $t = 1$ the state $x(1)$ belongs to the set of controllable positions then the decision makers fix the vector of control parameter $u(t) \in U_t(x(t))$ and we obtain the state $x(2)$. If $x(1)$ belongs to the set of uncontrollable positions then the system passes to the next state in the random way and so on. In this dynamic process the final state may be reached at a given moment of time with a probability which depends on the control of the system in the controllable states as well as the expectation of integral time cost by a trajectory depends on the control of the system in these states. Therefore, our main concentration will be addressed on studying and solving the following classes of problems:

Problem 4.1 Determine for given vectors of control parameters $u(t) \in U_t(x(t))$, $x(t) \in Z^C$, the probability that the dynamical system $\mathbb{L}$ with a given starting state $x_0 = x(0)$ will reach the final state x_f at the moment of time $t(x_f)$ such that $t_1 \le t(x_f) \le t_2$. We denote this probability by $P_{x_0}(u(t), x_f, t_1 \le t(x_f) \le t_2)$; if $t_1 = t_2 = \bar{t}$ then we use the notation $P_{x_0}(u(t), x_f, \bar{t})$.

Problem 4.2 Find the vectors of control parameters $u^*(t) \in U_t(x(t))$, $x(t) \in Z^C$ for which the probability in Problem 4.1 is maximal. We denote this probability by $P_{x_0}(u^*(t), x_f, t_1 \le t(x_f) \le t_2)$; in the case $t_1 = t_2 = \bar{t}$ we shall use the notation $P_{x_0}(u^*(t), x_f, \bar{t})$.

Problem 4.3 Determine for given vectors of control parameters $u(t) \in U_t(x(t))$, $x(t) \in Z^C$ and given $\bar{t}$ the expected total cost during $\bar{t}$ state's transitions of the system $\mathbb{L}$ if it starts transitions in the state $x_0 = x(0)$ at the moment of time $t = 0$. We denote this value by $\sigma_{x_0}(u(t), \bar{t})$.

Problem 4.4 Determine the vectors of control parameters $u^*(t) \in U_t(x(t)), x(t) \in Z^C$ for which the expected total cost in Problem 4.3 is minimal. We denote this value by $\sigma_{x_0}(u^*(t), \bar{t})$.

Problem 4.5 For given vectors of control parameters $u(t) \in U_t(x(t)),\ x(t) \in Z^C$, determine the expected total cost of state's transitions from a starting state x_0 to a final state x_f if x_f is reached at the time-moment $t(x_f)$ such that $t_1 \leq t(x_f) \leq t_2$. We denote this expected cost by $\sigma_{x_0}(u(t),\ x_f,\ t_1 \leq t(x_f) \leq t_2)$; we denote in the case $t_1 = t_2 = \bar{t}$ this value by $\sigma_{x_0}(u(t),\ x_f,\ \bar{t})$.

Problem 4.6 Determine the vectors of control parameters $u^*(t) \in U_t(x(t)), x(t) \in Z^C$ for which the expected total cost in Problem 4.5 is minimal. We denote this minimal expected cost by $\sigma_{x_0}(u^*(t),\ x_f,\ t_1 \leq t(x_f) \leq t_2)$; in the case $t_1 = t_2 = \bar{t}$ we denote this value by $\sigma_{x_0}(u^*(t),\ x_f,\ \bar{t})$.

We define the probability $P_{x_0}(u(t),\ x,\ \bar{t})$ for the dynamical system $\mathbb{L}$ in the considered problem in the following way:

Let $u'(t)$ be a fixed feasible control. Then on each feasible set $U_t(x(t)$ in the controllable states $x(t) \in Z^C$ we can define the following probability functions: $p(u(t)) = 1$ if $u(t) = u'(t)$; otherwise put $p(u(t)) = 0$.

So, the vector $u'(t)$ in a given controllable state $x = x(t)$ means that the dynamical system makes a transition from $x = x(t)$ to the state $y = g_t(x(t), u'(t))$ with the probability equal to 1 and the rest of probability transitions from x to another state are equal to zero. Thus, a fixed control $u'(t)$ generates a finite Markov process for which the probability of system's transition from a starting state x_0 to a final state x by using $\bar{t}$ units of time can be defined. We denote this probability by $P_{x_0}(u'(t),\ x,\ \bar{t})$.

We define the probability $P_{x_0}(u'(t),\ x,\ t_1 \leq t(x) \leq t_2)$ for given t_1 and t_2 and fixed feasible control $u'(t)$ as the probability of the dynamical system $\mathbb{L}$ to reach the state x at least at one of the moment of times $t_1, t_1 + 1, \dots, t_2$ if the system starts transitions in the state x_0 (at the moment of time $t = 0$).

In an analogues way we define the expected total cost in the problems 4.3–4.6 using the corresponding notion for Markov processes with transition costs from Chap. 1. We define for a fixed control $u'(t)$ the expected total cost $\sigma_{x_0}(u'(t), \bar{t})$ of system $\mathbb{L}$ in Problem 4.3 as the expected total cost during $\bar{t}$ state's transitions of the dynamical system in the Markov process generated by the control $u'(t)$ and the corresponding transition costs. We will precise the expected total cost $\sigma_{x_0}(u(t),\ x,\ t_1 \leq t(x) \leq t_2)$ in the Problems 4.5 and 4.6 in a more detailed form in Sect. 4.3.

For an additional characterization of the finite stochastic processes with transition costs we shall use also the notion of the variance of the total cost for the dynamical system and the corresponding problem of determining the variance in such processes will be considered.

The problems formulated above comprises a class of deterministic and stochastic dynamic problems from [6, 11, 47]. The problems from [47] related to finite Markov processes became Problems 4.1–4.3 in the case if $Z^C = \emptyset$, $t_1 = t_2 = \bar{t}$ and if the probabilities $p(u^i_{x(t)})$ do not depend on time but depend only on the states.

The discrete optimal control problems from [6,11] became Problems 4.4–4.6 in the case $Z^N = \emptyset$.

The considered problem can de studied and solved separately, however the combined joint solution for some of them also may be discussed and justified. For example, if we solve Problem 4.2 and find the control with the maximal probability of system's transitions from the starting state to the final one then after that may be expected the optimal control estimate the expected total cost, i.e. we have to solve additionally Problem 4.5. If we solve Problem 4.6 and find the control which provides the minimal expected total cost of states' transitions of the system from x_0 to x_f then after that the optimal control estimates the probability of states' transition of the system from x_0 to x_f, i.e., we have to solve additionally Problem 4.1.

We show that all considered characteristics and the optimal control in Problems 4.1–4.6 can be found using algorithms based on the direct and backward dynamic programming procedures.

4.2 Algorithms for Solving Stochastic Control Problems Using Time-Expanded Networks

In order to simplify the description of the main approach and to ground the algorithms for solving the problems formulated in Sect. 4.1 we shall use the network representation of the dynamics of the system and we will formulate these problems on networks. Note that in our control problems the probabilities and the costs of states' transitions of the system in general case depend on time. Therefore, here we develop a time-expanded network method from [76,79] for the stochastic versions of control problems and reduce them to the stationary case of the problems. This will allow us to describe dynamic programming algorithms for solving the problems on static networks [80]. At first we show how to construct the time-expanded network and how to solve the problems with a fixed number of stages, i.e. we consider the case $t_1 = t_2 = \overline{t}$.

4.2.1 Construction of the Time-Expanded Network with a Fixed Number of Transitions

If the dynamics of the discrete system $\mathbb{L}$ and the information related to the feasible sets $U_t(x(t))$ and the cost functions $c_t(x(t), g_t(x(t), u(t)))$ in the problems with $t_1 = t_2 = \overline{t}$ are known then our time-expanded network can be obtained in the following way: We identify each position (x, t) which corresponds to a dynamic state $x(t)$ with a vertex $z = (x, t)$ of the network.

So, the set of vertices Z of the network can be represented as follows

$$Z = Z_1 \cup Z_2 \cup \ldots \cup Z_{\overline{t}}$$

where

$$Z_t = \{(x,t) \mid x \in X\}, \quad t = 0, 1, 2, \dots, \bar{t}.$$

To each vector of control parameters $u(t) \in U_t(x(t)), t = 1, 2, \dots, \bar{t} - 1$ which provides a systems' transition from the state $x(t) = (x,t)$ to the state $x(t+1) = (y, t+1)$ we associate in our network a directed edge $e(z,w) = \big((x,t),(y,t+1)\big)$ from the vertex $z = (x,t) \in Z_t$ to the vertex $w = (y, t+1) \in Z_{t+1}$, i.e. the set of edges E of the network is determined by the feasible sets $U_t(x(t)), t = 0, 1, 2, \dots, \bar{t}-1$. To each directed edge $e = (z,w) = \big((x,t),(y,t+1)\big)$ originating in the uncontrollable positions (x,t) we put in correspondence the probability $p_e = p\big(u^i(t)\big)$, where $u^i(t)$ is a vector of control parameter which provides the transition of the system from the state $x = x(t)$ to the state $x(t+1) = (y, t+1)$. Thus, if we distinguish in E the subset of edges $E_N = \{e = (z,w) \in E \mid z \in Z^N\}$ originating in uncontrollable positions Z^N then on E_N we obtain the probability function $p : E \to \mathbb{R}$ which satisfies the condition

$$\sum_{e \in E(z)} p_e = 1, \quad z \in Z^N \backslash Z_{\bar{t}}$$

where $E(z)$ is the set of edges originating in z, i.e., $E(z) = \{e = (z,w) \mid e \in E, w \in Z\}$. In addition in the network we add to the edges $e = (z,w) = \big((x,t),(y,t+1)\big)$ the costs $c_{(z,w)} = c((x,t),(y,t+1)) = c_t\big(x(t), x(t+1)\big)$ which correspond to the costs of system's transition from the states $x(t)$ to the states $x(t+1)$. We denote the subset of edges of the graph G originating in vertices $z \in Z^C$ by E_C, i.e., $E_C = E \backslash E_N$.

So, our network is determined by the tuple $(G, Z^C, Z^N, c, p, \bar{t}, z_0, z_f)$, where $G = (Z, E)$ is the graph which describes the dynamics of the system; the vertices $z_0 = (x_0, 0)$ and $z_f = (x_f, \bar{t})$ correspond to the starting and the final states of the dynamical system, respectively; c represents the cost function defined on the set of edges E and p is the probability function defined on the set of edges E_N which satisfy condition (4.5). Thus, $Z = Z^C \cup Z^N$, where Z^C is a subset of vertices of G which corresponds to the set of controllable positions of the dynamical system and Z^N is a subset of vertices of G which corresponds to the set of uncontrollable positions of system $\mathbb{L}$. The subsets Z^C and Z^N can be represented as follows $Z^C = \bigcup_{t=0}^{\bar{t}} Z_t^C$, $Z^N = \bigcup_{t=0}^{\bar{t}} Z_t^N$, where $Z_t^C = \{(x,t) \in Z_t \mid (x,t) \in Z^C\}$ and $Z_t^N = \{(x,t) \in Z_t \mid (x,t) \in Z^N\}$. The graphical representation of the structure of this network is given on Fig. 4.1. For Problems 4.1 and 4.2 the information about the cost function c is not required, therefore this function in the notation above of the network can be omitted; for Problems 4.3 and 4.4 the final state is not given and therefore in the notation of the network we can also omit.

It is easy to observe that the Problem 4.1 in the case $t_1 = t_2 = \bar{t}$ can be formulated and solved on network $(G, Z^C, Z^N, p, \bar{t}, z_0, z_f)$. A control $u(t)$ of system $\mathbb{L}$ in the controllable state $x(t)$ on this network means a transitions from the vertex $z = (x,t)$ to $w = (y, t+1)$ through a leaving edges $e^u = (z,w) = ((x,t),(y,t+1))$ generated by $u(t)$.

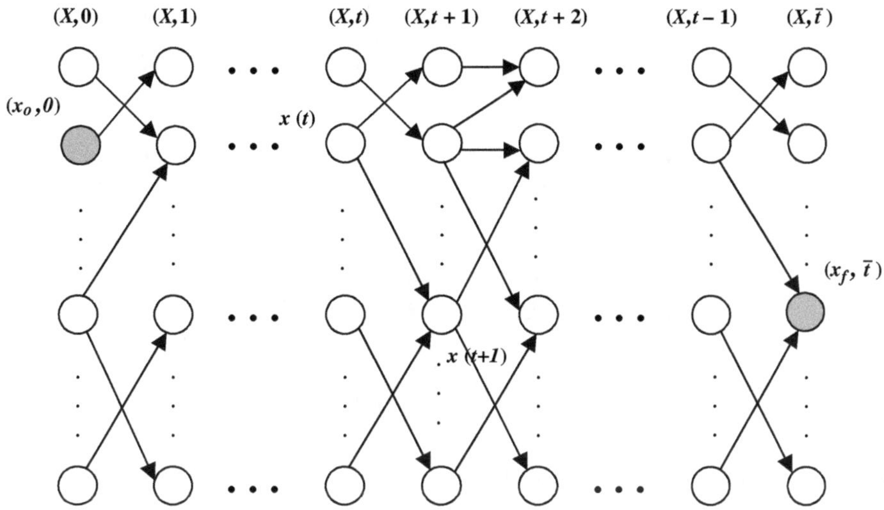

Fig. 4.1 The graphical representation of the structure of the time-expanded network

This is equivalent with an association to these leaving edges of the probability $p_{e^u} = 1$ of the states' transition from (x, t) to $(y, t + 1)$ considering $p_e = 0$ for the rest of the leaving directed edges. In other words a control on a time-expanded network means an extension of the probability function p from E_N to E according to the rule mentioned above. We denote this probability function on E by p^u and we will keep in mind that $p_e^u = p_e$ for $e \in E_N$ and this function satisfies on E_C the following property

$$p^u : E_C \to \{0, 1\}, \quad \sum_{e \in E_C(z)} p_e^u = 1 \quad \text{for} \quad z \in Z^C.$$

So, each feasible control $u(t)$ in the problems from Sect. 4.1 uniquely defines the function p_e^u on E and vice versa. Therefore if $u(t)$ is given for the control problems then we denote the time-expanded network by $(G, Z^C, Z^N, c, p^u, \bar{t}, z_0, z_f,)$. In the case if the control $u(t)$ in the Problems 4.1–4.6 is not fixed then for the time-expanded network we shall use the notation $(G, Z^C, Z^N, c, p, \bar{t}, z_0, z_f)$. For the probability transition from a starting position z_0 to a final position z_f of the system $\mathbb{L}$ on this network we shall use a similar notation $P_{z_0}(u(t), z_f, \bar{t})$.

4.2.2 An Example of Constructing the Time-Expanded Network

We describe an example how to construct the time-expanded network for the stochastic control problem which we will use for the illustration of the calculation procedures of the state-time probabilities and the expected total cost. We consider the dynamical system with the set of states $X = \{1, 2, 3, 4\}$ and $\bar{t} = 5$. The starting state of the system is $x(0) = 2$ and the final state is $x_f = x(5)$.

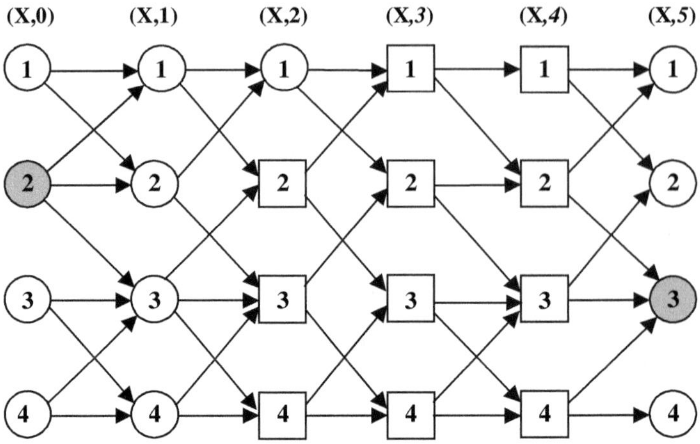

Fig. 4.2 The structure of the time-expanded network

All necessary information related to the dynamics and the control parameters of the problem are given in Table 4.1. The structure of the time-expanded network is represented on Fig. 4.2. In this figure the controllable positions of the dynamical system are represented by circles and the uncontrollable ones are represented by squares.

If the feasible sets $U_t(x(t))$ are known for all dynamical state $x(t)$ then the time-expanded network can be constructed by using $O\big(\sum_{t=0}^{\overline{t}} \sum_{x(t)\in X} |U_t(x(t))|\big)$ elementary operations. In general the complexity of the construction procedure of the time-expanded network depends on input data of the dynamics of the system and on the corresponding parameters of the problem. In our example we assume that the input data of the problem is presented by the list of the dynamical states $x(t) = (x, t)$, the feasible sets $U_t(x(t))$ in the controllable states $x(t)$, the values of the cost functions $c_t(x(t), x(t+1))$ and the corresponding transition probabilities $p_{x(t),\ x(t+1)}$ in the uncontrollable states $x(t)$ for every discrete moment of time t. Therefore we will estimate the complexity of the proposed algorithms for our problems with respect to this information.

If for the dynamical system we fix the control

$$
\begin{aligned}
&u\colon (1,0) \to (2,1) \ \text{ for } \ x(0) = (1,0),\\
&u\colon (2,0) \to (3,1) \ \text{ for } \ x(0) = (2,0),\\
&u\colon (3,0) \to (3,1) \ \text{ for } \ x(0) = (3,0),\\
&u\colon (4,0) \to (3,1) \ \text{ for } \ x(0) = (4,0),\\
&u\colon (1,1) \to (2,2) \ \text{ for } \ x(1) = (1,1),\\
&u\colon (2,1) \to (3,2) \ \text{ for } \ x(1) = (2,1),\\
&u\colon (3,1) \to (3,2) \ \text{ for } \ x(1) = (3,1),\\
&u\colon (4,1) \to (3,2) \ \text{ for } \ x(1) = (4,1),\\
&u\colon (1,2) \to (2,3) \ \text{ for } \ x(2) = (1,2),
\end{aligned}
$$

Table 4.1 Information related to the dynamics and the control parameters of the problem

t	x	$X(t)$ (x,t)	$U_t(x(t)) = \{u^i\}$ $u^i: (x,t) \to (y,t+1)$	Distribution functions for uncontrollable states	Cost functions $c_t(x(t), u(t))$
0	1	(1,0)	u^1: (1,0) → (1,1)		2
			u^2: (1,0) → (2,1)		4
	2	(2,0)	u^1: (2,0) → (1,1)		4
			u^2: (2,0) → (2,1)		6
			u^3: (2,0) → (3,1)		3
	3	(3,0)	u^1: (3,0) → (3,1)		6
			u^2: (3,0) → (4,1)		2
	4	(4,0)	u^1: (4,0) → (3,1)		2
			u^2: (4,0) → (4,1)		5
1	1	(1,1)	u^1: (1,1) → (1,2)		6
			u^2: (1,1) → (2,2)		2
	2	(2,1)	u^1: (2,1) → (1,2)		7
			u^2: (2,1) → (3,2)		2
	3	(3,1)	u^1: (3,1) → (2,2)		2
			u^2: (3,1) → (3,2)		5
			u^3: (3,1) → (4,2)		3
	4	(4,1)	u^1: (4,1) → (3,2)		7
			u^2: (4,1) → (4,2)		4
2	1	(1,2)	u^1: (1,2) → (1,3)		8
			u^2: (1,2) → (2,3)		1
	2	(2,2)	u^1: (2,2) → (1,3)	$p(u^1) = 0.6$	3
			u^2: (2,2) → (3,3)	$p(u^2) = 0.4$	7
	3	(3,2)	u^1: (3,2) → (2,3)	$p(u^1) = 0.3$	3
			u^2: (3,2) → (4,3)	$p(u^2) = 0.7$	8
	4	(4,2)	u^1: (4,2) → (3,3)	$p(u^1) = 0.5$	2
			u^2: (4,2) → (4,3)	$p(u^2) = 0.5$	4
3	1	(1,3)	u^1: (1,3) → (1,4)	$p(u^1) = 0.4$	1
			u^2: (1,3) → (2,4)	$p(u^2) = 0.6$	5
	2	(2,3)	u^1: (2,3) → (2,4)	$p(u^1) = 0.5$	1
			u^2: (2,3) → (3,4)	$p(u^2) = 0.5$	4
	3	(3,3)	u^1: (3,3) → (3,4)	$p(u^1) = 0.6$	3
			u^2: (3,3) → (4,4)	$p(u^2) = 0.4$	4
	4	(4,3)	u^1: (4,3) → (3,4)	$p(u^1) = 0.7$	5
			u^2: (4,3) → (4,4)	$p(u^2) = 0.3$	4

(continued)

Table 4.1 (continued)

t	x	$X(t)$ (x, t)	$U_t(x(t)) = \{u^i\}$ $u^i: (x, t) \to (y, t+1)$	Distribution functions for uncontrollable states	Cost functions $c_t(x(t), u(t))$
4	1	(1,4)	u^1: (1,4) → (1,5)	$p(u^1) = 0.2$	7
			u^2: (1,4) → (2,5)	$p(u^2) = 0.8$	2
	2	(2,4)	u^1: (2,4) → (1,5)	$p(u^1) = 0.3$	6
			u^2: (2,4) → (3,5)	$p(u^2) = 0.7$	3
	3	(3,4)	u^1: (3,4) → (2,5)	$p(u^1) = 0.5$	2
			u^2: (3,4) → (3,5)	$p(u^2) = 0.5$	4
	4	(4,4)	u^1: (4,4) → (3,5)	$p(u^1) = 0.6$	5
			u^2: (4,4) → (4,5)	$p(u^2) = 0.4$	3

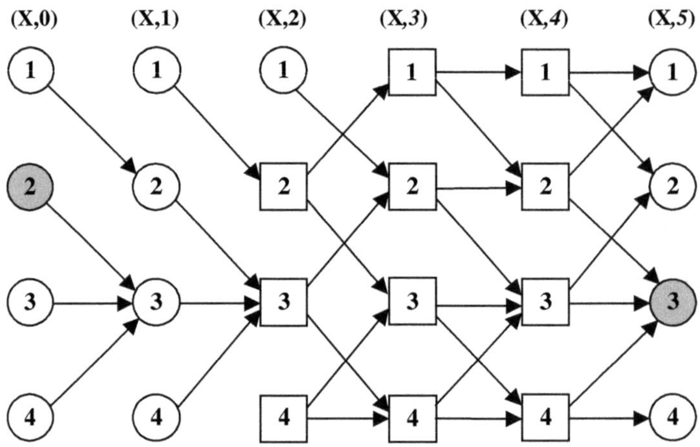

Fig. 4.3 The corresponding time-expanded network induced by the control $u(t)$

then we obtain the network $(G, Z^C, Z^P, c, p^u, \bar{t}, z_0, z_f)$ represented on Fig. 4.3. For this network the edges originating in controllable positions (vertices) get the following probabilities:

$$p_{((1,0),(2,1))} = 1; \quad p_{((2,0),(3,1))} = 1; \quad p_{((3,0),(3,1))} = 1; \quad p_{((4,0),(3,1))} = 1;$$

$$p_{((1,1),(2,2))} = 1; \quad p_{((2,1),(3,2))} = 1; \quad p_{((3,1),(3,2))} = 1; \quad p_{((4,1),(3,2))} = 1;$$

$$p_{((1,2),(2,3))} = 1.$$

The rest of transition probabilities on the edges originating in controllable positions are equal to zero. The transition probabilities on the edges originating in uncontrollable positions (vertices) are determined by distribution function given in Table 4.1.

4.2.3 Algorithms for Determining the State-Time Probabilities and the Optimal Control for the Problem with a Fixed Number of Transitions

In this section we describe algorithms for determining the solutions of Problems 4.1 and 4.2 using the direct and backward dynamic programming algorithms. At first we use the direct procedure for the calculation of the probabilities $P_{z_0}(u(t), z, \tau)$, of system transitions from z_0 to $z \in Z_\tau$, $\tau = 1, 2, \ldots, \overline{t}$ for the Problem 4.1 on the network $(G, Z^C, Z^N, p^u, \overline{t}, z_0, z_f)$.

Algorithm 4.7 Determining the State-Time Probabilities Using a the Direct Dynamic Programming Procedure

Preliminary step (Step 0): Put $P_{z_0}(u(t), z_0, 0) = 1$ for $z_0 = (x_0, 0)$ and $P_{z_0}(u(t), z, t) = 0$ for every $z \in Z \backslash \{z_0\}$.

General step (Step τ, $\tau \geq 1$): For every $z \in Z_\tau$ calculate

$$P_{z_0}(u(t), z, \tau) = \sum_{(w,z) \in E^-(z)} P_{z_0}(u(t), w, \tau - 1) p^u_{(w,z)}$$

where $E^-(z) = \{(w, z) \in E \mid w \in Z_{\tau-1}\}$. If $\tau = \overline{t}$ then STOP; otherwise go to the next step.

The correctness of the algorithm is evident. Algorithm 4.7 is based on the recursive formula 1.1 from Sect. 1.1 for the calculation of the state-time probabilities of the system on the time-expanded network.

Lemma 4.8 *If for every $x(t) \in X$ and $t = 0, 1, 2, \ldots, \overline{t} - 1$ the feasible sets $U_t(x(t))$ are nonempty then the state-time probabilities $P_{z_0}(u(t), z, \tau)$ calculated according to Algorithm 4.7 satisfy the condition*

$$\sum_{z \in Z_\tau} P_{z_0}(u(t), z, \tau) = 1, \quad \tau = 0, 1, 2, \ldots, \overline{t}.$$

Proof We prove this lemma using an induction principle on the number of stages τ. In the case $\tau = 0$ the lemma is evident. Assume that lemma holds for an arbitrary $\tau \geq 0$ and let us show that it is true for $\tau + 1$. Indeed,

$$\sum_{z \in Z_{\tau+1}} P_{z_0}(u(t), z, \tau + 1) = \sum_{z \in Z_{\tau+1}} \sum_{(w,z) \in E^-(z)} P_{z_0}(u(t), w, \tau) p^u_{(w,z)}.$$

If for every (w, z) which does not belong to the edges' set E we consider $p^u_{(w,z)} = 0$ then we can introduce in the last sum all terms $P_{z_0}(u(t), w, \tau) p^u_{(w,z)}$.

Thus,

$$\sum_{z \in Z_{\tau+1}} P_{z_0}(u(t),\, z,\, \tau+1) = \sum_{z \in Z_{\tau+1}} \sum_{w \in Z_\tau} P_{z_0}(u(t),\, w,\, \tau) p^u_{(w,z)}$$

$$= \sum_{w \in Z_\tau} \sum_{z \in Z_{\tau+1}} P_{z_0}(u(t),\, w,\, \tau) p^u_{(w,z)}$$

$$= \sum_{w \in Z_\tau} P_{z_0}(u(t),\, w,\, \tau) \sum_{z \in Z_{\tau+1}} p^u_{(w,z)} = \sum_{w \in Z_\tau} P_{z_0}(u(t),\, w,\, \tau) = 1,$$

i.e., lemma holds. □

In the following for the control problems from Sect. 4.1 we will assume that the condition $U_t(x(t)) \neq \emptyset$ for every $x(t) \in X$, $t = 0, 1, 2, \ldots, t_2 - 1$ holds.

Algorithm 4.7 allows us for a given control $u(t)$ to find the probabilities $P_{x_0}(u(t),\, x,\, \tau)$ of state's transition from x_0 to an arbitrary x for every $\tau = 0, 1, 2, \ldots, \bar{t}$. Using the network interpretation of Problem 4.1 we can develop an algorithm for the calculation of probabilities $P_{x(\bar{t}-\tau)}(u(t),\, x_f,\, \bar{t})$ of states' transition from an arbitrary dynamical state $x(\bar{t} - \tau)$ to the final state x_f for every $\tau = 0, 1, 2, \ldots, \bar{t}$ starting from a final state x_f if the control $u(t)$ is given. This algorithm is based on a backward dynamic programming procedure.

Algorithm 4.9 Determining the State-Time Probabilities of the System Using the Backward Dynamic Programming Procedure

Preliminary step (Step 0): Put $P_{z_f}(u(t),\, z_f,\, \bar{t}) = 1$ for $z_f = (x_f, \bar{t})$ and $P_z(u(t),\, z,\, \bar{t}) = 0$ for every $z \in Z_{\bar{t}} \setminus \{z_f\}$.

General step (Step $\tau,\ \tau \geq 1$*)*: For every $z \in Z_{\bar{t}-\tau}$ calculate

$$P_z(u(t),\, z_f,\, \bar{t}) = \sum_{(z,w) \in E(z)} P_w(u(t),\, z_f,\, \bar{t}) p^u_{(z,w)}$$

where $E(z) = \{(z, w) \in E \mid w \in Z_{\tau+1}\}$. If $\tau = \bar{t}$ then STOP; otherwise go to next step.

Algorithm 4.9 finds the values $P_{x(\bar{t}-\tau)}(u(t),\, x_f,\, \bar{t}) = P_{(x,\bar{t}-\tau)}(u(t), (x_f, \bar{t}), \bar{t})$ which represent the probability of system's transition from the state x at the moment of time $\bar{t} - \tau$ to the state x_f using τ units of time. Finally if we fix $\tau = \bar{t}$ then we obtain the probabilities $P_x(u(t),\, x_f,\, \bar{t})$ of system's transition from every x to x_f by using $\bar{t}$ units of time.

Theorem 4.10 *For a given control* $u(t)$ *Algorithm 4.9 correctly finds the state-time probabilities* $P_{(x,\bar{t}-\tau)}(u(t),\, x_f,\, \bar{t})$ *for every* $x \in X$ *and* $\tau = 0, 1, 2, \ldots, \bar{t}$*. The running time of the algorithm is* $O(|X|^2 \bar{t})$.

Proof The preliminary step of the algorithm is evident. The correctness of the general step of the algorithm follows from a recursive formula for the calculation of the state-time probabilities of the dynamical system in the Markov processes on the network. In order to estimate the running time of the algorithm it is sufficient to estimate the number of elementary operations of the general step of the algorithm. It is easy to see that the number of elementary operations for the tabulation of the state-time probabilities at the general step is $O(|X|^2)$. Taking into account that the number of steps of the algorithms is equal to $\overline{t}$ we obtain that the running time of the algorithm is $O(|X|^2\overline{t})$. □

Remark 4.11 The direct and the backward dynamic algorithm are valid for determining the probabilities $P_x(u(t), x_f, \overline{t})$ also in the case if the condition (4.5) may not take place, i.e. $\sum_{i=1}^{k(x(t))} p(u^i_{x(t)}) \leq 1$.

Below we illustrate the calculation procedures of the direct and backward dynamic programming algorithms for the stochastic control problem from Sect. 4.2.2.

Example Let the discrete system $\mathbb{L}$ with the dynamics described in Sect. 4.2.2 be given and consider the problem of determining the state-time probabilities $P_{x_0}(u(t),\ y,\ \tau)$ and $P_{x(T-\tau)}(u(t),\ x_f,\ \overline{t})$ for the following control $u(t)$:

$$
\begin{aligned}
u&: (1,0) \to (2,1) \quad \text{for } x(0) = (1,0),\\
u&: (2,0) \to (3,1) \quad \text{for } x(0) = (2,0),\\
u&: (3,0) \to (3,1) \quad \text{for } x(0) = (3,0),\\
u&: (4,0) \to (3,1) \quad \text{for } x(0) = (4,0),\\
u&: (1,1) \to (2,2) \quad \text{for } x(1) = (1,1),\\
u&: (2,1) \to (3,2) \quad \text{for } x(1) = (2,1),\\
u&: (3,1) \to (3,2) \quad \text{for } x(1) = (3,1),\\
u&: (4,1) \to (3,2) \quad \text{for } x(1) = (4,1),\\
u&: (1,2) \to (2,3) \quad \text{for } x(2) = (1,2).
\end{aligned}
$$

Assume that $\overline{t} = 5$, $x(0) = 2$ and $x_f = x(5) = 3$. The corresponding time-expanded network induced by the control $u(t)$ is represented on Fig. 4.3.

If we apply algorithm Algorithms 4.7 and 4.9, respectively, then we obtain the state-time probabilities $P_{x_0}(u(t),\ y,\ \tau)$ and $P_{x(\overline{t}-\tau)}(u(t),\ x_f,\ \overline{t}), \tau = 0, 1, 2, 3, 4, 5$ given in Table 4.2.

We can see that the direct dynamic programming algorithm finds the probabilities of the system transitions from a fixed starting state $x(0) = 2$ to an arbitrary state $x(t),\quad t = 0, 1, 2, 3, 4, 5$; the backward dynamic programming algorithm finds the probabilities of the system's transitions from an arbitrary dynamical state $x(\overline{t} - \tau) \in X, \tau = 0, 1, 2, 3, 4, 5$ to a fixed final state $x(5) = 3$. All these probabilities are given in Table 4.2.

Now let us show that the backward dynamic procedure of the calculation of the state probabilities described above can be extended for determining the optimal control $u^*(t)$ in Problem 4.1 with $t_1 = t_2 = \overline{t}$.

Table 4.2 State-time probabilities $P_{x_0}(u(t),\ y,\ \tau)$ and $P_{x(\bar{t}-\tau)}(u(t),\ x_f,\ \bar{t})$, $\tau = 0, 1, 2, 3, 4, 5$

Direct Dynamic Programming Algorithm				Backward Dynamic Programming Algorithm		
τ	x	$x(t) = (x, t)$	$P_{x_0}(u(t), y, \tau)$	$\bar{t} - \tau$	$x(\bar{t} - \tau) = (x, \bar{t} - \tau)$	$P_{x(\bar{t}-\tau)}(u(t), x_f, \bar{t})$
0	1	(1,0)	0	5	(1,5)	0
	2	(2,0)	1		(2,5)	0
	3	(3,0)	0		(3,5)	1
	4	(4,0)	0		(4,5)	0
1	1	(1,1)	0	4	(1,4)	0
	2	(2,1)	0		(2,4)	0,7
	3	(3,1)	1		(3,4)	0,5
	4	(4,1)	0		(4,4)	0,6
2	1	(1,2)	0	3	(1,3)	0,42
	2	(2,2)	0		(2,3)	0,6
	3	(3,2)	1		(3,3)	0,54
	4	(4,2)	0		(4,3)	0,53
3	1	(1,3)	0	2	(1,2)	0,6
	2	(2,3)	0,3		(2,2)	0,468
	3	(3,3)	0		(3,2)	0,551
	4	(4,3)	0,7		(4,2)	0,535
4	1	(1,4)	0	1	(1,1)	0,468
	2	(2,4)	0,15		(2,1)	0,551
	3	(3,4)	0,64		(3,1)	0,551
	4	(4,4)	0,21		(4,1)	0,551
5	1	(1,5)	0,045	0	(1,0)	0,551
	2	(2,5)	0,32		(2,0)	0,551
	3	(3,5)	0,551		(3,0)	0,551
	4	(4,5)	0,084		(4,0)	0,551

Algorithm 4.12 Determining the Optimal Control for Problem 4.2 **with** $\mathbf{t_1 = t_2 = \bar{t}}$

We describe the algorithm for finding the optimal control $u^*(t)$ and the probabilities $P_{x(\bar{t}-\tau)}(u^*(t),\ x_f,\ \bar{t})$ of the systems' transitions from the states $x \in X$ at the moment of times $\bar{t} - \tau$ to the state x_f by using τ units of time for $\tau = 0, 1, 2, \ldots, \bar{t}$. The algorithm consists of the preliminary, the general and the final steps.

The preliminary and the general steps of the algorithm determines the transition probabilities $P_{x(\bar{t}-\tau)}(u^*(t),\ x_f,\ \bar{t})$ from the states $x(\bar{t} - \tau) \in X$ at the moment of time $\bar{t} - \tau$ to the state $x_f = x(\bar{t}) \in X$ at the moment of time $\bar{t}$ if the optimal control $u^*(t)$ is applied. At the end of the last iteration of the general step the algorithm determines the subset of edges E_C^* of E_C which determines the optimal controls. The final step of the algorithm constructs an optimal control $u^*(t)$ of the problem.

Preliminary step (Step 0): Put $P_{z_f}(u^*(t),\ x_f,\ \bar{t}) = 1$ for $z_f = (x_f, \bar{t})$ and $P_z(u^*(t),\ x_f,\ \bar{t}) = 0$ for every $z \in Z \setminus \{z_f\}$. In addition set $E_C^* = \emptyset$.

General step (Step $\tau \geq 1,\ \tau \geq 1$*):* For given τ do the following items a) and b):

(a) For each uncontrollable position $z = (x, \bar{t} - \tau) \in Z^N_{\bar{t}-\tau}$, i.e., $z = x(\bar{t} - \tau)$, calculate

$$P_z(u^*(t),\ x_f,\ \bar{t}) = \sum_{(z,w)\in E(z)} P_w(u^*(t),\ x_f,\ \bar{t})p_{(z,w)},$$

where $E(z) = \left\{(z,\ w) \in E \mid w = x(\bar{t} - \tau + 1) = (x, \bar{t} - \tau + 1) \in Z^N_{\bar{t}-\tau+1}\right\}$;

(b) For each controllable position $z = (x, \bar{t} - \tau) \in Z^C_\tau$ calculate

$$P_z(u^*(t),\ x_f,\ \bar{t}) = \max_{(z,w)\in E(z)} P_w(u^*(t),\ x_f,\ \bar{t})$$

and include in the set E_C^* the edges $e^* = (z,\ w)^* \in E(x, \bar{t} - \tau)$ that satisfy the condition

$$(z,\ w)^* = \underset{(z,w)\in E(z)}{\text{argmax}}\ \{P_w(u^*(t),\ x_f,\ \bar{t})\}.$$

If $\tau = \bar{t}$ then go to Final step; otherwise go to step $\tau + 1$.

Final step: Form the graph $G^* = (Z,\ E_C^* \cup (E \setminus E_C))$ and fix on E_C^* an arbitrary map

$$u^* : (x, t) \to (y, t + 1) \in X_{G^*}(x, t) \ \text{ for } \ (x, t) \in Z^C$$

where $X_{G^*}(x, t) = \{(y, t + 1) \in Z \mid ((x, t),\ (y, t + 1)) \in E_C^*\}$. The map u^* corresponds to an optimal control for Problem 4.2 and $P_{x(0)}(u^*(t),\ x_f,\ \bar{t})$ for $x(0) \in X$ represent the probability transitions of the system from the starting states $x(0) \in X$ to x_f by using $\bar{t}$ units of time if the optimal control is applied.

Theorem 4.13 *Algorithm 4.12 correctly finds the optimal control $u^*(t)$ and the state-time probability $P_{x(0)}(u^*(t),\ x_f,\ \bar{t})$ for an arbitrary starting position $x(0) \in X$ in Problem 4.1 in the case $t_1 = t_2 = \bar{t}$. The running time of the algorithm is $O(|X|^2\bar{t})$.*

Proof The general step of the algorithm reflects the optimality principle of the dynamic programming for the problem of determining the control with maximal probabilities $P_{x(\bar{t}-\tau)}(u^*(t),\ x_f,\ \bar{t})$ of system's transition from the states $x \in X$ at the moment of time $\bar{t} - \tau$ to the state x_f at the moment of time $\bar{t}$.

For each controllable position $(x, \bar{t} - \tau) \in Z$ the values $P_{x(\bar{t}-\tau)}(u^*(t),\ x_f,\ \bar{t})$ are calculated on the time-expanded network in consideration that for a given moment of time $\bar{t} - \tau$ and a given state $x \in X$ the optimal control $u^*(\bar{t} - \tau) \in U_t(x(\bar{t} - \tau))$ is applied. Therefore, the values $P_{(x,\bar{t}-\tau)}(u^*(t),\ (x_f, \bar{t}), \bar{t})$ are calculated correctly for every $x \in X$ and $\tau = 0, 1, 2, \dots, \bar{t}$. In the case $\tau = \bar{t}$ we obtain the state-time probabilities $P_x(u^*(t), x_f, \bar{t}) = P_{(x,0)}(u^*(t), x_f, \bar{t})$ for $x \in X$. At each step of the algorithm an directed edge e^* corresponds to an optimal control and therefore at the final step we obtain the set of edges E^* which determines the optimal controls for an arbitrary state x and at every moment of time $\bar{t} - \tau$. The computational complexity of the algorithm can be estimated in the same way as in Algorithm 4.9. The algorithm needs $\bar{t}$ steps and at each step it uses $O(|X|^2)$ elementary operations. So, the running time of the algorithm is $O(|X|^2\bar{t})$. □

The following example illustrates the calculation procedure of Algorithm 4.12.

Example Consider the problem of determining the optimal control $u^*(t)$ for the example from Sect. 4.2.2 where $\bar{t} = 5$, $x_f = 3$; the corresponding time-expanded network is represented in Fig. 4.2.

If we apply Algorithm 4.12 then we obtain:

Step 0

$P_{(1,5)}(u^*(t), 3, 5) = 0$; $P_{(2,5)}(u^*(t), 3, 5) = 0$; $P_{(3,5)}(u^*(t), 3, 5) = 1$;
$P_{(4,5)}(u^*(t), 3, 5) = 0$; $E_C^* = \emptyset$.

Step 1

$P_{(1,4)}(u^*(t), 3, 5) = 0$; $P_{(2,4)}(u^*(t), 3, 5) = 0.7$;
$P_{(3,4)}(u^*(t), 3, 5) = 0.5$; $P_{(4,4)}(u^*(t), 3, 5) = 0.6$; $E_C^* = \emptyset$.

Step 2

$P_{(1,3)}(u^*(t), 3, 5) = 0.42$; $P_{(2,3)}(u^*(t), 3, 5) = 0.6$;
$P_{(3,3)}(u^*(t), 3, 5) = 0.54$; $P_{(4,3)}(u^*(t), 3, 5) = 0.53$; $E_C^* = \emptyset$.

Step 3

$P_{(1,2)}((u^*(t), 3, 5) = 0.6$; $P_{(2,2)}(u^*(t), 3, 5) = 0.468$;
$P_{(3,2)}(u^*(t), 3, 5) = 0.551$; $P_{(4,2)}(u^*(t), 3, 5) = 0.5035$;

$$E_C^* = \{((1,2), (2,3))\}.$$

Step 4

$P_{(1,1)}(u^*(t), 3, 5) = 0.6$; $P_{(2,1)}(3, 5) = 0.6$;
$P_{(3,1)}(u^*(t), 3, 5) = 0.551$; $P_{(4,1)}(u^*(t), 3, 5) = 0.551$;

$$\begin{aligned} E_C^* =& \{((1,2), (2,3))\} \\ & \cup \{((1,1), (1,2)),\ ((2,1), (1,2)),\ ((3,1), (3,2)),\ ((4,1), (3,2))\}. \end{aligned}$$

Step 5

$P_{(1,0)}(u^*(t), 3, 5) = 0.6$; $P_{(2,0)}(u^*(t), 3, 5) = 0.6$;
$P_{(3,0)}(u^*(t), 3, 5) = 0.551$; $P_{(4,0)}(u^*(t), 3, 5) = 0.551$;

$$\begin{aligned} E_C^* = & \{((1,2), (2,3)),\ ((1,1), (1,2)),\ ((2,1), (1,2)),\ ((3,1), (3,2)), \\ & ((4,1), (3,2))\} \cup \{((1,0), (1,1)),\ ((1,0), (2,1)),\ ((2,0), (1,1)),\ ((2,0), \\ & (2,1)),\ ((3,0), (3,1)),\ ((3,0), (4,1)),\ ((4,0), (3,1)),\ ((4,0), (4,1))\}. \end{aligned}$$

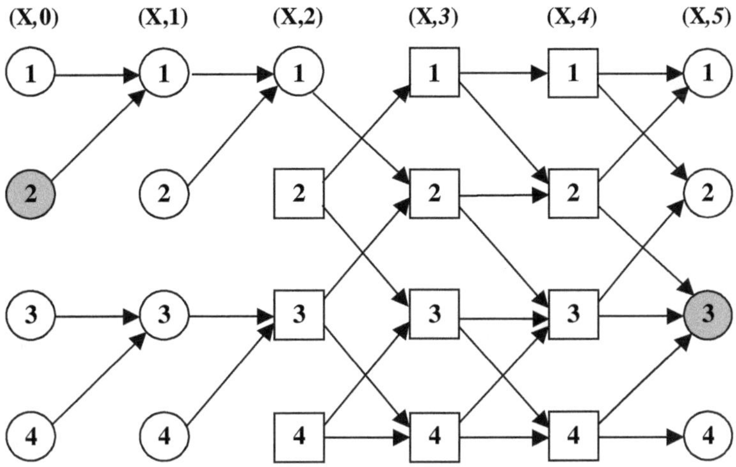

Fig. 4.4 The network which corresponds to the optimal control u^*

After steps 1–5 we fix the map

$$u^* : (1,0) \to (1,1); \quad (2,0) \to (1,1); \quad (3,0) \to (3,1); \quad (4,0) \to (3,1);$$

$$(1,1) \to (1,2); \quad (2,1) \to (1,2); \quad (3,1) \to (3,2); \quad (4,1) \to (3,2);$$

$$(1,2) \to (2,3).$$

which determines the optimal control. The network which corresponds to this optimal control is represented in Fig. 4.4.

4.2.4 Algorithms for Determining the State-Time Probabilities and the Optimal Control for the Problem with the Number of Transitions from a Given Interval

Now we consider the problems of determining the state-time probabilities of the dynamical system in the case $t_1 \neq t_2$. We show how to solve this problem applying the time-expanded network method. In this case we shall use the time-expanded network $(G, Z^C, Z^N, c, p, \overline{t}, z_0, z_f)$ with $\overline{t} = t_2$ and we will make some minor modification in its construction. We delete all directed edges originating in the vertices (x_f, t) for $t = t_1, t_1 + 1, \ldots, t_2 - 1$ preserving all remaining edges of the network. The structure of the time-expanded network after this additional transformation is represented in Fig. 4.5. We denote the time-expanded network in this case by $(G^0, Z^C, Z^N, c, p, t_1, t_2, z_0, Y)$, where $Y = \{(x_f, t_1), (x_f, t_1+1), \ldots, (x_f, t_2)\}$ and $G^0 = (Z, E^0)$ is the graph obtained from G by deleting all edges which originate in vertices from Y, i.e., $E^0 = E \backslash \{(z, w) \in E \mid z \in Y\}$.

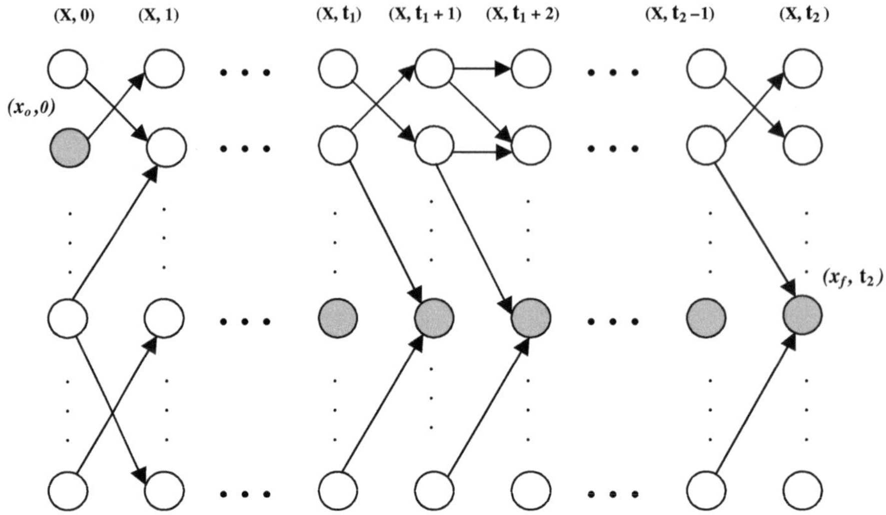

Fig. 4.5 The time-expanded network for the problem with $t_1 \neq t_2$

Based on the construction above we obtain the following property of the probability $P_{x_0}(u(t),\ x_f,\ t_1 \leq t(x_f) \leq t_2)$.

Lemma 4.14 *For an arbitrary feasible control* $u(t)$ *and a given starting state* x_0 *of the dynamical system* $\mathbb{L}$ *the following formula holds*

$$P_{x_0}(u(t),\ x_f,\ t_1 \leq t(x_f) \leq t_2) = \sum_{k=0}^{t_2-t_1} P_{z_0}(u(t),\ (x_f, t_1+k),\ t_1+k), \qquad (4.6)$$

where $P_{z_0}(u(t),\ (x_f, t_1+k),\ t_1+k),\ k = 0, 1, 2, \ldots, t_2 - t_1$ *represents the time-state probabilities calculated on the time-expanded network* $(G^0, Z^C, Z^N, c, p, t_1, t_2, z_0, Y)$.

Based on Lemma 4.14 we can calculate $P_{x_0}(u(t),\ x_f,\ t_1 \leq t(x_f) \leq t_2)$ as follows. We apply Algorithm 4.7 on the time-expanded network $(G^0, Z^C, Z^N, c, p^u, t_1, t_2, z_0, Y)$ and determine the state-time probabilities $P_{z_0}(u(t),\ (x, \tau),\ \tau)$ for every $(x, \tau) \in Z$ and $\tau = 0, 1, 2, \ldots, t_2$. Then we find the probability $P_{x_0}(u(t),\ x_f,\ t_1 \leq t(x_f) \leq t_2)$ using formula (4.6).

Example Let us calculate the probability $P_2(u(t), 3 \leq t(3) \leq 5)$ with the control $u(t)$ given in the example from Sect. 4.2.2. The corresponding network induced by the control $u(t)$ is represented in Fig 4.6. If we apply formula (4.6) then we obtain

$$\begin{aligned} &P_2(u(t),\ 3,\ 3 \leq t(3) \leq 5) \\ &\quad = P_2(u(t),\ (3,3),\ 3)) + P_2(u(t),\ (3,4),\ 4) + P_2(u(t),\ (3,5),\ 5), \end{aligned}$$

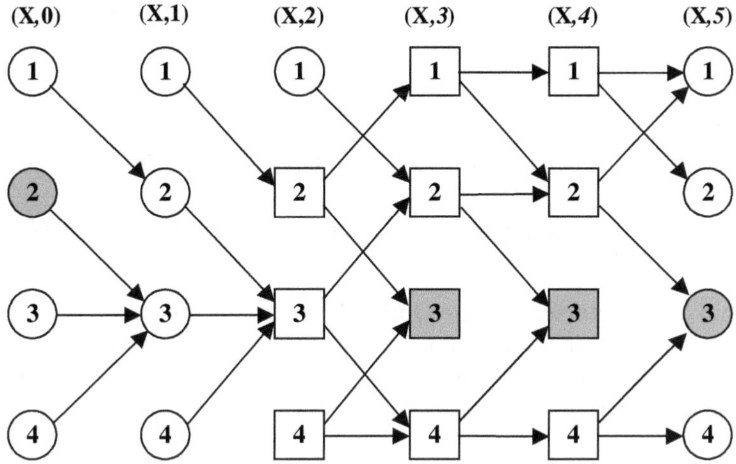

Fig. 4.6 The network induced by the control $u(t)$

where $P_{(2,0)}(u(t), (3,3), 3))$, $P_{(2,0)}(u(t), (3,4), 4)$ and $P_{(2,0)}(u(t), (3,5), 5)$ are calculated on the time-expanded network given in Fig. 4.6. So,

$$P_{(2,0)}(u(t), 3, 3 \le t(3) \le 5) = 0 + 0.64 + 0.231 = 0.871 .$$

The state-time probabilities $P_{(2,0)}(u(t), (3,3), 3)$, $P_{(2,0)}(u(t), (3,4), 4)$, and $P_{(2,0)}(u(t), (3,5), 5)$ on the auxiliary network can be calculated applying Algorithm 4.7 or Algorithm 4.9. We obtain a more suitable algorithm for the tabulation of the values $P_{x(0)}(u(t), x_f, t_1 \le t(x_f) \le t_2)$ for a fixed control in Problem 4.1 if we use the backward induction principle on the time-expanded network. Below we describe such an algorithm.

Algorithm 4.15 Determining the State-Time Probabilities for the Problem 4.1 **with $t_1 \neq t_2$ based on a Backward Induction Principle**

Preliminary step (Step 0): Put $P_{(x_f, t_2-\tau)}(u(t), x_f, t_1 \le t(x_f) \le t_2) = 1$, for $\tau = 0, 1, 2, \dots, t_2 - t_1$ and set $P_{(x, t_2-\tau)}(u(t), x_f, t_1 \le t(x_f) \le t_2) = 0$, for $(x, t_2 - \tau) \in Z \backslash Y$, $t = 0, 1, 2, \dots, t_2$.

General step (Step τ, $\tau \ge 1$): For every $(x, t_2 - \tau) = z \in Z^N_{t_2-\tau} \backslash Y$ calculate

$$\begin{aligned} &P_{(x,t_2-\tau)}(u(t), x_f, t_1 \le t(x_f) \le t_2) \\ &\quad = \sum_{(z,w)\in E^0(z)} P_w(u(t), x_f, t_1 \le t(x_f) \le t_2) p^u_{(z,w)}, \end{aligned}$$

where $w = (y, t_2 - \tau + 1) \in Z_{t_2-\tau+1}$, $E^0(z) = \{(z, w) \in E^0 \mid w \in Z_{\tau+1}\}$.

If $\tau = t_2$ then go to the final step; otherwise go to step $\tau + 1$.

Final step: Set $P_{x(t_2-\tau)}(u(t),\ x_f,\ t_1 \le t(x_f) \le t_2) = P_{(x,t_2-\tau)}(u(t),\ x_f,\ t_1 \le t(x_f) \le t_2)$ for every $(x, t_2-\tau) = z \in Z_{t_2-\tau}$, $\tau = 0, 1, 2, \dots, t_2$ and fix $P_{x(0)}(u(t),\ x_f,\ t_1 \le t(x_f) \le t_2) = P_{(x,0)}(u(t),\ x_f,\ t_1 \le t(x_f) \le t_2)$ for $x(0) \in X$.

It easy to check that in the worst case the algorithm uses $O(|X|^2 t_2)$ elementary operations.

The backward dynamic programming algorithm described above can be extended for the control Problem 4.2.

Algorithm 4.16 Determining the Optimal Control for Problem 4.2
The algorithm determines the optimal control $u*(t)$ and the probabilities $P_{x(t_2-\tau)}(u^*(t),\ x_f,\ t_1 \le t(x_f) \le t_2)$ for $\tau = 0, 1, 2, \dots, t_2$. These probabilities are tabulated using the time-expanded network $(G^0,\ Z^C,\ Z^N,\ c,\ p,\ t_1, t_2,\ z_0,\ Y)$ in the following way:

Preliminary step (Step 0): Put $P_{(x_f,t_2-\tau)}(u^*(t),\ x_f,\ t_1 \le t(x_f) \le t_2) = 1$, for $\tau = 0, 1, 2, \dots, t_2 - t_1$ and set $P_{(x,t_2-\tau)}(u^*(t),\ x_f,\ t_1 \le t(x_f) \le t_2) = 0$, for $(x, t_2-\tau) \in Z \setminus Y$, $t = 0, 1, 2, \dots, t_2$; in addition set $E_C^* = \emptyset$.

General step (Step τ, $\tau \ge 1$): For given τ do items (a) and (b):

(a) For every $(x, t_2-\tau) = z \in Z^N_{t_2-\tau} \setminus Y$ calculate

$$P_{(x,t_2-\tau)}(u^*(t),\ x_f,\ t_1 \le t(x_f) \le t_2) = \sum_{(z,w)\in E(z)} P_w(u^*(t),\ x_f,\ t_1 \le t(x_f) \le t_2)\, p_{(z,w)},$$

where $w = (y, t_2-\tau+1) \in Z_{t_2-\tau+1}$;

(b) For every $(x, t_2-\tau) = z \in Z^C_\tau \setminus Y$ calculate

$$P_{(x,t_2-\tau)}(u^*(t), x_f, t_1 \le t(x_f) \le t_2) = \max_{(z,w)\in E(z)} P_w(u^*(t), x_f, t_1 \le t(x_f) \le t_2)$$

and include in E_C^* the edges $e^* = (z, w)^*$ for which

$$(z, w)^* = \operatorname*{argmax}_{(z,w)\in E(z)} \{P_w(u^*(t), x_f, t_1 \le t(x_f) \le t_2)\}.$$

If $\tau = t_2$ then go to the final step; otherwise go to step $\tau + 1$.

Final step: Form the graph $G^* = (Z,\ E_C^* \cup (E^* \setminus E_C))$ and fix on E_C^* an arbitrary map

$$u^* : (x, t) \to (y,\ t+1) \in X_{G^*}(x, t) \quad \text{for} \quad (x, t) \in Z^C$$

where $X_{G^*} = \{(y,\ t+1) \in Z \mid ((x, t),\ (y, t+1)) \in E^*\}$.

4.2.5 Determining the State-Time Probabilities and the Optimal Control Using a Modified Time-Expanded Network

The algorithm from the previous section can be modified using a simple modification of the time-expanded network that allows us to reduce Problem 4.1 with $t_1 \neq t_2$ to Problem 4.1 with fixed number of stages $\overline{t}$ were $\overline{t} = t_2$. We ground this reduction procedure using a new time-expanded network which we denote by $(G^+,\ Z^C,\ Z^N,\ c^+,\ p^{u+},\ t_2,\ z_0,\ z_f)$. This network is obtained from $(G^0,\ Z^C,\ Z^N,\ c,\ p^u,\ t_1,\ t_2,\ z_0,\ Y)$ in the following way:

Graph $G^+ = (Z, E^+)$ is obtained from $G^0 = (Z, E^0)$ by adding the directed edges

$$((x_f, t_1), (x_f, t_1+1)), ((x_f, t_1+1), (x_f, t_1+2)), \dots, ((x_f, t_2-1), (x_f, t_2)),$$

i.e.,

$$E^+ = E \cup \{((x_f, t_1+i),\ (x_f, t_1+i+1)),\quad i = 0, 1, 2, \dots, t_2 - t_1 - 1\}.$$

To each directed edge $e_i = ((x_f, t_1+i),\ (x_f,\ t_1+i+1)), i = 0, 1, 2, \dots, t_2-t_1-1$ we associate the transition probability $p_{(e_i)} = 1$ and the transition cost $c^*_{(e_i)} = 0$; for the rest of the edges in G^* we preserve the same probabilities and the same costs as in G^0. We denote the network obtained after this construction by $(G^+,\ Z^C,\ Z^N,\ c^+,\ p^{u+},\ t_2,\ z_f)$. The graphical representation of this network is given in Fig. 4.7.

It is easy to observe that if for a fixed control $u(t)$ we consider the problem of determining the probabilities $P_{(x_f,t_2-\tau)}(u(t),\ (x_f, t_2),\ t_2)$ on the network $(G^+,\ Z^C,\ Z^N,\ c^+,\ p^{u+},\ t_2,\ z_f)$ then

$$P_{(x_f,t_2-\tau)}(u(t),\ (x_f, t_2),\ t_2) = P_{(x_f,t_2-\tau)}(u(t),\ Y,\ t_1 \le t(Y) \le t_2).$$

These probabilities can be found by using Algorithm 4.9 on the network $(G^+,\ Z^C,\ Z^N, c^+,\ p^{u+},\ T_2,\ z_f)$. Thus, Algorithm 4.15 on the network $(G^0,\ Z^C,\ Z^N,\ c,\ p^u, t_1, t_2,\ Y)$ can be replaced by Algorithm 4.9 on the network $(G^+,\ Z^C,\ Z^N,\ c^+, p^{u+}, t_2,\ z_f)$.

Example 1 Consider the problem of calculation the probability $P_2(u(t), 3,\ 3 \le t(3) \le 5)$ for the example from Sect. 4.2.4. We can calculate this probability using the modified network represented on Fig. 4.8, because

$$P_2(u(t),\ 3,\ 3 \le T(3) \le 5) = P_{(2,0)}(u(t),\ (3,5),\ 5),$$

where $P_{(2,0)}(u(t),\ (3,5),\ 5)$ is calculated on this network.

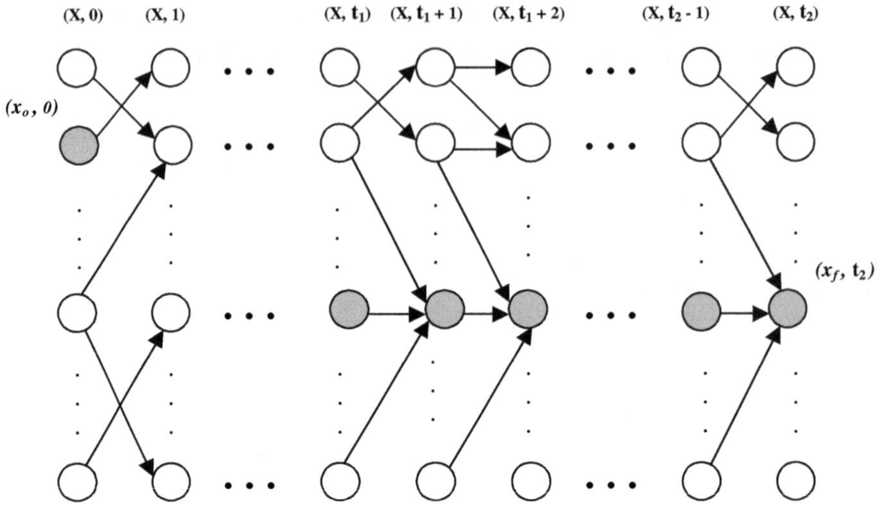

Fig. 4.7 The graphical representation of modified time-expanded network

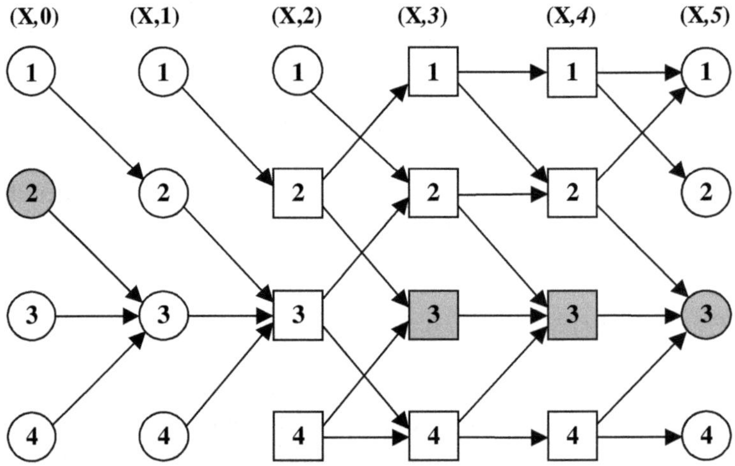

Fig. 4.8 The network which corresponds to the control $u(t)$

The modified network is obtained from the network on Fig. 4.6 by adding directed edges $((3,3),(3,4))$, $((3,4),(3,5))$ with the probabilities $p_{((3,3),(3,4))} = p_{((3,4),(3,5))} = 1$. If we apply Algorithm 4.7 or Algorithm 4.9 then we obtain

$$P_{(2,0)}(u(t),\ (3,5),\ 5) = 0.871\,.$$

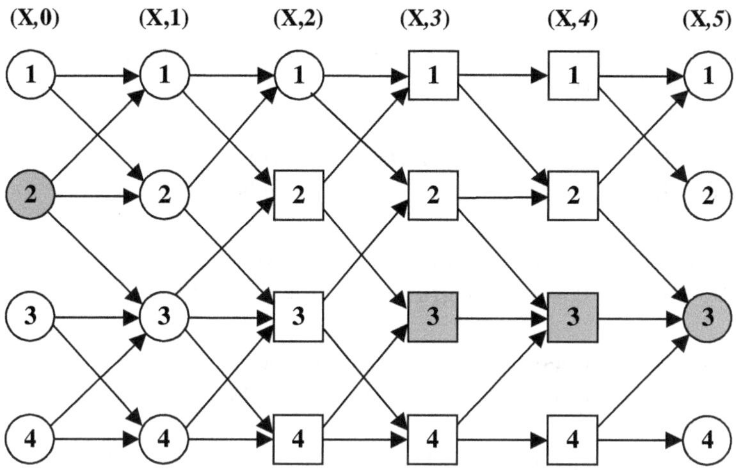

Fig. 4.9 The time-expanded network for the control problem

Algorithm 4.17 Determining the Optimal Control for Problem 4.2 **with $t_1 \neq t_2$**
The optimal control $u^*(t)$ and the transition probabilities $P_{x(t_2-\tau)}(u^*(t), x_f, t_1 \leq t(x_f) \leq t_2)$ for Problem 4.2 can be found using Algorithm 4.12 on the time-expanded network $(G^+, Z^C, Z^N, c^+, p^{u+}, t_2, z_0, z_f)$, i.e. we determine $u^*(t)$ and $P_{x(t_2-\tau)}(u^*(t), x_f, t_1 \leq t(x_f) \leq t_2)$ as follows:

(1) Construct the time-expanded network $(G^+, Z^C, Z^N, c^+, p^{u+}, t_2, z_0, z_f)$;
(2) Find the optimal control $u^*(t)$ and the transition probabilities $P_{x(t_2-\tau)}(u^*(t), x_f, t_2)$ for $x(t_2 - \tau) \in X$, $\tau = 0, 1, 2, \ldots, t_2$ using Algorithm 4.12.
(3) Fix the optimal control $u^*(t)$ and the transition probabilities $P_{x(t_2-\tau)}(u^*(t), x_f, t_1 \leq t(x_f) \leq t_2) = P_{x(t_2-\tau)}(u^*(t), x_f, t_2)$ for $x(t_2 - \tau) \in X$, $\tau = 0, 1, 2, \ldots, t_2$.

Example 2 Consider the problem of determining the optimal control $u^*(t)$ and the state-time probabilities $P_{x(5-\tau)}(u^*(t), 3, 3 \leq t(3) \leq 5)$, $\tau = 0, 1, 2, 3, 4, 5$ in Problem 4.2 for the example from Sect. 4.2.2.

The time-expanded network for this example is represented in Fig. 4.9. In this network $p_{((3,3),(3,4))} = p_{((3,4),(3,5))} = 1$; the rest of the probability transitions are the same as in the example from Sect. 4.2.2.

We determine the optimal control $u^*(t)$ and the probabilities $P_{(x,t_2-\tau)}(u^*(t), 3, 5)$ on the network $(G^+, Z^C, Z^N, c^+, p^{u+}, t_2, z_0, z_f)$ using Algorithm 4.12:

Step 0

$P_{(1,5)}(u^*(t), 3, 5) = 0;$ $P_{(2,5)}(u^*(t), 3, 5) = 0;$ $P_{(3,5)}(u^*(t), 3, 5) = 1;$
$P_{(4,5)}(u^*(t), 3, 5) = 0;$ $E_C^* = \emptyset.$

Step 1
$P_{(1,4)}(u^*(t), 3, 5) = 0; \quad P_{(2,4)}(u^*(t), 3, 5) = 0.7; \quad P_{(3,4)}(u^*(t), 3, 5) = 1;$
$P_{(4,4)}(u^*(t), 3, 5) = 0.6; \quad E_C^* = \emptyset.$

Step 2
$P_{(1,3)}(u^*(t), 3, 5) = 0.42; \quad P_{(2,3)}(u^*(t), 3, 5) = 0.85;$
$P_{(3,3)}(u^*(t), 3, 5) = 1; \quad P_{(4,3)}(u^*(t), 3, 5) = 0.88; \quad E_C^* = \emptyset.$

Step 3
$P_{(1,2)}(u^*(t), 3, 5) = 0, 85; \quad P_{(2,2)}(u^*(t), 3, 5) = 0.652;$
$P_{(3,2)}(u^*(t), 3, 5) = 0.871; \quad P_{(4,2)}(u^*(t), 3, 5) = 0.94; \quad E_C^* = \{((1, 2), (2, 3))\}.$

Step 4
$P_{(1,1)}(u^*(t), 3, 5) = 0.85; \quad P_{(2,1)}(u^*(t), 3, 5) = 0.871;$
$P_{(3,1)}(u^*(t), 3, 5) = 0.94; \quad P_{(4,1)}(3, 5) = 0.85;$

$$E_C^* = \{((1, 2), (2, 3))\} \cup \{((1, 1), (1, 2)), \ ((2, 1), (3, 2)), \ ((3, 1), (4, 2))\}.$$

Step 5
$P_{(1,0)}(u^*(t), 3, 5) = 0.871; \quad P_{(2,0)}(u^*(t), 3, 5) = 0.94;$
$P_{(3,0)}(u^*(t), 3, 5) = 0.94; \quad P_{(4,0)}(u^*(t), 3, 5) = 0.94;$

$$\begin{aligned} E_C^* = \{&((1, 2), (2, 3)), \ ((1, 1), (1, 2)), \ ((2, 1), (3, 2)), \ ((3, 1), (4, 2))\} \\ &\cup\{((1, 0), (1, 1)), \ ((1, 0), (2, 1)), \ ((2, 0), (3, 1)), \ ((3, 0), (3, 1)), \\ &((3, 0), (4, 1)), \ ((4, 0), (3, 1)), \ ((4, 0), (4.1))\}. \end{aligned}$$

After steps 0–5 we fix the map

$$\begin{aligned} u^*\colon &(1, 0) \to (2, 1); \quad (2, 0) \to (3, 1); \quad (3, 0) \to (3, 1); \quad (4, 0) \to (4, 1); \\ &(1, 1) \to (1, 2); \quad (2, 1) \to (3, 2); \quad (3, 1) \to (4, 2); \quad (4, 1) \to (4, 2); \\ &(1, 2) \to (2, 3) \end{aligned}$$

which determines the optimal control. In addition we fix

$$P_{(1,0)}(u^*(t), \ 3, \ 3 \le t(3) \le 5) = P_{(1,0)}(u^*(t), \ 3, \ 5) = 0.871;$$
$$P_{(2,0)}(u^*(t), \ 3, \ 3 \le t(3) \le 5) = P_{(2,0)}(u^*(t), \ 3, \ 5) = 0.94;$$
$$P_{(3,0)}(u^*(t), \ 3, \ 3 \le t(3) \le 5) = P_{(3,0)}(u^*(t), \ 3, \ 5) = 0.94;$$
$$P_{(4,0)}(u^*(t), \ 3, \ 3 \le t(3) \le 5) = P_{(4,0)}(u^*(t), \ 3, \ 5) = 0.94.$$

The network which corresponds to the optimal control $u^*(t)$ is represented in Fig. 4.10.

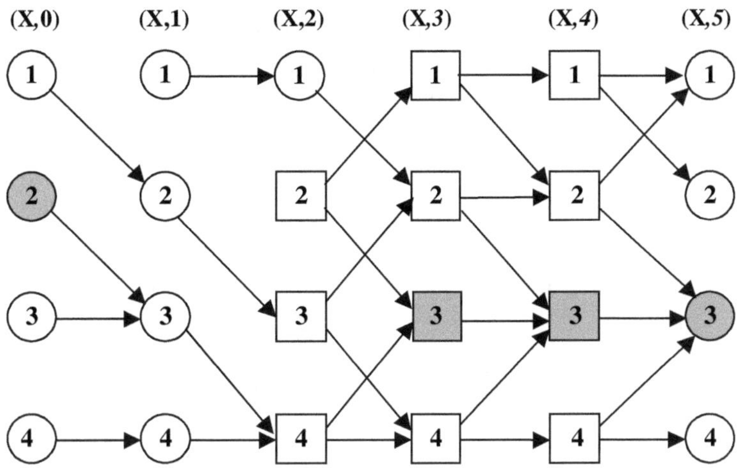

Fig. 4.10 The network which corresponds to the optimal control $u^*(t)$

4.3 Algorithms for Determining the Expected Total Cost and the Optimal Control Using a Time-Expanded Network

In this section we describe dynamic programming algorithms for the calculation of the expected total cost of the dynamical system in Problems 4.3–4.6. For the calculation of these values we shall use the time-expanded networks from the previous section.

4.3.1 Determining the Expected Total Cost and the Optimal Control in the Problems with a Fixed Number of Transitions

To develop a backward dynamic programming algorithm for the calculation of the expected total cost $\sigma_{x(0)}(u(t), \overline{t})$ in problems 4.3–4.6 we introduce the values $\sigma_{x(\overline{t}-\tau)}(u(t), \overline{t})$ for $\tau = 0, 1, 2 \ldots, \overline{t}$.

The value $\sigma_{x(\overline{t}-\tau)}(u(t), \overline{t})$ for a fixed τ expresses the cost to be expected during τ transitions if the system starts transitions in a state x at the moment of time $\overline{t}-\tau$ and finish transitions at the moment of time $\overline{t}$. It is evident that the following recursive formula holds:

$$\sigma_{x(\overline{t}-\tau)}(u(t), \overline{t}) = \sum_{x(\overline{t}-\tau+1)\in X} p_{x(\overline{t}-\tau),x(\overline{t}-\tau+1)} \left(c_{x(\overline{t}-\tau),x(\overline{t}-\tau+1)} + \sigma_{x(\overline{t}-\tau+1)}(u(t), \overline{t})\right).$$

If we set $\sigma_{x(\overline{t})}(u(t), \overline{t}) = 0$ and apply this formula for $\tau = 1, 2, \ldots, \overline{t}$ then we determine $\sigma_{x(0)}(u(t), \overline{t})$. For the time-expanded network $(G, Z^C, Z^N, c, p^u, \overline{t}, z_0)$ this recursive formula can be written as follows

$$\sigma_z(u(t), \overline{t}) = \sum_{(z,w)\in E(z)} p^u_{(z,w)}(c_{(z,w)} + \sigma_w(u(t), \overline{t})),$$
$$z \in Z_{\overline{t}-\tau}, \quad \tau = 1, 2, \ldots, \overline{t},$$

where $\sigma_z(u(t), \overline{t}) = 0$ for $z \in Z_T$; $E(z) = \{(z, w) \in E \mid w \in Z_{\overline{t}-\tau+1}\}$. Here $z = (x, \overline{t} - \tau) \in Z_{\overline{t}-\tau}$, $w = (x, \overline{t} - \tau + 1) \in Z_{\overline{t}-\tau+1}$ and the value

$$\sigma_z(u(t), \overline{t}) = \sigma_{(x,\overline{t}-\tau)}(u(t), \overline{t})$$

for a given control $u(t)$ expresses the expected total cost during τ transitions on G if the system starts transitions in $z = (x, \overline{t} - \tau)$ at the moment of time $\overline{t} - \tau$ and finish transitions at the moment of time $\overline{t}$ in a position $(y, \overline{t}) \in Z_{\overline{t}}$.

Thus, the values $\sigma_z(u(t), \overline{t}) = \sigma_{x(\overline{t}-\tau)}(u(t), \overline{t})$ for $\tau = 0, 1, 2, \ldots, \overline{t}$ can be tabulated using the backward dynamic programming algorithm. A similar procedure can be used for the tabulation of the values for the variance $D_{x(\overline{t}-\tau)}(u(t), \overline{t})$.

In the following we show how to determine the optimal control and the optimal value of the expected total cost for Problem 4.4 on the time-expanded network $(G, Z^C, Z^N, c, \ p, \overline{t}, z_0)$.

Algorithm 4.18 Determining the Optimal Control for Problem 4.4

The algorithm consists of the preliminary, the general and the final steps. At the preliminary and general steps the algorithm determines the optimal values of the expected total costs $\sigma_{(x,\overline{t}-\tau)}(u^*(t), \overline{t})$ of the system during τ transitions from the states $x(\overline{t} - \tau) = (x, \overline{t} - \tau)$ at the moment of time $\overline{t} - \tau$ to the final state x_f if the optimal control $u^*(t)$ is applied. The final step of the algorithm determines an optimal control for Problem 4.4.

Preliminary step (Step 0): Put $\sigma_{(x,\overline{t}-\tau)}(u^*(t), \overline{t}) = 0$ for every $(x, \overline{t}) \in Z_{\overline{t}}$ and set $E^*_C = \emptyset$;

General step (Step τ, $\tau \geq 1$): For given τ do items *a*) and *b*):

(a) For each uncontrollable position $z = (x, \overline{t} - \tau) \in Z^N_{\overline{t}-\tau}$ calculate

$$\sigma_z(u^*(t), \overline{t}) = \sum_{(z,w)\in E(z)} p_{(z,w)}(c_{(z,w)} + \sigma_w(u^*(t), \overline{t}));$$

(b) For each controllable position $z \in Z^C_{\overline{t}-\tau}$ calculate

$$\sigma_z(\overline{t}) = \min_{(z,w)\in E(z)} (c_{(z,w)} + \sigma_w(u^*(t), \overline{t}))$$

and include in the set E^*_G the edges $e^* = (z, w)^*$ for which

$$(z, w)^* = \underset{(z,w)\in E(z)}{\operatorname{argmin}} \{c_{(z,w)} + \sigma_w(u^*(t), \bar{t})\}.$$

If $\tau = T$ then go to the final step; otherwise go to step $\tau + 1$.

Final step: Form the graph $G^* = (Z,\ E^*_G \cup (E \backslash E_C))$ and fix in G^* an arbitrary map

$$u^* : (x, t) \to (y, t + 1) \in X_{G^*}(x, t) \quad \text{for} \quad (x, t) \in Z^C.$$

Algorithm 4.18 uses the backward induction principle and therefore it finds correctly the optimal control $u^*(t)$ and the expected total costs $\sigma_{x(0)}(\bar{t})$ arbitrary starting position $x = x(0) \in X$ in Problem 4.4. The running time of the algorithm is $O(|X|^2\bar{t})$.

4.3.2 Determining the Expected Total Cost and the Optimal Control with a Given Number of Transitions and a Fixed Final State

In this section we describe an algorithm for the calculation of the expected total cost $\sigma_{x_0}(u(t),\ x_f,\ \bar{t}_1 \le t(x_f) \le \bar{t}_2)$ in Problem 4.5. This value expresses the cost to be expected if the system starts a transition in the state x_0 and the final state x_f is reached at the moment of time $t(x)$ such that $t_1 \le t(x) \le t_2$.

At first we show how to calculate $\sigma_{x_0}(u(t),\ x_f, \bar{t}_1 \le t(x_f) \le \bar{t}_2)$ for a given control $u(t)$ in the case $t_1 = t_2 = \bar{t}$.

We construct the time-expanded network $(G,\ Z^C,\ Z^N,\ z_0,\ z_f,\ c,\ p^u,\ \bar{t})$ and assume that $P_{x_0}(u(t),\ x_f,\ \bar{t}) \neq 0$ (if $P_{x_0}(u(t),\ x_f,\ \bar{t}) = 0$ then we set $\sigma_{x_0}(u(t),\ x_f,\ \bar{t}) = 0$).

If $P_{x_0}(u(t),\ x_f,\ \bar{t}) = 1$ then $P_{z_0}(u(t),\ z_f,\ \bar{t}) = 1$ and therefore the expected total cost $\sigma_{z_0}(u(t),\ z_f,\ \bar{t})$ can be calculated as the expected total cost $\sigma'_{x_0}(u(t),\ x_f,\ \bar{t})$ of the system $\mathbb{L}$ during next $\bar{t}$ transitions if it starts transitions in x_0, i.e., $\sigma_{z_0}(u(t),\ z_f,\ \bar{t}) = \sigma_{z_0}(u(t), \bar{t})$.

If $0 < P_{x_0}(u(t),\ x_f,\ \bar{t}) < 1$ then we will calculate the expected total cost $\sigma_{z_0}(u(t),\ z_f,\ \bar{t})$ with the condition that after $\bar{t}$ transitions the final state x_f is reached. This means that the probabilities p^u_e on the edges $e \in E$ of the time-expanded network should be redefined or transformed in such way that the probability of the system's transition from the starting position z_0 to final position z_f during $\bar{t}$ transitions is equal to 1. We denote these redefined values by p'_e and call them the conditional probabilities. The conditional probabilities $p'_e,\ e \in E$ provide a transfer of the dynamical system $\mathbb{L}$ from the starting state x_0 to the final state x_f at the time moment $\bar{t}$ with the probability $P'_{x_0}(u(t),\ x_f,\ \bar{t}) = 1$. It is evident that if the system couldn't pass through a directed edge $e \in E$ during transitions from $(x, 0)$ to z_f then the conditional probability p'_e of this edge is equal zero. So, the subnetwork of the time-expanded network, generated by edges with nonzero probabilities p'_e has the structure of a directed graph with a sink vertex z_f. Thus, if from the graph $G = (Z, E)$ we delete all directed edges which do not belong to a directed path from z_0 to z_f then we obtain a new graph $G' = (Z, E')$. It is evident that in the obtained new network for some positions $z \in Z$ the condition $\sum_{(z,w)\in E'(z)} p_{(z,w)} = 1$ is not satisfied (here $E'(z)$ represents the subset of edges from E' which originate in the vertex z, i.e. $E'(z) = \{(z, w) \mid (z, w) \in E'\}$). Therefore for the new network we define the conditional probabilities p'_e in such way that the condition $\sum_{(z,w)\in E'(z)} p_{(z,w)'} = 1$ holds for an arbitrary vertex that contains at least one leaving directed edge. More precisely, we define the probabilities p'_e for $e \in E'$ in the following way:

For each position $z \in Z$ we determine the value $\theta(z) = \sum_{(z,w)\in E'(z)} p^u_{(z,w)}$ and after that for an arbitrary position $z \in Z$ with $\theta(z) \neq 0$ we make for every $(z, w) \in E'(z)$ the transformation

$$p'_{(z,w)} = \frac{1}{\theta(z)} p^u_{(z,w)}.$$

After these transformations we can apply Algorithm 4.18 on a stochastic network $(G', Z^C, Z^N, z_0, c, p', \bar{t})$ with conditional probabilities p'_e on edges $e \in E$, where $p'_e = 0$ for $e \in E\backslash E'$. If for this network we determine $\sigma'_{z0}(u(t), \bar{t})$ then we determine $\sigma_{x_0}(u(t), x_f, \bar{t}) = \sigma'_{z0}(u(t), \bar{t})$, i.e., this value represents the expected total cost for the dynamical system in Problem 4.5 if it starts transition in x_0 in the case $t_1 = t_2 = t$. It is easy to observe that the expected total cost $\sigma_{z_0}(u(t), z_f, \bar{t})$ defined and calculated in such a way satisfies the condition

$$C^{\min}(z_0, z_f) \leq \sigma_{z_0}(u(t), z_f, \bar{t}) \leq C^{max}(z_0, z_f),$$

where $C^{\min}(z_0, z_f)$ and $C^{\max}(z_0, z_f)$ represent, respectively, the minimal and the maximal total costs of the directed paths from z_0 to z_f in the time-expanded network.

In the case $t_1 \neq t_2$ the expected total cost $\sigma_{x_0}(u(t), x_f, t_1 \leq t(x) \leq t_2)$ can be defined and calculated in an analogues way if we consider Problem 4.5 on a stochastic network $(G^+, Z^C, Z^N, c^+, p^+, t_2, z_0, z_f)$ with a fixed number of stages $\bar{t} = t_2$ and if we make the similar transformations of the probability on the edges of this network. After these transformations we obtain the time-expanded network with new probabilities p'_e, $e \in X$ for which the expected total cost $\sigma_{z_0}(u(t), z_f, T_2) = \sigma'_{z0}(u(t), t_2)$ can be defined. Using the network $(G^+, Z^C, Z^N, c^+, p^+, t_2, z_0, z_f)$ we show that Algorithm 4.18 can be modified in such a way that it finds simultaneously the probabilities p'_e and the expected total costs $\sigma_{x(t_2-\tau)}(u(t), x_f, t_1 \leq t(x_f) \leq t_2) = \sigma_{x(t_2-\tau)}(u(t), x_f, t_2)$ for $t = 0, 1, 2, \ldots, t_2$.

Algorithm 4.19 Determining the Expected Total Cost and the Conditional probabilities for Problem 4.5

Preliminary step (Step 0): Put $\theta(z_f) = \theta'(z_f) = 1$ and $\theta(z) = \theta'(z) = 0$ for $z \in Z_{t_2}\backslash\{z_f\}$; in addition fix $\sigma'_z(u(t), z_f) = 0$ for every $z \in Z$.

General step (Step τ, $\tau \geq 1$): For each position $(x, t_2 - \tau) \in Z_{t_2-\tau}$ find

$$\theta(z) = \sum_{(z,w)\in E^*(z)} p^u_{(z,w)}\theta'(w),$$

where $E^+(z) = \{e = (z, w) \mid (z, w) \in E^+\}$, i.e., $E^+(z)$ represents the subset of directed edges from E^+ which originate in z. Then calculate

$$p'_{(z,w)} = \frac{1}{\theta(z)} p^u_{(z,w)}\theta'(w)$$

for every $(z, w) \in E^*(z)$ and put

$$\theta'(z) = \sum_{w\in E^*(z)} p'_{(z,w)}$$

for $z \in Z_{t_2-\tau}$.

After that for every $(x, t_2 - \tau) \in Z_{t_2-\tau}$ calculate

$$\sigma'_{(x,t_2-\tau)}(u(t),\ t_2) = \sum_{(z,w)\in E'(z)} (c_{(z,w)} + \sigma'_w(u(t),\ t_2))p'_{(z,w)},$$

where $w = (y,\ t_2 - \tau + 1) \in Z_{t_2-\tau+1}$.

Check if $\tau < t_2$? If $\tau < t_2$ then go to the next step; otherwise fix

$$\sigma_{x(0)}(u(t),\ x_f,\ t_1 \le t(x_f) \le t_2) = \sigma_{(x,0)}(u(t), t_2)$$

for every $x \in X$ and STOP.

Remark 4.20 In Algorithm 4.19 the values $\theta(z)$ and $\theta(z)'$ for $z \in Z$ satisfy the conditions $0 \le \theta(z) \le 1,\quad \theta(z)' \in \{0, 1\}$. The value $\theta(z)'$ in the algorithm is equal to 1 if and only if the vertex z in G' contains at least one leaving directed edge.

Remark 4.21 In the case $t_1 = t_2 = \bar{t}$ the algorithm determines the expected total cost $\sigma_{x(0)}(u'^*(t),\ x_f,\ \bar{t})$ for the problem with a fixed number of stages.

Example Consider the problem of determining the expectation $\sigma_{x_0}(u(t),\ x_f,\ t_1 \le t(x_f) \le t_2)$ for the example from Sect. 4.2.2 with a given control $u(t)$, where $x_f = 3,\ t_1 = 3,\ t_2 = 5$.

We reduce the considered problem to the problem with a fixed number of stages on the auxiliary network represented in Fig. 4.2. So, on this approach we have to find the expected total cost $E_{z_0}(u(t),\ z_f,\ T_2)$, where $z_f = (3, 5)$; $t_2 = 5$. Note that in this network

$$p_{((3,3),(3,4))} = p_{((3,4),(3,5))} = 1, \quad c_{((3,3),(3,4))} = c_{((3,4),(3,5))} = 0;$$

the probabilities and the costs for the rest of edges are the same as in the network from Fig. 4.2.

If we apply Algorithm 4.19 then we obtain:

Step 0

$\theta(1,5) = 0;\quad \theta(2,5) = 0;\quad \theta(3,5) = 1;\quad \theta(4,5) = 0;$
$\theta'(1,5) = 0;\quad \theta'(2,5) = 0;\quad \theta'(3,5) = 1;\quad \theta'(4,5) = 0;$
$\sigma'_{(1,5)}(u(t),5) = 0;\quad \sigma'_{(2,5)}(u(t),5) = 0;$
$\sigma'_{(3,5)}(u(t),5) = 0;\quad \sigma'_{(4,5)}(u(t),5) = 0.$

Step 1

$p'_{((1,4),(1,5))} = 0;\quad p'_{((1,4),(2,5))} = 0;\quad p'_{((2,4),(1,5))} = 0;\quad p'_{((2,4),(3,5))} = 1;$
$p'_{((3,4),(3,5))} = 1;\quad p'_{((4,4),(3,5))} = 1,\quad p'_{((4,4),(4,5))} = 0;$
$\theta(1,4) = 0;\quad \theta(2,4) = 0.7;\quad \theta(3,4) = 1;\quad \theta(4,4) = 0.6;$
$\theta'(1,4) = 0;\quad \theta'(2,4) = 1;\quad \theta'(3,4) = 1;\quad \theta'(4,4) = 1;$
$\sigma'_{(1,4)}(u(t),5) = 0;\quad \sigma'_{(2,4)}(u(t),5) = 3;$
$\sigma'_{(3,4)}(u(t),5) = 0;\quad \sigma'_{(4,4)}(u(t),5) = 5.$

Step 2

$p'_{((1,3),(1,4))} = 0;\quad p'_{((1,3),(2,4))} = 1;\quad p'_{((2,3),(2,4))} = 0.5;$
$p'_{((2,3),(3,4))} = 0.5;\quad p'_{((3,3),(3,4))} = 1;\quad p'_{((4,3),(3,4))} = 0.7;$
$p'_{((4,3),(4,4))} = 0.3;$
$\theta(1,3) = 0.6;\quad \theta(2,3) = 1;\quad \theta(3,3) = 1;\quad \theta(4,3) = 1;$
$\theta'(1,3) = 1;\quad \theta'(2,3) = 1;\quad \theta'(3,3) = 1;\quad \theta'(4,3) = 1;$
$\sigma'_{(1,3)}(u(t),5) = 5 + \sigma'_{(2,4)}(u(t),5) = 8;$
$\sigma'_{(2,3)}(u(t),5) = (1 + \sigma'_{(2,4)}(u(t),5))0.5 + (4 + \sigma'_{(3,4)}(u(t),5))0.5 = 4;$
$\sigma'_{(3,3)}(u(t),5) = 0;$
$\sigma'_{(4,3)}(u(t),5) = (5 + \sigma'_{(3,4)}(u(t),5))0.7 + (4 + \sigma'_{(4,4)}(u(t),5))0.3 = 6.2\,.$

Step 3

$p'_{((1,2),(2,3))} = 1;\quad p'_{((2,2),(1,3))} = 0.6;\quad p'_{((2,2),(3,3))} = 0.4;$
$p'_{((3,2),(2,3))} = 0.3;\quad p'_{((3,2),(4,3))} = 0.7;\quad p'_{((4,2),(3,3))} = 0.5;$
$p'_{((4,2),(4,3))} = 0.5;$
$\theta(1,2) = 1;\quad \theta(2,2) = 1;\quad \theta(3,2) = 1;\quad \theta(4,2) = 1;$
$\theta'(1,2) = 1;\quad \theta'(2,2) = 1;\quad \theta'(3,2) = 1;\quad \theta'(4,2) = 1;$
$\sigma'_{(1,2)}(u(t),5) = 1 + \sigma'_{(2,3)}(u(t),5) = 5;$
$\sigma'_{(2,2)}(u(t),5) = (3 + \sigma'_{(1,3)}(u(t),5))0.6 + (7 + \sigma'_{(3,3)}(u(t),5))0.4 = 9.4;$
$\sigma'_{(3,2)}(u(t),5) = (3 + \sigma'_{(2,3)}(u(t),5))0.3 + (8 + \sigma'_{(4,3)}(u(t),5))0.7 = 12.04;$
$\sigma'_{(4,2)}(u(t),5) = (2 + \sigma'_{(3,3)}(u(t),5))0.5 + (4 + \sigma'_{(4,3)}(u(t),5))0.5 = 6.1\,.$

Step 4

$p'_{((1,1),(2,2))} = 1;\quad p'_{((2,1),(3,2))} = 1;\quad p'_{((3,1),(3,2))} = 1;\quad p'_{((4,1),(3,2))} = 1;$
$\theta(1,1) = 1;\quad \theta(2,1) = 1;\quad \theta(3,1) = 1;\quad \theta(4,1) = 1;$
$\theta'(1,1) = 1;\quad \theta'(2,1) = 1;\quad \theta'(3,1) = 1;\quad \theta'(4,1) = 1;$
$\sigma'_{(1,1)}(u(t),5) = 2 + \sigma'_{(2,2)}(u(t),5) = 11.4;$
$\sigma'_{(2,1)}(u(t),5) = 2 + \sigma'_{(3,2)}(u(t),5) = 14.04;$
$\sigma'_{(3,1)}(u(t),5) = 5 + \sigma'_{(3,2)}(u(t),5) = 17.04;$
$\sigma'_{(4,1)}(u(t),5) = 7 + \sigma'_{(3,2)}(u(t),5) = 19.04\,.$

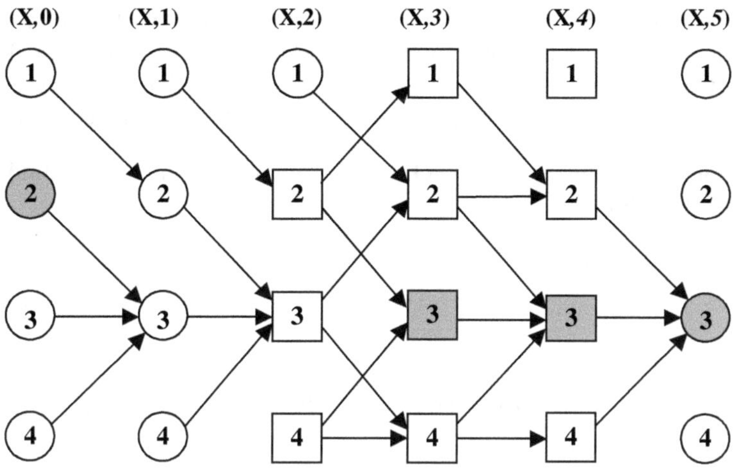

Fig. 4.11 The network induced by the control $u(t)$

Step 5

$p'_{((1,0),(2,1))} = 1;\quad p'_{((2,0),(3,1))} = 1;\quad p'_{((3,0),(3,1))} = 1;\quad p'_{((4,0),(3,1))} = 1;$
$\theta(1,0) = 1;\quad \theta(2,0) = 1;\quad \theta(3,0) = 1;\quad \theta(4,0) = 1;$
$\theta'(1,0) = 1;\quad \theta'(2,0) = 1;\quad \theta'(3,0) = 1;\quad \theta'(4,0) = 1;$
$\sigma'_{(1,0)}(u(t), 5) = 4 + \sigma'_{(2,1)}(u(t), 5) = 18.04;$
$\sigma'_{(2,0)}(u(t), 5) = 3 + \sigma'_{(3,1)}(u(t), 5) = 20.04;$
$\sigma'_{(3,0)}(u(t), 5) = 6 + \sigma'_{(3,1)}(u(t), 5) = 23.04;$
$\sigma'_{(4,0)}(u(t), 5) = 2 + \sigma'_{(3,1)}(u(t), 5) = 19.04\,.$

So, after the final step of the algorithm we obtain

$$\sigma_1(u(t),\ 3,\ 3 \le t(3) \le 5) = \sigma'_{(1,0)}(u(t), 5) = 18.04;$$
$$\sigma_2(u(t),\ 3,\ 3 \le t(3) \le 5) = \sigma'_{(2,0)}(u(t), 5) = 20.04;$$
$$\sigma_3(u(t),\ 3,\ 3 \le t(3) \le 5) = \sigma'_{(3,0)}(u(t), 5) = 23.04;$$
$$\sigma_4(u(t),\ 3,\ 3 \le t(3) \le 5) = \sigma'_{(3,0)}(u(t), 5) = 19.04.$$

The corresponding network generated by edges $e \in E$ with nonzero probabilities p'_e is represented in Fig. 4.11.

4.3.3 Determining the Optimal Control and the Expected Total Cost for the Control Problem with a Restriction on the Number of Transitions

We extend now the previous algorithm for determining the optimal control in Problem 4.6. The algorithm determines the optimal expected total cost $\sigma_{x(0)}(u^*(t),$

$x_f, t_1 \leq t(x_f) \leq t_2$) for an arbitrary starting state $x(0) \in X$, the optimal control $u^*(t)$ and the conditional probabilities p'_e that correspond to the optimal control $u^*(t)$. In the same way as in the previous algorithm we shall use the time-expanded network $(G^+, Z^C, Z^N, c^+, p^+, t_1, t_2, z_0, z_f)$ and the backward induction principle.

Algorithm 4.22 Determining the Optimal Control and the Conditional Probabilities for Problem 4.6

Preliminary step (Step 0)

Set:

$$\theta(z_f) = \theta'(z_f) = 1; \;\; \theta(z) = \theta'(z) = 0, \;\; \forall z \in Z_{t_2} \setminus \{z_f\};$$
$$\sigma'_{z_f}(u^*(t), t_2) = 0; \;\;\; \sigma'_z(u^*(t), t_2 = M, \;\; \forall z \in Z \;\; \left(M > \sum_{e \in E^+} |c_e/p_e|\right);$$
$$E^* = \emptyset.$$

General step (Step $\tau, \; \tau \geq 1$)

For each position $z \in Z_{t_2-\tau}$ find

$$\theta(z) = \sum_{(z,w) \in E^+(z)} p_{(z,w)} \theta'(w),$$

where $E^+(z) = \{e = (z, w) \,|\, (z, w) \in E^+\}$, i.e., $E^+(z)$ represents the subset of directed edges from E^+ which originate in z. Then for each $z \in Z_{T_2-\tau}$ do items (a), (b) and (c):

(a) If $z \in Z^N$ and $E^+(z) \neq \emptyset$ then calculate

$$p'_{(z,w)} = \frac{1}{\theta(z)} p_{(z,w)} \theta'(w)$$

for $(z, w) \in E^*(z)$ and put

$$\theta'(z) = \sum_{w \in E^*(z)} p'_{(z,w)}$$

for $z \in Z_{t_2-\tau}$. Then for $z = (x, t_2 - \tau) \in Z_N$ find

$$\sigma'_{(x,t_2-\tau)}(u^*(t), T_2) = \sum_{(z,w) \in E'(z)} (c_{(z,w)} + \sigma'_w(u^*(t), t_2)) p'_{(z,w)},$$

where $w = (y, t_2 - \tau + 1) \in Z_{t_2-\tau+1}$.

If $E^+(z) = \emptyset$ then put

$$\sigma'_{(x,t_2-\tau)}(u^*(t), t_2) = M$$

and fix $\theta(z) = \theta'(z) = 0$.

(*b*) If $z \in Z^C$ and $E^+(z) \neq \emptyset$ then calculate

$$\sigma'_{(x,t_2-\tau)}(u^*(t), t_2) = \min_{(z,w)\in E^*(z)} (c_{(z,w)} + \sigma(w)).$$

After that include in E^* the directed edges $(z, w)^*$ for which

$$(z, w)^* = \underset{(z,w)\in E^*(z)}{\text{argmin}} \{c_{(z,w)} + \sigma_w(u^*(t), t_2)\}$$

and put $\theta(z) = \theta'(w) = 1$. If $E^*(z) = \emptyset$ then put

$$\sigma'_{(x,t_2-\tau)}(u^*(t), t_2) = M$$

and fix $\theta(z) = \theta'(z) = 0$.

(*c*) If $\tau \leq T$ then go to the next step; otherwise fix an arbitrary map

$$u^*:(x, t) \to (y, t + 1), \quad t = 0, 1, 2, \dots, t_2$$

on the set of controllable positions such that $((x, t), (y, t + 1)) \in E_{u^*}$. Then set

$$\sigma_{x(0)}(u'^*(t), x_f, t_1 \leq t(x_f) \leq t_2) = \sigma_{(x,0)}(u^*, t_2)$$

for $x(0) \in X$ and STOP.

Example Consider the Problem 4.6 of determining the optimal control $u^*(t)$ and the expected total cost $\sigma_{x_0}(u^*(t), x_f, t_1 \leq t(x_f) \leq t_2)$ for the example from Sect. 4.2.2, where $x_f = 3$, $t_1 = 3$, $t_2 = 5$ (Problem 4.6).

We reduce this problem to the control problem with a fixed number of transitions on the time-expanded network represented in Fig. 4.9. Thus, we obtain the problem of finding the optimal control $u^*(t)$ with the minimal expected total cost $\sigma_{z_0}(u^*(t), z_f, t_2)$, where $z_f = (3, 5)$; $t_2 = 5$. In this network

$$p_{((3,3),(3,4))} = p_{((3,4),(3,5))} = 1, \quad c_{((3,3),(3,4))} = c_{((3,4),(3,5))} = 0;$$

the probabilities and the costs for the remaining edges are the same as in the network from Fig. 4.2.

If we apply Algorithm 4.18 then we obtain:

Step 0

$\theta(1, 5) = 0; \quad \theta(2, 5) = 0; \quad \theta(3, 5) = 1; \quad \theta(4, 5) = 0;$
$\theta'(1, 5) = 0; \quad \theta'(2, 5) = 0; \quad \theta'(3, 5) = 1; \quad \theta'(4, 5) = 0;$

$\sigma'_{(1,5)}(u^*(t), 5) = M; \quad \sigma'_{(2,5)}(u^*(t), 5) = M; \quad \sigma'_{(3,5)}(u^*(t), 5) = 0;$
$\sigma'_{(4,5)}(u^*(t), 5) = M; \quad E^+ = \emptyset.$

Step 1
The set Z_4 contains only uncontrollable positions, therefore, we obtain:

$$p'_{((1,4),(1,5))} = 0;\ p'_{((1,4),(2,5))} = 0;\ p'_{((2,4),(1,5))} = 0;\ p'_{((2,4),(3,5))} = 1;$$
$$p'_{((3,4),(3,5))} = 1;\ p'_{((4,4),(3,5))} = 1;\ p'_{((4,4),(4,5))} = 0;$$
$$\varepsilon(1,4) = 0;\quad \varepsilon(2,4) = 0.7;\quad \varepsilon(3,4) = 1;\quad \varepsilon(4,4) = 0.6;$$
$$\sigma'(1,4) = M;\quad \sigma'(2,4) = 3;\quad \sigma'(3,4) = 0;\quad \sigma'(4,4) = 5;$$
$$E^+ = \emptyset.$$

Step 2
The set Z_3 contains only uncontrollable positions, therefore we have:

$$p'_{((1,3),(1,4))} = 0;\ p'_{((1,3),(2,4))} = 0.7;\ p'_{((2,3),(2,4))} = 0.5;\ p'_{((2,3),(3,4))} = 0.5;$$
$$p'_{((3,3),(3,4))} = 1;\ p'_{((4,3),(3,4))} = 0.7;\ p'_{((4,3),(4,4))} = 0.3;$$
$$\theta(1,3) = 0.6;\quad \theta(2,3) = 1;\quad \theta(3,3) = 1;\quad \theta(4,3) = 1;$$
$$\theta'(1,3)= 0.6;\quad \theta'(2,3)= 1;\quad \theta'(3,3)= 1;\quad \theta'(4,3)= 1;$$
$$\sigma'_{(1,3)}(u^*(t),5) = 5 + \sigma'_{(2,4)}(u^*(t),5) = 8;$$
$$\sigma'_{(2,3)}(u^*(t),5) = (1 + \sigma'_{(2,4)}(u^*(t),5))0.5 + (4 + \sigma'_{(3,4)}(u^*(t),5))0.5 = 4;$$
$$\sigma'_{(3,3)}(u^*(t),5) = 0;$$
$$\sigma'_{(4,3)}(u^*(t),5) = (5 + \sigma'_{(3,4)}(u^*(t),5))0.7 + (4 + \sigma'_{(4,4)}(u(t),5))0.3 = 6.2.$$
$$E^+ = \emptyset.$$

Step 3
The set Z_2 contains the controllable position $(1,2)$, therefore, we calculate

$$\begin{aligned}\sigma'_{(1,2)}&(u^*(t),5)\\ &= \min\{c_{((1,2),(2,3))} + \sigma'_{(2,3)}(u^*(t),5),\ c_{((1,2),(1,3))} + \sigma'_{(1,3)}(u^*(t),5)\}\\ &= c_{((1,2),(2,3))} + \sigma(2,3) = 1 + \sigma(2,3) = 5\end{aligned}$$

and include in E^+ the edge $((1,2),(2,3))$, i.e., $E^{opt} = \{((1,2),(2,3))\}$. After that we find

$$p'_{((1,2),(1,3))} = 0;\quad p'_{((1,2),(2,3))} = 1;\quad p'_{((2,2),(1,3))} = 0.6;\ p'_{((2,2),(3,3))} = 0.4;$$
$$p'_{((3,2),(2,3))} = 0.3;\ p'_{((3,2),(4,3))} = 0.7;\ p'_{((4,2),(3,3))} = 0.5;\ p'_{((4,2),(4,3))} = 0.5;$$
$$\theta(1,2) = 1;\quad \theta(2,2) = 1;\quad \theta(3,2) = 1;\quad \theta(4,2) = 1;$$
$$\theta'(1,2) = 1;\quad \theta'(2,2) = 1;\quad \theta'(3,2) = 1;\quad \theta'(4,2) = 1;$$

$$\sigma'_{(2,2)}(u^*(t),5) = (3 + \sigma'_{(1,3)}(u^*(t),5))0.6 + (7 + \sigma'_{(3,3)}(u^*(t),5))0.4 = 9.4;$$
$$\sigma'_{(3,2)}(u^*(t),5) = (3 + \sigma'_{(2,3)}(u^*(t),5))0.3 + (8 + \sigma'_{(4,3)}(u^*(t),5))0.7 = 12.04;$$
$$\sigma'_{(4,2)}(u^*(t),5) = (2 + \sigma'_{(3,3)}(u^*(t),5))0.5 + (4 + \sigma'_{(4,3)}(u^*(t),5))0.5 = 6.1\,.$$

Step 4
The set Z_1 contains only controllable positions, therefore we calculate

$$\begin{aligned}
\sigma'_{(1,1)}(u^*(t),5) &= \min\{6+\sigma'_{(1,2)}(u^*(t),5), 2+\sigma'_{(2,2)}(u^*(t),5)\}\\
&= \min\{6+5, 2+9.4\} = 11;\\
\sigma'_{(2,1)}(u^*(t),5) &= \min\{7+\sigma'_{(1,2)}(u^*(t),5), 2+\sigma'_{(3,2)}(u^*(t),5)\} = 7+5\\
&= \min\{7+5,\ 2+12.04\} = 12;\\
\sigma'_{(3,1)}(u^*(t),5) &= \min\{2+\sigma'_{(2,2)}(u^*(t),5), 5+\sigma'_{(3,2)}(u^*(t),5), 3+\sigma'_{(4,2)}(u^*(t),5)\}\\
&= \min\{2+9.4,\ 5+12.04, 3+6.1\} = 9.1;\\
\sigma'_{(4,1)}(u^*(t),5) &= \min\{7+\sigma'_{(3,2)}(u^*(t),5), 4+\sigma'_{(4,2)}(u^*(t),5)\}\\
&= \min\{7+12.04,\ 4+6.1\} = 10.1
\end{aligned}$$

and find

$$\begin{aligned}
E^* = &\{((1,2),(2,3))\}\\
&\cup\{((1,1),(1,2)),\ ((2,1),(1,2)),\ ((3,1),(4,2)),\ ((4,1),(4,2))\};
\end{aligned}$$

$$\begin{aligned}
&p'_{((1,1),(2,2))} = 1;\ p'_{((2,1),(3,2))} = 1;\ p'_{((3,1),(3,2))} = 1;\ p'_{((4,1),(3,2))} = 1;\\
&\theta(1,1) = 1;\quad \theta(2,1) = 1;\quad \theta(3,1) = 1;\quad \theta(4,1) = 1;\\
&\theta'(1,1) = 1;\quad \theta'(2,1) = 1;\quad \theta'(3,1) = 1;\quad \theta'(4,1) = 1.
\end{aligned}$$

Step 5
The set Z_1 contains only controllable positions, therefore we calculate

$$\begin{aligned}
\sigma'_{(1,0)}(u^*(t),5) &= \min\{2+\sigma'_{(1,1)}(u^*(t),5),\ \ 4+\sigma'_{(2,1)}(u^*(t),5)\}\\
&= \min\{2+11,\ 4+12\} = 13;\\
\sigma'_{(2,0)}(u^*(t),5) &= \min\{4+\sigma'_{(1,1)}(u^*(t),5),\ \ 6+\sigma'_{(2,1)}(u^*(t),5), 3+\sigma'_{(3,1)}(u^*(t),5)\}\\
&= \min\{4+11,\ 6+12,\ 3+9.1\} = 12.1;\\
\sigma'_{(3,0)}(u^*(t),5) &= \min\{6+\sigma'_{(3,1)}(u^*(t),5),\ 2+\sigma'_{(4,1)}(u^*(t),5)\}\\
&= \min\{6+9.1,\ 2+10.1\} = 12.1;\\
\sigma'_{(4,0)}(u^*(t),5) &= \min\{2+\sigma'_{(3,1)}(u^*(t),5),\ 5+\sigma'_{(4,1)}(u^*(t),5)\}\\
&= \min\{2+9.1,\ 5+10.1\} = 11.1
\end{aligned}$$

and find

$$\begin{aligned}
E^* = &((1,2),(2,3)),\ ((1,1),(1,2)),\ ((2,1),(1,2)),\ ((3,1),(4,2)),\\
&((4,1),(4,2)),\ ((1,0),(1,1)),\ ((2,0),(3,1)),\ ((3,0),(4,1)),\\
&((4,0),(3,1));
\end{aligned}$$

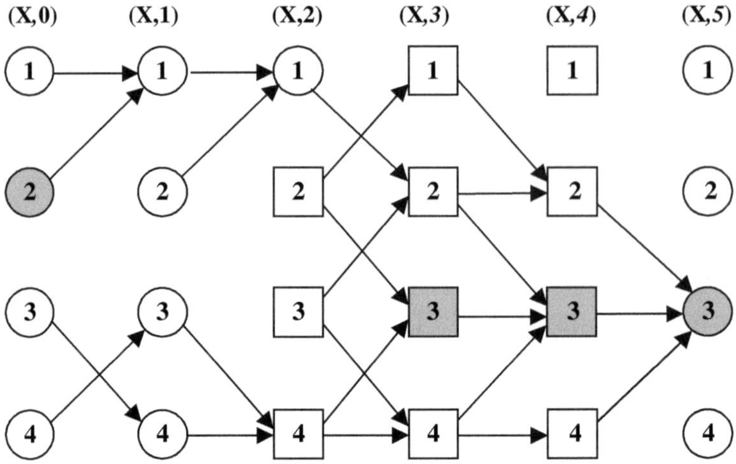

Fig. 4.12 The network which corresponds to the optimal control $u^*(t)$

$$\begin{array}{llll} p'_{((1,0),(2,1))} = 1; & p'_{((2,0),(3,1))} = 1; & p'_{((3,0),(3,1))} = 1; & p'_{((4,0),(3,1))} = 1; \\ \varepsilon(1,0) = 1; & \varepsilon(2,0) = 1; & \varepsilon(3,0) = 1; & \varepsilon(4,0) = 1. \end{array}$$

At the final step of the algorithm we fix the map

$$\begin{array}{llll} u^* : (1,2) \to (2,3); & (1,1) \to (1,2); & (2,1) \to (1,2); & (3,1) \to (4,2); \\ \quad (4,1) \to (4,2); & (1,0) \to (1,1); & (2,0) \to (3,1); & (3,0) \to (4,1); \\ \quad (4,0) \to (3,1) & & & \end{array}$$

that determine the optimal control $u^*(t)$. Additionally, we obtain the expected total costs

$$\sigma_1(u^*(t),\ 3,\ 3 \le t(3) \le 3) = \sigma'_{(1,0)}(u^*(t), 5) = 13;$$
$$\sigma_2(u^*(t),\ 3,\ 3 \le t(3) \le 3) = \sigma'_{(2,0)}(u^*(t), 5) = 14;$$
$$\sigma_3(u^*(t),\ 3,\ 3 \le t(3) \le 3) = \sigma'_{(3,0)}(u^*(t), 5) = 14.1;$$
$$\sigma_4(u^*(t),\ 3,\ 3 \le t(3) \le 3) = \sigma'_{(3,0)}(u^*(t), 5) = 14.$$

The time-expanded network generated by the edges $e \in E$ with nonzero probabilities p'_e is represented in Fig. 4.12; the directed edges originating in the controllable positions correspond to the optimal control of the considered problem.

4.4 Discrete Control Problems with Varying Time of States' Transitions of the Dynamical System

In the previous sections we have studied the control models with a fixed unit time of states' transitions of the dynamical system by a trajectory. Now we extend these models and consider the discrete control problems if the time of systems' transitions from one state to another may be different from 1 and may vary in the control process. We assume that the time of states' transitions depends on the vectors of control parameters belonging to a feasible set defined for an arbitrary state at every discrete moment of time. In this section we show that the time-expanded network method and the dynamic programming algorithms can be extended for the problems with varying time of states' transitions of the dynamical system. We describe these results for the deterministic control problem and show that the time-expanded network method for the stochastic cases of the problem can be applied in a similar way.

4.4.1 Deterministic Control Problem with Varying Time of States' Transitions

Consider the dynamical system $\mathbb{L}$ with the finite set of states X, where at every discrete moment of time $t = 0, 1, 2, \ldots$ the state of $\mathbb{L}$ is $x(t) \in X$. The starting state $x_0 = x(0)$ and the final state x_f are fixed. Assume, that the dynamical system should reach the final state x_f at the time moment $t(x_f)$ such that

$$t_1 \leq t(x_f) \leq t_2$$

where t_1 and t_2 are given.

The control of the time-discrete system $\mathbb{L}$ at each time-moment $t = 0, 1, 2, \ldots$ for an arbitrary state $x(t)$ is made by using the vector of control parameters $u(t)$ for which a feasible set $U_t(x(t))$ is given, i.e., $u(t) \in U_t(x(t))$. In addition we assume that for an arbitrary t and $x(t)$ on $U_t(x(t))$ it is defined an integer function

$$\tau_{x(t)} : U_t(x(t)) \rightarrow \mathbb{N}$$

which gives to each control $u(t) \in U_t(x(t))$ an integer value $\tau_{x(t)}(u(t))$. This value represents the time of the system's passage from the state $x(t)$ to the state $x\big(t + \tau_{x(t)}(u(t))\big)$ if the control $u(t) \in U_t(x(t))$ has been applied at the moment t for a given state $x(t)$.

Assume that the dynamics of the system is described by the following system of difference equations

$$\begin{cases} t_{j+1} &= t_j + \tau_{x(t_j)}\big(u(t_j)\big); \\ x(t_{j+1}) &= g_{t_j}\big(x(t_j), u(t_j)\big); \\ & (t_j) \in U_{t_j}\big(x(t_j)\big); \\ & j = 0, 1, 2, \ldots, \end{cases} \tag{4.7}$$

where

$$t_0 = 0, \quad x(t_0) = x_{t_0} \tag{4.8}$$

is a starting representation of the dynamical system $\mathbb{L}$.

We suppose that the functions g_{t_j} and $\tau_{x(t_j)}$ in (4.7) are known and t_{j+1} and $x(t_{j+1})$ are determined uniquely by $x(t_j)$ and $u(t_j)$ at every step $j = 0, 1, 2, \ldots$.

Let $u(t_j)$, $j = 0, 1, 2, \ldots$, be a control, which generates the trajectory

$$x(0), x(t_1), x(t_2), \ldots, x(t_k), \ldots .$$

Then either this trajectory passes through the final state x_f and $t(x_f) = t_k$ represents the time-moment if the final state x_f is reached or this trajectory does not pass trough x_f.

For an arbitrary control we define the quantity

$$F_{x_0,x_f}(u(t)) = \sum_{j=0}^{k-1} c_{t_j}\big(x(t_j),\ g_{t_j}\big(x(t_j),\ u(t_j)\big)\big) \tag{4.9}$$

if the trajectory $x(0), x(t_1), x(t_2), \ldots, x(t_k), \ldots$ passes through the final state x_f at the time-moment $t_k = t(x_f)$ such that $t_1 \le t(x_f) \le t_2$; otherwise we put

$$F_{x_0,x_f}(u(t)) = \infty.$$

Here $c_{t_j}\big(x(t_j),\ g_{t_j}\big(x(t_j), u(t_j)\big)\big) = c_{t_j}\big(x(t_j), x(t_{j+1})\big)$ represents the cost of the system $\mathbb{L}$ to pass from the state $x(t_j)$ to the state $x(t_{j+1})$ at the stage $[j, j+1]$.

We consider the following control problem:

Problem 4.23 To find time-moments $t = 0, t_1, t_2, \ldots, t_{k-1}$ and vectors of control parameters $u(t_0), u(t_1), u(t_2), \ldots, u(t_{k-1})$ which satisfy the conditions (4.7), (4.8) and minimize the functional (4.9).

It is evident that for this problem there exists the optimal control $u^*(t)$ if there exists a feasible control which provides a trajectory from a starting state x_0 to a final state x_f. The optimal solution of the problem can be found by using direct or backward dynamic programming algorithms. Both algorithms can be grounded on the basis of the time expanded method. A specification of this method for the control problem with varying time of states' transition of the dynamical system is given in the next section.

4.4.2 The Time-Expanded Network Method for a Deterministic Control Problem with Varying Time of States' Transitions

At first we describe how to construct the auxiliary time-expanded network if $t_2 = t_1 = \bar{t}$ and then we show that the general case of the problem with $t_2 > t_1$ can be reduced to the case with fixed $\bar{t}$.

Assume that $t_2 = t_1 = \bar{t}$ and construct a time-expanded network with the structure of an acyclic directed graph $G = (Z, E)$, where Z consists of $\bar{t} + 1$ copies of the set of states X corresponding to the time moments $t = 0, 1, 2, \ldots, \bar{t}$. So,

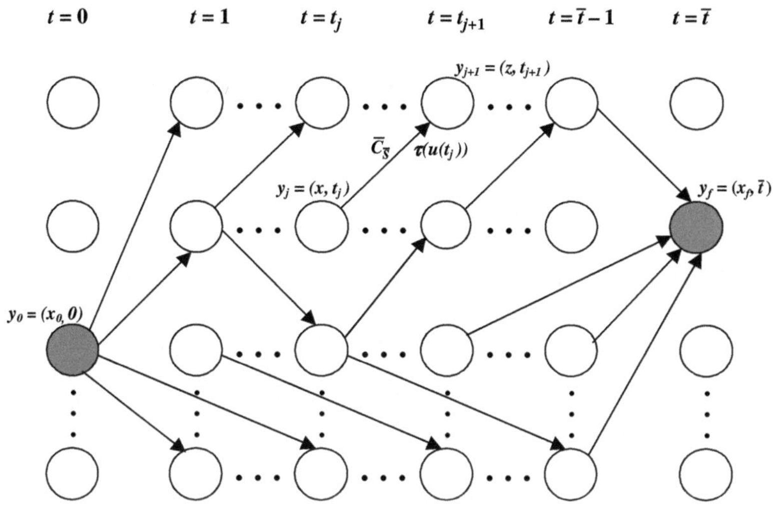

Fig. 4.13 The time-expanded network for the problem in general case

$$Z = Z^0 \cup Z^1 \cup Z^2 \cup \ldots \cup Z^{\overline{t}} \quad (Z^t \cap Z^l = \emptyset,\ t \neq l),$$

where $Z^t = (X, t)$ corresponds to the set of states of the dynamical system at the time moment $t = 0, 1, 2, \ldots, \overline{t}$. This means that $Z^t = \{(x, t) \mid x \in X\}$, $t = 0, 1, 2, \ldots, \overline{t}$. The graph G is represented in Fig. 4.13, where at each moment of time $t = 0, 1, 2, \ldots, \overline{t}$ we can see all copies of the vertex set X.

We define the set of edges E of the graph G in the following way:

If at a given moment of time $t_j \in [0, \overline{t}]$ for a given state $x = x(t_j)$ of the dynamical system there exists a vector of control parameters $u(t_j) \in U_{t_j}\left(x(t_j)\right)$ such that

$$z = x(t_{j+1}) = g_{t_j}\left(x(t_j), u(t_j)\right),$$

where

$$t_{j+1} = t_j + \tau_{x(t_j)}\left(u(t_j)\right),$$

then $\left((x, t_j), (z, t_{j+1})\right) \in E$, i.e., in G we connect the vertex $y_j = (x, t_j) \in Y^{t_j}$ with the vertex $y_{j+1} = (z, t_{j+1})$ (see Fig. 4.13). To this edge $e = \left((x, t_j), (z, t_{j+1})\right)$ we associate in G the cost $c_e = c_{t_j}\left(x(t_j), x(t_{j+1})\right)$.

The following lemma holds

Lemma 4.24 *Let $u(t_0), u(t_1), u(t_2), \ldots, u(t_{k-1})$ be a control of the dynamical system in Problem 4.23, which generates a trajectory*

$$x_0 = x(t_0),\ x(t_1),\ x(t_2),\ \ldots,\ x(t_k) = x_f$$

from x_0 to x_f, where

$$\begin{aligned} &t_0 = 0, \ \ t_{j+1} = t_j + \tau_{x(t_j)}\left(u(t_j)\right), \ j = 0, 1, 2, \dots, k-1; \\ &u(t_j) \in U_t\left(x(t_j)\right), \ \ j = 0, 1, 2, \dots, k-1; \\ &t_k = T. \end{aligned}$$

Then in G there exists a directed path

$$P_G(z_0, z_f) = \{z_0 = (x_0, 0), (x_{t_1}, t_1), (x_{t_2}, t_2), \dots, (x_{t_k}, \overline{t}) = z_f\}$$

from y_0 to z_f, where $x_{t_j} = x(t_j)$, $j = 0, 1, 2, \dots, k$, and $x(t_k) = x_f$, i.e., $t(x_f) = t_k = \overline{t}$. So, between the set of states of the trajectory $x_{t_0} = x(t_0), x(t_1), x(t_2), \dots, x(t_k) = x_f$ and the set of vertices of the directed path $P_G(z_0, z_f)$ there exists a bijective mapping

$$(x_{t_j}, t_j) \Longleftrightarrow x(t_j), \quad j = 0, 1, 2, \dots, k$$

such that $x_{t_j} = x(t_j)$, $j = 0, 1, 2, \dots, k$, and

$$\sum_{j=0}^{k-1} c_{t_j}\left(x(t_j), x(t_{j+1})\right) = \sum_{j=0}^{k-1} c_{\left((x_{t_j}, t_j), (x_{t_{j+1}}, t_{j+1})\right)},$$

where $t_0 = 0$, $x_{t_0} = x(t_0)$ and $x_f = x(t_k)$, $t_k = \overline{t}$.

Proof In Problem 4.23 an arbitrary control $u(t_j)$ at a given moment of time t_j for a given state $x(t_j) \in U_{t_j}\left(x(t_j)\right)$ uniquely determines the next state $x(t_{j+1})$. So, $u(t_j)$ can be identified with a unique transition $\left(x(t_j),\ x(t_{j+1})\right)$ from the state $x(t_j)$ to the state $x(t_{j+1})$. In $G = (Z, E)$ this transition corresponds to a unique directed edge $\left((x_{t_j}, t_j),\ (x_{t_{j+1}},\ t_{j+1})\right)$ which connects vertices $(x_{t_j},\ t_j)$ and $(x_{t_{j+1}},\ t_{j+1})$; the cost of this edge is $c\left((x_{t_j},\ t_j),\ (x_{t_{j+1}},\ t_{j+1})\right) = c_{t_j}\left(x(t_j),\ x(t_{j+1})\right)$. This one-to-one correspondence between the control $u(t_j)$ at a given moment of time and the directed edge $e = \left((x_{t_j}, t_j),\ (x_{t_{j+1}}, t_{j+1})\right) \in E$ implies the existence of a bijective mapping between the set of trajectories from the starting state x_0 to the final state x_f in Problem 4.23 and the set of directed paths from z_0 to z_f in G, which preserve the integral-time costs. □

Corollary 4.25 *If $u^*(t_j)$, $j = 0, 1, 2, \dots, k-1$ is the optimal control of the dynamical system in Problem 4.23, which generates a trajectory*

$$x_{t_0} = x^*(0),\ x^*(t_1),\ x^*(t_2),\ \dots,\ x^*(t_k) = x_f$$

from x_0 to x_f, then in G the corresponding directed path

$$P_G^*(z_0, z_f) = \{z_0 = (x_{t_0}, 0),\ (x^*_{t_1}, t_1),\ (x^*_{t_2}, t_2),\ \ldots,\ (x^*_{t_k}, t_k) = z_T\}$$

is the minimal integral cost directed path from z_0 to z_f and vice-versa.

So, the control problem with varying time of states' transitions can be reduced to the problem of finding the minimum cost directed path with fixed starting and final vertices in the time-expanded network. For this problem direct and backward dynamic programming algorithms an be applied. Note that if we apply direct dynamic programming algorithms then we determine the optimal trajectories from a starting state x_0 to every state $x(t)$ with $t = 0, 1, 2, \ldots, T$. If we apply backward dynamic programming algorithms then we obtain the optimal trajectories with starting states $x(t)$ for $t = 0, 2, \ldots, T$ to a final state $x_f = x(T)$.

Based on the results mentioned above we can determine the solution of Problem 4.23 by using the time-expanded network method in the following way:

(a) *Determining the optimal solution of Problem* 4.23 *with a fixed number of transitions.*

1. We construct the auxiliary time-expanded network consisting of a directed acyclic graph $G = (Z, E)$, a cost function $c : E \to \mathbb{R}^1$ and given starting and final vertices z_0 and z_f.
2. We find in G the directed path $P_G^*(z_0, y_f)$ from a starting vertex y_0 to a final vertex y_f with a minimal sum of the edge's costs.
3. We determine the control $u^*(t_j)$, $j = 0, 1, 2, \ldots, k-1$, which corresponds to a directed path $P_G^*(z_0, z_f)$ from z_0 to z_f. Then $u^*(t_j)$, $j = 0, 1, 2, \ldots, k-1$, is a solution of Problem 4.23.

(b) *Determining the optimal solution of Problem* 4.23 *with a fixed number of transitions from a given interval.*

In the case $t_1 \le t(x_f) \le t_2$ $(t_2 > t_1)$ Problem 4.23 can be reduced to $t_2 - t_1 + 1$ problems fixing each time $\bar{t} = t_1, t_1 + 1, \ldots, t_2$. If we solve these problems and find the best solution then we determine the solution of the problem in the case $t_1 \ne t_2$. In this case a more suitable procedure to solve the problem is to reduce it to the minimum cost path problem on an auxiliary graph $\overline{G} = (Z, \overline{E})$ which is obtained from $G = (Z, E)$ by adding a new vertex $\bar{z}_f$ and directed edges $((x, t), \bar{z}_f)$, $\bar{t} = t_1, t_1 + 1, \ldots, t_2$, where $c((x, t), \bar{z}_f) = 0$. An arbitrary directed path from a starting vertex $(x_0, 0)$ to a final vertex $\bar{z}_f$ in $\overline{G}$ corresponds to the optimal trajectory of the control problem with the starting state x_0 and the final state x_f.

Remark 4.26 The time-expanded network can be simplified if we delete from G all vertices $z \in Z$ which are not attainable from z_0 as well as vertices $y \in Y$ for which there is no directed path from z to z_f.

(c) *Determining the solution of the control problem with a varying time of states' transitions.*

In the considered control problem it is assumed that the cost function $c_t\big(x(t),\ g_t (x(t), u(t)) = c_t\ (x(t),\ x(t+1)\big)$ of the system's transition from the state $x = x(t)$ to the state $y = x(t+1)$ depends on time t, on the state $x = x(t)$ and on the vector of control parameters $u(t)$. In general we may consider that the cost function of the system's transition from the state $x(t)$ to the state $x(t+1)$ depends also on the transition time $\tau_{x(t)}(u(t))$, i.e., the cost function $c_t\big(x(t),\ g_t\ (x(t),\ u(t))\ ,\ \tau_{x(t)}(u(t))\big) = c_t\ \big(x(t),\ x(t+1),\ \tau_{x(t)}(u(t))\big)$ depends on $t,\ x(t),\ u(t)$ and $\tau_{x(t)}(u(t))$.

It is easy to observe that the problem in such a general form can be solved in an analogous way as the previous problem by using the time-expanded network method with a simple specification of the costs on the edges of the network. In the auxiliary time-expanded network we define the cost functions $\overline{c}_{\overline{e}}$ on the edges $\overline{e}$ as follows:

$$\overline{c}_{\overline{e}} = c_{t_j}\big(x(t_j),\ x(t_{j+1}),\ \tau_{x(t_j)}\ (u(t_j))\big).$$

So, the problem with the cost functions that depend on the transition time of the states' transitions of the system can be solved by using the proposed algorithms from above. The dynamic programming algorithms based on the time-expanded network method for solving the control problem with a varying time of states' transitions have been developed in [87].

4.4.3 The Stochastic Discrete Control Problem with a Varying Time of States' Transitions

We formulate the stochastic version of the control problem with a varying time of states' transitions in a similar way as the problem with unit state's transitions of the system. We assume that in the control process the dynamical system may admit the states $x(t_j)$ in which the vector of control parameters $u(t) \in U_{t_j}(x(t)$ is changed in a random way according to a given distribution functions of the probabilities

$$p : U_t(x(t)) \to [0, 1], \quad \sum_{i=1}^{k(x(t_j))} p(u^i_{x(t_j)}) = 1 \tag{4.10}$$

on the corresponding dynamical feasible sets $U_t(x(t))$. If we regard each dynamical state $x(t)$ of the system $\mathbb{L}$ for $t \in \{0, 1, 2, \dots, t_2\}$ as a position (x, t) then the set of positions

$$Z = \{(x, t)\,|\, x \in X,\ t = 0, 1, 2, \dots, t_2\}$$

can be divided into two disjoint subsets

$$Z = Z^C \cup Z^N \quad (Z^C \cap Z^N = \emptyset),$$

such that Z^C represents the set of controllable positions and Z^N represents the set of positions $(x, t) = x(t)$ for which the distribution function (4.10) of the vectors of control parameters $u(t) \in U_t(x(t))$ are given. So, in this case we have the following behavior of the dynamical system: If the starting state $x_0 = x(t_0)$ belongs to the controllable positions then the decision maker fixes a vector of the control parameters and the system passes to a new state $x(t_1)$ which is reached at the moment of time t_1. In the case if the state $x(t_0)$ belongs to the set of uncontrollable positions the system passes to a state $x(t_1) \in X$ in a random way. If the state $x(t_1)$ belongs to the set of controllable positions then the decision maker fixes a vector of the control parameter $u(t_1) \in U_{t_1}(x(t_1))$ and the system passes to the state $x(t_2)$ and reaches it at the moment of time t_2. In the case if $x(t_1)$ belongs to the set of uncontrollable positions then the system passes to a state $x(t_2) \in X$ in a random way and so on. In this dynamic process the final state may be reached at a given moment of time with a probability which depends on the control of the decision maker as well as on the probability distribution functions in the uncontrollable states. The total cost in this process also is a random variable that depends on the probabilities in controllable positions. Therefore, for this control model we can consider the Problems 4.1–4.6. The probability transitions and the expected total cost in these problems can be defined and calculated in a similar way as for the unit states' transitions with some minor modifications.

4.5 Dynamic Programming Algorithms for Finite Horizon Markov Decision Problems

The stochastic control problems from Sect. 4.1 can be extended for Markov decision processes with a finite time horizon, and similar dynamic programming algorithms for determining their solutions can be derived. Below we formulate the corresponding decision problems for finite horizon Markov processes and show how to apply the time-expanded method for solving them.

4.5.1 Problem Formulations

A non-stationary discrete Markov decision process is determined by a tuple (X, A, p, c) where:

- X is a finite set of states;
- A is a finite set of actions;
- $p : A \times X \times X \times \{0, 1, 2, \dots, \overline{t}\} \to [0, 1]$ is a probability function which for an arbitrary state $x \in X$, a fixed action $a \in A(x)$ and every discrete moment of time $t \in \{0, 1, 2, \dots, \overline{t}\}$ gives the transition probabilities $p^a_{x,y}(t)$ from the state x to $y \in X$ such that $\sum_{y\in X} p^a_{x,y}(t) = 1$;

- $c : A \times X \times X \times \{0, 1, 2, \ldots, \bar{t}\} \to \mathbb{R}$ is a cost function which for an arbitrary state $x \in X$, an arbitrary action $a \in A(x)$ and an arbitrary discrete moment of time $t \in \{0, 1, 2, \ldots, \bar{t}\}$ gives the transition costs $c^a_{x,y}(t)$ from the state x to the states $y \in X$.

We consider Markov decision processes with a finite time horizon, i.e., $t = 0, 1, 2, \ldots, \bar{t}$, where $\bar{t}$ is given. Assume that the starting state $x_0 = x(0)$ is known and the decision maker fixes the actions $a \in A$ in the states $x(t) \in X$, for $t = 0, 1, 2, \ldots, \bar{t} - 1$. Then we obtain a Markov process in which we can determine the probability of the states' transitions from a starting state x_0 to an arbitrary other state and the corresponding expected total cost with a fixed number of transitions.

In the considered Markov process the decision maker may use the stationary as well as the non-stationary strategies of fixing the actions in the dynamical states. We define the non-stationary strategy as a map

$$s : (x, t) \to a \in A(x) \text{ for } (x, t) \in X \times \{0, 1, 2, \ldots, \bar{t} - 1\},$$

where $A(x)$ is the set of actions in the state x. Here (x, t) has the same meaning as the notation $x(t)$, i.e., $(x, t) = x(t)$.

Let $\mathbb{S}$ be the set of strategies in the finite Markov decision process (X, A, p, c). An arbitrary strategy $s \in \mathbb{S}$ determines the control in the states $x = x(t)$ and, therefore, we can formulate the following decision problems that are similar to the stochastic control problems 4.1–4.6:

Problem 4.27 For a given strategy $s \in \mathbb{S}$, determine the probability that the dynamical system $\mathbb{L}$ with a given starting state $x_0 = x(0)$ will reach the final state x_f at the moment of time $t(x_f)$ such that $t_1 \leq t(x_f) \leq t_2$. We denote this probability by $P_{x_0}(s, x_f, t_1 \leq t(x_f) \leq t_2)$; if $t_1 = t_2 = \bar{t}$ then we use the notation $P_{x_0}(s, x_f, \bar{t})$.

Problem 4.28 Find the strategy $s^* \in \mathbb{S}$, for which the probability in Problem 4.27 is maximal. We denote this probability by $P_{x_0}(s^*, x_f, t_1 \leq t(x_f) \leq t_2)$; in the case $t_1 = t_2 = \bar{t}$ we shall use the notation $P_{x_0}(s^*, x_f, \bar{t})$.

Problem 4.29 For a given strategy $s \in \mathbb{S}$, and given T determine the expected total cost during $\bar{t}$ states' transitions of the system $\mathbb{L}$ if it starts transitions in the state $x_0 = x(0)$ at the moment of time $t = 0$. We denote this value by $\sigma_{x_0}(s, \bar{t})$.

Problem 4.30 Determine the strategy $s^* \in S$ for which the expected total cost in Problem 4.29 is minimal. We denote this value by $\sigma_{x_0}(s^*, x_f, \bar{t})$.

Problem 4.31 For a given stationary strategy $s \in \mathbb{S}$, determine the expected total cost of the states' transitions from a starting state x_0 to a final state x_f if x_f is reached at the time-moment $t(x_f)$ such that $t_1 \leq t(x_f) \leq t_2$. We denote this expected total cost by $\sigma_{x_0}(s, x_f, t_1 \leq t(x_f) \leq t_2)$; in the case $t_1 = t_2 = \bar{t}$ we denote this value by $\sigma_{x_0}(s, x_f, \bar{t})$.

Problem 4.32 Determine the strategy $s^* \in S$ for which the expected total cost in Problem 4.5 is minimal. We denote this minimal expected total cost by $\sigma_{x_0}(s^*, x_f, t_1 \leq t(x_f) \leq t_2)$; in the case $t_1 = t_2 = \bar{t}$ we denote this value by $\sigma_{x_0}(s^*, x_f, \bar{t})$.

Obviously, the probabilities $P_{x_0}(s, x_f, t_1 \leq t(x_f) \leq t_2)$, $P_{x_0}(s, x_f, \bar{t})$ and the expected total cost $\sigma_{x_0}(s, \bar{t})$ in the Problems 4.27 and 4.29 for a fixed strategy $s \in \mathbb{S}$ can be found by using the algorithms from previous sections. The optimal strategy s^* in Problem 4.30 can be obtained by using the backward dynamic programming algorithm from [115]. For the Problems 4.28, 4.31 and 4.32 the backward dynamic algorithms can be derived if we apply the time-expanded method in a similar way as for the stochastic control problems. In general we can see that all considered problems can be reduced to the stochastic control Problems 4.27–4.32 on an auxiliary time-expanded network.

4.5.2 Construction of the Time-Expanded Network for a Finite Horizon Markov Decision Process

The structure of the time-expanded network is determined by the directed graph $G = (Z, E)$, where the set of vertices Z and the set of edges E are defined as follows: Z consists of $2\bar{t} + 1$ disjoint subsets $Z_0^C, Z_1^C, Z_2^C, \dots, Z_{\bar{t}}^C$ and $Z_0^N, Z_1^N, \dots, Z_{\bar{t}-1}^N$, where Z_t^C for $t \in \{0, 1, 2, \dots, \bar{t}\}$ corresponds to the set of states of the system at the moment of time t and Z_t^N for $t \in \{0, 1, 2, \dots, \bar{t} - 1\}$ corresponds to the set of possible fixed actions in different states for the dynamical system at the moment of time t.

Thus,

$$Z = \left(Z_0^C \cup Z_0^N\right) \cup \left(Z_1^C \cup Z_1^N\right) \cup \left(Z_2^C \cup Z_2^N\right) \cup \dots \cup \left(Z_{\bar{t}-1}^C \cup Z_{\bar{t}-1}^N\right) \cup Z_{\bar{t}}^N,$$

where $Z_t^C = \{(x, t) \,|\, x \in X\}, t = 0, 1, 2, \dots, \bar{t}$ and $Z_t^N = \{((x, a), t) | x \in X, a \in A\}, t = 0, 1, 2, \dots, \bar{t} - 1$. In the time-expanded network

$$Z^C = Z_0^C \cup Z_1^C \cup Z_2^C \cup \dots \cup Z_{\bar{t}}^C$$

represents the set of controllable states and

$$Z^N = Z_0^N \cup Z_1^N \cup Z_2^N \cup \dots \cup Z_{\bar{t}-1}^N,$$

represents the set of uncontrollable states.

The set of edges E consists of $2\bar{t}$ disjoint subsets $E_0^C, E_1^C, E_2^C, \dots, E_{\bar{t}-1}^C$ and $E_0^N, E_1^N, \dots, E_{\bar{t}-1}^N$, where

$$E_t^C = \{e = ((x, t), ((x, a), t)) \,|\, x \in X,\ a \in A\}, \quad t = 0, 1, 2, \dots, \bar{t} - 1$$

is the set of directed edges $e \in E$ that originate in the vertices $(x, t) \in Z_t^C$ and end in $((x, a), t) \in Z_t^N$, and

$$E_t^N = \{e = (((x,a),t),(y,t+1)) \mid a \in A,\ y \in X\},\quad t = 0,1,2,\ldots,\bar{t}-1$$

is the set of directed edges $e \in E$ that originate in vertices $((x,a),t) \in Z_t^N$ and end in $(y,t+1) \in Z_{t+1}^C$.

So,

$$E^C = E_0^C \cup E_1^C \cup E_2^C \cup \ldots \cup E_{\bar{t}-1}^C$$

represents the set of directed edges that originate in the controllable states and

$$E^N = E_0^N \cup E_1^N \cup E_2^N \cup \ldots \cup E_{\bar{t}-1}^N$$

represents the set of directed edges that originate in the uncontrollable states.

To each directed edge $e = (((x,a),t),(y,t+1)) \in E^N$ we associate the transition probability $p_e = p_{x,y}^a(t)$ and the transition cost $c_e = c_{x,y}^a(t)$. To each directed edge $e = ((x,t),((x,a),t)) \in E^C$ we put in correspondence the transition cost $c_e = 0$.

In such a way we obtain the time-expanded network on which we can formulate the problem of determining the probabilities $P_{x_0}(u(t),\ x_f,\ \bar{t}')$, $P_{x_0}(u(t),\ x_f,\ t_1' \le t(x_f) \le t_2')$ for a given control $u(t)$, the control $u^*(t)$ which provides the corresponding maximal transition probabilities $P_{x_0}(u^*(t),\ x_f,\ \bar{t}')$, $P_{x_0}(u^*(t),\ x_f,\ t_1' \le t(x_f) \le t_2')$, and the optimal control $u^*(t)$ that provides the corresponding maximal expected total costs $\sigma_{x_0}(u^*(t),\ x_f,\ \bar{t}')$, $\sigma_{x_0}(u^*(t),\ x_f, t_1' \le t(x_f) \le t_2')$.

It is easy to observe that if we solve these problems (Problems 4.1–4.6) on the time-expanded network with $\bar{t}' = 2\bar{t}$, $t_1' = 2t_1$ and $t_2' = 2t_2$ then we obtain the solutions of the Markov decision Problems 4.27–4.32.

4.5.3 Backward Dynamic Programming Algorithms for Finite Horizon Markov Decision Problems

The backward dynamic programming algorithms for the Markov decision problems 4.28 and 4.30 can be grounded without their reduction to stochastic control problems and the construction of the auxiliary time-expanded network. Below we describe such dynamic programming algorithms for the Problems 4.28 and 4.30.

Algorithm 4.33 Determining the Optimal Strategies for the Decision Problem 4.28

Let s be a strategy that determines the actions $a = s(x(t)) \in A(x(t))$ in the dynamical states $x(t) \in X$, $t = 0,1,2,\ldots,t_2$. For a given strategy s we denote by $P_{x(t_2-\tau)}(s),\ x_f,\ t_1 \le t(x_f) \le t_2)$ the probability transition from the state $x(t_2-\tau)$ at the moment of time $t_2-\tau$ to the state x_f if the state x_f is reached at the moment of time $t(x_f)$ such that $t_1 \le t(x_f) \le t_2$.

Then the optimal strategy $s^*(x(t))$ and the corresponding values

$$P_{x(t_2-\tau)}(s^*, x_f, t_1 \le t(x_f) \le t_2) \quad \text{for} \quad t = 0, 1, 2, \ldots, t_1$$

can be found by using the following algorithm:

Preliminary step (Step 0)
Fix $s^*(x_f(t)) = \emptyset$, $P_{x_f(t)}(s^*,\ x_f,\ t_1 \le t(x_f) \le t_2) = 1$ for $t = t_1, t_1 + 1, \ldots, t_2$, and set $s^*(x(t_2)) = \emptyset$, $P_{x(t_2)}(s^*(x(t_2)),\ x_f,\ t_1 \le t(x_f) \le t_2) = 0$ for $x(t_2) \in X \backslash \{x_f\}$.

General step (Step $\tau,\ \tau \ge 1$)
If $1 \le \tau \le t_2 - t_1$ then for every $x(t_2 - \tau) \in X \backslash \{x_f\}$ find

$$P_{x(t_2-\tau)}(s(x(t_2-\tau)), x_f, t_1 \le t(x_f) \le t_2)$$
$$= \max_{a \in A(x(t_2-\tau))} \left\{ p^a_{x(t_2-\tau), y(t_2-\tau+1)}(t_2-\tau) \cdot P_{x(t_2-\tau+1)}(s^*, x_f, t_1 \le t(x_f) \le t_2) \right\}$$

and fix $s^*(x(t_2 - \tau)) = a^* \in A(x(t_2 - \tau))$, where a^* is the action for which this maximum is reached.

If $t_2 - t_1 < \tau \le t_2$ then for every $x(t_2 - \tau) \in X$ find

$$P_{x(t_2-\tau)}(s(x(t_2-\tau)), x_f, t_1 \le t(x_f) \le t_2)$$
$$= \max_{a \in A(x(t_2-\tau))} \left\{ p^a_{x(t_2-\tau), y(t_2-\tau+1)}(t_2 - \tau) \cdot P_{x(t_2-\tau+1)}(s^*, x_f, t_1 \le t(x_f) \le t_2) \right\}$$

and fix $s^*(x(t_2 - \tau)) = a^* \in A(x(t_2 - \tau))$, where a^* is the action for which this maximum is reached.

Check if $\tau < t_2$?

If $\tau < t_2$ then go to the next step; otherwise STOP.

After t_2 steps Algorithm 4.33 determines the optimal strategy s^* and the corresponding probabilities $P_{x(0)}(s^*, x_f, t_1 \le t(x_f) \le t_2)$ for an arbitrary starting state $x(0) \in X$.

Algorithm 4.34 Determining the Optimal Strategies for the Decision Problem 4.30

For a given strategy s we denote by $\sigma_{x(\bar{t}-\tau)}(s,\bar{t})$ the expected total cost for the dynamical system if it starts a transition in the state $x(\bar{t}-\tau)$ at the moment of time $\bar{t}-\tau$ and finishes transitions at the moment of time $\bar{t}$. The optimal strategy s^* and the corresponding values

$$\sigma_{x(\bar{t}-\tau)}(s^*,\bar{t}), \quad \tau = 0, 1, 2, \dots, \bar{t}$$

can be found by using the following algorithm:

Preliminary step (Step 0)
Fix $s^*(x(\bar{t})) = \emptyset$, $\sigma_{x(\bar{t})}(s^*,\bar{t}) = 0$ for every $x(\bar{t}) \in X$.
General step (Step τ, $\tau \geq 1$)
Find

$$\sigma_{x(\bar{t}-\tau)}(s^*,\bar{t}) = \min_{a \in A(x(\bar{t}-\tau))} \Bigg\{ \mu^a_{x(\bar{t}-\tau)}(t_2-\tau) + \sum_{y(\bar{t}-\tau+1)\in X} p^a_{x(\bar{t}-\tau),y(\bar{t}-\tau+1)}(t_2-\tau)\cdot \sigma^a_{y(\bar{t}-\tau+1)} \Bigg\},$$

where

$$\mu^a_{x(\bar{t}-\tau)}(t_2-\tau) = \sum_{y(\bar{t}-\tau+1)\in X} p^a_{x(\bar{t}-\tau),y(\bar{t}-\tau+1)}(t_2-\tau)\cdot c^a_{x(\bar{t}-\tau),y(\bar{t}-\tau+1)}(t_2-\tau),$$

and fix $s^*(x(\bar{t}-\tau)) = a^* \in A(x(\bar{t}-\tau))$ where a^* is the action for which this maximum is reached.

Check if $\tau < \bar{t}$?

If $\tau < \bar{t}$ then go to the next step; otherwise STOP.

After $\bar{t}$ steps Algorithm 4.34 determines the optimal strategy s^* and the corresponding expected total costs $\sigma_{x(\bar{t}-\tau)}(s^*,\bar{t})$ for an arbitrary starting state $x(\bar{t}-\tau)$, $\tau = 0, 1, 2, \dots, \bar{t}$. In the case $\tau = \bar{t}$ we obtain the optimal total cost $\sigma_{x(0)}(s^*,\bar{t})$ during $\bar{t}$ transitions if the system starts transitions in $x(0) \in X$.

Errata to: Optimization of Stochastic Discrete Systems and Control on Complex Networks

Dmitrii Lozovanu and Stefan Pickl

Errata to:
D. Lozovanu and S. Pickl, *Optimization of Stochastic Discrete Systems and Control on Complex Networks*, Advances in Computational Management Science, DOI 10.1007/978-3-319-11833-8

Below listed corrections need to be incorporated in published volume:

The online version of the original book can be found under
DOI 10.1007/978-3-319-11833-8

D. Lozovanu (✉)
Institute of Mathematics and Informatics, Academy of Sciences of Moldova,
Chisinau, Moldova
e-mail: lozovanu@math.md

S. Pickl
Institut für Theoretische Informatik, Mathematik und Operations Research,
Universität der Bundeswehr München, Munchen, Bayern, Germany
e-mail: stefan.pickl@unibw.de

D. Lozovanu and S. Pickl, *Optimization of Stochastic Discrete Systems and Control on Complex Networks*, Advances in Computational Management Science 12,
DOI 10.1007/978-3-319-11833-8_5

Page No	Published content	Replace with
Copy right page: Author city name	"Munchen"	"Munich"
Page 124: Equation (2.27)	$\begin{cases} h_x = \min\limits_{y \in X(x)} h_y, & \forall x \in X_C; \\ h_x = \sum\limits_{y \in X(x)} p_{x,y} h_x, & \forall x \in X_N. \end{cases}$ (2.27)	$\begin{cases} h_x = \min\limits_{y \in X(x)} h_y, & \forall x \in X_C; \\ h_x = \sum\limits_{y \in X(x)} p_{x,y} h_y, & \forall x \in X_N. \end{cases}$ (2.27)
Page 164: Equation (2.75)	$\omega_x = \min\limits_{a \in A(x)} \left\{ \sum\limits_{y \in X} p_{x,y}^a \omega_x \right\}, \quad \forall x \in X,$ (2.75)	$\omega_x = \min\limits_{a \in A(x)} \left\{ \sum\limits_{y \in X} p_{x,y}^a \omega_y, \right. \quad \forall x \in X,$ (2.75)
Page 165: Equation	$\varepsilon_x + \omega_x \le \mu_{x,a} + \sum\limits_{y \in X} p_{x,y}^a \varepsilon_y, \quad \forall x \in X, \quad \forall a \in A(x);$ $\omega_x \le \sum\limits_{y \in X} p_{x,y}^a \omega_x, \quad \forall x \in X, \quad \forall a \in A(x).$	$\varepsilon_x + \omega_x \le \mu_{x,a} + \sum\limits_{y \in X} p_{x,y}^a \varepsilon_y, \quad \forall x \in X, \quad \forall a \in A(x);$ $\omega_x \le \sum\limits_{y \in X} p_{x,y}^a \omega_y, \quad \forall x \in X, \quad \forall a \in A(x).$
	$\begin{cases} \varepsilon_x + \omega_x \le \mu_{x,a} + \sum\limits_{y \in X} p_{x,y}^a \varepsilon_y, & \forall x \in X, \quad \forall a \in A(x); \\ \omega_x \le \sum\limits_{y \in X} p_{x,y}^a \omega_x, & \forall x \in X, \quad \forall a \in A(x). \end{cases}$ (2.78)	$\begin{cases} \varepsilon_x + \omega_x \le \mu_{x,a} + \sum\limits_{y \in X} p_{x,y}^a \varepsilon_y, & \forall x \in X, \quad \forall a \in A(x); \\ \omega_x \le \sum\limits_{y \in X} p_{x,y}^a \omega_y, & \forall x \in X, \quad \forall a \in A(x) \end{cases}$ (2.78)
	$\begin{cases} \varepsilon_x + \omega_x = \mu_{x,s(x)} + \sum\limits_{y \in X} p_{x,y}^{s(x)} \varepsilon_y, & \forall x \in X; \\ \omega_x = \sum\limits_{y \in X} p_{x,y}^{s(x)} \omega_x, & \forall x \in X \end{cases}$ (2.79)	$\begin{cases} \varepsilon_x + \omega_x = \mu_{x,s(x)} + \sum\limits_{y \in X} p_{x,y}^{s(x)} \varepsilon_y, & \forall x \in X; \\ \omega_x = \sum\limits_{y \in X} p_{x,y}^{s(x)} \omega_y, & \forall x \in X \end{cases}$ (2.79)
Page 235: Equation	$\begin{cases} \omega_x = \max\limits_{a \in A(x)} \left\{ \sum\limits_{y \in X} p_{x,y}^a \omega_x \right\}, & \forall_x \in X_1; \\ \omega_x = \min\limits_{a \in A(x)} \left\{ \sum\limits_{y \in X} p_{x,y}^a \omega_x \right\}, & \forall_x \in X_2, \end{cases}$ (3.18)	$\begin{cases} \omega_x = \max\limits_{a \in A(x)} \left\{ \sum\limits_{y \in X} p_{x,y}^a \omega_y \right\}, & \forall_x \in X_1; \\ \omega_x = \min\limits_{a \in A(x)} \left\{ \sum\limits_{y \in X} p_{x,y}^a \omega_y \right\}, & \forall_x \in X_2, \end{cases}$ (3.18)
	$s^{1*}(x) \in \left(\arg\max\limits_{a \in A(x)} \left\{ \sum\limits_{y \in X} p_{x,y}^a \omega_x^* \right\} \right) \bigcap \left(\arg\max\limits_{a \in A(x)} \left\{ \mu_{x,a} + \sum\limits_{y \in X} p_{x,y}^a \varepsilon_y^* \right\} \right), \quad \forall x \in X_1$ (3.20) and	$s^{1*}(x) \in \left(\arg\max\limits_{a \in A(x)} \left\{ \sum\limits_{y \in X} p_{x,y}^a \omega_y^* \right\} \right) \bigcap \left(\arg\min\limits_{a \in A(x)} \left\{ \mu_{x,a} + \sum\limits_{y \in X} p_{x,y}^a \varepsilon_y^* \right\} \right). \quad \forall x \in X_1$ (3.20) and

(continued)

(continued)

Page No	Published content	Replace with
	$s^{2*}(x) \in \left(\underset{a\in A(x)}{\arg\min}\left\{\sum_{y\in X} p^a_{x,y}\omega^*_x\right\}\right) \bigcap \left(\underset{a\in A(x)}{\arg\min}\left\{\mu_{x,a} + \sum_{y\in X} p^a_{x,y}\varepsilon^*_y\right\}\right), \quad \forall x \in X_2$ (3.21)	$s^{2*}(x) \in \left(\underset{a\in A(x)}{\arg\min}\left\{\sum_{y\in X} p^a_{x,y}\omega^*_y\right\}\right) \bigcap \left(\underset{a\in A(x)}{\arg\min}\left\{\mu_{x,a} + \sum_{y\in X} p^a_{x,y}\varepsilon^*_y\right\}\right). \quad \forall x \in X_2$ (3.21)
Page 236: Equation	$\begin{cases} \varepsilon_x + \omega_x \geq \mu_{x,a} + \sum_{y\in X} p^a_{x,y}\varepsilon_y, & \forall x \in X_1, a = \bar{s}^1(x); \\ \varepsilon_x + \omega_x = \mu_{x,a} + \sum_{y\in X} p^a_{x,y}\varepsilon_y, & \forall x \in X_2, a = \bar{s}^2(x); \\ \omega_x = \sum_{y\in X} p^a_{x,y}\omega_x, & \forall x \in X_1, a = \bar{s}^1(x); \\ \omega_x = \sum_{y\in X} p^a_{x,y}\omega_x, & \forall x \in X_2, a = \bar{s}^2(x) \end{cases}$ (3.22)	$\begin{cases} \varepsilon_x + \omega_x = \mu_{x,a} + \sum_{y\in X} p^a_{x,y}\varepsilon_y, & \forall x \in X_1, a = \bar{s}^1(x); \\ \varepsilon_x + \omega_x = \mu_{x,a} + \sum_{y\in X} p^a_{x,y}\varepsilon_y, & \forall x \in X_2, a = \bar{s}^2(x); \\ \omega^*_x = \sum_{y\in X} p^a_{x,y}\omega_y, & \forall x \in X_1, a = \bar{s}^1(x); \\ \omega^*_x = \sum_{y\in X} p^a_{x,y}\omega_y, & \forall x \in X_2, a = \bar{s}^2(x) \end{cases}$ (3.22)
	$\begin{cases} \varepsilon^*_x + \omega^*_x \geq \mu_{x,a} + \sum_{y\in X} p^a_{x,y}\varepsilon^*_y, & \forall x \in X_1, a \in A(x); \\ \varepsilon^*_x + \omega^*_x = \mu_{x,a} + \sum_{y\in X} p^a_{x,y}\varepsilon^*_y, & \forall x \in X_2, a = \bar{s}^2(x); \\ \omega^*_x \geq \sum_{y\in X} p^a_{x,y}\omega^*_x, & \forall x \in X_1, a \in A(x); \\ \omega^*_x = \sum_{y\in X} p^a_{x,y}\omega^*_x, & \forall x \in X_2, a = \bar{s}^2(x) \end{cases}$	$\begin{cases} \varepsilon^*_x + \omega^*_x \geq \mu_{x,a} + \sum_{y\in X} p^a_{x,y}\varepsilon^*_y, & \forall x \in X_1, a \in A(x); \\ \varepsilon^*_x + \omega^*_x = \mu_{x,a} + \sum_{y\in X} p^a_{x,y}\varepsilon^*_y, & \forall x \in X_2, a = \bar{s}^2(x); \\ \omega^*_x \geq \sum_{y\in X} p^a_{x,y}\omega^*_y, & \forall x \in X_1, a \in A(x); \\ \omega^*_x = \sum_{y\in X} p^a_{x,y}\omega^*_y, & \forall x \in X_2, a = \bar{s}^2(x) \end{cases}$
	$\begin{cases} \varepsilon^*_x + \omega^*_x = \mu_{x,a} + \sum_{y\in X} p^a_{x,y}\varepsilon^*_y, & \forall x \in X_1, a = \bar{s}^1(x); \\ \varepsilon^*_x + \omega^*_x \leq \mu_{x,a} + \sum_{y\in X} p^a_{x,y}\varepsilon^*_y, & \forall x \in X_2, a \in A(x); \\ \omega^*_x = \sum_{y\in X} p^a_{x,y}\omega^*_x, & \forall x \in X_1, a = \bar{s}^1(x); \\ \omega^*_x \leq \sum_{y\in X} p^a_{x,y}\omega^*_x, & \forall x \in X_2, a \in A(x) \end{cases}$	$\begin{cases} \varepsilon^*_x + \omega^*_x = \mu_{x,a} + \sum_{y\in X} p^a_{x,y}\varepsilon^*_y, & \forall x \in X_1, a = \bar{s}^1(x); \\ \varepsilon^*_x + \omega^*_x \leq \mu_{x,a} + \sum_{y\in X} p^a_{x,y}\varepsilon^*_y, & \forall x \in X_2, a \in A(x); \\ \omega^*_x = \sum_{y\in X} p^a_{x,y}\omega^*_y, & \forall x \in X_1, a = \bar{s}^1(x); \\ \omega^*_x \leq \sum_{y\in X} p^a_{x,y}\omega^*_y, & \forall x \in X_2, a \in A(x) \end{cases}$
Page 237: Equation	$\begin{cases} \varepsilon_x + \omega_x \geq \mu_{x,a} + \sum_{y\in X} p^a_{x,y}\varepsilon_y, & \forall x \in X_1, a \in A(x); \\ \varepsilon_x + \omega_x \leq \mu_{x,a} + \sum_{y\in X} p^a_{x,y}\varepsilon_y, & \forall x \in X_2, a \in A(x); \\ \omega_x \geq \sum_{y\in X} p^a_{x,y}\omega_x, & \forall x \in X_1, a \in A(x)(x); \\ \omega_x \leq \sum_{y\in X} p^a_{x,y}\omega_x, & \forall x \in X_2, a \in A(x) \end{cases}$	$\begin{cases} \varepsilon_x + \omega_x \geq \mu_{x,a} + \sum_{y\in X} p^a_{x,y}\varepsilon_y, & \forall x \in X_1, a \in A(x); \\ \varepsilon_x + \omega_x \leq \mu_{x,a} + \sum_{y\in X} p^a_{x,y}\varepsilon_y, & \forall x \in X_2, a \in A(x); \\ \omega_x \geq \sum_{y\in X} p^a_{x,y}\omega_y, & \forall x \in X_1, a \in A(x)(x); \\ \omega_x \leq \sum_{y\in X} p^a_{x,y}\omega_y, & \forall x \in X_2, a \in A(x) \end{cases}$

(continued)

(continued)

Page No	Published content	Replace with
Page 238: Equation	$s_k^1(x) \in \arg\max_{a\in A(x)}\left\{\sum_{y\in X} p_{x,y}^a \omega_x^{s_{k-1}^1}\right\}, \quad \forall x \in X_1;$ $s_k^2(x) \in \arg\max_{a\in A(x)}\left\{\sum_{y\in X} p_{x,y}^a \omega_x^{s_{k-1}^1}\right\}, \quad \forall x \in X_2$ and set $s_k = s_{k-1}$ if $s_{k-1}^1(x) \in \arg\max_{a\in A(x)}\left\{\sum_{y\in X} p_{x,y}^a \omega_x^{s_{k-1}^1}\right\}, \quad \forall x \in X_1;$ $s_{k-1}^2(x) \in \arg\max_{a\in A(x)}\left\{\sum_{y\in X} p_{x,y}^a \omega_x^{s_{k-1}^1}\right\}, \quad \forall x \in X_2.$	$s_k^1(x) \in \arg\max_{a\in A(x)}\left\{\sum_{y\in X} p_{x,y}^a \omega_y^{s_{k-1}^1}\right\}, \quad \forall x \in X_1;$ $s_k^2(x) \in \arg\max_{a\in A(x)}\left\{\sum_{y\in X} p_{x,y}^a \omega_y^{s_{k-1}^1}\right\}, \quad \forall x \in X_2$ and set $s_k = s_{k-1}$ if $s_{k-1}^1(x) \in \arg\max_{a\in A(x)}\left\{\sum_{y\in X} p_{x,y}^a \omega_y^{s_{k-1}^1}\right\}, \quad \forall x \in X_1;$ $s_{k-1}^2(x) \in \arg\max_{a\in A(x)}\left\{\sum_{y\in X} p_{x,y}^a \omega_y^{s_{k-1}^1}\right\}, \quad \forall x \in X_2.$

Conclusion

The stochastic dynamic programming problems considered in this book generalize classical ones and have large applications for studying and solving many practical dynamic decision problems from various areas. New results concerned with determining the optimal stationary strategies for the stochastic discrete control problems and Markov decision processes with an average and expected total discounted costs' optimization criteria are derived. Based on these results and classical numerical methods new algorithms for solving the dynamic decision problems with finite and infinite time horizon are elaborated and grounded.

The most important results of the book are related to complex decision processes, where the dynamics of the system is controlled by several actors. The corresponding decision models for such processes are formulated and studied applying the game-theoretical concept to Markov decision processes. Following this concept a new class of stochastic positional games and multi-objective decision problems that extend deterministic positional games and multi-criteria control problems to networks have been studied. Nash equilibria conditions for average and discounted stochastic positional games are gained and algorithms for determining the optimal stationary strategies of the players are grounded. Furthermore, the concept of multi-objective control is applied to classical discrete optimal control problems and new classes of multi-criteria decision problems are studied. Necessary and sufficient conditions for determining Pareto optima and Stackelberg strategies in the considered multi-criteria control models are obtained. The dynamic programming techniques for such a class of problems are developed and new polynomial time algorithms for determining Nash equilibria and Pareto optima are elaborated. Polynomial time algorithms for determining the optimal strategies of the players for the dynamic c-games and for the game control problems in positional form are suggested.

D. Lozovanu and S. Pickl, *Optimization of Stochastic Discrete Systems and Control on Complex Networks*, Advances in Computational Management Science 12,
DOI 10.1007/978-3-319-11833-8

References

1. Bartlett, M.S.: An Introduction to Stochastic Processes. Cambridge University Press, London (1956)
2. Basar, T., Olsder, G.: Dynamic Noncooperative Game Theory. SIAM, London (1999)
3. Bellman, R.: Dynamic Programming. Princeton University Press, Princeton (1957)
4. Bellman, R.: A Markovian decision process. J. Math. Mech. **6**, 679–684 (1957)
5. Bellman, R.: Functional equations in the theory of dynamic programming. XI-Limit theorems. Rand. Circolo Math. Palermo **8**(3), 343–345 (1959)
6. Bellman, R., Kalaba, R.: Dynamic Programming and Modern Control Theory. Academic Press, New York (1965)
7. Berman, A., Plemmons, R.J.: Nonnegative Matrices in the Mathematical Sciences. Academy, New York (1979)
8. Blackwell, D.: Discounted dynamic programming. Ann. Math. Stat. **36**, 226–235 (1965)
9. Boliac, R., Lozovanu, D., Solomon, D.: Optimal paths in network games with p players. Discrete Appl. Math. **99**(3), 339–346 (2000)
10. Bollobas, B.: Random Graphs, 2nd edn. Cambridge University Press, Cambridge (2001)
11. Boltjanski, W.G.: Optimale Steuerung Diskreter Systeme. Leipzig Akademische Verlagsgesellschaft Geest & Portig K.-G, Leipzig (1976)
12. Boros, E., Gurvich, V.: On Nash-solvability in pure stationary strategies of finite games with perfect information which may have cycles. DIMACS Tech. Rep. **18**, 1–32 (2002)
13. Boulogne, T., Altman, E., Kameda, H., Pourtallier, O.: Mixed equilibrium for multiclass routing games. IEEE Trans. Autom. Control 47(6), 903–916 (2002) (Special issue on control issues in telecommunication networks).
14. Brucher, P.: Discrete parameter optimization problems and essential efficient points. Oper. Res. **16**(5), 189–197 (1972)
15. Burkard, R., Krarup, J., Pruzan, P.: A relationship between optimality and efficiency in multicriteria 0–1 programming problems. Comp. Oper. Res. **8**(2), 241–247 (1981)
16. Burkard, R., Krarup, J., Pruzan, P.: Efficiency and optimality in minisum, minimax 0–1 programming problems. J. Oper. Res. Soc. **33**(2), 137–151 (1982)
17. Butenko, S., Murphey, R., Pardalos, P.M. (eds.): Cooperative Control: Models, Applications, and Algorithms. Kluwer Academic Publishers, New York (2003)
18. Butkobic, P., Cuninghame-Green, R.A.: An $O(n^2)$ algorithm for the maximum cycle meean of an $n \times n$ bivalent matrix. Discrete Appl. Math. **35**, 157–162 (1992)
19. Christofides, N.: Graph Theory: An Algorithmic Approach. Academic Press, New York (1975)
20. Chung, K.L.: Markov Chains with Stationary Transitions Probabilities. Springer, New York (1960)
21. Cinlar, E.: Introduction to Stochastic Processes. Prentice-Hall, Englewood Cliffs (1975)

D. Lozovanu and S. Pickl, *Optimization of Stochastic Discrete Systems and Control on Complex Networks*, Advances in Computational Management Science 12,
DOI 10.1007/978-3-319-11833-8

22. Condon, A.: The complexity of stochastic games. Inf. Comput. **96**(2), 203–224 (1992)
23. Dasgupta, P., Maskin, E.: The existence of equilibrium in discontinuous economic games. Rev. Econ. Stud. **53**, 1–26 (1986)
24. Debreu, G.: A social equilibrium existence theorem. Proc. Nat. Acad. Sci. **38**, 386–393 (1952)
25. Denardo, E.V.: On linear programming in a Markov decision processes. Man. Sci. **16**, 281–226 (1986)
26. Derman, C.: Finite State Markov Decision Processes. Academic Press, New York (1970)
27. Durinovic, S., Lee, H.M., Katehakis, M.N., Filar, J.A.: Multiobjective Markov decision process with average reward criterion. Large Scale Syst. **100**, 215–226 (1986)
28. Dynkin, E.B., Yushkevich, A.A.: Controlled Markov Processes. Springer, Berlin (1975) (Stanford University).
29. Ehrenfeucht, A., Mycielski, J.: Positional strategies for mean payoff games. Int. J. Game Theory **8**, 109–113 (1979)
30. Emelichev, V., Perepelitsa, V.: The complexity of discrete multicriterion problems. Discrete Math. Appl. **6**(1), 5–33 (1994)
31. Fabrikant, A., Papadimitrio, C., Talwar, K.: The complexity of pure Nash equilibria. In: Proceedings of the 36th Annual ACM Symposium, on Theory of Computing (STOC'04), IL, USA, pp. 604–612, (2005).
32. Federgruen, A., Schweitzer, P.J.: Discounted and undiscounted value-iteration in Markov decision problems. In: Puterman, M.L (ed.) Dynamic Programming and its Applications pp. 23–52. Academic Press, New York (1978).
33. Feller, W.: An Introduction to Probability Theory and its Applications, vol. 1, 2nd edn. Wiley, New York (1957).
34. Filar, J.A., Kallenberg, L.C.M., Lee, H.M.: Variance penalized Markov decision processes. Math. Oper. Res. **14**, 147–161 (1989)
35. Filar, J.A., Vrieze, K.: Competitive Markov Decision Processes. Springer, New York (1997)
36. Fleming, W.H., Rishel, R.: Deterministic and Stochastic Optimal Control. Springer, New York (1975)
37. Fox, B.D., Landi, D.M.: An algorithm for identification the ergodic subchains and transient states of a stochastic matrix. Comm. ACM **2**, 619–621 (1968)
38. Garey, M., Johnson, D.: Computers and Intractability. Freeman, San Francisco (1979)
39. Georghe, A.V.: Decision Processes in Dynamic Probabilistic Systems. Kluver Academic Publishers, Dordrecht (1990)
40. Ghosh, M.K.: Markov decision processes with multiple costs. Oper. Res. Lett. **9**(4), 257–260 (1990)
41. Gillette, D.: Stochastic games with zero stop probabilities. Contribution to the Theory of Games, vol. III, pp. 179–187. Princeton University Press, Princeton (1957).
42. Gottlob, G., Greco, G., Scarcello, F.: Pure nash equilibria: hard and easy games. J. Artif. Intell. Res. **24**, 357–406 (2005)
43. Gurvich, V.A., Karzanov, A.V., Khachian, L.G.: Cyclic games and an algorithm to find minimax cycle means in directed graphs. USSR Comput. Math. Math. Phys. **28**, 85–91 (1988)
44. Helmberg, G., Voltkamp, G.: On Fadeev-Leverrier's method for the computation of the characteristic polynomial of the matrix and of eigenvectors. Linear Algebra Appl. **185**, 219–233 (1993)
45. Hordijk, A., Kallenberg, L.C.M.: Linear programming and Markov decision chains. Man. Sci. **25**, 352–362 (1979)
46. Hordijk, A., Kallenberg, L.C.M.: On solving Markov decision problems by linear programming. In: Hartley, R., Thomas, L.C., White, D.J. (eds.) Recent Developments in Markov Decision Processes, pp. 127–143. Academic, New York (1980)
47. Howard, R.A.: Dynamic Programming and Markov Processes. Wiley, New York (1960)
48. Howard, R.A.: Semi-Markov Decision Processes. In: Proceedings of the Internal Statistician Institute, Ottawa, Canada (1963).
49. Howard, R.A.: Dynamic Probabilistic Systems. Wiley, New York (1972)

50. Hu, Q., Yue, W.: Markov Decision Processes with their Applications. Springer, New York (2008)
51. Kallenberg, L.S.: Linear programming and finite Markovian control problems. Mathematical Centre Tracts, 148, Mathematisch Centrum, Amsterdam (1983).
52. Karlin, S.: A First Course in Stochastic Processes. Academic, New York (1969)
53. Karp, R.: A characterization of the minimum cycle mean in a digraph. Discrete Mathematics **23**(3), 309–311 (1978)
54. Karzanov, A., Lebedev, V.: Cyclical games with prohibitions. Mathematical Programming **60**, 277–293 (1993)
55. Kemeny, J.G., Snell, J.L.: Finite Markov Chains. Van Nostrand-Reinhold, New York (1960)
56. Kemeny, J.G., Snell, J.L., Knapp, A.W.: Denumerable Markov Chains. Springer, New York (1976)
57. Khachian, L.G.: Polynomial time algorithm in linear programming. USSR Comput. Math. Math. Phys **20**, 51–58 (1980)
58. Khachian, L.: On exact solution of the system of linear inequalities and linear programming problem. USSR Comput. Math. Math. Phys. **22**, 999–1002 (1982)
59. Krabs, W., Pickl, S.: Controllability of a time-discrete dynamical system with the solution of an approximation problem. Int. J. Control Theory Cybern. **32**(1), 57–74 (2003)
60. Krabs, W., Pickl, S.: Analysis, Controllability and Optimization of Time-Discrete Systems and Dynamical Games. Lecture Notes in Economics and Mathematical Systems. Springer, Berlin (2003).
61. Lal, K., Sinha, S.: Zero-sum two person semi-Markov games. J. Appl. Prob. **29**, 56–72 (1992)
62. Lamond, B.F., Puterman, M.L.: Generalized inverses in discrete time Markov processes. SIAM J. Matrix Anal. Appl. **10**, 118–134 (1989)
63. Lawler, E.: Optimal cycles in doubly weighted directed linear graphs. In: International Symposium on Theory of Graphs, p. 209. Dunod, Paris (1996).
64. Lazari, A.: Algorithms for determining the transient and differential matrices in finite Markov processes. Bull. Acad. Sci. RM Ser. Mathem. **2**(63), 84–99 (2010).
65. Lozovanu, D.: Extremal-Combinatorial Problems and Algorithms for their Solving. Kishinev, Stiinta (1991). (in Russian)
66. Lozovanu, D.: Algorithms to solve some classes of network minmax problems and their aplications. Cybernetics **29**, 93–100 (1991)
67. Lozovanu, D.: Polynomial time algorithm for determining optimal strategies in cyclic games. In: Proceedings of the 10th IPCO Conference, pp. 74–85. New York, NY (2004).
68. Lozovanu, D.: Multiobjective Control of Time-Discrete Systems and Dynamic Games on Networks. In: Chinchulun, A., Migdalas, A., Pardalos, P., Pitsoulis, L. (eds.) Chapter in the Book "Pareto Optimality, Game Theory and Equilibria", pp. 665–757. Springer, New York (2008)
69. Lozovanu, D.: The game-theoretical approach to Markov decision problems and determining nash equilibria for stochastic positional games. Int. J. Mathematical Modelling and Numerical. Optimization **2**(2), 162–164 (2011)
70. Lozovanu, D., Fonoberova, M.: Optimal dynamic multicommodity flows in networks. Electron. Notes Discrete Math. **13**, 154–158 (2003)
71. Lozovanu, D., Fonoberova, M.: Optimal Dynamic Flows in Networks and Algorithms for Finding Them. In: Dehmer, M., Emmert-Sterb, F. (eds.) Chapter in the Book "Analysis of Complex Networks", pp. 377–400. Wiley-VGH Verlag GmbH and Co. KGaA, Weinheim (2009)
72. Lozovanu, D., Lazari, A.: An approach for determining the matrix of limiting state probabilities in discrete Markov processes. Bull. Acad. Sci. RM Ser. Mathem. **1**(62), 17–32 (2010)
73. Lozovanu, D., Petic, C.: Algorithms for finding the minimum cycle mean in the weighted directed graph. Comput. Sci. J. Moldova **6**(1), 27–34 (1998)
74. Lozovanu, D., Pickl, S.: Nash equilibria for multiobjective control of time-discrete systems and polynomial-time algorithm for k-partite networks. Cent. Eur. J. Oper. Res. **13**(2), 127–146 (2005)

75. Lozovanu, D., Pickl, S.: Nash equilibria conditions for cyclic games with p players. Eletronic Notes Discrete Math. **25**, 117–124 (2006)
76. Lozovanu, D., Pickl, S.: Algorithms and the calculation of nash equilibria for multi-objective control of time-discrete systems and polynomial-time algorithms for dynamic c-games on networks. Eur. J. Oper. Res. **181**, 1214–1232 (2007)
77. Lozovanu, D., Pickl, S.: Algorithms for solving multiobjective discrete control problems and dynamic c-games on networks. Discrete Appl. Math. **155**(14), 158–180 (2007)
78. Lozovanu, D., Pickl, S.: Multiobjective hierarchical control of time-discrete systems and determining stackelberg strategies. In: Proceedings CTW-2007, pp. 111–114. (2007).
79. Lozovanu, D., Pickl, S.: Optimization and Multiobjective Control of Time-Discrete Systems. Springer, Berlin (2009)
80. Lozovanu, D., Pickl, S.: Dynamic programming algorithms for solving stochastic discrete control problems. Bull. Acad. Sci. RM Ser. Mathem. **2**(2009), 17–34 (2009)
81. Lozovanu, D., Pickl, S.: An algorithmic solution of discrete control problems on stochastic networks. In: Proceedings CTW-2009, pp. 221–224 (2009).
82. Lozovanu, D., Pickl, S.: Algorithms for solving discrete optimal control problems with infinite time horizon and determining minimal mean cost cycles in a directed graph as decision support tool. Cent. Eur. J. Oper. Res. **17**(3), 255–264 (2009)
83. Lozovanu, D., Pickl, S.: Discrete control and algorithms for solving antagonistic dynamic games on networks. Optimization **58**(6), 665–683 (2009)
84. Lozovanu, D., Pickl, S.: An extension of a polynomial time algorithm for the calculation of the limit state matrix in a random graph. In: Dubois, D. (ed.) International Journal of Computing Anticipatory Systems, pp. 92–97. CHAOS, Liege (2010)
85. Lozovanu, D., Pickl, S.: Determining optimal stationary strategies for discounted stochastic optimal control problem on networks. In: Faigle, U., Schrader, R., Herrmann, D. (eds.) Proceedings CTW-2010, pp. 115–118 (2010).
86. Lozovanu, D., Pickl, S.: Optimal stationary control of discrete processes and a polynomial time algorithm for stochastic control problem on networks. In: Proceedings of ICCS 2010 Conference, vol. CS(1), pp. 1417–1426. Amsterdam, Elsevier, Procedia (2010).
87. Lozovanu, D., Pickl, S.: Algorithms for solving discrete optimal control problems with varying time of states' transitions of dynamical systems. Dyn. Continuous Discrete Impulsive Syst. Ser. B: Appl. Algorithms 17, 101–111 (2010).
88. Lozovanu, D., Pickl, S.: Algorithms for determining the state-time probabilities and the limit matrix in discrete Markov chains. Bull. Acad. Sci. RM Ser. Mathem. **1**(65), 66–82 (2011)
89. Lozovanu, D., Pickl, S.: Discounted Markov decision processes and algorithms for solving stochastic control problems on networks. In: Proceedings CTW-2011, pp. 194–197 (2011).
90. Lozovanu, D., Pickl, S.: Determining the optimal strategies for antagonistic positional games in Markov decision processes. Operation Research Proceedings-2011, Selected Paper of the International Conference on Operations Research GOR, pp. 229–244. Springer (2012).
91. Lozovanu, D., Pickl, S., Kropat, E.: Markov decision processes and determining Nash equilibria for stochastic positional games. In: Proceedings of the 18th World Congress IFAC-2011, 13398–13493 (2011).
92. Lozovanu, D., Solomon, D., Zelikovsky, A.: Multiobjective games and determining Pareto-Nash equilibria. Bull. Acad. Sci. Moldova Math. **3**(49), 115–122 (2005)
93. Lozovanu, D., Stratila, D.: The minimum cost flow problem on dynamic networks and algorithms for its solving. Bull. Acad. Sci. RM Ser. Mathem. **3**(37), 38–56 (2001)
94. Lozovanu, D., Stratila, D.: Optimal flows in dynamic networks with nonlinear cost functions on the edges. In: Barbu, V., Lasiecka, I. (eds.) Analysis and Optimization of Differential Systems, pp. 247–258. Kluwer Academic Publishers, Dordrecht (2003)
95. Mertens, J.F., Neyman, A.: Stochastic games. Int. J. Game Theory **10**, 53–66 (1981)
96. Meyn, S.P.: Control Techniques for Complex Networks. Cambridge University Press, Cambridge (2007)
97. Mine, H., Osaki, S.: Some remarks on a Markovian decision process with an absorbing state. J. Math. Anal. Appl. **18**, 362–380 (1968)

98. Mine, H., Osaki, S.: Markovian Decision Processes. Elsevier, New York (1970)
99. Moulin, H.: Prolongement des jeux a deux joueurs de somme nulle. Bull. Soc. Math. France. Mem **45**, 5–111 (1976).
100. Moulin, H.: Theory des Jeux pour l'Economie et la Politique. Hermann, Paris (1981)
101. Murphey, R., Pardalos, P.M. (eds.): Cooperative Control and Optimization. Kluwer Academic Publishers, Boston (2002)
102. Nash, J.F.: Non cooperative games. Ann. Math. **2**, 286–295 (1951)
103. Neumann, J., Morgenstern, O.: Theory of Games and Economic Behavior. Princeton University Press, Princeton (1953)
104. Neyman, A., Sorin, S.: Stochastic Games and Applications. Kluwer Academic Press, Dordrecht (2003)
105. Papadimitriou, C.H.: Algorithms, games, and the internet. In: Proceedings of the 33rd Annual ACM Symposium on the Theory of Computing, pp. 749–753. ACM, NY (2001).
106. Papadimitriou, C.H., Tsitsiklis, J.N.: The complexity of Markov decision processes. Math. Oper. Res. **12**, 441–450 (1987)
107. Pareto, V.: Manual d'Eeconomie Politique. Giard, Paris (1904)
108. Pickl, S., Lozovanu, D.: Stochastic discrete control problems and dynamic programming algorithms for solving them. In: Abstract of the International Conference: Actual Problems of Stability and Control Theory. Ekaterinburg, pp. 174–176 (2009).
109. Pickl, S., Lozovanu, D.: A linear programming approach for solving the discounted stochastic optimal control problem on certain networks. In: Proceedings of the 2010 Networking and Electronic Commerce Research Conference NAEC 2010, Riva del Garda, Italy, pp. 1–3 (2010).
110. Pickl, S., Lozovanu, D.: A dynamic programming approach for finite Markov processes and algorithms for the calculation of the limit matrix in Markov chains. Optimization **60**(10–11), 1339–1358 (2011)
111. Podinovsky, V., Noghin, V.: Pareto Optimal Solution for Multicriterion Problems. Nauka, Moscow (1982)
112. Porteus, E.: Overview of iterative methods for discounted finite Markov and semi-Markov decision chains. In: Hartley, R., Thomas, L.S., White, D.J. (eds.) Recent Developments in Markov Decision Processes pp. 1–20. Academic, New York (1980).
113. Puterman, M.: Markov Decision Processes. In: Heyman, D., Sobel, M. (eds.) Handbook of Operations Research, vol. II, pp. 331–432. Noth Holland Publishers, New York (1990).
114. Puterman, M.: Markov Decision Processes: Discrete Stochastic Dynamic Programming. Wiley, New York (1994)
115. Puterman, M.: Markov Decision Processes: Discrete Stochastic Dynamic Programming. Wiley, New Jersey (2005)
116. Reny, F.P.: On the existence of pure and mixed strategy Nash equilibria in discontinuous games. Econometrica **67**, 1029–1056 (1999)
117. Romanovski, I.: Optimization of stationary control of discrete deterministic processes. Cybernetics **2**, 66–78 (1967)
118. Romanovski, I.: On the solvability of Bellman's equations for a Markovian decision process. J. Math. Anal. Appl. **42**, 485–498 (1973)
119. Saad, Y.: Iterative Methods for Linear Systems, 2nd edn. SIAM, Philadelphia (2000)
120. Schweitzer, P.J.: Iterative solutions of the functional equations of undiscounted Markov renewall programming. J. Math. Anal. Appl. **34**, 405–501 (1971)
121. Schweitzer, P.J., Federgruen, A.: The asymptotic behavior of the value-iteration in Markov decision problems. Math. Oper. Res. **2**, 360–381 (1977)
122. Schweitzer, P.J., Federgruen, A.: Foolproof convergence in multichain policy iteration. J. Math. Anal. Appl. **64**, 360–368 (1978)
123. Schweitzer, P.J., Federgruen, A.: Geometric convergence of value-iteration in multichain Markov decision problems. Adv. Appl. Prob. **11**, 188–217 (1979)
124. Shapley, L.: Stochastic games. Proc. Natl. Acad. Sci. USA **39**, 1095–1100 (1953)
125. Simon, L.: Games with discontinuous payoffs. Rev. Econ. Stud. **54**, 569–597 (1987)

126. Stackelberg, H.: Marktform and Gleichgewicht. Springer, Vienna (1934), English translation, entitled The Theory of Market Economy. Oxford University Press, Oxford (1952).
127. Stewart, W.J.: Introduction to the Numerical Solution of Markov Chains. Princeton University Press, Princeton (1995)
128. Stoer, J., Bulirsch, R.: Introduction to Numerical Analysis, 2nd end. Springer, New York (1993)
129. Van der Wal, J.: Stochastic dynamic programming. Mathematical Centre Tracts, 139, Mathematisch Centrum, Amsterdam (1981).
130. Vieille, N.: Stochastic games: recent results. Handbook of Game Theory, pp. 1833–1850. Elsevier Science, Amsterdam (2002).
131. Voge, J., Jurdzinsky, M.: A discrete strategy improvement algorithm for solving parity games. In: Emerson, E.A., Sistla, A.P. (eds.) Computer Aided Verification, 12th International Conference. LNCS, vol. 1855, pp. 202–215. Springer, Heidelberg (2000)
132. Weber, G.W.: Optimal control theory: on the global structure and connection with optimization. J. Comput. Technol. **4**(2), 3–25 (1999)
133. White, C.C., Kim, K.: Solution procedures for vector criterion Markov decision processes. J. Large Scale Syst. **1**, 129–140 (1980)
134. White, C.C., White, D.J.: Markov decision processes. Eur. J. Oper. Res. **39**, 1–16 (1989)
135. White, D.J.: Finite Dynamic Programming. Wiley, Chichester (1978)
136. White, D.J.: Monotone value iteration for discounted finite Markov processes. J. Math. Anal. Appl. **109**, 311–324 (1981)
137. White, D.J.: A Multi-objective version of Bellman's inventory problem. J. Math. Anal. Appl. **87**, 219–227 (1982)
138. White, D.J.: Multi-objective infinite-horizon discounted Markov decision processes. J. Math. Anal. Appl. **89**, 639–644 (1982)
139. White, D.J.: Decomposition in multi-item inventory control. J. Optim. Theory Appl. **54**, 383–401 (1987)
140. White, D.J.: Markov Decision Processes. Wiley, New York (1993)
141. Yushkevich, A.A.: On semi-Markov controlled models with an average criterion. Theory Prob. Appl. **26**, 796–803 (1981)
142. Zwik, U., Paterson, M.: The complexity of mean payoff games on graphs. TCS **158**, 344–359 (1996)

Index

D. Lozovanu and S. Pickl, *Optimization of Stochastic Discrete Systems and Control on Complex Networks*, Advances in Computational Management Science 12,
DOI 10.1007/978-3-319-11833-8

MIX
Papier aus verantwortungsvollen Quellen
Paper from responsible sources
FSC® C105338

If you have any concerns about our products,
you can contact us on
ProductSafety@springernature.com

In case Publisher is established outside the EU,
the EU authorized representative is:
Springer Nature Customer Service Center GmbH
Europaplatz 3, 69115 Heidelberg, Germany

Printed by Libri Plureos GmbH
in Hamburg, Germany